Epigenetics and Human Health

Volume 14

This books series concerns the quickly expanding field of epigenetics. It evokes novel research questions and how these are linked to human health. Each of its volumes covers a particular topic that is important to clinical practice and research, and to researchers in epigenetics who study the relevant medical issues of today. The individual books of the series complement each other and underline the importance of epigenetics in health-related disciplines such as toxicology, reproduction, psychiatry, and nutritional and environmental medicine.

Epigenetics and Human Health was founded in 2011. Prof. Dr. Robert Feil, PD Dr. Mario Noyer-Weidner and Prof. Dr. Jörn Walter were the founding editors and were responsible as series editors for volumes 1-10 until 2016. In 2023, the series was relaunched together with Prof. Rasime Kalkan and Dr. Luis M. Vaschetto, who are the current series editors.

Epigenetics and Human Health is indexed in SCOPUS, EMBASE, Reaxys, EMCare, EMBiology and SCImago.

Luis M. Vaschetto
Editor

Transposable Elements, Transcriptomics, and Diseases

Springer

Editor
Luis M. Vaschetto
National Coalition of Independent Scholars
South Bend, IN, USA

ISSN 2191-2262 ISSN 2191-2270 (electronic)
Epigenetics and Human Health
ISBN 978-3-032-04124-1 ISBN 978-3-032-04125-8 (eBook)
https://doi.org/10.1007/978-3-032-04125-8

This Springer imprint is published by the registered company Springer Nature Switzerland AG
The registered company address is: Gewerbestrasse 11, 6330 Cham, Switzerland

A mi hija María Inés

Preface

Transposable elements (TEs), once dismissed as mere "selfish components" uniquely focused on their own replication and inheritance, are now understood to play a far more intricate and significant role in the host genome. Far from being parasitic, early hypotheses proposed TEs as positive control elements, working in synergy with the host to confer beneficial adaptations. This perspective has gained substantial ground, as it is now widely recognized that the movement and insertion of TEs into the host genome represent a crucial evolutionary mechanism. TE dynamics and activity can profoundly impact an organism's adaptive fitness by creating novel gene regulatory networks, altering gene expression patterns, and even generating new genes. The interplay between TEs and transcriptomics, the study of the complete set of RNA transcripts in a cell, is particularly critical. TE insertions or their aberrant activity can disrupt normal gene transcription, leading to altered RNA profiles that underpin various diseases. Conversely, the precise regulation of TE activity through epigenetic mechanisms and host defense systems is essential for maintaining genomic stability and preventing disease. Understanding the complex relationship between TEs, their impact on the transcriptome, and their direct or indirect roles in disease pathogenesis is a growing field, offering new avenues for therapeutic intervention and a deeper appreciation of the genome plasticity.

Transposable Elements, Transcriptomics, and Diseases is a reading piece that examines the diverse and multifaceted nature of mobile genetic elements. This comprehensive book covers their origins, the intricate mechanisms controlling their activity, their expression patterns, and their profound functional significance in various disease states. Ultimately, the book explores potential therapeutic interventions aimed at understanding the exceptional, yet often underestimated, potential of TEs within the human genome. We are sincerely grateful for the input of the reviewers, Dr. Dass, Dr. Nian Liu, Dr. Xinyi Lu, Jess Bell, Dr. Tomás Carrasco-Valenzuela, Serafina Turner, Prof. Yejing Ge, Dr. Miriam Konkel, Prof. Peter J. Hamilton, Dr. Sabri Saeed Sanabani, Prof. Manyuan Long, Dr. Mayra Martinez, Dr. Boshi Wang, Prof. Rabindra Nath Chatterjee, Dr. Cheng-Zhong Zhang, and Dr. Juliane Glaser, whose insightful comments significantly improved the quality of this book.

Alta Gracia, Córdoba, Argentina — Luis M. Vaschetto

Contents

1 **Transposable Elements and Genome Evolution in Vertebrates** 1
Vahid Nikoonejad Fard and Marc Tollis

2 **Transposable Elements and Gene Duplication** 37
Izabela Makałowska and Natalia Ryczek

3 **Transposable Elements and Their Roles in Genome Architecture** ... 73
Sampriti Majumder, Snata Pandey, and Sugopa Sengupta

4 **Epigenetic Control of Transposable Elements** 109
Zhengyan Liang, Chengyu Jia, Xinyi Yao, and Yin Zhang

5 **Transposable Element-Derived Noncoding RNAs** 153
Francesco Panariello, Alen Stambolliu, Michele Panepuccia, Filippo Vittorio Burattin, and Beatrice Bodega

6 **Transposable Elements: The Role of the piRNA Pathway** 191
Jiao-Yan Ma and Xin Zhiguo Li

7 **The Role of Transposable Elements in Development** 221
John R. McCoy and Tara T. Doucet-O'Hare

8 **The Impact of Transposable Elements on Immune Pathways** 249
Mónica Salinas-Pena, Pau Homs-Aguadé, and Albert Jordan

9 **Transposable Elements and Hereditary Diseases** 289
Sulev Kõks

10 **Transposable Elements in Cancer** 311
Sulev Kõks

11 **Transposable Elements as Drivers of Age-Related Neurodegenerative Disease** 341
Chiara Scopa, Megan Muench, and Davide Trotti

12 Retrotransposons: Silent Architects of Vascular Aging and Disease 375
Sudip Kumar Paul and Naoya Takayama

13 The Role of Transposable Elements in Therapeutics 389
Kathryn E. Howe and Tara T. Doucet-O'Hare

Editor and Contributors

About the Editor

Luis M. Vaschetto obtained his Ph.D. in Biological Sciences from the Universidad Nacional de Córdoba (Argentina). He was a postdoctoral researcher at the Consejo Nacional de Investigaciones Científicas y Técnicas (CONICET) and a research associate at the South Dakota State University (SDSU). He has first-authored over ten academic articles in top-tier journals such as *WIREs RNA*, *Briefings in Functional Genomics*, *Journal of Cellular Biochemistry*, *Molecular Breeding*, *Current Gene Therapy*, *RNA Biology*, and *Current Genetics*, and also published several book chapters. Currently, he is Series Editor for the book series Epigenetics and Human Health (EHH), published by Springer Nature (Switzerland). He has edited and authored eight international books, including two for the EHH series and three for Methods in Molecular Biology (MiMB, Springer Nature, USA). His research mainly focuses on understanding the epigenetic mechanisms involved in plant and animal development, as well as their effects on human health and diseases.

Contributors

Beatrice Bodega INGM, Istituto Nazionale di Genetica Molecolare "Romeo ed Enrica Invernizzi", Milan, Italy

Department of Clinical Sciences and Community Health, University of Milan, Milan, Italy

Filippo Vittorio Burattin INGM, Istituto Nazionale di Genetica Molecolare "Romeo ed Enrica Invernizzi", Milan, Italy

Tara T. Doucet-O'Hare Department of Genetics and Biochemistry, and Center for Human Genetics, Clemson University, Greenwood, SC, USA

Vahid Nikoonejad Fard School of Informatics, Computing, and Cyber Systems, Northern Arizona University, Flagstaff, AZ, USA

Pau Homs-Aguadé Molecular Biology Institute of Barcelona (IBMB-CSIC), Barcelona, Spain

Kathryn E. Howe Department of Genetics and Biochemistry, and Center for Human Genetics, Clemson University, Greenwood, SC, USA

Chengyu Jia Guangdong Provincial Key Laboratory of Malignant Tumor Epigenetics and Gene Regulation, Sun Yat-sen Memorial Hospital, Sun Yat-sen University, Guangzhou, Guangdong, China

Albert Jordan Molecular Biology Institute of Barcelona (IBMB-CSIC), Barcelona, Spain

Sulev Kõks Medical Genomics Laboratory, Murdoch University, Personalised Medicine Centre, Perron Institute, Murdoch, WA, Australia

Genomic Therapeutics Pty Ltd, Prion OÜ, Perth, WA, Australia

Xin Zhiguo Li Center for RNA Medicine, the Fourth Affiliated Hospital of School of Medicine, and International School of Medicine, International Institutes of Medicine, Zhejiang University, Yiwu, China

Zhengyan Liang Guangdong Provincial Key Laboratory of Malignant Tumor Epigenetics and Gene Regulation, Sun Yat-sen Memorial Hospital, Sun Yat-sen University, Guangzhou, Guangdong, China

Guangdong Pharmaceutical University, Guangzhou, Guangdong, China

Jiao-Yan Ma Center for RNA Medicine, the Fourth Affiliated Hospital of School of Medicine, and International School of Medicine, International Institutes of Medicine, Zhejiang University, Yiwu, China

Sampriti Majumder Department of Life Sciences, Presidency University Kolkata, Kolkata, India

Izabela Makałowska Institute of Human Biology and Evolution, Faculty of Biology, Adam Mickiewicz University, Poznań, Poland

John R. McCoy Department of Genetics and Biochemistry, and Center for Human Genetics, Clemson University, Greenwood, SC, USA

Megan Muench Weinberg ALS Center, Vickie and Jack Farber Institute for Neuroscience, Department of Neuroscience, Thomas Jefferson University, Philadelphia, PA, USA

Francesco Panariello INGM, Istituto Nazionale di Genetica Molecolare "Romeo ed Enrica Invernizzi", Milan, Italy

Snata Pandey Department of Life Sciences, Presidency University Kolkata, Kolkata, India

Michele Panepuccia INGM, Istituto Nazionale di Genetica Molecolare "Romeo ed Enrica Invernizzi", Milan, Italy

SEMM, European School of Molecular Medicine, Milan, Italy

Sudip Kumar Paul Department of Regenerative Medicine, Graduate School of Medicine, Chiba University, Chiba, Japan

Natalia Ryczek Institute of Human Biology and Evolution, Faculty of Biology, Adam Mickiewicz University, Poznań, Poland

Mónica Salinas-Pena Molecular Biology Institute of Barcelona (IBMB-CSIC), Barcelona, Spain

Centre for Genomic Regulation (CRG), The Barcelona Institute of Science and Technology (BIST), Barcelona, Spain

Chiara Scopa Weinberg ALS Center, Vickie and Jack Farber Institute for Neuroscience, Department of Neuroscience, Thomas Jefferson University, Philadelphia, PA, USA

Sugopa Sengupta Department of Life Sciences, Presidency University Kolkata, Kolkata, India

Alen Stambolliu INGM, Istituto Nazionale di Genetica Molecolare "Romeo ed Enrica Invernizzi", Milan, Italy

SEMM, European School of Molecular Medicine, Milan, Italy

Naoya Takayama Department of Regenerative Medicine, Graduate School of Medicine, Chiba University, Chiba, Japan

Marc Tollis School of Informatics, Computing, and Cyber Systems, Northern Arizona University, Flagstaff, AZ, USA

Davide Trotti Weinberg ALS Center, Vickie and Jack Farber Institute for Neuroscience, Department of Neuroscience, Thomas Jefferson University, Philadelphia, PA, USA

Xinyi Yao Guangdong Provincial Key Laboratory of Malignant Tumor Epigenetics and Gene Regulation, Sun Yat-sen Memorial Hospital, Sun Yat-sen University, Guangzhou, Guangdong, China

Yin Zhang Guangdong Provincial Key Laboratory of Malignant Tumor Epigenetics and Gene Regulation, Sun Yat-sen Memorial Hospital, Sun Yat-sen University, Guangzhou, Guangdong, China

Department of Cellular and Molecular Diagnostics Center, Sun Yat-Sen Memorial Hospital, Sun Yat-Sen University, Guangzhou, Guangdong, China

Chapter 1
Transposable Elements and Genome Evolution in Vertebrates

Vahid Nikoonejad Fard and Marc Tollis

Abstract Transposable elements (TEs) are pervasive genomic components that have profoundly influenced vertebrate genome evolution. Once regarded as junk DNA, TEs are now recognized as dynamic agents of genetic innovation, shaping genome architecture and gene regulation, while their hyperactivity and aberrant expression can contribute to the onset and progression of various diseases. This chapter provides an overview of TE classification, structure, and transposition mechanisms, highlighting their intra- and interspecies transmission through vertical inheritance and horizontal transfer (HT). We explain the evolutionary forces that shape the fate of TE insertions, their contribution to expanding genomic diversity, and their capacity to drive functional and regulatory innovations. We emphasize TE-mediated exonization, gene duplication, *cis*- and *trans*-regulatory effects, and their impact on evolutionary divergence and speciation. We also discuss the coevolutionary "arms race" between TEs and host genomes, detailing silencing mechanisms and TE counterstrategies. Ultimately, we explore how TE-driven genome expansion and contraction govern genome size dynamics under the "accordion model" and outline emerging research frontiers enabled by long-read sequencing, single-cell technologies, and comparative genomics. Altogether, TEs are not merely relics of past invasions but central architects of genome plasticity, innovation, and evolutionary adaptation in vertebrates.

Keywords Transposable elements · Genome evolution · Genome plasticity · Horizontal transfer · Structural variation · Gene regulation · Epigenetic reprogramming · Speciation · TE-host coevolution · Genome size

---

V. Nikoonejad Fard · M. Tollis (✉)
School of Informatics, Computing, and Cyber Systems, Northern Arizona University, Flagstaff, AZ, USA
e-mail: marc.tollis@nau.edu

L. M. Vaschetto (ed.), *Transposable Elements, Transcriptomics, and Diseases*, Epigenetics and Human Health 14, https://doi.org/10.1007/978-3-032-04125-8_1

Abbreviations

TE	Transposable element
HT	Horizontal transfer
SINE	Short interspersed nuclear element
LINE	Long interspersed nuclear element
L1	Long interspersed nuclear element type-1
nLTRs	Non-long terminal repeat retrotransposons
TAD	Topologically associating domain
SVA	SINE-R/VNTR/Alu element
VNTR	Variable number tandem repeat
ORF	Open reading frame
Indel	Insertion and deletion
TF	Transcription factor
TFBS	Transcription factor binding site
ERV	Endogenous retrovirus
HERV	Human endogenous retrovirus
SV	Structural variation
CNV	Copy number variation
UTR	Untranslated region
Pol	RNA polymerase
Mya	Million years ago
ncRNA	Noncoding RNA
miRNA	MicroRNA
lncRNA	Long noncoding RNA
piRNA	Piwi-interacting RNA
NOR	Nucleolar organizer region

1.1 Introduction

The vertebrate genome is complex and dynamic, mainly encompassing repetitive noncoding sequences with only a tiny fraction composed of protein-coding genes (Feschotte and Pritham 2007). Transposable elements (TEs), also termed "jumping genes," "selfish DNA," or "genome parasites," are the predominant type of repetitive elements with the ability to relocate and proliferate within the genome, offering both advantages and disadvantages to the host. TEs typically range from 100 bp to 10 kb in length and comprise between 6% and 85% of vertebrate genomes (Arkhipova and Yushenova 2019; Wells and Feschotte 2020), accounting, on average, for 45% of the mammalian genome (Osmanski et al. 2023) and approximately 50% of the human genome (Lander et al. 2001; Nurk et al. 2022).

TEs were first discovered by Barbara McClintock in 1944 in the maize genome. She termed them "*transposable controlling elements*," noting that "*they can appear at new locations, disappear from previously determined locations, and alter the expression of nearby genes, leading to phenotypic variations such as changes in kernel color*" (McClintock 1950, 1956; Feschotte 2023). These insights gained broader recognition after the 1970s, as scientists rediscovered TEs and their profound impact on shaping the genomes and critical phenotypes in various organisms (Doolittle and Sapienza 1980; Orgel and Crick 1980; Kazazian et al. 1988). This resurgence in TE research transformed their perception from genomic parasites to key regulators of genome architecture, function, and evolution. As highlighted by Shapiro (2022), McClintock's seminal work revolutionized our conventional paradigm that genome evolution is driven solely by small, incremental mutations or localized genetic errors. Instead, she provided the first evidence that species have a remarkable inherent capacity to restructure their genomes in a dynamic, dispersed, and regulated manner, a process in which mobile elements play a fundamental role.

It is now fully understood that TE activity can be a major force in driving structural rearrangements and, occasionally, facilitating the emergence of novel functional elements that could result in critical evolutionary adaptations (Chalopin et al. 2015; Wells and Feschotte 2020; Fueyo et al. 2022). TEs influence gene expression by inserting regulatory sequences, rewiring transcriptional networks, and serving as enhancers, silencers, and alternative promoters (Babaian and Mager 2016; Ricci et al. 2018; Senft and Macfarlan 2021; Shapiro 2022; Fueyo et al. 2022). Additionally, TE insertions play an essential role in epigenome remodeling, where their sequences interact with DNA methylation, histone modifications, and small RNA pathways to regulate chromatin accessibility and transcriptional states. Their ability to modify chromatin structure contributes to three-dimensional genome organization, facilitating long-range interactions between regulatory elements and influencing higher-order nuclear architecture (Slotkin and Martienssen 2007; Pehrsson et al. 2019; Choudhary et al. 2023; Matsuo and Cristofari 2025). In addition to these potentially fitness-enhancing changes to the genomes, uncontrolled TE activity can disrupt genome stability, contributing to mutagenesis, deleterious chromosomal rearrangements, and genetic disorders. TE-mediated disruptions have been implicated in the onset and progression of various diseases, including cancers, neurodegenerative disorders, autoimmune conditions, and many other age-related diseases, where pervasive TE derepression alters gene expression and genomic integrity (Poduri et al. 2013; Hancks and Kazazian 2016; De Cecco et al. 2019; Payer and Burns 2019; Gorbunova et al. 2021; Chénais 2022; Liu et al. 2023).

This chapter explores the classification, structure, and transposition mechanisms of TEs, their colonization strategies, and population dynamics. We review how TE activity drives genomic variation, structural innovation, gene regulation, and epigenetic remodeling. Additionally, we discuss TE-host coevolution, regulatory mechanisms controlling TE activity, their impact on evolutionary divergence and speciation, and their influence on genome size dynamics. Finally, we highlight emerging research directions in TE biology and genome evolution.

1.2 TE Classification, Structure, and Mechanisms of Transposition

TEs are categorized based on whether they rely on self-encoded enzymes or require other elements or host-encoded factors for mobilization (Finnegan 1989). Autonomous TEs are those equipped with the essential tools to relocate within the genome. In contrast, nonautonomous TEs do not have endogenous equipment and hijack that machinery from the host or other elements for transposition. Nonautonomous elements commonly emerge and evolve as parasites of autonomous TEs, such as short interspersed nuclear elements (SINEs) or miniature inverted repeats (Dewannieux and Heidmann 2005; Ohshima and Okada 2005). TEs are also classified into two major classes based on their replication mechanisms, enzymatic properties, and sequence similarities (Wicker et al. 2007; Kapitonov and Jurka 2008; Arkhipova 2018). The first mechanism is called replicative transposition (also known as copy-and-paste), which mimics the replication of a viral retroelement (Eickbush and Malik 2007; Grandi and Tramontano 2018). Class I elements, or retrotransposons, use this mechanism in which the original element remains in the same place while a copy is inserted at the new position. Consequently, this transposition mode results in the proliferation of the element within the genome (i.e., an increase in TE's copy numbers). This mechanism is usually facilitated by an RNA intermediate, transcribed from an endogenous sequence, and a reverse transcriptase.

The second mechanism, the conservative process or cut-and-paste transposition, is characteristic of most Class II elements or DNA transposons. It involves excision of the element from its primary position and reintegration elsewhere via a DNA intermediate. A few DNA transposon families like cryptons, helitrons, and polintons (also called mavericks) recruit distinctive replicative mechanisms other than the cut-and-paste mechanism (Kapitonov and Jurka 2006; Kojima and Jurka 2011). For instance, helitrons transpose via a unique process known as rolling-circle transposition, using replication initiator helicase protein, which combines both the rolling-circle endonuclease and helicase domains (Kapitonov and Jurka 2001; Thomas and Pritham 2015). Although the cut-and-paste mechanism does not result in the propagation of the DNA transposons within the genome, their copy number can still increase if transposition occurs during DNA replication from replicated to unreplicated sites (Ros and Kunze 2001; Spradling et al. 2011; Skipper et al. 2013; Fricker and Peters 2014) or if excised sites are repaired through homologous recombination using an unexcized chromosomal copy (Engels et al. 1990; Hsia and Schnable 1996; Rubin and Levy 1997). Furthermore, DNA transposons occasionally mediate the amplification of nonautonomous elements that have lost their ability to transpose due to internal deletions (Hartl et al. 1992; Yang et al. 2009; Almojil et al. 2021).

Generally, important Class I elements, such as long- and short-interspersed nuclear elements (LINEs & SINEs), have less stringent insertion site preferences than Class II elements. For example, LINE-1 (L1) is almost randomly distributed across the mammalian genome, and this relaxed insertion site preference is one reason it represents the most frequent class of TEs in this taxon (Flasch et al. 2019; Sultana et al. 2019). Retrotransposons are a common feature of eukaryotic genomes and are less common in prokaryotes (Ramakrishnan et al. 2022). They are further subdivided into two prominent subclasses—the long terminal repeat (LTR) and the non-long terminal repeat (nLTR) elements—which are further categorized into different superfamilies, families, and subfamilies based on the phylogenetic origins and sequence homology (Wicker et al. 2007; Kapitonov and Jurka 2008; Arkhipova 2018).

LTRs typically range from 6 to 11 kb and are abundant in plants, invertebrates, and some vertebrates (Petersen et al. 2019). Across mammals, they cover 2 to 17.8% of the genome, averaging 7.8% (Osmanski et al. 2023). The most common LTRs in the human genome are endogenous retroviruses (ERVs), typically categorized into three classes based on their ancient retroviral origins (Gifford et al. 2018; Jansz and Faulkner 2021). Two 100–300 bp long terminal repeats flank the proviral sequences in ERVs, which contain the group-specific antigen (*GAG*), protease (*PRO,* also known as *PEP*), polymerase (*POL*), and envelope (*ENV*) genes of retroviral origin. The LTR has a dual function. It acts as a regulatory element, with promoter and enhancer functions, while also playing a pivotal role in the reverse transcription of the element into DNA (Kojima et al. 2023). The *POL* encodes essential enzymatic components, including reverse transcriptase and integrase, while the *GAG* produces structural proteins that form viral capsid particles. The *PRO* encodes a protease crucial in processing polyproteins into their functional forms. Ultimately, the *ENV*, which is lost in most ERVs, is vital for reinfection and integrating the element into new cells (Wilhelm and Wilhelm 2001; Hughes 2015; Wells and Feschotte 2020).

The majority of the human ERVs (HERVs) were inserted in the human genome >25 million years ago (Mya) (Lander et al. 2001; Mills et al. 2007). They are approximately 9.5 kb in length (i.e., the length of the consensus sequence) and cover nearly 9% (i.e., <400 k loci) of the human genome (Pace and Feschotte 2007; Mager and Stoye 2015). Even though HERVs appear to have lost their transposition ability in the human genome (Bannert and Kurth 2004), their remnants still play critical regulatory roles in early embryonic development, particularly in totipotent and pluripotent cells (e.g., human- and chimpanzee-specific HERVK/LTR5) (Hirata et al. 2022). Recent evidence suggests that some transcriptionally active HERVs, such as HERVH, demarcate topologically associating domains (TADs) in pluripotent stem cells (Zhang et al. 2019b). TADs are regions of the genome that interact more frequently with each other than with other regions. These interactions are determined by how the DNA supercoil is twisted and organized in its 3D structure (Fueyo et al. 2022). In differentiated cells, HERV regulatory sequences still act as both canonical and cryptic promoters, upregulating the nearby genes, such as the *SLC4A8* gene, which regulates intracellular pH and electrolyte balance, and the *IFT172* gene, which is essential for cilia formation and function (Gogvadze et al. 2009; Babaian and Mager 2016; Grandi and Tramontano 2018; Deniz et al. 2020).

HERVK is another prominent HERV subfamily implicated in the regulation of gene expression and immune system modulation. Its aberrant activation can contribute to the development of neurodegenerative diseases like amyotrophic lateral sclerosis (ALS) and susceptibility to autoimmune diseases, such as rheumatoid arthritis in humans (Jönsson et al. 2019; Burn et al. 2022). A human-specific HERVK, known as HERVK HML-2, is the most recent HERVK to have integrated into the human genome after the human–chimpanzee divergence of approximately 0.5 Mya (Burn et al. 2022). The HERVK HML-2 insertion upstream of the schizophrenia-associated *PRODH* gene enhances its expression in the hippocampus. This gene regulates proline catabolism, which is crucial for central nervous system function (Suntsova et al. 2013). In many other mammalian species, however, ERVs are relatively more frequent than in humans and other hominoids (e.g., great apes and gibbons) with potential mutagenic and broader regulatory capacity (Mager and Stoye 2015; Magiorkinis et al. 2015).

The nLTR retrotransposons, on the other hand, have a more streamlined structure and do not contain long terminal repeats. The most active and well-documented subfamilies of human nLTRs are LINEs, SINEs, and SINE-R/VNTR/Alu (SVAs) (Kapitonov and Jurka 2008; Bourque et al. 2018; Fueyo et al. 2022). With over 1.7 million copies, SINEs have the highest copy number for any interspersed repetitive DNA, covering roughly 14% of the human genome. They also range between 0.4 and 32.1% (i.e., 10.5%, on average) of the genomes in other mammals. LINEs are less frequent, with approximately one million copies (e.g., including ~0.5 million copies of L1), yet they span over 17% of the human genome, as they are longer than SINEs. They also occupy 8.2 and 52.8% of the genomes in various mammals (i.e., 22.6% on average) (Lander et al. 2001; Chénais 2022; Osmanski et al. 2023). LINEs are thought to have evolved from bacterial organelle group II introns (Malik et al. 1999) and encompass three subfamilies, including L1, LINE-2, and LINE-3, of which the latter two have almost entirely degenerated and lack transposition ability (Cordaux and Batzer 2009).

L1 is an autonomous TE of approximately 6.1 kb and is flanked by an untranslated region (UTR) at its 5′ end, which is composed of a promoter responsive to RNA polymerase II (Pol II) and a series of A-T base pairs at its 3′ end, truncated in different lengths among copies across the genome (Babushok and Kazazian 2007; Beck et al. 2011). Having been active for around one billion years, L1 is perhaps the most ancient TE in eukaryotes and the most successful TE family across therian mammals (Ivancevic et al. 2018; Percharde et al. 2020). Only a tiny fraction of L1 elements in the human genome remain active, yet their evolutionary history of transposition and activity has played a crucial role in genome organization and function (Brouha et al. 2003; Betancourt et al. 2024). The full-length active L1 uses an endonuclease and a reverse transcriptase for transposition. Primarily, Pol II transcribes an intact L1 to a capped polyadenylated mRNA, which is further exported into the cytoplasm and translated into ORF1p and ORF2p (Beck et al. 2011). Subsequently, ORF1p facilitates the integration of both proteins with the mRNA to assemble an RNP complex, which will then be shuttled to the nucleus and bind with the A-T-rich bundle, nicked by an endonuclease. This final complex is a substrate for reverse transcriptase, facilitated by ORF2p, which has both endonuclease and reverse transcriptase activities (Luan et al. 1993; Doucet et al. 2010; Richardson et al. 2015). A primate-specific ORF0 is located in the 5′UTR of the L1 transcript on the opposite strand of the ORF1 and ORF2 structural genes. ORF0 boosts the L1 retrotransposition by stabilizing the ORF1p and facilitating its interaction with RNA (Denli et al. 2015).

Nonautonomous SINEs are typically derived from noncoding molecules such as tRNAs, 7SL, and 5S RNAs transcribed by Pol III in eukaryotic cells (Kriegs et al. 2007; Kramerov and Vassetzky 2011). They are generally between 100 and 400 bp in length and do not contain any genes, meaning they do not synthesize reverse transcriptase. Instead, they rely on the L1s' and other autonomous elements' reverse transcriptase for transposition (Dewannieux and Heidmann 2005). The most common SINE element in primates is Alu, with more than one million copy numbers in the human genome (i.e., ~10% of the human genome) (Lander et al. 2001). Alu elements consist of two halves (i.e., two ancient monomeric 7SL-derived SINEs), each of which is made up of a similar 120 bp sequence, with a 31–32 bp insertion in the right half (Kriegs et al. 2007; Chénais 2022). Alu insertions are responsible for many differences between human and chimpanzee genomes (Hedges et al. 2004). They are highly active and have been found inside the sequence of numerous genes, such as *BRCA1* and *BRCA2*, which have been implicated in the development of ovarian and breast cancer (Batzer and Deininger 2002; Teugels et al. 2005), and other types of genes, such as the *TBXT* (Brachyury), which resulted in tail loss in hominids and apes (Xia et al. 2024). The B1 element, which is equivalent to half of an Alu sequence in primates, originates from a common 7SL RNA ancestor and remains one of the most active SINEs in rodent genomes (Kriegs et al. 2007; Chénais 2022).

L1s and SINEs can be found in almost all intergenic and intragenic regions of the mammalian genome, and their widespread distribution and persistent activity have allowed them to be co-opted (also known as domestication, exaptation, or adaptation) for various functional and regulatory roles (Richardson et al. 2015). The term refers to the process by which previously inactive or nonfunctional elements are recruited for new regulatory or functional purposes. This can lead to adaptive innovation, trait modification, or even pathological consequences, such as the co-option of inactive TEs by tumor cells to reprogram gene expression patterns, a process referred to as oncoexaptation. Accordingly, L1s and SINEs have contributed to the evolution of human and other mammalian genomes by facilitating gene structural modifications, modulating gene expression, and influencing chromatin architecture (Richardson et al. 2015; Kordyukova et al. 2018; Matsuo and Cristofari 2025). Notably, the profile and activity of these elements vary across tissues, and they may be co-opted for various roles throughout an individual's lifetime (Philippe et al. 2016), yet many of the resulting modifications cannot be vertically transmitted unless they occur in the germline or stem cells (Platt et al. 2018; Kofler et al. 2018).

SVA is another prominent active hominid-specific retroelement, named based on its structural components—an Alu-like sequence, a variable number tandem repeat (VNTR), and an LTR-derived fragment (Wang et al. 2005). Like SINEs, SVA activity relies on the L1 reverse transcription machinery, although it is relatively longer, spanning approximately 1 to 2 kb. SVAs are among the youngest elements in the human genome with active retrotransposition ability. Evidence suggests that intronic SVA insertions are enriched in human genes associated with neurodevelopment (Nadler et al. 2023). While typically being repressed by the KRAB-ZNF pathway in primed embryonic stem cells and the epiblast, derepressed SVAs act as enhancers in the fetal brain, pineal gland, and hippocampus, regulating the nearby genes (Trizzino et al. 2018; Pontis et al. 2019). They are also capable of causing deleterious mutations and contributing to diseases. For instance, alternative splicing patterns caused by SVA insertions are responsible for the development of Fukuyama muscular dystrophy (FCMD) (i.e., disrupting the *Fukutin* gene) and Lynch syndrome (Taniguchi-Ikeda et al. 2011; Yamamoto et al. 2021).

Active DNA transposons, alongside cryptons, helitrons, and polintons, are more prevalent in plants and, to a greater extent, in prokaryotes than in humans and other mammals (Kojima and Jurka 2011). These elements constitute approximately 4% of mammalian genomes, with DNA transposons ranging from 0.5% to 8.4% and rolling-circle elements varying between 0.01% and 19.7% across species (Osmanski et al. 2023). These elements remained active throughout the early evolutionary history of primates until approximately 37 Mya, yet have since lost their transpositional activity in humans due to missing or mutated terminal sequences required for transposition (Feschotte and Pritham 2007; Pace and Feschotte 2007). Roughly 2.8% of the human genome (i.e., ~294,000 copies, equivalent to ~6% of total human TEs) is made up of DNA transposons of various types, notably TIR-containing transposons, including hATs, piggyBacs, and TcMariners, all of which are entirely transpositionally inactive (Kapitonov and Jurka 2006; Chénais 2022).

To date, limited evidence indicates the functionality of a few DNA transposons in the human genome (Fugmann et al. 2000; Gellert 2002; Pace and Feschotte 2007; Majumdar et al. 2013). Some beneficial insertions have been identified, such as the retention of the P-element transposase activity in the human *THAP9* gene (Majumdar et al. 2013) or the ancient domestication of a Transib family of DNA transposons, which occurred nearly 500 Mya in jawed vertebrates. The latter event led to the transformation of the ancestral RAG transposase into the modern RAG recombinase, which enabled the evolution of a diverse repertoire of antibodies and T cell receptors, significantly enhancing immune recognition of pathogens in this clade (Kapitonov and Jurka 2005; Zhang et al. 2019a). Nevertheless, DNA transposons are more associated with disease implications than beneficial effects, mainly due to their potential to disrupt genes and regulatory elements. For instance, an oncogenic role has been suggested for the human piggyBac-derived-5 gene (*PGBD5*), contributing to pediatric cancers (Henssen et al. 2017).

1.3 TE Dispersal Across Generations and Species

Over evolutionary time, TEs have successfully occupied vast portions of eukaryotic genomes, shaping their structure, function, and regulation (Wells and Feschotte 2020). The primary strategy for long-term propagation within the populations is through vertical inheritance, which requires de novo insertions of active elements within germlines or stem cells during early embryogenesis, ensuring the integration of novel insertions into the genetic material passed on to offspring. TEs exhibit elevated activity in the germline and during early embryonic development due to relaxed epigenetic constraints, such as widespread DNA hypomethylation and accessible chromatin configuration, which transiently relieve transcriptional repression and facilitate TE mobilization (Zamudio and Bourc'his 2010; DiRusso and Clark 2023; Oomen and Torres-Padilla 2024; Betancourt et al. 2024). This means that even without exogenous input elements, TEs can still increase their copy numbers, securing their persistence throughout extended timescales (Shapiro 2022).

Beyond vertical inheritance, TEs have the ability to invade new hosts through horizontal transfer (HT). HT refers to the passage of genetic material between organisms through mechanisms other than conventional reproduction strategies (Gilbert and Feschotte 2018; Zhang et al. 2020). Several biological agents have been implicated in this process. Bacteria, including their endosymbiotic forms, are among the well-established vectors capable of harboring TEs and integrating them into the genomes of host organisms (Arnold et al. 2022). Similarly, proxiviruses can excise and encapsulate TEs along with their own viral sequences, subsequently introducing them into new hosts (Piskurek and Okada 2007). Pathogens that infect gonadal tissues are another potential route for HT, as they introduce novel TEs into the germline (Gilbert et al. 2014). Beyond microbial vectors, parasitic organisms are also important agents of HT. Blood-feeding arthropods, such as triatomine bugs and ticks, as well as endoparasitic nematodes, have been suggested as mediators of TE transfer across species (Gilbert et al. 2010; Walsh et al. 2013; Dunemann and Wasmuth 2019). For instance, L1 elements are believed to have originated in bacteria and were horizontally transferred to eukaryotic hosts through specific parasites (Kordis and Gubensek 1998; Ivancevic et al. 2018). Another mode of HT involves predator–prey interactions, such as HT of RTE-BovBs from snakes to frogs (Kambayashi et al. 2022) or potential DNA transposons from arthropods and non-mammalian vertebrates to carnivorous mammals, where diet might have served as a mediator (Osmanski et al. 2023).

The rate of HT is highly influenced by geographical proximity, as overlapping ecological niches or habitats increase physical contact and shared vectors. Generally, DNA transposons are more frequently horizontally transferred between species than retroelements (Wallau et al. 2018; Zhang et al. 2020). Upon generating or rewiring a critical beneficial function, the HT is often followed by a rapid proliferation of the element due to selection, as each new insertion is a source for more transpositions (Gilbert and Feschotte 2018; Galbraith et al. 2021). HT also appears to be more prevalent in aquatic animals, such as ray-finned fishes, than terrestrial species (Galbraith et al. 2021), yet it is not just restricted to closely related species but also occurs between distant taxa, such as animals and plants, or between vertebrates and invertebrates (Gilbert and Feschotte 2018; Liu et al. 2023). Multiple examples exist to demonstrate HT between distant taxa, such as the ancient transfer of Penelope-like retroelements from arthropods to conifers (Lin et al. 2016) or the transfer of the RTE-BovB elements between squamates and bovids (Kordis and Gubensek 1998).

Despite numerous examples, however, identifying HT events is difficult, primarily due to the challenges in inferring TE homology across species and the degeneration of TEs over time, which obscures their evolutionary history. Current methods use three main criteria to detect HT of TEs: (i) discrepancies between host and TE phylogenies, (ii) similarity of TEs found in distantly related species, and (iii) uneven distribution of TEs across phylogenies (Wallau et al. 2018; Zhang et al. 2020).

1.4 The Evolutionary Fate of TEs: Selection Versus Drift

The genome is a blueprint containing the foundational molecular instructions for the entire organism's ontogeny. Evolution represents the continual modification of this blueprint across generations, enabling species to adapt to ongoing environmental changes. Genome evolutionary dynamics, including those of TEs, are primarily driven by the forces of natural selection and genetic drift. The interplay of these forces with TE transposition and deterioration rates, which are shaped by mutation accumulation, the HT rates, and the complexity of host defense mechanisms, ultimately determines the content, dynamics, and evolutionary trajectory of TEs in host populations and the eventual fixation or elimination of TEs within the genome (Charlesworth and Charlesworth 1983; Charlesworth et al. 1994).

Most TE insertions are considered "dead on arrival," meaning they generally do not promote functional or regulatory modifications. Accordingly, they neither significantly harm nor benefit the host organism. The immediate impact of these neutral insertions is an expansion of genome size, which may impose a subtle genomic burden by increasing replication demands and transcriptional noise (Beck et al. 2011; Fueyo et al. 2022). Selectively neutral TE insertions also add to the repository of potential coding and regulatory variants that can be co-opted during the genome's evolutionary trajectory (Shapiro 2022). If co-option does not occur, these neutral insertions gradually accumulate mutations, leading to their degradation and eventual inactivation, leaving behind nonfunctional molecular remnants often referred to as TE fossils. Over time, these nonfunctional TEs may be purged from the genome through recombination-based mechanisms, notably through unequal crossing over or large-scale deletions, helping to restore genome efficiency, maintain stability, and decrease genome size (Payer and Burns 2019; Chénais 2022; Pascarella et al. 2022).

De novo TE insertions can disrupt proper gene function, influence nearby regulatory networks, and alter gene expression and chromatin dynamics, potentially leading to highly penetrant phenotypes and disease (Boissinot et al. 2006; Hancks and Kazazian 2016; Payer and Burns 2019). Subsequently, intense selective pressure against these highly deleterious insertions often results in the rapid elimination of carriers from populations. However, some insertions may persist if they exert complex pleiotropic effects that impact multiple traits or if functional redundancy within the affected genes or pathways buffers their immediate phenotypic consequences. In some cases, cellular mechanisms such as apoptosis or DNA repair pathways act swiftly to mitigate damage, preventing the transmission of harmful insertions.

Notably, while germline harmful insertions are subject to purifying selection at the population level, deleterious somatic insertions can accumulate over an organism's lifetime, contributing to diseases such as cancer (Hancks and Kazazian 2012; Gorbunova et al. 2021; Grundy et al. 2022). Slightly deleterious insertions, in contrast, can persist in the populations for extended periods. While these insertions may reduce fitness, their effects are often too mild, occur later in life, or arise post-reproduction, preventing their immediate elimination by selection. In small populations where the effective population size is smaller than the minimum viable population size, genetic drift plays a dominant role in shaping the population dynamics of TEs, allowing slightly deleterious insertions to persist and even increase in frequency by chance despite their adverse effects. In large populations, however, these elements are more effectively purged because genetic drift does not overpower selection (Lynch and Conery 2003).

TE insertions can also confer beneficial effects that can be co-opted by the host, contributing to genomic innovations that give rise to constitutive or tissue-specific functional or regulatory elements (Brosius 1991; Fukuda 2024; Cagliani et al. 2024). In rare cases, these insertions become essential for organismal survival or maintenance (Gilbert and Feschotte 2018; Galbraith et al. 2021). Although rare, beneficial germline insertions can rise rapidly in frequency, as each new insertion may seed further transpositions. The probability of such insertions reaching fixation is proportional to their impact on fitness (Tollis and Boissinot 2012; Fueyo et al. 2022). Recent TE insertions are often detected as polymorphic alleles that segregate within a population, present in some individuals but not yet fixed, providing insight into ongoing TE dynamics and activity (Fueyo et al. 2022). While the evolutionary forces governing TE fate vary across populations, the broader impact of beneficial insertions contributing to structural and functional innovations can extend beyond individual lineages. Collectively, the interplay between genetic drift, selection, and the diverse effects of TE insertions shapes the genomic landscape, driving continuous adaptation in both TEs and their hosts (Osmanski et al. 2023). Lastly, like deleterious insertions, somatic beneficial insertions (also known as local exaptations) can enhance individual fitness without being passed on to the next generation. The long-term imprint of the heritable beneficial TEs on genome architecture and evolutionary diversification will be explored in the following section.

1.5 TE Insertions as Catalysts of Genomic Diversity

Genetic diversity forms the essential foundation upon which evolutionary forces can operate, with broader variation expanding the opportunities for more innovation and adaptation. TE-induced variation represents a significant source of structural rearrangements and functional and regulatory innovations within the genome. Given their remarkable ability to move and proliferate within genomes, TEs are not only key contributors to genome plasticity but also pivotal players in shaping the structure and evolution of eukaryotic genomes (Bennett 2004; O'Donnell and Burns

2010; Tollis and Boissinot 2012; Chalopin et al. 2015). Animals, for example, exhibit a vast array of distinguished traits and phenotypes that have evolved throughout millions of years of stringent evolutionary forces. However, when comparing the genomes of various species, most variation resides not within the differences in coding sequences but rather in the noncoding and repetitive elements (Shapiro 2022). This variation occurs despite each species exhibiting a distinct repertoire of transposons and retrotransposons (Britten and Kohne 1968; Wu and Lu 2019), yet it is so extensive that almost all organisms have a unique inherent capacity to evolve dispersed genomic networks encoding complex cellular and multicellular adaptations (Shapiro 2022). For example, ~63% of the differences between the genomes of humans and chimpanzees are attributed to TEs, which influence the differential expression of almost identical genes (Polavarapu et al. 2011).

TEs exhibit considerable diversity in their sequences and structures, resulting in variations in their mutagenic and regulatory effects. Even within the same TE families, sequences and their associated impacts can differ between populations and sometimes even within populations of the same species. Consequently, each insertion has the potential to deliver unique information capable of modifying genes and cellular functions in various ways (Bourque et al. 2018; Betancourt et al. 2024). Furthermore, the insertion patterns among individuals vary significantly, as the most active elements (i.e., nLTR retrotransposons) exhibit minimal site preferences, leading to predominantly random insertion patterns. For example, de novo TE insertions occur in approximately 6% of human births, and any two human haploid genomes, on average, differ by approximately one thousand TE insertions, mainly due to L1 and SINE/Alu activities (Bourque et al. 2018). Consistently, the frequency of genomic novelty due to L1 insertions is approximately 1 in 100 human births (Hancks and Kazazian 2016) and 1 in 8 mouse births (Richardson et al. 2017). These estimates are conservative, as there are many ways by which TEs can impact the genome (Payer and Burns 2019; Percharde et al. 2020). Thus, such massive diversity, upon improving fitness and surviving evolutionary forces like bottlenecks and genetic drift, can be subject to natural selection and develop morphological alterations that facilitate the emergence of novel evolutionary adaptations (Percharde et al. 2020; Colonna Romano and Fanti 2022; Shapiro 2022).

1.6 TE Activities Shape the Structural, Functional, and Regulatory Landscape of Genomes

The unique transpositional capability of TEs makes them key contributors to shaping the genome landscape, architecture, and function. TEs are major drivers of SVs within the genome, including duplications, insertions, deletions (collectively termed indels), inversions, and even large-scale chromosomal rearrangements and neochromosome formations, as observed in cancer genomes and certain rapidly evolving lineages in plants and vertebrates (e.g., chromoanagenesis events) (Moran et al.

1999; Hancks and Kazazian 2016; Janoušek et al. 2016; Senft and Macfarlan 2021; Kim et al. 2022). As a result, they can be observed across all intergenic and intragenic landscapes (Wicker et al. 2007; Richardson et al. 2015), including exons, introns (Gotea and Makałowski 2006), pseudogenes, transcription factor binding sites (TFBSs) (Roman et al. 2008; Sundaram et al. 2014), promoters (Batut et al. 2013), enhancers (Jacques et al. 2013), silencers (Garcia-Perez et al. 2016; Chuong et al. 2017), splicing sites (Attig et al. 2018), and polyadenylation signals (Tang et al. 2000).

TE-induced exonization is the process by which TE sequences become incorporated into mature transcripts as novel exons and can lead to alternative splicing patterns or give rise to new protein isoforms with altered or expanded functions (Hancks and Kazazian 2010; Etchegaray et al. 2021). LTR genes such as *ENV*, *PRO*, *GAG*, *POL*, and ORFs from nLTRs widely contribute to these new functionalities at insertion sites. Two illustrative examples are the exonization of an L1-derived ORF1 exon into the *MYL4* gene, which encodes myosin and plays critical roles across various organs in several vertebrates (Kitao et al. 2024), and the syncytin proteins encoded by the *ENV* gene, which contribute to trophoblast cell fusion and immunosuppression in the mammalian placenta (Mi et al. 2000; Mangeney et al. 2007). Similarly, TEs facilitate exon shuffling by mediating the rearrangement of exonic sequences and recombination, generating hybrid genes or new domain architectures (Makałowski et al. 2019), such as a MIR-b SINE element insertion in the fifth exon of insulin-like growth factor 1s (IGF-1Eb and IGF-1Ec) in mammals (Annibalini et al. 2016). In the human genome, over 3000 exons (i.e., >1% of all known exons) are derived from TEs, with LINEs, Alu/SINEs, MIRs, and LTRs contributing the most (Piriyapongsa et al. 2007; Sela et al. 2010).

Beyond incorporating their endogenous genes into novel sites, TEs contribute to novel gene functionalities through deletion, inversion, and transduction mechanisms (Gasparotto et al. 2023). Transduction occurs when all or part of the sequences of nearby genes are inadvertently "hitchhiked" along with the TE sequence during its movement. Thus, transposition can result in the partial or complete duplication of protein-coding genes, increasing copy number variations (CNVs), which are occasionally coupled with novel gene restructuring, rearrangements, and regulatory shifts (the interplay between TEs and gene duplication is discussed in Chap. 2). This phenomenon is common in the transcription of L1 and SVA elements, where bypassing the 3′ termination signals can extend transcription into downstream genomic regions. A striking example is the acyl-malonyl condensing enzyme 1 (*AMAC1*) gene family, which originated through multiple SVA-mediated 3′ transductions, leading to dispersed copies across several chromosomes (Xing et al. 2006). Another interesting example from other species is the retroduplication of the *CDKN2C* gene (also known as *CDKN2CRTG*) in bowhead whales. *CDKN2C*, which functions as a cell cycle control gene, is embedded within a cetacean-specific L1 element and may be responsible for long lifespan and cancer resistance in these species (Vazquez et al. 2022).

Occasionally, TE exaptations can go beyond a simple exonization and result in massive evolutionary adaptations. For instance, TE insertions have been widely implicated in developing advanced adaptive immunities in mammals (Koonin and Krupovic 2015) and other vertebrates (Zhang et al. 2019a), such as the development of the modern RAG recombinase as a diverse repertoire of antibodies and T cells due to the domestication of a Transib family of DNA transposons in jawed vertebrates, as mentioned earlier. Equally important, TE insertions have shaped critical genes involved in placental and embryonic development in therians (Oomen and Torres-Padilla 2024). The *ENV*-derived *syncytin* and *SUPYN* genes and *GAG*- and *POL*-derived *PEG10* and *PNMA* families, for example, are necessary for normal placental development in mammals and marsupials (Kim et al. 2004; Campillos et al. 2006; Lavialle et al. 2013; Frank et al. 2022; Keighley et al. 2023; Kaneko-Ishino and Ishino 2023). Interestingly, many of these genes are species-specific, arising from independent exaptation events across different lineages that lead to convergent evolution (Chuong et al. 2013).

Nonetheless, the primary legacy of TEs lies not in their capability to generate functional innovations but in their profound impact on modulating gene regulation and expression networks across the genome via two main mechanisms. Primarily, TEs create an extensive repository of *cis*-regulatory elements (Jacques et al. 2013; Trizzino et al. 2017; Senft and Macfarlan 2021), of which nearly 47% can be in an active state in normal, fully differentiated human cells—the proportion that tends to increase with age or during the advanced stages of some diseases, such as cancer (Pehrsson et al. 2019). It has been estimated that approximately 20% of *cis*-regulatory elements in the human genome have been co-opted from TEs (i.e., L1 and LTRs in particular) (Lowe and Haussler 2012). These elements have been instrumental in shaping the lineage-specific gene expression patterns that underlie the evolution of the placenta (Chuong et al. 2013; Du et al. 2023; Frost et al. 2023), second plate (Nishihara et al. 2016), mammary gland (Nishihara 2019), and innate immunity (Chuong et al. 2016) in mammals. For instance, specific endogenous sequences within HERV and L1 can function as alternative promoters, initiating the transcription of nearby genes (e.g., L1 5′ UTR-derived bidirectional promoter). Some other sequences, such as in LTRs and SINEs, can act as enhancers or insulators by facilitating TF binding or acting as boundary elements that restrict enhancer–promoter interactions (Roman et al. 2008; Sundaram et al. 2014; Chuong et al. 2017). DNA transposons also serve as significant sources of regulatory regions, including remarkably conserved noncoding elements (CNEs), which have been co-opted into homeodomain TFs. Since the dawn of animal evolution, these elements have been subjected to millions of years of purifying selection and remained highly conserved across both vertebrates and invertebrates (Gonzalez et al. 2024).

Such a pervasive regulatory capacity also enables TEs to shape nearly a quarter of the human epigenome (Pehrsson et al. 2019). TEs influence chromatin state, contribute to TAD demarcations, and modulate DNA methylation and histone modification patterns, thereby regulating chromatin folding and three-dimensional genome organization (Zhang et al. 2019b; Pehrsson et al. 2019; Fueyo et al. 2022; Choudhary et al. 2023). It has been shown that most CpG methylation sites in the human genome are located within TE sequences (Ohtani et al. 2018; Jansz 2019). TEs also modulate ribosomal DNA activity within nucleolar organizer regions

(NOR), thereby affecting nucleolar organization and function (Ataei et al. 2025). The nucleolus is a key nuclear structure responsible for rRNA production and ribosome assembly, essential for protein synthesis, particularly during early development (Yu et al. 2021; Xie et al. 2022). For instance, a hominin-specific L1 element within NORs is actively transcribed in naive human embryonic stem cells. Deletion or silencing of this L1 element disrupts nucleolar organization and function, leading to alterations in cellular identity (Matsuo and Cristofari 2025).

The second mechanism through which TEs participate in modulating gene regulatory networks is by contributing functional sequences to noncoding RNAs (ncRNAs), which act as *trans*-regulatory elements, primarily at the transcriptional and post-transcriptional levels. Approximately 80% of the human genome is transcribed into RNA molecules, while less than 2% encodes proteins (Birney et al. 2007; Park et al. 2022). Accordingly, the remaining RNA transcripts that are not translated into proteins are called ncRNAs. TEs encode a large portion of ncRNAs, such as long noncoding RNAs (lncRNAs) (Johnson and Guigó 2014) and Piwi-interacting RNAs (piRNAs) (i.e., predominantly active in germlines) (Betel et al. 2007), thereby playing critical roles in regulating gene expression in both transcribed and non-transcribed regions of the genome (Birney et al. 2007; Park et al. 2022). These molecules participate in various cell functional and developmental regulations, making their biological significance equal to proteins in eukaryotes (Oo et al. 2022; Andergassen and Rinn 2022; Shapiro 2022). For instance, the depletion of the SVA-containing lncRNA AK057321, present in humans, chimpanzees, bonobos, and gorillas, leads to reduced expression of several critical genes involved in neuronal maturation that harbor intronic SVA elements, including *CHAF1B*, *KCNJ6*, *POFUT2*, *HTT*, *CDK5RAP2*, and *SCN8A* (Nadler et al. 2023). The miRNAs can also originate from TEs. These molecules, called TE-derived miRNAs, are implicated in various cellular functions. Their upregulations are also responsible for the progression of infectious diseases and cancer. However, while ncRNAs, in general, have been the subject of numerous research studies, the functional significance of TE-derived ncRNAs remains largely understudied (Park et al. 2022).

TEs also facilitate many other developmental and regulatory processes in animals, such as the compensation of X chromosome dosage in Drosophila (Ellison and Bachtrog 2013), sex determination in fish (Chalopin et al. 2015), and general early embryonic (Xiang and Liang 2021), uterine (Lynch et al. 2011), placental (Chuong 2018), skeletal (Kubota et al. 2020), and nervous system development and regulation (Ferrari et al. 2021) in mammals. TE remnants are expressed at high levels in undifferentiated human embryonic cells (e.g., LINE1- and ERV-mutated genes in stem cells), regulating various essential mechanisms that are necessary for embryonic development (Garcia-Perez et al. 2016; Xiang and Liang 2021), viral infection defense (Frank et al. 2022), and maternal immune tolerance in the fetus (Liang et al. 2010; Zhu et al. 2021). TEs have also been an exciting research topic in human neurobiology, as numerous studies have emphasized their ambivalent regulatory roles in the development of the nervous system and the onset and progression of neurological disorders (Fukuda 2024). By establishing centromere identity and structure, TEs contribute to key meiotic processes and shape the three-dimensional organization of the genome (Henikoff et al. 2001; Klein and O'Neill 2018). For example, specific SINE subfamilies have been shown to facilitate synapsis during mammalian meiosis (Johnson and Giles 2013). In addition, TEs (i.e., LINE subfamilies) can regulate telomeres and preserve them from degradation due to multiple cell divisions and assemble the centromeres' structural identity (Kordyukova et al. 2018; Kalmykova and Sokolova 2023).

1.7 TEs as Drivers of Evolutionary Divergence and Speciation

All the aforementioned TE-driven restructuring and regulatory modifications may contribute to rapid genome evolution (Platt et al. 2018; Arkhipova 2018), differentiation (Huff et al. 2016), divergence, and, ultimately, speciation (Zeh et al. 2009; Jurka et al. 2011; Oliver and Greene 2012). Speciation refers to the evolutionary process by which new, reproductively isolated species arise from ancestral populations, typically driven by genetic divergence or ecological adaptation that establishes barriers to gene flow (Coyne and Orr 2004). As potent agents of genetic divergence, TEs can promote diverse forms of SVs, including CNVs and large-scale chromosomal rearrangements. These alterations can lead to genomic incompatibilities, disrupted meiotic pairing, and reduced hybrid viability, ultimately contributing to reproductive isolation through mechanisms such as meiotic drive, hybrid incompatibility genes, genomic shock, behavioral shifts, and ecological niche adaptations (Almojil et al. 2021; Betancourt et al. 2024).

Hybridization is a powerful catalyst for the emergence of many eukaryotic species (Shapiro 2022). TE insertions, for example, can induce genomic incompatibilities through hybrid dysgenesis, an epigenetic phenomenon in which mismatched parental TE landscapes trigger transpositional bursts in hybrid offspring, resulting in developmental abnormalities and sterility. This phenomenon is well-documented in Drosophila, where active P-element transposons in one parental lineage cause sterility when introduced into a TE-deficient genome (Spradling et al. 2011). On some occasions, however, even a simple critical insertion can potentially drive divergence, such as in the Chinese Yangtze River fish (*Coilia nasus*), where a SINE insertion in a critical olfactory gene has contributed to the divergence of a resident ecotype from its migratory counterpart (Liu et al. 2020).

TE-induced regulatory rewiring can generate species-specific transcriptional landscapes, leading to differential gene expression patterns and the emergence of novel morphological, physiological, and behavioral traits essential for reproductive isolation. Additionally, failures in TE silencing mechanisms can lead to TE hyperactivity, indirectly contributing to the genetic incompatibilities described above (Parhad et al. 2017; Wang et al. 2020). These multifaceted mechanisms may explain the observed strong correlation between speciation rates and bursts of TE amplification in mammals (Ricci et al. 2018), notably in specific lineages, such as rodents

(Verneau et al. 1998), bats (Ray et al. 2008), and woodpeckers (Manthey et al. 2018). As such, TEs not only influence genome evolution at the intra-species level but also drive long-term macroevolutionary changes, reinforcing their role as key contributors to speciation.

1.8 The Evolutionary Arms Race: TEs Versus Host Defenses

As discussed earlier, selfish elements of diverse origins have persistently invaded and colonized host genomes throughout evolution. Such persistent pressure has driven a dynamic "arms race" between TEs and hosts, with hosts evolving intricate defensive mechanisms, such as RNA interference, DNA methylation, and histone modifications, to mitigate the genome instability caused by TE activities (Slotkin and Martienssen 2007). In response, TEs have developed counterstrategies, including sequence diversification and host-regulatory emulation, to evade suppression (Fu et al. 2013; Ellison and Bachtrog 2013). This dynamic interplay highlights the complex coevolution of TEs and their hosts, shaping both genome and TE evolutionary trajectories.

Some key considerations must be taken into account when examining this evolutionary arms race. First, the relationship between TE activity and abundance within and across species is not always linear. For instance, a species with fewer active TE copies may exhibit higher TE activity than a species with a more abundant but largely inactive TE repertoire. Such discrepancies often reflect the differences in the effectiveness of host defense mechanisms in controlling TE activity (Tollis and Boissinot 2012). Second, selection operates at two levels in this conflict. While it generally favors robust suppression of TE activity at the host genome level to maintain stability, it promotes selfish replication and activity at the TE level. Additionally, random genetic drift can influence this balance unpredictably, potentially supporting either enhanced TE activity or strengthened host suppression, with outcomes heavily dependent on specific genomic and environmental contexts (Betancourt et al. 2024).

TEs are typically repressed in normal, fully differentiated cells, given their profound capacity to induce mutagenesis and rewire genomic regulatory networks. This repression occurs at the epigenetic, transcriptional, or post-transcriptional levels through processes commonly referred to as "TE silencing" events (Slotkin and Martienssen 2007; Grow et al. 2015; Grundy et al. 2022; Almeida et al. 2022). The primary TE silencing mechanisms (i.e., the host's first line of defense) are heterochromatinization and epigenetic modifications, including methylation and histone modifications. Reflecting this, most of the methylated cytosines in the human genome are within the TE sequences (Ohtani et al. 2018; Jansz 2019), restricting access of the transcriptional machinery to those particular genome fragments. In humans, the methylation levels, particularly in ERVs, are inversely correlated with TE ages, meaning younger elements undergo more methylation than their older counterparts. This pattern likely reflects the fact that older elements have lost their ability to transpose due to accumulated mutations in their replicative regions. This age-related methylation pattern also underscores the impact of TE-induced de novo mutations on the overall methylation profile of the DNA. Histone modifications represent another key silencing mechanism capable of acting as either transcriptional activators or repressors, depending on context. Compared with younger elements, older elements, particularly ERVs, are mainly regulated through histone modifications (Simonti et al. 2017; Ohtani et al. 2018).

Transcriptional silencing of TEs, in contrast, is primarily mediated by sequence-specific transcription factors, such as Krüppel-associated box zinc finger proteins (KRAB-ZFPs), which are the central TE repressors in mammals, mainly targeting L1s (Bruno et al. 2019; Senft and Macfarlan 2021). RNA-mediated mechanisms also suppress TE activity at the post-transcriptional level, primarily targeting retrotransposon-derived transcripts and proteins to prevent successful mobilization. Strong evidence exists for the role of siRNAs (Yang and Kazazian 2006), miRNAs (Hamdorf et al. 2015), and piRNAs (i.e., in germlines) (Aravin et al. 2007; Newkirk et al. 2017) in the repression of L1 and other retroelements in mammals. Notably, whether these processes represent host-adapted strategies to suppress TE activity, selfish strategies by TEs to optimize their propagation, or a combination of both remains unclear. However, it is critical for TEs to maintain this balance, as triggering host defenses risks both their own suppression and detrimental effects on host genome stability, which could, in turn, result in their extinction (Cosby et al. 2019). Overall, the suppression pressure of hosts' genomes on different TE families varies depending on the structure and impact of the elements. For example, terminal sequences such as LTRs, UTRs, and ITRs have experienced less evolutionary pressure by the hosts than ERVs, LINEs, and DNA transposons, which comprise internal protein-coding genes (Wells and Feschotte 2020).

Studies across various species have shown that some TEs have evolved counter-defense strategies to evade host silencing mechanisms (Fu et al. 2013; Ellison and Bachtrog 2013). One common adaptation is sequence diversification, which reduces recognition by host sequence-specific silencing pathways, such as piRNA or KRAB-ZFPs. Some TEs mimic host-regulatory motifs, such as TFBSs, to integrate into active chromatin regions and evade repression (Roman et al. 2008; Sundaram et al. 2014; Chuong et al. 2017). Another strategy involves inducing pleiotropic effects, where TEs confer beneficial functions that may be subject to positive selection, enabling them to evade suppression and persist across evolutionary timescales. Furthermore, host silencing mechanisms often serve additional roles, particularly in the germline, or may inadvertently affect nearby host genes, making repression in these regions potentially harmful. TEs may exploit this constraint by inserting into such functionally sensitive regions, where host defenses are less likely to act robustly, or into other "safe havens" within the genome, such as heterochromatic regions or introns, to minimize the likelihood of detection or elimination by the host (Levin and Moran 2011; Choi and Lee 2020). Other strategies include actively interfering with silencing mechanisms. For instance, L1 elements produce antisense transcripts that can disrupt RNA-mediated silencing pathways (Goodier et al. 2013). TEs also actively capitalize on host stress conditions, such as chronic inflammation and DNA damage, to reactivate and transpose (Miousse et al. 2015). Finally, HT of TEs between species also enables them to invade new genomes and bypass co-adapted silencing systems (Gilbert et al. 2010). These adaptive capabilities highlight the dynamic arms race between TEs and their hosts, underscoring the evolutionary ingenuity of TEs in countering genomic surveillance.

1.9 Balancing DNA Gain and Loss: TEs Mediate Genome Size Evolution

The variation in TE types, abundance, and activity across species (Kidwell 2002; Tollis and Boissinot 2012; Kapusta et al. 2017), in addition to TE-mediated DNA gain and loss processes (Petrov 2002; Lynch and Conery 2003; Bourque et al. 2018), is a major factor contributing to differences in haploid genome size among eukaryotes. The "C-value enigma" refers to the observation that genome size does not exhibit a straightforward correlation with an organism's complexity (Elliott and Gregory 2015). The most widely accepted explanation of the C-value enigma is the lineage-specific accumulation of noncoding DNA, particularly TEs. Interestingly, there is often a significant positive correlation between one or a combination of the aforementioned TE attributes and genome size in most species (Wu and Lu 2019; Wells and Feschotte 2020; Almojil et al. 2021; Osmanski et al. 2023).

The mechanisms through which TEs drive DNA gain and genome expansion are well-established and outlined in earlier sections. TEs drive genome expansion through insertional duplication, exonization, and transduction. Epigenetic modifications can also contribute to DNA gain. For instance, a recent study involving 53 species revealed an interplay between TE insertions, methylations, and genome size. The researchers observed a negative correlation between genome size and methylation of CpG sites within TEs. This association implies that elevated hypomethylation, resulting from a common spontaneous deamination of cytosines to thymines in CpG sites, leads to the derepression and reactivation of TEs and, eventually, to genome size expansion (Zhou et al. 2020). TE insertions can also alter chromatin organization, increasing the likelihood of replication errors that result in additional genomic material. Furthermore, TE sequences may become integrated into the repair process during double-strand break repair, leading to sequence insertions that further expand genome size (Cavalier-Smith 2005; Gupta et al. 2016).

While TEs are often associated with genome expansion, they also play a critical role in DNA loss and genome size contraction. The primary mechanism by which TEs contribute to genomic reduction is recombination-mediated deletion. The high degree of homology among both active and inactive TEs makes them hotspots for ectopic recombination events during crossing over (Rubin and Levy 1997; Nam and Ellegren 2012; Kent et al. 2017; Bourgeois and Boissinot 2019), which often results in misalignments that lead to the loss of intervening sequences. Although recombination-mediated deletion is more prevalent among younger TE insertions, older elements, which are typically more fragmented and degenerated, are not entirely excluded from these mechanisms. Notably, L1 and Alu elements are frequently implicated in this process in the human genome, with some studies linking their activity to neurodegenerative disorders such as Parkinson's and Alzheimer's diseases (Sen et al. 2006; Payer and Burns 2019; Pascarella et al. 2022). In addition, deleterious insertions that inactivate critical genes, promote chromosome breakage, and induce DNA damage, along with imprecise excisions and ligations that leave unequal gaps (e.g., non-homologous end-joining), misrepair events, and host defense mechanisms, can lead to TE deletion and DNA loss (Gilbert et al. 2002; Callinan et al. 2005; Cordaux and Batzer 2009).

At the population level, bottlenecks and small population sizes can also elevate TE load and promote DNA gain by allowing the accumulation of mildly deleterious insertions where genetic drift overpowers selection, and random fluctuations in allele frequencies (i.e., haplotype blocks) can facilitate the stochastic accumulation of TEs. Conversely, purifying selection tends to be more effective in populations with large effective sizes, which may lead to more efficient deletion of TE insertions that otherwise may have mildly deleterious effects. This mechanism may explain why populations with smaller effective sizes tend to have larger genomes (Lynch and Conery 2003).

Despite these trends, the mechanisms underlying genome size variation remain largely unknown. For example, while birds (Kapusta et al. 2017), squamates (Pasquesi et al. 2018; Gable et al. 2024), and mammals (Tollis and Boissinot 2012) exhibit relatively streamlined genomes, the underlying reasons may differ. In birds and mammals, a more uniform TE content (i.e., retrotransposons) may lead to higher DNA loss through mechanisms like increased ectopic recombination and higher efficiency of host defenses against a narrow repertoire of active TE families. Consistent with this hypothesis, amphibian genomes harbor a diverse array of Class I and II TEs, contributing to their substantially larger genome sizes due to the inefficiency of host defenses against various TE types and lower recombination rates (Chalopin et al. 2015). However, such a hypothesis does not fully account for genome size patterns in squamates, as they maintain uniform genome sizes despite high TE abundance, diversity, and activity (Kapusta et al. 2017; Gable et al. 2024). Collectively, these conflicting observations highlight the inherent complexity of genome size variation across species, suggesting an interplay among multiple factors, including TE types, abundance, and activity; the efficacy of host defense mechanisms and potentially broader genetic and environmental contexts.

Taken together, these opposing forces, including TE-driven genome expansion and host-mediated DNA loss, are well described by the "accordion model," which explains how genome size fluctuates over evolutionary time due to cycles of TE proliferation and genomic deletions. The model suggests that genome size is not static but evolves through cycles of expansion via TE insertions, duplications, and proliferation, as well as contraction via deletions, recombination, and genome streamlining mechanisms explained above. This model provides a framework for understanding why genome size varies so dramatically across species despite nearly similar gene content, reinforcing the idea that genome size evolution is driven more by TE dynamics and DNA loss mechanisms than by organismal complexity (Kapusta et al. 2017; Anderson et al. 2019; Wu and Lu 2019; Blommaert 2020). The increasing affordability of high-throughput telomere-to-telomere sequencing technologies, combined with novel TE detection and annotation methods, can provide further insights into the mechanisms underlying genome size variation.

1.10 Emerging Frontiers in TE Research

The future of TE research is set for discoveries driven by advancements in high-throughput sequencing technologies and integrative computational approaches. While TEs are the most common polymorphic SVs, short-read sequencing and current annotation tools fall short in detecting a substantial portion of TEs, especially in heterochromatic and repeat-rich regions, limiting the development of robust hypotheses in population genomics and precise evolutionary association analyses (Jain et al. 2018; Kojima et al. 2023). Today, the availability of more affordable telomere-to-telomere long-read sequencing technologies, novel computational and bioinformatic tools, and machine-learning algorithms has enabled precise annotation of complex genomic regions and polymorphic variants that were previously intractable (Peona et al. 2018; Shahid and Slotkin 2020).

Combining long-read data with transcriptomics and epigenomics can further unravel the activity status of TEs, their influence on gene expression patterns, and DNA and RNA modifications, providing insights into the mechanisms through which TEs contribute to phenotypic variation and disease predisposition. These technologies not only facilitate the identification of lineage-specific TE insertions and SVs but also discover novel TE subfamilies, enriching our understanding of genome architecture and function. Besides, emerging approaches, including pangenomics and population-scale sequencing, promise to shed light on TE diversity and evolutionary trajectories across species. Moreover, single-cell sequencing and epigenomic profiling provide unprecedented insights into the dynamic regulatory roles of TEs in tissue-specific gene expression, developmental plasticity, and disease susceptibility. These methods advance our knowledge of how TEs influence host epigenetic defenses, such as DNA methylation, histone modifications, and ncRNA pathways, revealing critical interactions that govern early development and cell fate (Laszlo et al. 2013; Sakamoto et al. 2020; Shahid and Slotkin 2020).

Comparative genomics and phylogenomic studies also have expanded our understanding of TE-driven innovations in species diversification, genome size evolution, and regulatory network expansion (Kapusta et al. 2017; Serrato-Capuchina and Matute 2018; Zhou et al. 2020; Shapiro 2022). Investigating the evolutionary arms race between TEs and host genomes, particularly through piRNA pathways and other silencing mechanisms, can shed light on how genomes maintain stability while accommodating TE proliferation (Platt et al. 2018; Luo et al. 2020; Betancourt et al. 2024). Additionally, longitudinal studies, including those using ancient DNA, fossil genomes, or reconstructed ancestral genomes, hold great potential for uncovering the long-term evolutionary dynamics of TEs. Such approaches can reveal how ancient and degenerate TEs have contributed to developmental processes and lineage-specific adaptations over macroevolutionary timescales (Matsushima et al. 2024). Collectively, these future directions underscore the pivotal role of TEs in genome evolution, bridging cutting-edge sequencing technologies, epigenetic research, and evolutionary biology to deepen our understanding of how TEs shape genome landscapes over the course of evolution.

1.11 Conclusion

TEs exemplify the dynamic and multifaceted nature of genomes, acting as both architects of genetic innovation and potential disruptors of genomic stability. Once considered parasitic, TEs are now understood as pivotal drivers of genome evolution. They shape genome architecture, modulate regulatory complexities, and, when aberrantly expressed, contribute to the onset and progression of various diseases. Their ability to transpose, recombine, and mediate exon shuffling underpins their substantial influence on genomic diversity and adaptation. TEs propagate through both vertical inheritance and HT, contributing to genetic diversity within populations and facilitating gene flow across species boundaries. Their evolutionary fate is shaped by the interplay of natural selection and genetic drift, where beneficial insertions are co-opted for adaptive functions, while deleterious or neutral elements are subject to purifying selection or gradual recombination-mediated decay. Through SVs, TEs drive genomic innovation while their co-opted regulatory sequences shape alternative enhancers, silencers, and insulators, modulating transcriptional networks and fine-tuning gene expression and chromatin architecture.

Beyond their canonical regulatory roles, TEs are integral to epigenome dynamics, influencing DNA methylation patterns, histone modifications, and the three-dimensional organization of the genome. These activities are critical in developmental processes, cell fate decisions, and maintaining cellular homeostasis. TEs also play a fundamental role in speciation by promoting genetic divergence, contributing to reproductive isolation, and driving the evolution of novel phenotypic traits. The evolutionary arms race between TEs and host genomes, where hosts evolve silencing mechanisms while TEs develop counterstrategies, highlights the coevolutionary forces that shape genome stability and integrity. Ultimately, TEs significantly influence genome size evolution by balancing DNA gain via transpositional activity and DNA loss through recombination-mediated deletions. Such a dynamic "accordion model" of genome size regulation illustrates how TEs contribute to genome plasticity, allowing organisms to adapt to ever-changing environments and evolutionary pressures. Altogether, TEs are not just remnants of ancient genomic invasions but are dynamic elements that contribute to genomes' complexity, modulation, and adaptability. As research continues to unveil the multifaceted roles of TEs, our understanding of genome evolution will undoubtedly deepen, offering new insights into genetic and evolutionary biology and opening new avenues for exploring the complexity of life at the molecular level.

Acknowledgments We gratefully acknowledge Dr. Simone Gable, whose research has enriched our understanding of transposable elements and genome evolution, and guided aspects of this chapter.

Funding No external funding was received for this work.

Compliance with Ethical Standards

Conflict of Interest The authors declare no conflicts of interest.

Ethical Approval This chapter does not contain any studies involving human participants or animals performed by the authors.

References

Almeida MV, Vernaz G, Putman ALK, Miska EA (2022) Taming transposable elements in vertebrates: from epigenetic silencing to domestication. Trends Genet 38:529–553. https://doi.org/10.1016/j.tig.2022.02.009

Almojil D, Bourgeois Y, Falis M et al (2021) The structural, functional and evolutionary impact of transposable elements in eukaryotes. Genes 12:918. https://doi.org/10.3390/genes12060918

Andergassen D, Rinn JL (2022) From genotype to phenotype: genetics of mammalian long non-coding RNAs in vivo. Nat Rev Genet 23:229–243. https://doi.org/10.1038/s41576-021-00427-8

Anderson SN, Stitzer MC, Brohammer AB et al (2019) Transposable elements contribute to dynamic genome content in maize. Plant J 100:1052–1065. https://doi.org/10.1111/tpj.14489

Annibalini G, Bielli P, De Santi M et al (2016) MIR retroposon exonization promotes evolutionary variability and generates species-specific expression of IGF-1 splice variants. Biochim Biophys Acta 1859:757–768. https://doi.org/10.1016/j.bbagrm.2016.03.014

Aravin AA, Hannon GJ, Brennecke J (2007) The Piwi-piRNA pathway provides an adaptive defense in the transposon arms race. Science 318:761–764. https://doi.org/10.1126/science.1146484

Arkhipova IR (2018) Neutral theory, transposable elements, and eukaryotic genome evolution. Mol Biol Evol 35:1332. https://doi.org/10.1093/molbev/msy083

Arkhipova IR, Yushenova IA (2019) Giant transposons in eukaryotes: is bigger better? Genome Biol Evol 11:906–918. https://doi.org/10.1093/gbe/evz041

Arnold BJ, Huang I-T, Hanage WP (2022) Horizontal gene transfer and adaptive evolution in bacteria. Nat Rev Microbiol 20:206–218. https://doi.org/10.1038/s41579-021-00650-4

Ataei L, Zhang J, Monis S et al (2025) LINE1 elements at distal junctions of rDNA repeats regulate nucleolar organization in human embryonic stem cells. Genes Dev 39:280–298. https://doi.org/10.1101/gad.351979.124

Attig J, Agostini F, Gooding C et al (2018) Heteromeric RNP assembly at LINEs controls lineage-specific RNA processing. Cell 174:1067. https://doi.org/10.1016/j.cell.2018.07.001

Babaian A, Mager DL (2016) Endogenous retroviral promoter exaptation in human cancer. Mob DNA 7:24. https://doi.org/10.1186/s13100-016-0080-x

Babushok DV, Kazazian HH (2007) Progress in understanding the biology of the human mutagen LINE-1. Hum Mutat 28:527–539. https://doi.org/10.1002/humu.20486

Bannert N, Kurth R (2004) Retroelements and the human genome: new perspectives on an old relation. Proc Natl Acad Sci USA 101:14572–14579. https://doi.org/10.1073/pnas.0404838101
Batut P, Dobin A, Plessy C et al (2013) High-fidelity promoter profiling reveals widespread alternative promoter usage and transposon-driven developmental gene expression. Genome Res 23:169–180. https://doi.org/10.1101/gr.139618.112
Batzer MA, Deininger PL (2002) Alu repeats and human genomic diversity. Nat Rev Genet 3:370–379. https://doi.org/10.1038/nrg798
Beck CR, Garcia-Perez JL, Badge RM, Moran JV (2011) LINE-1 elements in structural variation and disease. Annu Rev Genomics Hum Genet 12:187–215. https://doi.org/10.1146/annurev-genom-082509-141802
Bennett PM (2004) Genome plasticity. In: Woodford N, Johnson AP (eds) Genomics, proteomics, and clinical bacteriology: methods and reviews. Humana Press, Totowa, pp 71–113
Betancourt AJ, Wei KH-C, Huang Y, Lee YCG (2024) Causes and consequences of varying transposable element activity: an evolutionary perspective. Annu Rev Genomics Hum Genet 25:1–25. https://doi.org/10.1146/annurev-genom-120822-105708
Betel D, Sheridan R, Marks DS, Sander C (2007) Computational analysis of mouse piRNA sequence and biogenesis. PLoS Comput Biol 3:e222. https://doi.org/10.1371/journal.pcbi.0030222
Birney E, Stamatoyannopoulos JA, Dutta A et al (2007) Identification and analysis of functional elements in 1% of the human genome by the ENCODE pilot project. Nature 447:799–816. https://doi.org/10.1038/nature05874
Blommaert J (2020) Genome size evolution: towards new model systems for old questions. Proc R Soc B Biol Sci 287:20201441. https://doi.org/10.1098/rspb.2020.1441
Boissinot S, Davis J, Entezam A et al (2006) Fitness cost of LINE-1 (L1) activity in humans. Proc Natl Acad Sci USA 103:9590–9594. https://doi.org/10.1073/pnas.0603334103
Bourgeois Y, Boissinot S (2019) On the population dynamics of junk: a review on the population genomics of transposable elements. Genes 10:419. https://doi.org/10.3390/genes10060419
Bourque G, Burns KH, Gehring M et al (2018) Ten things you should know about transposable elements. Genome Biol 19:199. https://doi.org/10.1186/s13059-018-1577-z
Britten RJ, Kohne DE (1968) Repeated sequences in DNA. Hundreds of thousands of copies of DNA sequences have been incorporated into the genomes of higher organisms. Science 161:529–540. https://doi.org/10.1126/science.161.3841.529
Brosius J (1991) Retroposons—seeds of evolution. Science 251:753–753. https://doi.org/10.1126/science.1990437
Brouha B, Schustak J, Badge RM et al (2003) Hot L1s account for the bulk of retrotransposition in the human population. Proc Natl Acad Sci USA 100:5280–5285. https://doi.org/10.1073/pnas.0831042100
Bruno M, Mahgoub M, Macfarlan TS (2019) The arms race between KRAB–zinc finger proteins and endogenous retroelements and its impact on mammals. Annu Rev Genet 53:393–416. https://doi.org/10.1146/annurev-genet-112618-043717
Burn A, Roy F, Freeman M, Coffin JM (2022) Widespread expression of the ancient HERV-K (HML-2) provirus group in normal human tissues. PLoS Biol 20:e3001826. https://doi.org/10.1371/journal.pbio.3001826
Cagliani R, Forni D, Mozzi A et al (2024) Evolution of virus-like features and intrinsically disordered regions in Retrotransposon-derived mammalian genes. Mol Biol Evol 41:msae154. https://doi.org/10.1093/molbev/msae154
Callinan PA, Wang J, Herke SW et al (2005) Alu retrotransposition-mediated deletion. J Mol Biol 348:791–800. https://doi.org/10.1016/j.jmb.2005.02.043
Campillos M, Doerks T, Shah PK, Bork P (2006) Computational characterization of multiple Gag-like human proteins. Trends Genet 22:585–589. https://doi.org/10.1016/j.tig.2006.09.006
Cavalier-Smith T (2005) Economy, speed and size matter: evolutionary forces driving nuclear genome miniaturization and expansion. Ann Bot 95:147–175. https://doi.org/10.1093/aob/mci010

Chalopin D, Naville M, Plard F et al (2015) Comparative analysis of transposable elements highlights mobilome diversity and evolution in vertebrates. Genome Biol Evol 7:567–580. https://doi.org/10.1093/gbe/evv005

Charlesworth B, Charlesworth D (1983) The population dynamics of transposable elements. Genet Res 42:1–27. https://doi.org/10.1017/S0016672300021455

Charlesworth B, Sniegowski P, Stephan W (1994) The evolutionary dynamics of repetitive DNA in eukaryotes. Nature 371:215–220. https://doi.org/10.1038/371215a0

Chénais B (2022) Transposable elements and human diseases: mechanisms and implication in the response to environmental pollutants. Int J Mol Sci 23:2551. https://doi.org/10.3390/ijms23052551

Choi JY, Lee YCG (2020) Double-edged sword: the evolutionary consequences of the epigenetic silencing of transposable elements. PLoS Genet 16:e1008872. https://doi.org/10.1371/journal.pgen.1008872

Choudhary MNK, Quaid K, Xing X et al (2023) Widespread contribution of transposable elements to the rewiring of mammalian 3D genomes. Nat Commun 14:634. https://doi.org/10.1038/s41467-023-36364-9

Chuong EB (2018) The placenta goes viral: Retroviruses control gene expression in pregnancy. PLoS Biol 16:e3000028. https://doi.org/10.1371/journal.pbio.3000028

Chuong EB, Rumi MAK, Soares MJ, Baker JC (2013) Endogenous retroviruses function as species-specific enhancer elements in the placenta. Nat Genet 45:325. https://doi.org/10.1038/ng.2553

Chuong EB, Elde NC, Feschotte C (2016) Regulatory evolution of innate immunity through co-option of endogenous retroviruses. Science 351:1083–1087. https://doi.org/10.1126/science.aad5497

Chuong EB, Elde NC, Feschotte C (2017) Regulatory activities of transposable elements: from conflicts to benefits. Nat Rev Genet 18:71–86. https://doi.org/10.1038/nrg.2016.139

Colonna Romano N, Fanti L (2022) Transposable elements: major players in shaping genomic and evolutionary patterns. Cells 11:1048. https://doi.org/10.3390/cells11061048

Cordaux R, Batzer MA (2009) The impact of retrotransposons on human genome evolution. Nat Rev Genet 10:691–703. https://doi.org/10.1038/nrg2640

Cosby RL, Chang N-C, Feschotte C (2019) Host-transposon interactions: conflict, cooperation, and cooption. Genes Dev 33:1098–1116. https://doi.org/10.1101/gad.327312.119

Coyne JA, Orr HA (2004) Speciation. Oxford University Press, Oxford, NY

De Cecco M, Ito T, Petrashen AP et al (2019) L1 drives IFN in senescent cells and promotes age-associated inflammation. Nature 566:73–78. https://doi.org/10.1038/s41586-018-0784-9

Deniz Ö, Ahmed M, Todd CD et al (2020) Endogenous retroviruses are a source of enhancers with oncogenic potential in acute myeloid leukaemia. Nat Commun 11:3506. https://doi.org/10.1038/s41467-020-17206-4

Denli AM, Narvaiza I, Kerman BE et al (2015) Primate-specific ORF0 contributes to retrotransposon-mediated diversity. Cell 163:583–593. https://doi.org/10.1016/j.cell.2015.09.025

Dewannieux M, Heidmann T (2005) LINEs, SINEs and processed pseudogenes: parasitic strategies for genome modeling. Cytogenet Genome Res 110:35–48. https://doi.org/10.1159/000084936

DiRusso J, Clark A (2023) Transposable elements in early human embryo development and embryo models. Curr Opin Genet Dev 81:102086. https://doi.org/10.1016/j.gde.2023.102086

Doolittle WF, Sapienza C (1980) Selfish genes, the phenotype paradigm and genome evolution. Nature 284:601–603. https://doi.org/10.1038/284601a0

Doucet AJ, Hulme AE, Sahinovic E et al (2010) Characterization of LINE-1 ribonucleoprotein particles. PLoS Genet 6:e1001150. https://doi.org/10.1371/journal.pgen.1001150

Du C, Jiang J, Li Y et al (2023) Regulation of endogenous retrovirus-derived regulatory elements by GATA2/3 and MSX2 in human trophoblast stem cells. Genome Res 33:197–207. https://doi.org/10.1101/gr.277150.122

Dunemann SM, Wasmuth JD (2019) Horizontal transfer of a retrotransposon between parasitic nematodes and the common shrew. Mob DNA 10:24. https://doi.org/10.1186/s13100-019-0166-3

Eickbush TH, Malik HS (2007) Origins and evolution of Retrotransposons. In: Mobile DNA II. Wiley, pp 1111–1144

Elliott TA, Gregory TR (2015) Do larger genomes contain more diverse transposable elements? BMC Evol Biol 15:69. https://doi.org/10.1186/s12862-015-0339-8

Ellison CE, Bachtrog D (2013) Dosage compensation via transposable element mediated rewiring of a regulatory network. Science 342:846–850. https://doi.org/10.1126/science.1239552

Engels WR, Johnson-Schlitz DM, Eggleston WB, Sved J (1990) High-frequency P element loss in Drosophila is homolog dependent. Cell 62:515–525. https://doi.org/10.1016/0092-8674(90)90016-8

Etchegaray E, Naville M, Volff J-N, Haftek-Terreau Z (2021) Transposable element-derived sequences in vertebrate development. Mob DNA 12:1. https://doi.org/10.1186/s13100-020-00229-5

Ferrari R, Grandi N, Tramontano E, Dieci G (2021) Retrotransposons as drivers of mammalian brain evolution. Life 11:376. https://doi.org/10.3390/life11050376

Feschotte C (2023) Transposable elements: McClintock's legacy revisited. Nat Rev Genet 24:797–800. https://doi.org/10.1038/s41576-023-00652-3

Feschotte C, Pritham EJ (2007) DNA transposons and the evolution of eukaryotic genomes. Annu Rev Genet 41:331–368. https://doi.org/10.1146/annurev.genet.40.110405.090448

Finnegan DJ (1989) Eukaryotic transposable elements and genome evolution. Trends Genet TIG 5:103–107. https://doi.org/10.1016/0168-9525(89)90039-5

Flasch DA, Macia Á, Sánchez L et al (2019) Genome-wide de novo L1 Retrotransposition connects endonuclease activity with replication. Cell 177:837–851.e28. https://doi.org/10.1016/j.cell.2019.02.050

Frank JA, Singh M, Cullen HB et al (2022) Evolution and antiviral activity of a human protein of retroviral origin. Science 378:422–428. https://doi.org/10.1126/science.abq7871

Fricker AD, Peters JE (2014) Vulnerabilities on the lagging-strand template: opportunities for mobile elements. Annu Rev Genet 48:167–186. https://doi.org/10.1146/annurev-genet-120213-092046

Frost JM, Amante SM, Okae H et al (2023) Regulation of human trophoblast gene expression by endogenous retroviruses. Nat Struct Mol Biol 30:527–538. https://doi.org/10.1038/s41594-023-00960-6

Fu Y, Kawabe A, Etcheverry M et al (2013) Mobilization of a plant transposon by expression of the transposon-encoded anti-silencing factor. EMBO J 32:2407–2417. https://doi.org/10.1038/emboj.2013.169

Fueyo R, Judd J, Feschotte C, Wysocka J (2022) Roles of transposable elements in the regulation of mammalian transcription. Nat Rev Mol Cell Biol 23:481–497. https://doi.org/10.1038/s41580-022-00457-y

Fugmann SD, Lee AI, Shockett PE et al (2000) The RAG proteins and V(D)J recombination: complexes, ends, and transposition. Annu Rev Immunol 18:495–527. https://doi.org/10.1146/annurev.immunol.18.1.495

Fukuda K (2024) The role of transposable elements in human evolution and methods for their functional analysis: current status and future perspectives. Genes Genet Syst 98:289–304. https://doi.org/10.1266/ggs.23-00140

Gable SM, Bushroe N, Mendez J et al (2024) Differential conservation and loss of CR1 retrotransposons in Squamates reveals lineage-specific genome dynamics across reptiles. BioRxiv Prepr Serv Biol:2024.02.09.579686. https://doi.org/10.1101/2024.02.09.579686

Galbraith JD, Ludington AJ, Sanders KL et al (2021) Horizontal transfer and subsequent explosive expansion of a DNA transposon in sea kraits (Laticauda). Biol Lett 17:20210342. https://doi.org/10.1098/rsbl.2021.0342

Garcia-Perez JL, Widmann TJ, Adams IR (2016) The impact of transposable elements on mammalian development. Dev Camb Engl 143:4101–4114. https://doi.org/10.1242/dev.132639

Gasparotto E, Burattin FV, Gioia VD et al (2023) Transposable elements co-option in genome evolution and gene regulation. Int J Mol Sci 24:2610. https://doi.org/10.3390/ijms24032610

Gellert M (2002) V(D)J recombination: RAG proteins, repair factors, and regulation. Annu Rev Biochem 71:101–132. https://doi.org/10.1146/annurev.biochem.71.090501.150203

Gifford RJ, Blomberg J, Coffin JM et al (2018) Nomenclature for endogenous retrovirus (ERV) loci. Retrovirology 15:59. https://doi.org/10.1186/s12977-018-0442-1

Gilbert C, Feschotte C (2018) Horizontal acquisition of transposable elements and viral sequences: patterns and consequences. Curr Opin Genet Dev 49:15–24. https://doi.org/10.1016/j.gde.2018.02.007

Gilbert N, Lutz-Prigge S, Moran JV (2002) Genomic deletions created upon LINE-1 retrotransposition. Cell 110:315–325. https://doi.org/10.1016/s0092-8674(02)00828-0

Gilbert C, Schaack S, Pace JK II et al (2010) A role for host–parasite interactions in the horizontal transfer of transposons across phyla. Nature 464:1347–1350. https://doi.org/10.1038/nature08939

Gilbert C, Chateigner A, Ernenwein L et al (2014) Population genomics supports baculoviruses as vectors of horizontal transfer of insect transposons. Nat Commun 5:3348. https://doi.org/10.1038/ncomms4348

Gogvadze E, Stukacheva E, Buzdin A, Sverdlov E (2009) Human-specific modulation of transcriptional activity provided by endogenous retroviral insertions. J Virol 83:6098–6105. https://doi.org/10.1128/JVI.00123-09

Gonzalez P, Hauck QC, Baxevanis AD (2024) Conserved noncoding elements evolve around the same genes throughout Metazoan evolution. Genome Biol Evol 16:evae052. https://doi.org/10.1093/gbe/evae052

Goodier JL, Cheung LE, Kazazian HH (2013) Mapping the LINE1 ORF1 protein interactome reveals associated inhibitors of human retrotransposition. Nucleic Acids Res 41:7401–7419. https://doi.org/10.1093/nar/gkt512

Gorbunova V, Seluanov A, Mita P et al (2021) The role of retrotransposable elements in ageing and age-associated diseases. Nature 596:43–53. https://doi.org/10.1038/s41586-021-03542-y

Gotea V, Makałowski W (2006) Do transposable elements really contribute to proteomes? Trends Genet TIG 22:260–267. https://doi.org/10.1016/j.tig.2006.03.006

Grandi N, Tramontano E (2018) Human endogenous Retroviruses are ancient acquired elements still shaping innate immune responses. Front Immunol 9:2039. https://doi.org/10.3389/fimmu.2018.02039

Grow EJ, Flynn RA, Chavez SL et al (2015) Intrinsic retroviral reactivation in human preimplantation embryos and pluripotent cells. Nature 522:221–225. https://doi.org/10.1038/nature14308

Grundy EE, Diab N, Chiappinelli KB (2022) Transposable element regulation and expression in cancer. FEBS J 289:1160–1179. https://doi.org/10.1111/febs.15722

Gupta A, LaBar T, Miyagi M, Adami C (2016) Evolution of genome size in asexual digital organisms. Sci Rep 6:25786. https://doi.org/10.1038/srep25786

Hamdorf M, Idica A, Zisoulis DG et al (2015) miR-128 represses L1 retrotransposition by binding directly to L1 RNA. Nat Struct Mol Biol 22:824–831. https://doi.org/10.1038/nsmb.3090

Hancks DC, Kazazian HH (2010) SVA retrotransposons: evolution and genetic instability. Semin Cancer Biol 20:234–245. https://doi.org/10.1016/j.semcancer.2010.04.001

Hancks DC, Kazazian HH (2012) Active human retrotransposons: variation and disease. Curr Opin Genet Dev 22:191–203. https://doi.org/10.1016/j.gde.2012.02.006

Hancks DC, Kazazian HH (2016) Roles for retrotransposon insertions in human disease. Mob DNA 7:9. https://doi.org/10.1186/s13100-016-0065-9

Hartl DL, Lozovskaya ER, Lawrence JG (1992) Nonautonomous transposable elements in prokaryotes and eukaryotes. Genetica 86:47–53. https://doi.org/10.1007/BF00133710

Hedges DJ, Callinan PA, Cordaux R et al (2004) Differential alu mobilization and polymorphism among the human and chimpanzee lineages. Genome Res 14:1068–1075. https://doi.org/10.1101/gr.2530404

Henikoff S, Ahmad K, Malik HS (2001) The centromere paradox: stable inheritance with rapidly evolving DNA. Science 293:1098–1102. https://doi.org/10.1126/science.1062939

Henssen AG, Koche R, Zhuang J et al (2017) PGBD5 promotes site-specific oncogenic mutations in human tumors. Nat Genet 49:1005–1014. https://doi.org/10.1038/ng.3866

Hirata M, Ichiyanagi T, Katoh H et al (2022) Sequence divergence and retrotransposon insertion underlie interspecific epigenetic differences in primates. Mol Biol Evol 39:msac208. https://doi.org/10.1093/molbev/msac208

Hsia AP, Schnable PS (1996) DNA sequence analyses support the role of interrupted gap repair in the origin of internal deletions of the maize transposon, MuDR. Genetics 142:603–618. https://doi.org/10.1093/genetics/142.2.603

Huff JT, Zilberman D, Roy SW (2016) Mechanism for DNA transposons to generate introns on genomic scales. Nature 538:533–536. https://doi.org/10.1038/nature20110

Hughes SH (2015) Reverse transcription of Retroviruses and LTR Retrotransposons. Microbiol Spectr 3:MDNA3-0027–2014. https://doi.org/10.1128/microbiolspec.MDNA3-0027-2014

Ivancevic AM, Kortschak RD, Bertozzi T, Adelson DL (2018) Horizontal transfer of BovB and L1 retrotransposons in eukaryotes. Genome Biol 19:85. https://doi.org/10.1186/s13059-018-1456-7

Jacques P-É, Jeyakani J, Bourque G (2013) The majority of primate-specific regulatory sequences are derived from transposable elements. PLoS Genet 9:e1003504. https://doi.org/10.1371/journal.pgen.1003504

Jain M, Koren S, Miga KH et al (2018) Nanopore sequencing and assembly of a human genome with ultra-long reads. Nat Biotechnol 36:338–345. https://doi.org/10.1038/nbt.4060

Janoušek V, Laukaitis CM, Yanchukov A, Karn RC (2016) The role of Retrotransposons in gene family expansions in the human and mouse genomes. Genome Biol Evol 8:2632–2650. https://doi.org/10.1093/gbe/evw192

Jansz N (2019) DNA methylation dynamics at transposable elements in mammals. Essays Biochem 63:677–689. https://doi.org/10.1042/EBC20190039

Jansz N, Faulkner GJ (2021) Endogenous retroviruses in the origins and treatment of cancer. Genome Biol 22:147. https://doi.org/10.1186/s13059-021-02357-4

Johnson PA, Giles JR (2013) The hen as a model of ovarian cancer. Nat Rev Cancer 13:432–436. https://doi.org/10.1038/nrc3535

Johnson R, Guigó R (2014) The RIDL hypothesis: transposable elements as functional domains of long noncoding RNAs. RNA N Y N 20:959–976. https://doi.org/10.1261/rna.044560.114

Jönsson ME, Ludvik Brattås P, Gustafsson C et al (2019) Activation of neuronal genes via LINE-1 elements upon global DNA demethylation in human neural progenitors. Nat Commun 10:3182. https://doi.org/10.1038/s41467-019-11150-8

Jurka J, Bao W, Kojima KK (2011) Families of transposable elements, population structure and the origin of species. Biol Direct 6:44. https://doi.org/10.1186/1745-6150-6-44

Kalmykova AI, Sokolova OA (2023) Retrotransposons and telomeres. Biochem Biokhimiia 88:1739–1753. https://doi.org/10.1134/S0006297923110068

Kambayashi C, Kakehashi R, Sato Y et al (2022) Geography-dependent horizontal gene transfer from vertebrate predators to their prey. Mol Biol Evol 39:msac052. https://doi.org/10.1093/molbev/msac052

Kaneko-Ishino T, Ishino F (2023) Retrovirus-derived RTL/SIRH: their diverse roles in the current Eutherian developmental system and contribution to Eutherian evolution. Biomolecules 13:1436. https://doi.org/10.3390/biom13101436

Kapitonov VV, Jurka J (2001) Rolling-circle transposons in eukaryotes. Proc Natl Acad Sci USA 98:8714–8719. https://doi.org/10.1073/pnas.151269298

Kapitonov VV, Jurka J (2005) RAG1 core and V(D)J recombination signal sequences were derived from Transib transposons. PLoS Biol 3:e181. https://doi.org/10.1371/journal.pbio.0030181

Kapitonov VV, Jurka J (2006) Self-synthesizing DNA transposons in eukaryotes. Proc Natl Acad Sci USA 103:4540–4545. https://doi.org/10.1073/pnas.0600833103

Kapitonov VV, Jurka J (2008) A universal classification of eukaryotic transposable elements implemented in Repbase. Nat Rev Genet 9:411–412. https://doi.org/10.1038/nrg2165-c1

Kapusta A, Suh A, Feschotte C (2017) Dynamics of genome size evolution in birds and mammals. Proc Natl Acad Sci 114:E1460–E1469. https://doi.org/10.1073/pnas.1616702114
Kazazian HH, Wong C, Youssoufian H et al (1988) Haemophilia A resulting from de novo insertion of L1 sequences represents a novel mechanism for mutation in man. Nature 332:164–166. https://doi.org/10.1038/332164a0
Keighley LM, Lynch-Sutherland CF, Almomani SN et al (2023) Unveiling the hidden players: the crucial role of transposable elements in the placenta and their potential contribution to preeclampsia. Placenta 141:57–64. https://doi.org/10.1016/j.placenta.2023.05.017
Kent TV, Uzunović J, Wright SI (2017) Coevolution between transposable elements and recombination. Philos Trans R Soc B Biol Sci 372:20160458. https://doi.org/10.1098/rstb.2016.0458
Kidwell MG (2002) Transposable elements and the evolution of genome size in eukaryotes. Genetica 115:49–63. https://doi.org/10.1023/a:1016072014259
Kim FJ, Battini J-L, Manel N, Sitbon M (2004) Emergence of vertebrate retroviruses and envelope capture. Virology 318:183–191. https://doi.org/10.1016/j.virol.2003.09.026
Kim WR, Park EG, Lee YJ et al (2022) Integration of TE induces cancer specific alternative splicing events. Int J Mol Sci 23:10918. https://doi.org/10.3390/ijms231810918
Kitao K, Ichiyanagi K, Nakagawa S (2024) Birth of protein-coding exons by ancient domestication of LINE retrotransposon. 2024.04.25.591049
Klein SJ, O'Neill RJ (2018) Transposable elements: genome innovation, chromosome diversity, and centromere conflict. Chromosom Res 26:5–23. https://doi.org/10.1007/s10577-017-9569-5
Kofler R, Senti K-A, Nolte V et al (2018) Molecular dissection of a natural transposable element invasion. Genome Res 28:824. https://doi.org/10.1101/gr.228627.117
Kojima KK, Jurka J (2011) Crypton transposons: identification of new diverse families and ancient domestication events. Mob DNA 2:12. https://doi.org/10.1186/1759-8753-2-12
Kojima S, Koyama S, Ka M et al (2023) Mobile element variation contributes to population-specific genome diversification, gene regulation and disease risk. Nat Genet 55:939–951. https://doi.org/10.1038/s41588-023-01390-2
Koonin EV, Krupovic M (2015) Evolution of adaptive immunity from transposable elements combined with innate immune systems. Nat Rev Genet 16:184–192. https://doi.org/10.1038/nrg3859
Kordis D, Gubensek F (1998) Unusual horizontal transfer of a long interspersed nuclear element between distant vertebrate classes. Proc Natl Acad Sci USA 95:10704–10709
Kordyukova M, Olovnikov I, Kalmykova A (2018) Transposon control mechanisms in telomere biology. Curr Opin Genet Dev 49:56–62. https://doi.org/10.1016/j.gde.2018.03.002
Kramerov DA, Vassetzky NS (2011) Origin and evolution of SINEs in eukaryotic genomes. Heredity 107:487–495. https://doi.org/10.1038/hdy.2011.43
Kriegs JO, Churakov G, Jurka J et al (2007) Evolutionary history of 7SL RNA-derived SINEs in Supraprimates. Trends Genet TIG 23:158–161. https://doi.org/10.1016/j.tig.2007.02.002
Kubota S, Ishikawa T, Kawata K et al (2020) Retrotransposons manipulating mammalian skeletal development in chondrocytes. Int J Mol Sci 21:1564. https://doi.org/10.3390/ijms21051564
Lander ES, Linton LM, Birren B et al (2001) Initial sequencing and analysis of the human genome. Nature 409:860–921. https://doi.org/10.1038/35057062
Laszlo AH, Derrington IM, Brinkerhoff H et al (2013) Detection and mapping of 5-methylcytosine and 5-hydroxymethylcytosine with nanopore MspA. Proc Natl Acad Sci 110:18904–18909. https://doi.org/10.1073/pnas.1310240110
Lavialle C, Cornelis G, Dupressoir A et al (2013) Paleovirology of 'syncytins', retroviral env genes exapted for a role in placentation. Philos Trans R Soc B Biol Sci 368:20120507. https://doi.org/10.1098/rstb.2012.0507
Levin HL, Moran JV (2011) Dynamic interactions between transposable elements and their hosts. Nat Rev Genet 12:615–627. https://doi.org/10.1038/nrg3030
Liang C-Y, Wang L-J, Chen C-P et al (2010) GCM1 regulation of the expression of Syncytin 2 and its Cognate Receptor MFSD2A in human Placenta1. Biol Reprod 83:387–395. https://doi.org/10.1095/biolreprod.110.083915

Lin X, Faridi N, Casola C (2016) An ancient Transkingdom horizontal transfer of Penelope-Like Retroelements from arthropods to Conifers. Genome Biol Evol 8:1252–1266. https://doi.org/10.1093/gbe/evw076

Liu D, Yang J, Tang W et al (2020) SINE Retrotransposon variation drives Ecotypic disparity in natural populations of Coilia nasus. Mob DNA 11:4. https://doi.org/10.1186/s13100-019-0198-8

Liu X, Liu Z, Wu Z et al (2023) Resurrection of endogenous retroviruses during aging reinforces senescence. Cell 186:287–304.e26. https://doi.org/10.1016/j.cell.2022.12.017

Lowe CB, Haussler D (2012) 29 mammalian genomes reveal novel Exaptations of Mobile elements for likely regulatory functions in the human genome. PLoS One 7:e43128. https://doi.org/10.1371/journal.pone.0043128

Luan DD, Korman MH, Jakubczak JL, Eickbush TH (1993) Reverse transcription of R2Bm RNA is primed by a nick at the chromosomal target site: a mechanism for non-LTR retrotransposition. Cell 72:595–605. https://doi.org/10.1016/0092-8674(93)90078-5

Luo S, Zhang H, Duan Y et al (2020) The evolutionary arms race between transposable elements and piRNAs in Drosophila melanogaster. BMC Evol Biol 20:14. https://doi.org/10.1186/s12862-020-1580-3

Lynch M, Conery JS (2003) The origins of genome complexity. Science 302:1401–1404. https://doi.org/10.1126/science.1089370

Lynch VJ, Leclerc RD, May G, Wagner GP (2011) Transposon-mediated rewiring of gene regulatory networks contributed to the evolution of pregnancy in mammals. Nat Genet 43:1154–1159. https://doi.org/10.1038/ng.917

Mager DL, Stoye JP (2015) Mammalian endogenous Retroviruses. Microbiol Spectr 3:MDNA3-0009–2014. https://doi.org/10.1128/microbiolspec.MDNA3-0009-2014

Magiorkinis G, Blanco-Melo D, Belshaw R (2015) The decline of human endogenous retroviruses: extinction and survival. Retrovirology 12:8. https://doi.org/10.1186/s12977-015-0136-x

Majumdar S, Singh A, Rio DC (2013) The human THAP9 gene encodes an active P-element DNA transposase. Science 339:446–448. https://doi.org/10.1126/science.1231789

Makałowski W, Gotea V, Pande A, Makałowska I (2019) Transposable elements: classification, identification, and their use as a tool for comparative genomics. In: Anisimova M (ed) Evolutionary genomics: statistical and computational methods. Springer, New York, pp 177–207

Malik HS, Burke WD, Eickbush TH (1999) The age and evolution of non-LTR retrotransposable elements. Mol Biol Evol 16:793–805. https://doi.org/10.1093/oxfordjournals.molbev.a026164

Mangeney M, Renard M, Schlecht-Louf G et al (2007) Placental syncytins: genetic disjunction between the fusogenic and immunosuppressive activity of retroviral envelope proteins. Proc Natl Acad Sci USA 104:20534–20539. https://doi.org/10.1073/pnas.0707873105

Manthey JD, Moyle RG, Boissinot S (2018) Multiple and independent phases of transposable element amplification in the genomes of Piciformes (Woodpeckers and Allies). Genome Biol Evol 10:1445–1456. https://doi.org/10.1093/gbe/evy105

Matsuo M, Cristofari G (2025) LINE-1, the NORth star of nucleolar organization. Genes Dev 39:183–185. https://doi.org/10.1101/gad.352583.124

Matsushima W, Planet E, Trono D (2024) Ancestral genome reconstruction enhances transposable element annotation by identifying degenerate integrants. Cell Genom 4:100497. https://doi.org/10.1016/j.xgen.2024.100497

McClintock B (1950) The origin and behavior of mutable loci in maize. Proc Natl Acad Sci USA 36:344–355. https://doi.org/10.1073/pnas.36.6.344

Mcclintock B (1956) Controlling elements and the gene. Cold Spring Harb Symp Quant Biol 21:197–216. https://doi.org/10.1101/sqb.1956.021.01.017

Mi S, Lee X, Li X et al (2000) Syncytin is a captive retroviral envelope protein involved in human placental morphogenesis. Nature 403:785–789. https://doi.org/10.1038/35001608

Mills RE, Bennett EA, Iskow RC, Devine SE (2007) Which transposable elements are active in the human genome? Trends Genet TIG 23:183–191. https://doi.org/10.1016/j.tig.2007.02.006

Miousse IR, Chalbot M-CG, Lumen A et al (2015) Transposable elements in response to environmental stressors. Mutat Res Rev Mutat Res 765:19–39. https://doi.org/10.1016/j.mrrev.2015.05.003

Moran JV, DeBerardinis RJ, Kazazian HH (1999) Exon shuffling by L1 Retrotransposition. Science 283:1530–1534. https://doi.org/10.1126/science.283.5407.1530

Nadler MJS, Chang W, Ozkaynak E et al (2023) Hominoid SVA-lncRNA AK057321 targets human-specific SVA retrotransposons in SCN8A and CDK5RAP2 to initiate neuronal maturation. Commun Biol 6:347. https://doi.org/10.1038/s42003-023-04683-8

Nam K, Ellegren H (2012) Recombination drives vertebrate genome contraction. PLoS Genet 8:e1002680. https://doi.org/10.1371/journal.pgen.1002680

Newkirk SJ, Lee S, Grandi FC et al (2017) Intact piRNA pathway prevents L1 mobilization in male meiosis. Proc Natl Acad Sci USA 114:E5635–E5644. https://doi.org/10.1073/pnas.1701069114

Nishihara H (2019) Retrotransposons spread potential cis-regulatory elements during mammary gland evolution. Nucleic Acids Res 47:11551–11562. https://doi.org/10.1093/nar/gkz1003

Nishihara H, Kobayashi N, Kimura-Yoshida C et al (2016) Coordinately co-opted multiple transposable elements constitute an enhancer for wnt5a expression in the mammalian secondary palate. PLoS Genet 12:e1006380. https://doi.org/10.1371/journal.pgen.1006380

Nurk S, Koren S, Rhie A et al (2022) The complete sequence of a human genome. Science 376:44–53. https://doi.org/10.1126/science.abj6987

O'Donnell KA, Burns KH (2010) Mobilizing diversity: transposable element insertions in genetic variation and disease. Mob DNA 1:21. https://doi.org/10.1186/1759-8753-1-21

Ohshima K, Okada N (2005) SINEs and LINEs: symbionts of eukaryotic genomes with a common tail. Cytogenet Genome Res 110:475–490. https://doi.org/10.1159/000084981

Ohtani H, Liu M, Zhou W et al (2018) Switching roles for DNA and histone methylation depend on evolutionary ages of human endogenous retroviruses. Genome Res 28:1147–1157. https://doi.org/10.1101/gr.234229.118

Oliver KR, Greene WK (2012) Transposable elements and viruses as factors in adaptation and evolution: an expansion and strengthening of the TE-Thrust hypothesis. Ecol Evol 2:2912–2933. https://doi.org/10.1002/ece3.400

Oo JA, Brandes RP, Leisegang MS (2022) Long non-coding RNAs: novel regulators of cellular physiology and function. Pflugers Arch 474:191–204. https://doi.org/10.1007/s00424-021-02641-z

Oomen ME, Torres-Padilla M-E (2024) Jump-starting life: balancing transposable element co-option and genome integrity in the developing mammalian embryo. EMBO Rep 25:1721. https://doi.org/10.1038/s44319-024-00118-5

Orgel LE, Crick FHC (1980) Selfish DNA: the ultimate parasite. Nature 284:604–607. https://doi.org/10.1038/284604a0

Osmanski AB, Paulat NS, Korstian J et al (2023) Insights into mammalian TE diversity through the curation of 248 genome assemblies. Science 380:eabn1430. https://doi.org/10.1126/science.abn1430

Pace JK, Feschotte C (2007) The evolutionary history of human DNA transposons: evidence for intense activity in the primate lineage. Genome Res 17:422–432. https://doi.org/10.1101/gr.5826307

Parhad SS, Tu S, Weng Z, Theurkauf WE (2017) Adaptive evolution leads to cross-species incompatibility in the piRNA transposon silencing machinery. Dev Cell 43:60–70.e5. https://doi.org/10.1016/j.devcel.2017.08.012

Park EG, Ha H, Lee DH et al (2022) Genomic analyses of non-coding RNAs overlapping transposable elements and its implication to human diseases. Int J Mol Sci 23:8950. https://doi.org/10.3390/ijms23168950

Pascarella G, Hon CC, Hashimoto K et al (2022) Recombination of repeat elements generates somatic complexity in human genomes. Cell 185:3025–3040.e6. https://doi.org/10.1016/j.cell.2022.06.032

Pasquesi GIM, Adams RH, Card DC et al (2018) Squamate reptiles challenge paradigms of genomic repeat element evolution set by birds and mammals. Nat Commun 9:2774. https://doi.org/10.1038/s41467-018-05279-1

Payer LM, Burns KH (2019) Transposable elements in human genetic disease. Nat Rev Genet 20:760–772. https://doi.org/10.1038/s41576-019-0165-8

Pehrsson EC, Choudhary MNK, Sundaram V, Wang T (2019) The epigenomic landscape of transposable elements across normal human development and anatomy. Nat Commun 10:5640. https://doi.org/10.1038/s41467-019-13555-x

Peona V, Weissensteiner MH, Suh A (2018) How complete are "complete" genome assemblies?—an avian perspective. Mol Ecol Resour 18:1188–1195. https://doi.org/10.1111/1755-0998.12933

Percharde M, Sultana T, Ramalho-Santos M (2020) What doesn't kill you makes you stronger: transposons as dual players in chromatin regulation and genomic variation. BioEssays 42:1900232. https://doi.org/10.1002/bies.201900232

Petersen M, Armisén D, Gibbs RA et al (2019) Diversity and evolution of the transposable element repertoire in arthropods with particular reference to insects. BMC Evol Biol 19:11. https://doi.org/10.1186/s12862-018-1324-9

Petrov DA (2002) Mutational equilibrium model of genome size evolution. Theor Popul Biol 61:531–544. https://doi.org/10.1006/tpbi.2002.1605

Philippe C, Vargas-Landin DB, Doucet AJ et al (2016) Activation of individual L1 retrotransposon instances is restricted to cell-type dependent permissive loci. elife 5:e13926. https://doi.org/10.7554/eLife.13926

Piriyapongsa J, Rutledge MT, Patel S et al (2007) Evaluating the protein coding potential of exonized transposable element sequences. Biol Direct 2:31. https://doi.org/10.1186/1745-6150-2-31

Piskurek O, Okada N (2007) Poxviruses as possible vectors for horizontal transfer of retroposons from reptiles to mammals. Proc Natl Acad Sci USA 104:12046. https://doi.org/10.1073/pnas.0700531104

Platt RN, Vandewege MW, Ray DA (2018) Mammalian transposable elements and their impacts on genome evolution. Chromosom Res 26:25–43. https://doi.org/10.1007/s10577-017-9570-z

Poduri A, Evrony GD, Cai X, Walsh CA (2013) Somatic mutation, genomic variation, and neurological disease. Science 341:1237758. https://doi.org/10.1126/science.1237758

Polavarapu N, Arora G, Mittal VK, McDonald JF (2011) Characterization and potential functional significance of human-chimpanzee large INDEL variation. Mob DNA 2:13. https://doi.org/10.1186/1759-8753-2-13

Pontis J, Planet E, Offner S et al (2019) Hominoid-specific transposable elements and KZFPs facilitate human embryonic genome activation and control transcription in Naive human ESCs. Cell Stem Cell 24:724–735.e5. https://doi.org/10.1016/j.stem.2019.03.012

Ramakrishnan M, Satish L, Sharma A et al (2022) Transposable elements in plants: recent advancements, tools and prospects. Plant Mol Biol Report 40:628–645. https://doi.org/10.1007/s11105-022-01342-w

Ray DA, Feschotte C, Pagan HJT et al (2008) Multiple waves of recent DNA transposon activity in the bat, Myotis lucifugus. Genome Res 18:717–728. https://doi.org/10.1101/gr.071886.107

Ricci M, Peona V, Guichard E et al (2018) Transposable elements activity is positively related to rate of speciation in mammals. J Mol Evol 86:303–310. https://doi.org/10.1007/s00239-018-9847-7

Richardson SR, Doucet AJ, Kopera HC et al (2015) The influence of LINE-1 and SINE Retrotransposons on mammalian genomes. Microbiol Spectr 3. https://doi.org/10.1128/microbiolspec.MDNA3-0061-2014

Richardson SR, Gerdes P, Gerhardt DJ et al (2017) Heritable L1 retrotransposition in the mouse primordial germline and early embryo. Genome Res 27:1395–1405. https://doi.org/10.1101/gr.219022.116

Roman AC, Benitez DA, Carvajal-Gonzalez JM, Fernandez-Salguero PM (2008) Genome-wide B1 retrotransposon binds the transcription factors dioxin receptor and Slug and regulates

gene expression in vivo. Proc Natl Acad Sci USA 105:1632–1637. https://doi.org/10.1073/pnas.0708366105
Ros F, Kunze R (2001) Regulation of activator/dissociation transposition by replication and DNA methylation. Genetics 157:1723–1733. https://doi.org/10.1093/genetics/157.4.1723
Rubin E, Levy AA (1997) Abortive gap repair: underlying mechanism for Ds element formation. Mol Cell Biol 17:6294–6302. https://doi.org/10.1128/MCB.17.11.6294
Sakamoto Y, Sereewattanawoot S, Suzuki A (2020) A new era of long-read sequencing for cancer genomics. J Hum Genet 65:3–10. https://doi.org/10.1038/s10038-019-0658-5
Sela N, Mersch B, Hotz-Wagenblatt A, Ast G (2010) Characteristics of transposable element exonization within human and mouse. PLoS One 5:e10907. https://doi.org/10.1371/journal.pone.0010907
Sen SK, Han K, Wang J et al (2006) Human genomic deletions mediated by recombination between Alu elements. Am J Hum Genet 79:41–53
Senft AD, Macfarlan TS (2021) Transposable elements shape the evolution of mammalian development. Nat Rev Genet 22:691–711. https://doi.org/10.1038/s41576-021-00385-1
Serrato-Capuchina A, Matute DR (2018) The role of transposable elements in speciation. Genes 9:254. https://doi.org/10.3390/genes9050254
Shahid S, Slotkin RK (2020) The current revolution in transposable element biology enabled by long reads. Curr Opin Plant Biol 54:49–56. https://doi.org/10.1016/j.pbi.2019.12.012
Shapiro JA (2022) What we have learned about evolutionary genome change in the past 7 decades. Biosystems 215–216:104669. https://doi.org/10.1016/j.biosystems.2022.104669
Simonti CN, Pavlicev M, Capra JA (2017) Transposable element exaptation into regulatory regions is rare, influenced by evolutionary age, and subject to Pleiotropic constraints. Mol Biol Evol 34:2856–2869. https://doi.org/10.1093/molbev/msx219
Skipper KA, Andersen PR, Sharma N, Mikkelsen JG (2013) DNA transposon-based gene vehicles - scenes from an evolutionary drive. J Biomed Sci 20:92. https://doi.org/10.1186/1423-0127-20-92
Slotkin RK, Martienssen R (2007) Transposable elements and the epigenetic regulation of the genome. Nat Rev Genet 8:272–285. https://doi.org/10.1038/nrg2072
Spradling AC, Bellen HJ, Hoskins RA (2011) Drosophila P elements preferentially transpose to replication origins. Proc Natl Acad Sci USA 108:15948–15953. https://doi.org/10.1073/pnas.1112960108
Sultana T, van Essen D, Siol O et al (2019) The landscape of L1 Retrotransposons in the human genome is shaped by pre-insertion sequence biases and post-insertion selection. Mol Cell 74:555–570.e7. https://doi.org/10.1016/j.molcel.2019.02.036
Sundaram V, Cheng Y, Ma Z et al (2014) Widespread contribution of transposable elements to the innovation of gene regulatory networks. Genome Res 24:1963–1976. https://doi.org/10.1101/gr.168872.113
Suntsova M, Gogvadze EV, Salozhin S et al (2013) Human-specific endogenous retroviral insert serves as an enhancer for the schizophrenia-linked gene PRODH. Proc Natl Acad Sci USA 110:19472–19477. https://doi.org/10.1073/pnas.1318172110
Tang W, Gunn TM, McLaughlin DF et al (2000) Secreted and membrane attractin result from alternative splicing of the human ATRN gene. Proc Natl Acad Sci USA 97:6025. https://doi.org/10.1073/pnas.110139897
Taniguchi-Ikeda M, Kobayashi K, Kanagawa M et al (2011) Pathogenic exon-trapping by SVA retrotransposon and rescue in Fukuyama muscular dystrophy. Nature 478:127–131. https://doi.org/10.1038/nature10456
Teugels E, De Brakeleer S, Goelen G et al (2005) De novo Alu element insertions targeted to a sequence common to the BRCA1 and BRCA2 genes. Hum Mutat 26:284–284. https://doi.org/10.1002/humu.9366
Thomas J, Pritham EJ (2015) Helitrons, the Eukaryotic rolling-circle transposable elements. Microbiol Spectr 3. https://doi.org/10.1128/microbiolspec.MDNA3-0049-2014

Tollis M, Boissinot S (2012) The evolutionary dynamics of transposable elements in eukaryote genomes. Genome Dyn 7:68–91. https://doi.org/10.1159/000337126
Trizzino M, Park Y, Holsbach-Beltrame M et al (2017) Transposable elements are the primary source of novelty in primate gene regulation. Genome Res 27:1623–1633. https://doi.org/10.1101/gr.218149.116
Trizzino M, Kapusta A, Brown CD (2018) Transposable elements generate regulatory novelty in a tissue-specific fashion. BMC Genomics 19:468. https://doi.org/10.1186/s12864-018-4850-3
Vazquez JM, Kraft M, Lynch VJ (2022) A CDKN2C retroduplication in Bowhead whales is associated with the evolution of extremely long lifespans and alerted cell cycle dynamics. 2022.09.07.506958
Verneau O, Catzeflis F, Furano AV (1998) Determining and dating recent rodent speciation events by using L1 (LINE-1) retrotransposons. Proc Natl Acad Sci USA 95:11284–11289
Wallau GL, Vieira C, Loreto ÉLS (2018) Genetic exchange in eukaryotes through horizontal transfer: connected by the mobilome. Mob DNA 9:6. https://doi.org/10.1186/s13100-018-0112-9
Walsh AM, Kortschak RD, Gardner MG et al (2013) Widespread horizontal transfer of retrotransposons. Proc Natl Acad Sci USA 110:1012–1016. https://doi.org/10.1073/pnas.1205856110
Wang H, Xing J, Grover D et al (2005) SVA elements: a hominid-specific retroposon family. J Mol Biol 354:994–1007. https://doi.org/10.1016/j.jmb.2005.09.085
Wang L, Barbash DA, Kelleher ES (2020) Adaptive evolution among cytoplasmic piRNA proteins leads to decreased genomic auto-immunity. PLoS Genet 16:e1008861. https://doi.org/10.1371/journal.pgen.1008861
Wells JN, Feschotte C (2020) A field guide to eukaryotic transposable elements. Annu Rev Genet 54:539–561. https://doi.org/10.1146/annurev-genet-040620-022145
Wicker T, Sabot F, Hua-Van A et al (2007) A unified classification system for eukaryotic transposable elements. Nat Rev Genet 8:973–982. https://doi.org/10.1038/nrg2165
Wilhelm M, Wilhelm FX (2001) Reverse transcription of retroviruses and LTR retrotransposons. Cell Mol Life Sci CMLS 58:1246–1262. https://doi.org/10.1007/PL00000937
Wu C, Lu J (2019) Diversification of transposable elements in arthropods and its impact on genome evolution. Genes 10:338. https://doi.org/10.3390/genes10050338
Xia B, Zhang W, Zhao G et al (2024) On the genetic basis of tail-loss evolution in humans and apes. Nature 626:1042–1048. https://doi.org/10.1038/s41586-024-07095-8
Xiang Y, Liang H (2021) The regulation and functions of endogenous Retrovirus in embryo development and stem cell differentiation. Stem Cells Int 2021:6660936. https://doi.org/10.1155/2021/6660936
Xie SQ, Leeke BJ, Whilding C et al (2022) Nucleolar-based Dux repression is essential for embryonic two-cell stage exit. Genes Dev 36:331–347. https://doi.org/10.1101/gad.349172.121
Xing J, Wang H, Belancio VP et al (2006) Emergence of primate genes by retrotransposon-mediated sequence transduction. Proc Natl Acad Sci USA 103:17608–17613. https://doi.org/10.1073/pnas.0603224103
Yamamoto G, Miyabe I, Tanaka K et al (2021) SVA retrotransposon insertion in exon of MMR genes results in aberrant RNA splicing and causes Lynch syndrome. Eur J Hum Genet EJHG 29:680–686. https://doi.org/10.1038/s41431-020-00779-5
Yang N, Kazazian HH (2006) L1 retrotransposition is suppressed by endogenously encoded small interfering RNAs in human cultured cells. Nat Struct Mol Biol 13:763–771. https://doi.org/10.1038/nsmb1141
Yang G, Nagel DH, Feschotte C et al (2009) Tuned for transposition: molecular determinants underlying the hyperactivity of a Stowaway MITE. Science 325:1391–1394. https://doi.org/10.1126/science.1175688
Yu H, Sun Z, Tan T et al (2021) rRNA biogenesis regulates mouse 2C-like state by 3D structure reorganization of peri-nucleolar heterochromatin. Nat Commun 12:6365. https://doi.org/10.1038/s41467-021-26576-2
Zamudio N, Bourc'his D (2010) Transposable elements in the mammalian germline: a comfortable niche or a deadly trap? Heredity 105:92–104. https://doi.org/10.1038/hdy.2010.53

Zeh DW, Zeh JA, Ishida Y (2009) Transposable elements and an epigenetic basis for punctuated equilibria. BioEssays News Rev Mol Cell Dev Biol 31:715–726. https://doi.org/10.1002/bies.200900026

Zhang Y, Cheng TC, Huang G et al (2019a) Transposon molecular domestication and the evolution of the RAG recombinase. Nature 569:79–84. https://doi.org/10.1038/s41586-019-1093-7

Zhang Y, Li T, Preissl S et al (2019b) Transcriptionally active HERV-H retrotransposons demarcate topologically associating domains in human pluripotent stem cells. Nat Genet 51:1380–1388. https://doi.org/10.1038/s41588-019-0479-7

Zhang H-H, Peccoud J, Xu M-R-X et al (2020) Horizontal transfer and evolution of transposable elements in vertebrates. Nat Commun 11:1362. https://doi.org/10.1038/s41467-020-15149-4

Zhou W, Liang G, Molloy PL, Jones PA (2020) DNA methylation enables transposable element-driven genome expansion. Proc Natl Acad Sci USA 117:19359–19366. https://doi.org/10.1073/pnas.1921719117

Zhu X, Fang H, Gladysz K et al (2021) Overexpression of transposable elements is associated with immune evasion and poor outcome in colorectal cancer. Eur J Cancer 157:94–107. https://doi.org/10.1016/j.ejca.2021.08.003

Chapter 2
Transposable Elements and Gene Duplication

Izabela Makałowska and Natalia Ryczek

Abstract Transposable elements are one of the most important players in genome evolution. The most obvious contribution of TEs to evolution is gene duplication, which provides new genetic material on which mutation, drift, and selection can act. There are several mechanisms involving TEs by which gene duplication can occur. These mechanisms can be divided into two main groups: DNA-based duplication and RNA-based duplication. In DNA-based duplication and transduction, the duplicated region can contain part of a gene, an entire gene, or multiple genes. In the latter two cases, promoters and introns may be conserved and present in duplicated genes. This is in sharp contrast to the results of RNA-based retroposition, where mature mRNA is retrotranscribed into cDNA and duplicated genes are intronless and stripped of regulatory sequences. There is increasing evidence that these 'dead on arrival' retrocopies can acquire active promoters and become functional. The evolutionary fate of duplicated genes varies. Many of them accumulate mutations and become pseudogenized. Others can either retain the function of their ancestors or undergo subfunctionalization or neofunctionalization. Gene duplications can have direct phenotypic consequences and contribute significantly to the ability of species to adapt to changing environments, but they are also associated with many human diseases.

Keywords Transposable elements · Gene duplication · Retroposition · Retrocopies · Recombination · Human diseases · Cancer · Pseudogenes

Abbreviations

AF	Atrial fibrillation
AML	Acute myeloid leukemia

I. Makałowska (✉) · N. Ryczek
Institute of Human Biology and Evolution, Faculty of Biology, Adam Mickiewicz University, Poznań, Poland
e-mail: izabela.makalowska@amu.edu.pl

L. M. Vaschetto (ed.), *Transposable Elements, Transcriptomics, and Diseases*, Epigenetics and Human Health 14, https://doi.org/10.1007/978-3-032-04125-8_2

Ap	Aspartic protease
ASD	Autism spectrum disorder
CMT	Charcot-Marie-Tooth disease
CMT1A	Charcot-Marie-Tooth disease type 1A
CVD	Cardiovascular disease
dsRNA	Double-stranded RNA
EDS VI	Ehlers-Danlos syndrome type VI
EH	Essential hypertension
EN	Endonuclease
endo-siRNA	Endogenous small interfering RNA
HD	Huntington's disease
HIV-1	Human immunodeficiency virus 1
HPRT	Hypoxanthine phosphoribosyltransferase
IN	Integrase
LDL	Low-density lipoprotein
LINE	Long interspersed nuclear element
LN	Lesch Nyhan syndrome
lncRNA	Long noncoding RNA
LTR	Long terminal repeats
miRNA	MicroRNA
MITE	Mariner transposon-like element
NAHR	Non-allelic homologous recombination
NAT	Natural antisense transcripts
OCD	Obsessive-compulsive disorder
ORF	Open reading frame
PD	Parkinson's disease
piRNA	Piwi-interacting RNA
PPT	Polypurine tract
RH	Rnase H
RT	Reverse transcriptase
SD	Segmental duplications
SINE	Short-interspersed nuclear element
TE	Transposable element
TPRT	Target-primed reverse transcription
TSD	Target site duplication
VLP	Virus-like particle
VNTR	Variable number of tandem repeats

2.1 Introduction

Transposable elements (TEs) are key players in the evolution of genomes, and they contribute to it in both direct and indirect ways. TEs are probably best known as recombination hotspots, but they also provide genomic material and contribute to the evolution of promoters (Fueyo et al. 2022; Rayan et al. 2016), transcriptomes, and proteomes (Kubiak et al. 2020). In the direct contribution, the coding potential of a TE is "domesticated" to perform a host cellular function (Kapitonov and Jurka 2004; Joly-Lopez and Bureau 2018). An example of a domesticated TE is the recombination activating protein RAG1 (Agrawal et al. 1998). Another example is the *gag* gene. It has been estimated that there may be as many as 85 *gag*-like genes in the human genome (Campillos et al. 2006). Sequences derived from TEs have also been exapted into a coding portion of an existing gene, generating novel variants of transcripts and proteins (Makalowski et al. 1994; Schmitz and Brosius 2011). In humans, *Alu* sequences are the major donors of TE-originating exons, but exons derived from other elements, such as LINEs (long interspersed nuclear elements), endogenous viruses, and DNA transposons, have also been identified (Makalowski et al. 2019). The functional roles of TEs in genome evolution are discussed in detail in Chap. 1.

Indirect contributions include gene duplication, which is one of the most important evolutionary forces because it creates material that can be further modified by natural selection (Ohno 1970; Nei 1969). This phenomenon, which contributes to adaptive evolutionary innovation, is very common in all eukaryotic organisms (Kaessmann 2010). TEs as mutagens catalyzing various structural changes in the genomes, including duplications, were already recognized by Barbara McClintock (Mc 1950). However, the first case of TE-mediated gene duplication in bacteria was published 25 years later (Kleckner et al. 1975). There are several mechanisms involving TEs by which these duplicates can be formed. These mechanisms can be divided into two main groups: DNA-based and RNA-based duplications (Fig. 2.1). DNA-based duplication was long thought to be the main mechanism for the evolution of new genes. However, subsequent studies have shown that many other mechanisms play an important role in the generation of new genes (Ma et al. 2023; Xia et al. 2025). The most important of these are retroduplication (Kaessmann et al. 2009) and transduction (Pickeral et al. 2000).

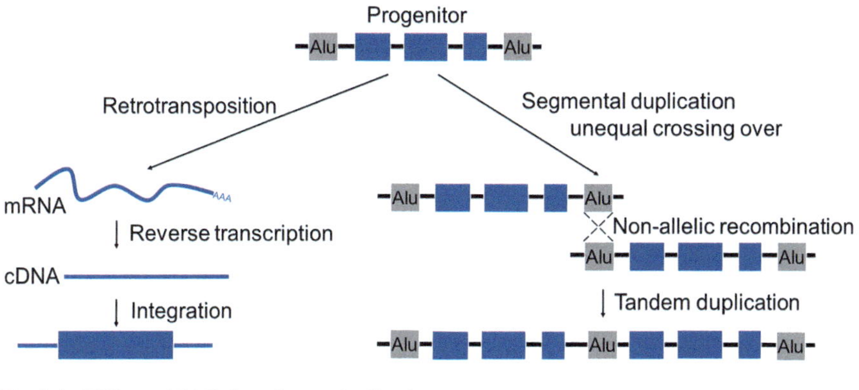

Fig. 2.1 RNA- and DNA-based gene duplication

2.2 DNA-Based Gene Duplication—Non-allelic Homologous Unequal Recombination

The human genome contains numerous blocks of DNA larger than 1 kb that share a high degree of sequence similarity, usually greater than 90%. These blocks of homologous DNA, called segmental duplications (SD) or low-copy repeats, are dispersed interchromosomally and intrachromosomally. They were originally estimated to make up 5%–6% of the human genome (Bailey et al. 2002), but this has been increased to 7% after the sequencing of the entire genome, from telomere to telomere (Vollger et al. 2022). These stretches of duplicated DNA result from unequal crossing-over, also known as non-allelic homologous recombination (NAHR). Once formed, segmental duplications promote further misalignments, rearrangements, and subsequent recombination (Stankiewicz and Lupski 2002). Depending on the position of crossing-over, the duplicated region may contain only part of a gene, an entire gene, or several genes. These structural variants contribute significantly to genomic variation between humans, and a high number of copy number polymorphic genes in humans map to SDs (Liao et al. 2023).

The association between TEs and recombination is quite strong, although it depends on the type of TE. While L1 LINEs are negatively correlated with the recombination rate (Sistrom et al. 2024), *Alu* elements are accumulated in gene-rich and highly recombining regions (Witherspoon et al. 2009). Our genome contains over a million *Alu* sequences. These nearly identical non-allelic regions trigger NAHR events, which among other things, lead to the duplication (Batzer and Deininger 2002). Highly significant enrichment of *Alu* sequences was found near or within DNA junctions. Almost 27% of all segmental duplications in the human genome are flanked with *Alu* sequences. Statistically significant enrichment of *AluS* and *AluY* elements was found at junctions of low-copy repeats, and it has been proposed that the burst of these two subfamilies triggered *Alu-Alu* recombination, and this in turn initiated the expansion of rich in genes segmental duplications (Bailey et al. 2003).

An excellent example of duplication by unequal recombination is the human growth hormone and human chorionic somatomammotropin (*HGH-HCS*) gene cluster. The cluster contains two growth hormone genes and three chorionic somatomammotropin genes. High sequence similarity and the presence of at least three classes of duplication units interspersed in the cluster suggest that the cluster evolved by a mechanism involving unequal homologous recombination between *Alu* sequences (Barsh et al. 1983; Hirt et al. 1987).

2.3 RNA-Based Gene Duplication

Various mobile genomic elements, such as *Ty* in yeast (Garfinkel et al. 1985) and LINE1 in humans (Mathias et al. 1991), were found to encode reverse transcriptase, which was quickly linked to their mobilization capabilities. Autonomous transposons, such as LTR (long terminal repeats) or LINE, encode their own mobilization machinery. This machinery can be used to proliferate nonautonomous retrotransposons such as *Alu* elements (Dewannieux et al. 2003) as well as protein-coding genes (Kaessmann et al. 2009; Kubiak and Makalowska 2017). The endogenous activity of reverse transcriptase and its ability to retrotranspose cellular non-retrotransposon RNAs was demonstrated in a minigene system in HeLa cells (Maestre et al. 1995). Further experiments revealed functional differences between the enzymes from retroviruses and those from LINE retrotransposons (Dhellin et al. 1997) and confirmed that LINE1 can act in trans and give rise to retroposed gene copies (Esnault et al. 2000) (Fig. 2.2a).

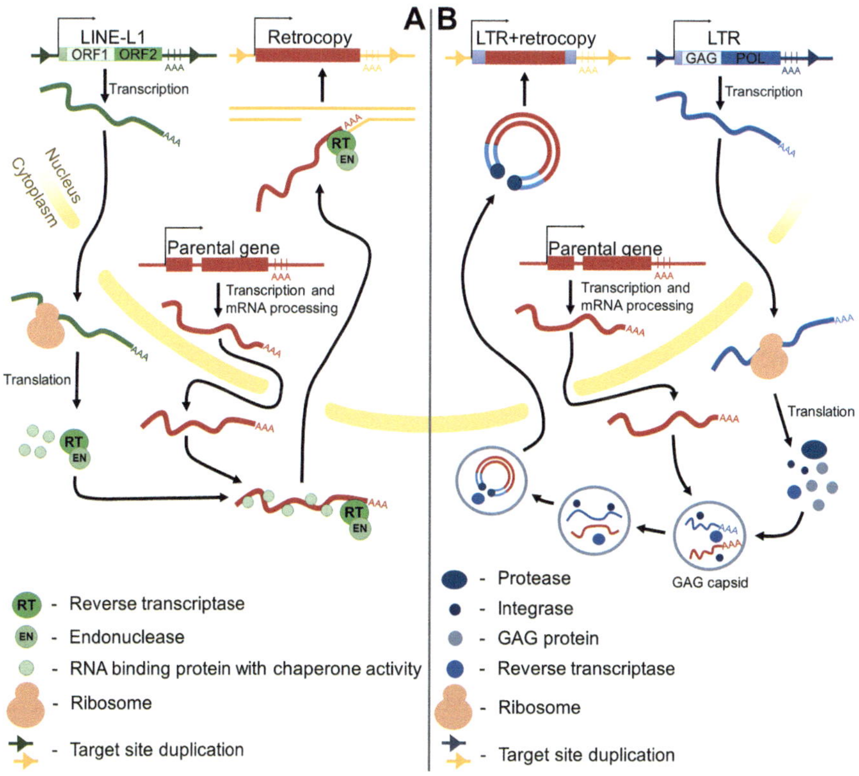

Fig. 2.2 Mechanisms of RNA-based gene retroposition. (**a**) LINE1-driven retroposition. (**b**) LTR-driven retroposition

While DNA-mediated duplication preserves the gene structure, i.e., exons and introns, retroposition mainly produces complete but deprived-of-introns copies. LINE-L1-mediated retrocopies are also characterized by the presence of a 3′ polyA tail and flanking target site duplications (TSDs). LTR retrotransposons trap host mRNA in their virus-like particles (VLPs), and when accidentally switched to this mRNA as a template, they generate chimeric cDNA. The copies generated by both LTR and non-LTR-mediated processes, and reintegrated into the genome, do not inherit regulatory elements and are "dead on arrival." These sequences are therefore often referred to as retropseudogenes or processed pseudogenes. Although the majority of these retrocopies are in a state of "relaxed" selection and remain "dormant," many acquire regulatory sequences and become functional.

2.3.1 LTR Retrotransposons-Based Genes Duplication

Autonomous LTR retrotransposons make up about 8% of the human genome. They are a diverse group, but they do share some common traits and are all characterized by long terminal repeats (LTRs) directly flanking an internal coding region. Functional LTR retrotransposons encode a minimum of two genes: *gag* and *pol*. *Gag* encodes a polyprotein that forms virus-like particles, and *pol* encodes various enzymatic domains, including aspartic protease (Ap), RNase H (RH), reverse transcriptase (RT), and integrase (IN) (Wilhelm and Wilhelm 2001). They also contain an RNA Pol II promoter in their 5′ LTR.

The mechanism of LTR transposition is similar to that observed in retroviruses, which undergo two rounds of template switching during viral DNA synthesis and use a low-affinity RT enzyme (Delviks-Frankenberry et al. 2011). It has been shown that retroviruses can capture host mRNA, and their RT can switch to mRNA and use it as a template (Luo and Taylor 1990; Hajjar and Linial 1993). Genome-wide studies in several taxa, including humans, show that LTR retrotransposons can drive retroposition (Tan et al. 2016). The process starts with transcription of the LTR retrotransposon by RNA polymerase II. The newly synthesized RNA is then transported to the cytoplasm, where translation and formation of VLPs takes place. Reverse transcriptase, integrase, and RNA molecules are packaged into VLPs. Cellular mRNA may also be accidentally encapsulated (Fig. 2.2b). Reverse transcription is primed by a tRNA that matches the primer binding site that is located at the 3′ end of the 5′ LTR. Reverse transcriptase begins synthesizing cDNA using the RNA of the LTR retrotransposon as the template. Induced by stretches of micro similarity between the retrotransposon RNA and cellular mRNA captured in the VLP, the enzyme switches templates, using the cellular mRNA as the new template. As a result, the mRNA is incorporated into emerging cDNA. Stretches of micro-similarity can occur several times, and template switching may also happen several times. However, usually only a limited portion of the mRNA is reverse transcribed and incorporated into the chimeric cDNA produced. Once the RT reaches the poly-purine tract (PPT) located in the 3′ LTR of the retrotransposon RNA template, synthesis of the second (plus) strand begins. After the minus-strand cDNA synthesis has copied the RNA strand beyond the polypurine tract, RNase H degrades most of the template but spares the PPT, which acts as a primer for plus-strand DNA synthesis. Next, the LTR ends of the newly synthesized cDNA are bound by integrase, and the VLP localizes to a nearby nucleus. The cDNA-IN complex is transported to the nucleus, where integrase cleaves the cellular DNA and joins the released ends to the LTRs of the retrotransposed copy (Tan et al. 2016; Hughes 2015; Wells and Feschotte 2020) (Fig. 2.2b).

The evidence for LTR-mediated retroposition has been found in multiple eukaryotic lineages and appears to be highly conserved. There is a striking consistency in the features that characterize LTR-mediated retroposition. However, its contribution to gene duplication varies between taxa (Tan et al. 2016).

2.3.2 LINE-L1-Based Genes Retroposition

Non-LTR retrotransposons are the most abundant TEs in the human genome. Among them, the LINE-L1 has been the most active during human evolution (Beck et al. 2010; Cordaux and Batzer 2009), which explains why the non-LTR gene retroposition is the best studied. Intact L1s are up to 6 kb in length. They consist of a 5′ end untranslated region (UTR) containing an internal RNA polymerase II promoter (Swergold 1990), two open reading frames (ORF) encoding well-characterized proteins, and a 3′ end UTR containing a polyadenylation signal terminating with an oligo A-rich tail (Babushok and Kazazian 2007). In the 5′ untranslated region is localized an internal promoter with sense and antisense activity (Speek 2001). The antisense promoter (ASP) drives transcription in the opposite direction, into adjacent cellular sequences, yielding chimeric transcripts (Nigumann et al. 2002). It produces transcripts at approximately 1/8 the rate of the sense promoter (Yang and Kazazian 2006). ORF1 encodes an RNA-binding protein, and ORF2 encodes a reverse transcriptase and endonuclease (EN) domain (Feng et al. 1996; Babushok and Kazazian 2007). The machinery encoded by L1 enables the retroposition process known as target-primed reverse transcription (TPRT). A third, primate-specific open reading frame has recently been found on the antisense strand (Denli et al. 2015; Boeke and Fenyo 2015). The biological function of the 71 amino-acid-long ORF0 protein has been proposed to be involved in L1 mobility (Boeke and Fenyo 2015). Other potential roles of the ORF0 transcript have also been suggested, such as L1 post-transcription degradation via the RNAi mechanism (Yang and Kazazian 2006; Martin et al. 2021).

Retroposition of L1 has been shown to occur at high frequency (Moran et al. 1996) and its activity is not restricted to transcripts of the transposon (Maestre et al. 1995). In the TPRT process, the L1 mRNA is transported to the cytoplasm and translated. In the L1 life cycle, the newly synthesized ORF1 and ORF2 proteins bind to the L1 mRNA in such a way that those with enzymatic domains bind to the 3′ end, while those with chaperone activity bind along the entire RNA molecule.

This ribonucleoprotein complex is then imported back into the nucleus (Kubo et al. 2006). The proteins encoded by L1 can also bind to other RNA molecules (Fig. 2.2a). LINE-1 elements are responsible for the *Alu* (Dewannieux et al. 2003), snRNA (Doucet et al. 2015a) and mRNA retrocopies (Esnault et al. 2000). LINEs can also mobilize primate-specific SVA retrotransposons (Hancks et al. 2011), hY RNA (Perreault et al. 2005) and even endogenous retroviruses (Pavlicek et al. 2002). However, the observation that removal of the polyadenylation signal results in a loss of retrotransposition indicated that the presence of a polyA may be required for mobilizing RNA via L1-derived mechanism (Doucet et al. 2015b).

When the ribonucleoprotein particle is imported into the nucleus, the ORF1 protein acts as a nuclear chaperone. The mechanism of nuclear import is not clear, but it has been proposed that the retroposition occurs during cell division when the nuclear membrane is disrupted (Pizarro and Cristofari 2016; Abyzov et al. 2013), although there is also evidence for retroposition in non-dividing cells (Kubo et al. 2006; Macia et al. 2017). In the nucleus, the complex needs to get close to the DNA, where the target-primed reverse transcription takes place. Encoded by the ORF2, endonuclease cleaves DNA at the degenerate consensus, which is often A/T rich (Rodic and Burns 2013; Jurka 1997). After cleavage, the 3′-end hydroxyl is left free for use as a primer. Reverse transcriptase then synthesizes a cDNA that is integrated back into the genome at the cut site (Kemp and Longworth 2015; Cost et al. 2002) (Fig. 2.2a). Non-classical L1 insertions into pre-existing gaps have also been reported (Sen et al. 2007). The second strand of the target sequence is then cleaved, typically 7–20 nucleotides downstream of the first nick (Kazazian and Moran 1998). The second strand of DNA is synthesized by RNA polymerase II or host repair mechanisms. There are four main canonical features of TPRT-derived duplications: absence of upstream promoter sequence, absence of introns, poly(A) tail at the 5-end, and flanking by direct repeats at the DNA target site.

Retroposed genes have been identified in most evolutionary clades of animal and plant species, although the intensity of retroposition varies significantly between taxa. While there are thousands of retroposed genes in placental mammals, only a handful of retrocopies were identified in birds' genomes, 25 in chicken and 29 in flycatcher, for example. In plants, the number varies from a few dozens to a few hundred retrocopies (Rosikiewicz et al. 2017).

2.3.3 Duplication by Transduction

Active transposable elements have the ability to transduce DNA sequences flanking their 3′ end into the new location. This occurs when the cellular processing machinery bypasses the transposon's termination signal and instead uses a downstream signal. Transduction is expected to be more common in TEs with a simple termination signal, a simple recognition sequence, and a simple transposition process. The most efficient in generation 3′ transductions are L1s, as they have a weak transcription termination signal—the polyA signal (Goodier et al. 2000; Pickeral et al. 2000;

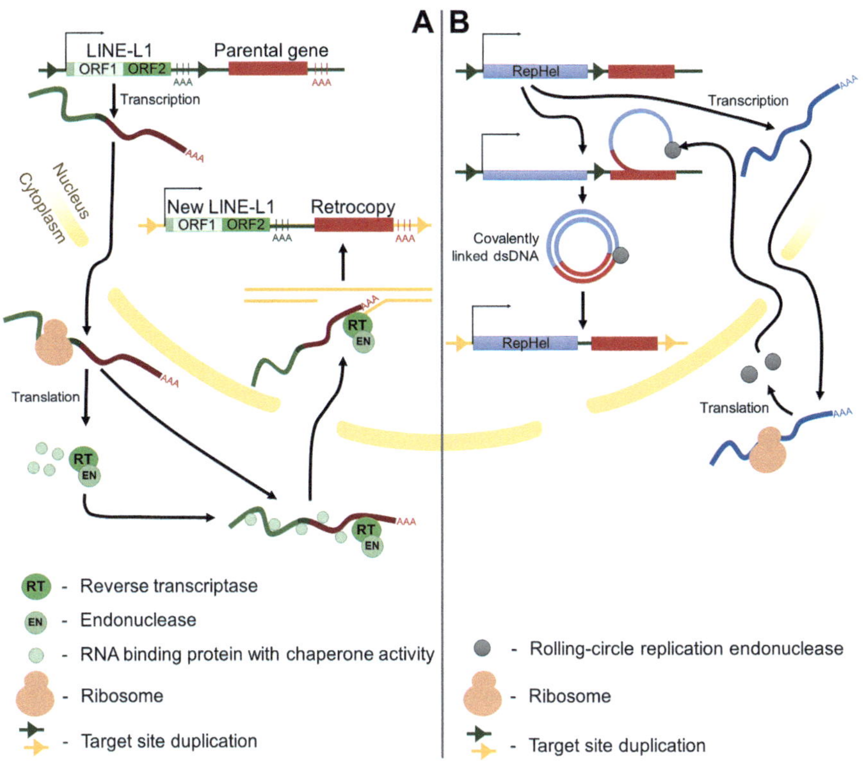

Fig. 2.3 Genes duplication by the transduction mechanisms. (**a**) Driven by the LINE-L1 element. (**b**) Driven by *Helitrons*

Moran et al. 1999). This mechanism can lead to the duplication of coding sequences located in the transduced genomic fragment (Fig. 2.3a). Exon shuffling of the L1 mediated transduction has been shown in cell culture assays (Moran et al. 1996, 1999) but not many examples of exon transduction have been demonstrated in the human genome. One of them is an amplification of exon 9 of the *CFTR* gene (Rozmahel et al. 1997).

Hominid-specific SVA elements have also been implicated in the transduction events (Wang et al. 2005; Ostertag et al. 2003). SVA elements, named after their components: SINE (short interspersed nuclear element), VNTR (variable number of tandem repeats), and *Alu* are thought to be transcribed by RNA polymerase II. These nonautonomous retroposons are mobilized by the L1 protein machinery (Raiz et al. 2012). It was found that at least 52 kb of human genomic sequences were duplicated by 142 different SVA-mediated transductions (Xing et al. 2006). A good example is the *AMAC1* gene (acryl-malonyl condensing enzymes), which has been three times duplicated through SVA-mediated transduction. These events took place before the divergence of humans and African great apes. All three copies contain an intact ORF in the human genome, and two of them have been found to be transcriptionally active, but it is unclear whether they have been conserved by natural selection (Xing et al. 2006).

Helitrons are another group of TEs that are reported to mediate gene duplication through the end bypass model (Kapitonov and Jurka 2007; Thomas and Pritham 2015). Although *Helitrons* occupy a substantial fraction of many genomes, they remained undiscovered until the early 2000s, more than 50 years after the discovery of TEs (Kapitonov and Jurka 2001). They are classified as DNA II class transposons, as they have no intermediate RNA involved in their transposition. However, unlike other DNA TEs, they lack terminal inverted repeats (TIRs) and do not generate target site duplications (TSDs) upon insertion. Their other characteristic features include 5′-TC and 3′-CTRR terminal motifs. They also have a short palindromic sequence of 16–20 nucleotides that forms a hairpin near the 3′ end (Kapitonov and Jurka 2001). Another feature distinguishing *Helitrons* from other DNA transposons is that they retain the original template unaltered during transposition. They have therefore been named peel-and-paste or rolling-circle elements (Kapitonov and Jurka 2001) (Fig. 2.3b). *Helitrons* are widespread in plants and invertebrates but have also been described in other organisms, including vertebrates (Thomas et al. 2010; Pritham and Feschotte 2007; Barro-Trastoy and Kohler 2024). The majority of *helitrons*-driven gene transposition events have been reported in plants. However, although the prevalence and activity of *helitrons* in mammals appears to be much rarer, one well-documented example of such activity has been reported. A study of the brown bat, *Myotis lucifugus*, where these TEs account for an exceptional 6% of the genome, provided evidence that *helitrons* have contributed exons as well as promoters of novel genes (Thomas et al. 2014). Many *helitrons* contain fragments of more than one gene. In the maze these TEs have captured and amplified tens of thousands of gene fragments, and certain *helitrons* have been observed to capture fragments of up to 12 genes (Dong et al. 2011). Several *helitrons* containing more than one gene fragment have also been reported in the brown bat (Thomas et al. 2014; Grabundzija et al. 2018).

2.4 Evolutionary Fate of Duplicated Genes

In all domains of life, a large proportion of genes have been generated by gene duplication, but the outcome depends on the mechanism of the duplication (Cerbin and Jiang 2018; Xia et al. 2025). In the DNA-based duplication and transduction, the duplicated region may contain part of a gene, an entire gene, or multiple genes. In the latter two cases, introns may be preserved and present in duplicated genes. The promoter region can also be included if the 5′ end or the whole gene has been duplicated. In some transduction events, such as those driven by L1 or SVA elements, which involve an intermediate RNA sequence and reverse transcription, promoters that are captured are transcribed into cDNA. This is in sharp contrast to the results of typical retroposition, where mature mRNA is retrotranscribed into cDNA. In this process, regulatory sequences are lost. Another difference is that in the retroposition it is impossible to duplicate blocks of genes together.

Due to the template switch mechanism, only fragments of the gene are duplicated in LTR-elements-driven retroposition, in contrast to the L1-driven process, in which the entire mRNA is often reverse transcribed and inserted into the genome. In the retrocopies originated by LTR-elements-driven mechanism, both 5′ and 3′ LTRs usually coexist, as they are essential for integrase recognition (Tan et al. 2016). In L1-driven retroposition, reverse transcription starts at the 3′ end of the RNA template and often aborts prematurely, which results in 5′ truncation (Kazazian and Moran 1998). It has also been reported that template switching can occur in the case of L1 transposition due to microsimilarity shared by the breakpoints of host sequences and L1, resulting in chimeric sequences (Buzdin et al. 2003).

DNA-based duplication was long thought to be the primary mechanism of the creation of new genes, as these duplications were the earliest observed (Ohno 1970). However, subsequent studies have shown that RNA-mediated duplications occur at a much higher rate than DNA-mediated duplications (Jun et al. 2009). In humans, it has been estimated that L1 has duplicated 1% of the genome (Pickeral et al. 2000), and some estimates of the total number of retrocopies are as high as over 18,000 (Casola and Betran 2017). This exceptionally high number of retrocopies in the human and other primate genomes led to the conclusion that there was a burst of retroposition in ancestral primates (Zhang et al. 2003; Ohshima et al. 2003). However, some studies suggest that an equally high rate of the gene retroposition occurred in the mouse genome (Zhang and Tautz 2022).

2.4.1 Promoter Acquisition

Genes duplicated by retrotransposition are stripped of their regulatory sequences, and the new copy often decays slowly and is silenced by the accumulation of degenerative mutations. For this reason, gene retrotransposition was long thought to produce non-functional gene copies called retropseudogenes or processed pseudogenes. It was also commonly thought that pseudogenes are DNA segments with high similarity to a functional gene but containing nucleotide changes such as frameshift or nonsense mutations that prevent their expression (Li et al. 1981). However, genome-wide searches have confirmed the importance of gene retrotransposition for the emergence of new genes in plants (Abdelsamad and Pecinka 2014; Zhu et al. 2016), invertebrates (Chen et al. 2010; Bai et al. 2007), and primates (Kaessmann et al. 2009; Kubiak et al. 2020).

Although L1-mediated retrocopies are "dead-on-arrival," a large number of transcripts derived from retrocopies were found in cells, suggesting the acquisition of active promoters. One way to achieve this is to "hitchhike" on the regulatory elements of nearby genes (Troskie et al. 2021). Indeed, many retrocopies are found close to or within other transcribed genes. Another scenario is that a retrocopy is inserted near a pre-promoter, a CpG-rich sequence, that has evolved into a functional element over time (Kaessmann et al. 2009). It was observed that many retrocopies acquired a distant promoter by gaining a new 5′ exon from the vicinity of the insertion site (Troskie et al. 2021; Fablet et al. 2009). In the human genome, core promoters are frequently recruited from transposable elements (Makalowski et al. 2019; Kellner and Makalowski 2019). In some cases, a retrogene can also acquire a promoter from its ancestor. This can happen when the parental gene is transcribed from the site upstream of the canonical transcription start site (Okamura and Nakai 2008; Carelli et al. 2016). A good example of such a case is the *PABP3* retrogene. Sequence analysis of this duplicate and its progenitor showed high conservation of the 5′ region, which suggests that a retrogene arose from a gene transcript containing a fragment of the promoter (Okamura and Nakai 2008).

2.4.2 Conservation of Gene Function and Subfunctionalization

Gene duplications generate functional redundancy, which may be preserved or eliminated during evolution, depending on whether it is deleterious or beneficial to the organism. Signatures of purifying selection, such as intact open reading frames or lower rates of non-synonymous to synonymous mutations, together with transcriptional activity, provide strong evidence for the functionality and beneficial outcome of functional redundancy. However, it has been suggested that maintaining both copies may be possible in the case of highly demanded genes like rRNA and histones. Evidence for dosage compensation was provided for some retrocopies. For example, two testicular-specific retrogenes, *RPL10L* and *RPL39L*, may compensate for their parental genes that are inactivated during spermatogenesis (Uechi et al. 2002). A similar mechanism has been proposed for the *HNRNP G-T* retrogene, which may functionally replace the parental protein in meiosis (Elliott et al. 2000).

It has been predicted that both duplicates can coexist stably if there are some differences in their function (Nowak et al. 1997; Lynch and Force 2000). This can occur through subfunctionalization, where the ancestral gene already has a dual function and each of the duplicates takes on parts of the original function. An example of subfunctionalization in human is the cell cycle gene *CDC14B* and its retrocopy *CDC14Bretro*. The retroposition of the *CDC14B* transcript occurred approximately 18–25 million years ago. The ancestral function was probably maintained in both copies until the separation of African and Asian apes. In the ancestral genome of the African apes, several mutations were fixed in the 5′ end of the retrogene sequence. As a result, the subcellular localization of the encoded protein was shifted from the microtubule to the endoplasmic reticulum (Rosso et al. 2008). However, it is debatable whether this is subfunctionalization or neofunctionalization, as the re-localization was associated with the novel function and specific testis/brain expression (Rastogi and Liberles 2005).

The analysis of human retrocopies led to the identification of 111 known-protein-coding genes that originated via retroposition (Rosikiewicz et al. 2017). Many of these provide good evidence for subfunctionalization. One example is *ARF6,* a retrocopy of *ARF3* gene. Both genes encode ribosylation factors, factor 6 and factor 3, respectively, and play important roles in cytokinesis (Hanai et al. 2016), both are also described to activate the cholera toxin (Welsh et al. 1994). Some functions have been ascribed specifically to the *ARF6* retrogene. It has been demonstrated that it plays an important role in cell-cell interactions (Taatjes and Roth 2024) and the encoded protein ARF6 is strongly associated with human immunodeficiency virus type1 (Wonderlich et al. 2011). The retrogene encoding connexin 43 (Cx43) protein, whose expression is restricted to breast cancer (Kandouz et al. 2004), has an intact open reading frame and encodes a protein of the same size as the parental gene. However, while a parental gene is involved in cell growth control and intercellular communication, the retrogene does not appear to be involved in the latter function (Bier et al. 2009). Recently, Amici et al. identified a functional paralog of *HAPSTR1* gene—a key component of a pathway that regulates the cellular response to various molecular stressors and regulates the expression of several stress signaling proteins such as p53, p21, and HO-1(Amici et al. 2023). Phylogenetic analysis shows that the identified retrogene, named *HAPSTR2*, originated in the early evolution of placental mammals. Analyses performed also show that, in contrast to the ubiquitous expression of *HAPSTR1*, the expression of *HAPSTR2* is limited to the testis and neural tissues. Interestingly, the retrocopy was also present in several tumors from tissues that do not normally express *HAPSTR2*. They also demonstrated that *HAPSTR2* encodes a functional protein and that this protein retains key biochemical features of its precursor. Overexpression of *HAPSTR2* led to a very interesting observation—a significant increase in HAPSTR1 protein abundance without any change in *HAPSTR1* mRNA level. The data collected suggest that *HAPSTR2* was inserted close to the proto-promoter and gained regulatory capabilities as a result of subtle evolutionary events (Amici et al. 2023).

A retrogene can also become a functional replacement after pseudogenization or deletion of its progenitor (Ciomborowska et al. 2013; Carelli et al. 2016). The first so-called “orphan” retrogenes were discovered as a result of comparative analysis of the worm, chicken, and human genomes. Twenty-five such cases have been found in the human genome, and all of them represent well-studied genes that had not previously been identified as retrogenes. Seven of these have been linked to various human diseases like diabetes, attention-deficit/hyperactivity disorder, congestive heart failure, and Huntington’s disease. One of them, *CHMP1B,* linked to hereditary spastic paraplegia, encodes chromatin-modifying protein 1B. The parental gene was pseudogenized in the ancestor of Old World and New World monkeys, but it is still active in rodents (Ciomborowska et al. 2013).

2.4.3 Neofunctionalization

For a long time, the novel functions of duplicated genes were mainly considered in terms of proteins. An acceleration of protein evolution with relaxed purifying selection after gene duplication was reported (Lynch and Conery 2000). For this reason, retrocopy functionality has been associated with intact ORFs (Kaessmann et al. 2009). Other retrocopies have been considered biologically insignificant because they harbor premature stop codons, insertions and deletions, frameshifts, and missense mutations. Nevertheless, their sequences remain largely similar to those of their progenitors, suggesting that there is selective pressure to maintain these genetic elements and that they may play important roles. The development of high-throughput sequencing methods has made it possible to study transcriptomes at an unprecedented level, and there is increasing evidence for novel functions of duplicated genes, especially retrocopies (Kubiak et al. 2020; Makalowska and Kubiak 2023).

2.4.3.1 Retrocopies as miRNA "Sponges"

One of these novel roles is regulatory, and the best-known example is probably the *PTENpg1* retrogene, which regulates the expression of its parental gene *PTEN* in several ways (Poliseno et al. 2010; Vitiello et al. 2020). *PTENpg1* has three different transcripts, two of which are antisense to the parental gene. One isoform, called *PTENpg1 as RNA alpha*, recruits the methyltransferase 3A to the parental promoter. As a result, expression of the parental gene is repressed by the addition of three methyl groups to histone H3 Lys27 (Johnsson et al. 2013). The sense transcript of *PTENpg1* shares many microRNA binding sites with the parental gene. It sequesters microRNAs, preventing the binding and subsequent degradation of *PTEN* transcripts (Poliseno et al. 2010). *PTEN* is a tumor suppressor gene that negatively regulates the phosphatidylinositol 3-kinase signaling pathway, which is involved in the control of cell proliferation (Stambolic et al. 1998). Downregulation of *PTEN*, which occurs in the absence of *PTENpg1* transcription, enhances cell growth, which promotes tumorigenesis. Oncosuppressive effects of *PTEN-PTENpg1* endogenous competition for shared microRNA molecules have been demonstrated in prostate and colon cancer (Poliseno et al. 2010), melanoma (Poliseno et al. 2011), renal cell carcinoma (Yu et al. 2014), and hepatocellular carcinoma (Chen et al. 2015).

Other examples of retrocopies "sponging" microRNAs are two retrogenes of the high mobility group A1 (*HMGA1*) gene, *HMGA1P6* and *HMGA1P7*. HMGA1 proteins are expressed at relatively low levels in adult normal cells, but they are overexpressed in cancers. Both retrogenes have a conserved microRNA target site from the parental gene. The ability of miRNA-retrogene interaction has been experimentally confirmed, and a significant reduction in the *HMGA1*, *HMGA1P6*, and *HMGA1P7* was observed after miRNA transfection in human breast adenocarcinoma cells (Esposito et al. 2014b). *HMGA1* retrogenes were also found to be highly expressed in human anaplastic thyroid carcinoma, one of the most aggressive tumors (Esposito et al. 2014a), and in pituitary tumors (Esnault et al. 2000). In contrast to the oncosuppressive function of *PTENpg1*, *HMGA1P6* and *HMGA1P7* exhibit oncogenic activity, and their high expression contributes to cancer progression.

Genome-wide analysis of human retrocopies revealed up to 181 retrocopies that could be regulating transcripts of 187 genes by sequestering microRNA molecules (Kubiak et al. 2020). This included a retrocopy of *NDUFV2* gene, *NDUFV2P1*, which had previously been shown to be upregulated in schizophrenia-derived cells. *NDUFV2P1* also showed a significant inverse correlation with the NDUFV2 protein levels. It has been suggested that this may indicate possible competition with the *NDUFV2* at the translational level (Bergman et al. 2018).

2.4.3.2 Natural Antisense Transcripts

As the number of transcriptomic studies increases, it is becoming clear that many pseudogenes, including retrocopies, are transcribed into long noncoding RNAs (lncRNAs) (Milligan and Lipovich 2014). Several genome-wide analyses of retrocopies have led to the identification of dozens of retrocopies that are transcribed in antisense and give rise to lncRNAs. These lncRNAs share sequence similarity with the parental genes of the retrocopies in a sense/antisense orientation, can form RNA:RNA duplexes, and often show a significant expression correlation. This makes them perfect candidates as natural antisense transcripts (NATs) acting in *trans* (Bryzghalov et al. 2016; Muro and Andrade-Navarro 2010; Kubiak et al. 2020). One of these lncRNAs derived from the *HNRNPA1P7* retrocopy may contribute to heterogeneous nuclear riboprotein A1 (hnRNPA1) pre-processing. The bioinformatics analysis strongly suggests that the antisense lncRNA may mask the 5′ splice site in the sixth intron of the parental gene transcript, allowing expression of an isoform with a longer sixth exon (Bryzghalov et al. 2016). Another example comes from the previously described *PTENpg1*. Its second antisense transcript, *PTENpg1 asRNA beta*, stabilizes the expression of the sense transcript (Johnsson et al. 2013).

2.4.3.3 Source of the Small RNA Molecules

Retroposed genes were also found to be a source of small regulatory RNA molecules. Retrocopy-derived small RNAs were identified in the analysis of the noncoding RNA repertoire in the human transcriptome. It was observed that the density of small RNAs was higher in retrocopies than in other duplicated pseudogenes or coding genes (Guo et al. 2014). Studies by Mercuri et al. revealed 17 miRs that emerged from retrocopies of mRNA sequences. All of these retro-miRs were primate-specific and two of them were human-specific (Mercuri et al. 2023). Interestingly, six of the identified miRs represented retroduplications of exonic miRNAs that were present in the parental genes. Five of these retro-miRs originated from the *PTMA* parental gene. MicroRNAs located within retrocopies were reported also by other groups. Devor, for example, found two primate-specific microRNAs, miR-220 and miR-492, that originated from retrotransposed sequences (Devor 2006). One of them, miR-492, which originated from the keratin 19 gene, has been proposed to play a role in hepatoblastoma (von Frowein et al. 2011) and was found to be a proto-oncogenic miRNA that acts as a proliferation promoter in breast cancer and was associated with shorter survival in patients with gastric cancer, ovarian cancer, oropharyngeal carcinoma, colorectal cancer, and hepatocellular carcinoma (Shen et al. 2015, 2023).

Retrocopies can also give rise to piwi-interacting RNA (piRNA), which is involved in the formation of silencing complexes in animal germlines. Bioinformatics analyses focused on the small RNA of human sperm showed that some piRNA overlap with retroposed genes. An example could be the *NPAP1P6* retrocopy, which originated from the primate-specific *NPAP1* gene encoding a nuclear pore-associated protein associated with Prader-Willi syndrome (Pantano et al. 2015). Other examples from the same study include piRNAs derived from *IMPDH1P5*, *TMX2P1*, and *RP11-545A16.3* retrocopies. PiRNA clusters overlapping with retrocopies were also found in other studies (Gebert et al. 2015; Milligan et al. 2016).

Retrocopy-derived lncRNAs are often transcribed from the opposite DNA strand and, as already mentioned, function as antisense transcripts that can pair with parental and other genes. These double-stranded RNA molecules can be processed into endogenous small interfering RNA (endo-siRNA). A study of mouse oocytes demonstrated that endo-siRNAs annotated in the retrocopy locus can regulate the expression of their parental genes. For example, 77 small RNAs were mapped to an expressed retrocopy, *Gm15681*, which originated from the phosphatase 4 regulatory subunit 1 (*Ppp4r1*) gene, and almost all of them were oriented antisense to the parental gene. This strongly suggests that the siRNA arose from the double-stranded RNA (dsRNA) region formed by the retrocopy and the parental gene (Watanabe et al. 2008).

2.4.4 Gene Fusion and Chimeric Transcripts

The first reported chimeric transcript generated by retrocopy insertion into a pre-existing gene was *jingwei* observed in *Drosophila* (Long and Langley 1993), but it took a decade to find a fusion gene in a vertebrate. The retroposition of cyclophilin A (*PPIA*) into the *TRIM5* gene, which occurred after the divergence of New World and Old World monkeys, resulted in the origin of a novel protein responsible for the post-entry restriction of HIV-1 (Human immunodeficiency virus 1) (Sayah et al. 2004).

Although a lot of chimeric transcripts have been reported, fusion events resulting from TEs-driven NAHR appear to be rare. Newman et al. identified 184 duplications from 170 unrelated individuals and analyzed genes at the 118 breakpoint junctions. Identified duplications resulted from recombination events, but only three of them were associated with *Alu-Alu* recombination. Only some of the identified duplications created fusion genes at breakpoints, and none of them was associated with *Alu*-driven events (Newman et al. 2015). However, there are some reports providing evidence for gene fusion after TE-driven NAHR events. A chimeric transcript resulting from *Alu-Alu* recombination was described in a case of the low-density lipoprotein receptor (*LDLR*). The tandem duplication of a 14-kilobase sequence containing exons 2 to 8 led to the incorporation of an additional (duplicated) exon, and larger-than-normal protein is causing familial hypercholesterolemia (Lehrman et al. 1987a).

Although thousands of retrocopies are found in the introns of other genes, the creation of chimeric transcripts resulting from gene retroposition also does not appear to be a widespread phenomenon. Baertch and colleagues identified only 34 cases as potential gene fusions (Baertsch et al. 2008). Three chimeric transcripts were found in the study focused on human-specific retrocopies with 5′ CpG islands. One is composed of exon 8 of the *RNF13* gene and the *TMEM183A-r* retrogene, and another is composed of the *HSF2BP* gene and the *H2BFS* retrogene. The third chimeric transcript identified in these studies, created between the *VRK2* gene and the *EIF3P3* retrogene, was expressed only in malignant prostate cancer cell lines (Mori et al. 2016). Carelli et al. identified 13 human retrocopies integrated into introns of another gene and expressed together with the upstream exons of the host gene as a new chimeric transcript (Carelli et al. 2016). Retrogenic sequences are also used as alternatively spliced exons. For example, the *BRCA1* gene has an internal retrogene-derived exon that generates a 22 amino acid cassette (Baertsch et al. 2008).

A retrocopy indicating a fusion of *PIP5K1A* and *PSD4* was identified by Akiva et al. While both genes are located on chromosome 1, the retrocopy is located on chromosome 10. The retrocopy has undergone considerable evolution when compared to the fused transcripts but encodes an uninterrupted PIP5K1A-PSD4 fusion protein (Akiva et al. 2006).

A fusion transcript containing exons of *ATR* and *EIF2AK1* was found in cancer cells. It has been proposed that this fusion results from a recombination event between chromosomes 7 and 3 (Wang et al. 2014). Apparently, the retrocopy of the *EIF2AK1* gene is located in the intron of the *ATR* gene, suggesting that this non-allelic recombination was stimulated by the high sequence similarity between the parental gene and its retrocopy embedded in the *ATR* gene (Kubiak et al. 2020). Studies of the data from FusionGDB (Kim et al. 2022) revealed nineteen fusion transcripts resulting from chromosomal rearrangements with evidence of retrocopy involvement. In fifteen cases, the breakpoint was in the body of the retrocopy; in the remaining four, it was at some distance from the retrocopy (Kubiak et al. 2020). Another example of genes forming fusion transcripts involving retrocopies is from the *FOXP2* gene associated with language and speech disorders. *FOXP2* hosts a retrocopy of the gene *RPL36* and forms a fusion transcript with this retrocopy parental gene (Herrero and Gitton 2018). All of these examples also show that, in addition to providing material for new genes, retroposed genes can also play an important role as recombination hotspots.

2.5 Duplication of Other Than mRNA Sequences

Although the majority of retrotransposed genes originated from mRNA sequences, there is also evidence for other retrotransposed RNA molecules. Pan et al. found several retrocopies that were derived from the mir-302-367 cluster gene. All precursors of these microRNAs were found to be highly expressed in tissues with elevated retrotransposon activity (Pan and Lin 2020). PiRNA sequences can originate from retroposed genes, but this class of small RNAs can also expand by segmental duplication. It has been shown that piRNA clusters were rapidly expanding via segmental duplication in primate and rodent genomes (Assis and Kondrashov 2009). This process, driven by intense positive selection, provides an efficient way for expanding the piRNA repertoire and allows the evolution of protection against the expansion of lineage-specific transposable elements.

2.6 Phenotypic Innovations

Gene duplications can have immediate phenotypic consequences. One of the most known examples is a retrocopy derived from the canine growth factor gene *fgf4*, which is responsible for the chondrodysplasia, a short-legged phenotype, in at least 19 dog breeds (Parker et al. 2009). The retrogene is located within the LINE element. In the upstream sequence there are also other LINEs as well as SINEs, and it has been proposed that these elements provided a regulatory sequence allowing expression of the retrogene (Parker et al. 2009). *fgf4* causes overactivation of the fibroblast growth factor receptor *FGFR3*, which is responsible for the majority of cases of achondroplasia, the most common form of dwarfism in humans (Horton and Lunstrum 2002). Another example comes also from canine studies. Poodles with red coats have been shown to carry the *SNNL1* retrocopy, which is located upstream of *GPR22* gene, an orphan G-protein-coupled receptor that is normally expressed exclusively in the brain and heart. It is likely that the *SNNL1* retrocopy disrupts the regulation of *GPR22*, leading to atypical expression in the skin (Batcher et al. 2022).

An interesting example is the rodent-specific retrocopy of the ribosomal protein gene *Rps23*. There are hundreds of retrocopies of ribosomal protein genes in mammalian genomes, but they are usually non-functional retropseudogenes. This particular retrocopy, *Rps23r1,* is transcribed from the reverse DNA strand. It has incorporated sequences flanking its insertion site as new exons, resulting in the emergence of a new protein-coding gene that encodes a protein unrelated to that encoded by its progenitor. This new protein appears to increase resistance to the formation of Alzheimer's-causing amyloid plaques and tau phosphorylation in mice (Zhang et al. 2009).

Studies on elephants show that they have a low rate of tumor transformation compared to other mammals. Elephants have 20 copies of the *TP53* gene, a known oncosuppressor, 19 of which are retrocopies. It has been proposed that the presence of these copies is associated with an increased apoptotic response in the elephant population (Abegglen et al. 2015). The presence of extra copies leads to an effective DNA damage response through the hyperactive TP53 pathway (Sulak et al. 2016). This is a good example of the "Peto's paradox," which indicates that the incidence of cancer in animals does not increase with body size and longevity.

In most mammals, glutamate hydrogenase (hGDH1 in humans) is encoded by a single functional gene, *GLUD1*. However, humans and other primates have acquired a second copy through retroposition. *GLUD1* was retroposed to chromosome X in the ape ancestor and gave rise to *GLUD2*, which encodes a very similar isoenzyme (hGDH2) expressed in the testes and brain. GDH is an enzyme central to glutamate metabolism and plays a key role in cellular energetics (Kanavouras et al. 2007). Two amino acid substitutions in *GLUD2* were fixed by positive selection and contributed to the adaptation of the *GLUD2* GDH to the unique metabolic needs of the neural tissue (Plaitakis et al. 2003; Burki and Kaessmann 2004) and to a function in the glutamate metabolism in the brain (Rosso et al. 2008).

Retroposition gives rise to considerable genetic variation between species, including between humans. The 1000 Genomes Project (Genomes Project et al. 2012) made it possible to explore the variation of retrocopies in human populations. A total of 208 polymorphic retrocopies were discovered (Richardson et al. 2014). In another study, focusing on retrocopy deletions, 214 indels affecting 190 retrocopies were identified. Out of these, 68 were found to be ancestral, and the polymorphism of these retrocopies clearly resulted from a deletion. This study also showed a variation in the expression level of the retrocopies (Kabza et al. 2015).

2.7 TE-Driven Duplications and Human Diseases

TE activity undoubtedly influences human biology in both health and disease. Efforts to understand exactly how TEs contribute to human disease have been ongoing for decades, and duplications have been found to play a central role in a large number of human diseases (Deininger and Batzer 1999; Ciomborowska-Basheer et al. 2021).

2.7.1 *Disorders Resulting from Alu-Driven NAHRs*

Alu-Alu recombination has been reported to cause many human diseases. Although the majority of these diseases are associated with gene deletions (Deininger and Batzer 1999), diseases caused by recombination resulting in duplications have also been reported. An example is the type VI variant of Ehlers-Danlos syndrome (EDS VI) caused by the duplication of seven exons in the lysyl hydroxylase (*PLOD*—procollagen-lysine, 2-oxoglutarate 5-dioxygenase) gene (Pousi et al. 1994). The recombination has occurred via two identical 44 nucleotide-long parts of the right subunit of the *Alu* sequences located in intron 9 and intron 16 of the *PLOD* gene (Yeowell et al. 1997). Duplication of these seven exons has been identified as a common cause of EDS VI. This mutation results in reduced activity of lysyl hydroxylase and consequently weakened cross-links in collagen (Yeowell and Walker 2000).

A tandem duplication of 1.5 MB of chromosome 17 (17p11.2-p12), encompassing the peripheral myelin protein 22 gene (*PMP22*), is the most frequent cause of Charcot-Marie-Tooth disease (CMT). This duplication is caused by the unequal crossing-over and subsequent recombination (Lupski et al. 1991). This duplication was found in approximately 50% of all CMT patients and in over 70% of patients who developed CMT1A (Charcot-Marie-Tooth disease type 1A) (Boerkoel et al. 2002), and it is generally accepted that the extra copy of the *PMP22* gene causes the majority of CMT1A. Although there are other genes in the duplicated region, *PMP22* appears to be the most sensitive to dose. The PMP22 protein plays an essential role in myelination. Altered expression of *PMP22* by gene dosage leads to the demyelinating neuropathy phenotype (Chance et al. 1992). Analysis of the DNA sequence surrounding the recombination hotspot revealed the presence of a *mariner* transposon-like element (MITE), and it was proposed that MITE stimulates homologous recombination (Lupski 1998; Kiyosawa and Chance 1996). Further studies showed that the CMT1A-REP repeat region is also flanked by inverted *Alu* sequences (Chance et al. 1992).

Another example comes from the Lesch-Nyhan syndrome (LN), a severe neurological disorder. It is caused by a deficiency of the enzyme hypoxanthine phosphoribosyltransferase (HPRT). The *HPRT* locus contains 49 *Alu* repeats, which make up 25% of the gene's DNA sequence. Although many reports demonstrate *HPRT* gene deletion, there are also known cases of LN caused by *HPRT* duplications. Yang et al. described a duplication of exons 2 and 3 (Yang et al. 1988), and Marcus et al. reported a duplication of exons 7 and 8 resulting from unequal homologous recombination between two *Alu* sequences in introns 6 and 8. The duplication resulted in a frameshift followed by a stop codon that disrupted protein functionality (Marcus et al. 1993).

Mutations within the LDL receptor caused by *Alu-Alu* recombination are associated with familial hypercholesterolemia. Although in the majority of cases, the disease results from partial deletions of the *LDLR* gene (Lehrman et al. 1987a), an *LDLR* duplication has also been reported. This is a 14-kilobase duplication involving exons 2 to 8. Although the resulting larger-than-normal protein binds reduced amounts of LDL, it undergoes internalization and recycling (Lehrman et al. 1987b). Abnormalities resulting from partial tandem duplications of an *ALL1* (also known as *MLL*) gene are associated with acute myeloid leukemia (AML). The homologous recombination leading to this duplication is mediated by *Alu* elements. The duplicated region spans exons 2–6 (Whitman et al. 2001). Another example is Duchenne muscular dystrophy, which is mostly associated with deletions of gene segments but can also be caused by a subset of duplicated exons of the dystrophin gene. At least one of these duplications is due to NAHR between two *Alu* elements (Hu et al. 1991). An exon 13 and more than 6 kb of intronic sequence were found to be duplicated in the *BRCA1* gene. The presence of consecutive exons 13 leads to a frameshift in the mutant mRNA. Breakpoints leading to the recombination were found to occur in a 23-bp region within two *AluSx* sequences (Puget et al. 1999).

2.7.2 Retroposed Genes in Cancer

A number of pathological mechanisms involving retrocopies have been recognized. This includes cancer, with numerous studies highlighting the potential of retrocopies to act as both oncogenes and tumor suppressors (Staszak and Makalowska 2021; Chen et al. 2020). For example, the noncoding retrocopies of the *POU5F1* gene (also known as *OCT4*), *POU5F1P4* (*OCT4-pg4*), and *POU5F1P5* (*OCT4-pg5*) have been implicated in cancer progression. Mechanistically, *POU5F1P4* acts as a natural microRNA sponge, sequestering miR-145 and thereby preventing the inhibition of *POU5F1* transcripts. This regulatory interaction facilitates the emergence of *POU5F1*-mediated cancer stem cell populations in hepatocellular carcinoma (Wang et al. 2013) and promotes progression of bladder cancer (Zhou et al. 2024). *POU5F1P5* also interacts with miR-145 in endometrial cancer, where it contributes significantly to cancer cell growth and proliferation (Bai et al. 2015). In MCF-7 breast adenocarcinoma cells, *POU5F1P5* forms a silencing complex with the factors G9a and Ezh2 that suppresses the transcription of the parental gene. This transcriptional repression can be inhibited when the antisense RNA is bound and sequestered by RNA-binding proteins, such as PURA and NCL (Hawkins and Morris 2010).

In human breast cancer and glioblastoma cell lines, the low expression of two *TUSC2* retrocopies, *TUSC2P1* and *TUSC2P2*, has been found to be associated with many cancer cell expansion processes. These retrocopies have binding sites for several microRNAs, including miR-17, miR-93, miR-299-3p, miR-520a, miR-608, and miR-661. The interaction between these small RNAs and retrocopies stimulates the translation of *TUSC2*, a known tumor suppressor. This results in the inhibition of cell proliferation, invasion, migration, and colony formation and consequently increased tumor cell death (Rutnam et al. 2014). It has been hypothesized that overexpression of the *UTP14C*, a protein-coding retrocopy of the *UTP14A* gene, in ovarian cancer leads to the proteolytic degradation of the p53 protein in both the nucleus and cytoplasm. This process is thought to contribute to the promotion of oncogenesis by reducing the ability of cells to undergo apoptosis and preventing cell cycle arrest (Rohozinski et al. 2012). The expression level of retrocopy *KRASP1* was found to correlate with prostate cancer phenotype. Its parental gene, *KRAS*, is one of the best-known oncogenes. Studies in cancer cell lines have shown that *KRASP1* overexpression causes increased parental gene expression and cell proliferation (Poliseno et al. 2010). Another example of a retrocopy acting as a microRNA "sponge" is *CTNNA1P1*. It was found to compete with its parental gene *CTNNA1* for miR-141. Low expression of *CTNNAP1* and high availability of miR-141 result in reduced *CTNNA1* mRNA levels and the manifestation of a malignant phenotype in colorectal cancer cell lines (Chen et al. 2016).

The oldest known cancer-related retrocopies are the protein-coding genes *RHOB* and *DIO3*, which arose during the early evolution of vertebrates (Staszak and Makalowska 2021). The conservation of ORFs in these quite ancient retrogenes may indicate that they acquired transcriptional capabilities very quickly and that the tendency to accumulate mutations typical of retrocopies was "locked" due to their functional importance. *RHOB* is an important oncosuppressor, and a decrease in its expression promotes cancer (Prendergast 2001). Increased expression of the *DIO3* retrocopy has been associated with tumor progression in papillary thyroid cancer and colon cancer (Goemann et al. 2020).

Other examples of retrocopies involved in cancers include *MSL3P1* (Chen et al. 2018), *ANXA2P2* (Wang et al. 2019), and *CSDAP1* (Xu et al. 2018). However, these are just selected examples. An analysis of 293 samples representing 13 cancer and normal tissue types revealed 218 pseudogenes that were expressed only in cancer samples. Of these, 178 were observed in multiple cancers, and 40 were identified only in a single cancer type (Kalyana-Sundaram et al. 2012). In another study, two cancer cell lines, chronic myelogenous leukemia (K562) and hepatocyte carcinoma (HepG2), were found to have the highest number of expressed retrocopies. A comparison of normal tissue and cancer cell lines also showed that three retrocopies are expressed in all cancer cell lines analyzed, but not in normal tissues (Kubiak et al. 2020).

2.7.3 Retrogenes in Cardiovascular, Neurodegenerative, Mental, and Other Diseases

Cardiovascular disease (CVD), which refers to a diverse group of disorders affecting the heart and blood vessels, has also been shown to be associated with retrocopies. A study in CVD has identified *VDAC2P2* as an lncRNA that may be involved in atrial fibrillation (AF) (Wu et al. 2020). AF is one of the most common and complex forms of arrhythmia and is associated with an increased risk of stroke and heart failure (Bidaoui et al. 2024). Although the exact function of *VDAC2P2* remains unclear, it is thought to be involved in homologous recombination and gene conversion or to act as an antisense RNA. This retrocopy was found to be significantly differentially expressed in atrial tissue from patients with atrial fibrillation and regulates the parental gene *VDAC2* as well as the neighboring gene *KLRG1* (Wu et al. 2020). *VDAC2* encodes a mitochondrial outer membrane protein, and its deregulation leads to AF-related mitochondrial dysfunction (Jeganathan et al. 2017). Another CVD in which retrocopies play a role is essential hypertension (EH). *LOC646616* and *LAP3P2* are two lncRNAs-acting retrocopies that have been recently identified as promising therapeutic targets for EH. They were found to be overexpressed in EH patients and act as a molecular sponge for miR-637, which regulates *WNT3* and *CAMK2N2* expression in human vascular smooth muscle cells during hypertension.

This interaction leads to the activation of the WNT/β-catenin signaling pathway, which contributes to the onset and progression of EH (Yin et al. 2020).

Retrocopies have also been reported to play a role in neurodegenerative disorders such as Huntington's disease (HD) and Parkinson's disease (PD). Retrocopies *HMGB1P1*, *HMGB1P5*, and *HMGB1P10* have been found to have altered expression in HD (Costa et al. 2012). They share microRNA binding sites with their parental gene *HMGB1*, which encodes a protein that interacts with huntingtin. Upregulation of *HMGB1* has been described to be associated with the inhibition of HD progression (Angelopoulou et al. 2020). Three other retrocopies, *PHC1P1*, *RBMXP2*, and *CHCHD2P2*, have been selected as potentially involved in the pathogenesis of Parkinson's disease (PD) (Costa et al. 2012). It is hypothesized that the *PHC1P1* retrocopy may act as a microRNA sponge, affecting the expression of the parental *PHC1* gene, which is involved in neuroinflammation (Nuzziello and Liguori 2019; Kubiak et al. 2020).

Mental diseases are also impacted by retrocopies. One of the genetic factors contributing to one of these disorders, autism spectrum disorder (ASD), is connected with the expression of the antisense transcript *MSNP1AS* of the *MSNP1* retrocopy (Tang et al. 2017; Luo et al. 2020). This transcript forms a secondary double-stranded structure that inhibits the expression of the parental *MSN* gene (Kerin et al. 2012; DeWitt et al. 2016). The *MSN* protein is a part of the PI3K/Akt, RhoA, and Rac1 signaling pathways involved in the proper functioning of neurons (Luo et al. 2020). In patients diagnosed with ASD, the upregulation of *MSNP1AS* has a significant effect on the length and number of neurites in neural progenitor cell lines (Kerin et al. 2012). Retrocopies are also involved in the development of obsessive-compulsive disorder (OCD). For example, *NANOGP8* retrocopy, which is known to be involved in the regulation of neuronal development (Nestadt et al. 2012).

The *HMGA1P8* retrocopy was found to be highly expressed in some patients with diabetes, a common metabolic disease. This was linked with the reduced expression of the parental *HMGA1* gene, which encodes a protein involved in the activation of the insulin receptor gene. *HMGA1P8* has been shown to compete for a trans-acting cytoplasmic protein that stabilizes *HMGA1* mRNA, leading to insulin resistance (Chiefari et al. 2010). Other retrocopies, *HK2P1* and *PGK1P2*, are associated with preeclampsia, which is characterized by proteinuria and hypertension in pregnant women. Both retrocopies regulate their parental genes, *HK2* and *PGK1*, respectively, through competition for microRNA. Downregulation of *HK2P1* and *HK2* leads to the poor proliferation of endometrial stromal cells, and low expression of *PGK1* adversely affects the glycolytic pathway, which is crucial for the endometrium (Lv et al. 2018). Deregulation of the *ANXA2P3* retrocopy, derived from the *ANXA2* gene, has been demonstrated to play a key role in the development of fibrosis in biliary atresia. Upregulation of both *ANXA2P3* and *ANXA2* has been shown to enhance cell proliferation, making them promising candidates as biomarkers for liver diseases (Nuerzhati et al. 2019).

Another example of a disease associated with a retrocopy is gelatinous drop-like corneal dystrophy. Mutation in a retrogene *TACSTD2*, tumor-associated calcium signal transducer 2, causes this autosomal recessive disorder characterized by severe corneal amyloidosis leading to blindness (Tsujikawa et al. 1999).

Similarly to the insertion of TEs (Deininger and Batzer 1999), a retrocopy insertion can disrupt a function of the gene and lead to the disease. Although not many of such instances are known, a paper by the Boer et al. describes an interesting case of the X-linked disorder, chronic granulomatous disease (de Boer et al. 2014). The disease is caused by the insertion of semi-processed pseudogene of the TATA element modulatory factor gene (*TMF1*) into intron 1 of *CYBB*. The insertion was in reverse orientation and truncated at the 3′ end. It contained exons 1 to 8 of *TMF1* along with intron 7 and most of intron 8. The inserted sequence contained an excellent donor site in exon 2 of *TMF1* and the newly created 117 bp exon contained a nonsense codon that rendered the *CYBB* gene to be non-functional (de Boer et al. 2014).

2.8 Conclusions

Multiple studies have shown that TEs contribute significantly to molecular evolution. They have been found to be factors that shape differences between species, individuals, or even tissues and cell types, and are therefore considered to be a source of genetic polymorphism. The most obvious contribution of TEs to evolution is gene duplication, which provides new genetic material on which mutation, drift, and selection can act. The results are genes with specialized or new functions (Xia et al. 2025; Brosius 2003). Without gene duplication, the ability of a genome or species to adapt to changing environments would be severely limited. Species-specific gene duplication often leads to species-specific gene functions and phenotypic traits that facilitate species-specific adaptation. Of course, there is also a price to pay, in the form of TEs-driven diseases, a tradeoff that affects all species and accompanies all evolutionary mechanisms.

Acknowledgments We would like to thank Wojciech Makalowski for valuable comments and suggestions.

Compliance with Ethical Standards

Ethical approval This chapter is a review of previously published accounts; as such, no animal or human studies were performed.

Conflict of Interest The authors have no conflict of interest to declare.

References

Abdelsamad A, Pecinka A (2014) Pollen-specific activation of Arabidopsis retrogenes is associated with global transcriptional reprogramming. Plant Cell 26(8):3299–3313. https://doi.org/10.1105/tpc.114.126011

Abegglen LM, Caulin AF, Chan A et al (2015) Potential mechanisms for cancer resistance in elephants and comparative cellular response to DNA damage in humans. JAMA 314(17):1850–1860. https://doi.org/10.1001/jama.2015.13134

Abyzov A, Iskow R, Gokcumen O et al (2013) Analysis of variable retroduplications in human populations suggests coupling of retrotransposition to cell division. Genome Res 23(12):2042–2052. https://doi.org/10.1101/gr.154625.113

Agrawal A, Eastman QM, Schatz DG (1998) Transposition mediated by RAG1 and RAG2 and its implications for the evolution of the immune system. Nature 394(6695):744–751. https://doi.org/10.1038/29457

Akiva P, Toporik A, Edelheit S et al (2006) Transcription-mediated gene fusion in the human genome. Genome Res 16(1):30–36. https://doi.org/10.1101/gr.4137606

Amici DR, Cingoz H, Alasady MJ et al (2023) The HAPSTR2 retrogene buffers stress signaling and resilience in mammals. Nat Commun 14(1):152. https://doi.org/10.1038/s41467-022-35697-1

Angelopoulou E, Paudel YN, Piperi C (2020) Exploring the role of high-mobility group box 1 (HMGB1) protein in the pathogenesis of Huntington's disease. J Mol Med (Berl) 98(3):325–334. https://doi.org/10.1007/s00109-020-01885-z

Assis R, Kondrashov AS (2009) Rapid repetitive element-mediated expansion of piRNA clusters in mammalian evolution. Proc Natl Acad Sci USA 106(17):7079–7082. https://doi.org/10.1073/pnas.0900523106

Babushok DV, Kazazian HH Jr (2007) Progress in understanding the biology of the human mutagen LINE-1. Hum Mutat 28(6):527–539. https://doi.org/10.1002/humu.20486

Baertsch R, Diekhans M, Kent WJ et al (2008) Retrocopy contributions to the evolution of the human genome. BMC Genomics 9(1):466. https://doi.org/10.1186/1471-2164-9-466

Bai Y, Casola C, Feschotte C et al (2007) Comparative genomics reveals a constant rate of origination and convergent acquisition of functional retrogenes in Drosophila. Genome Biol 8(1):R11. https://doi.org/10.1186/gb-2007-8-1-r11

Bai M, Yuan M, Liao H et al (2015) OCT4 pseudogene 5 upregulates OCT4 expression to promote proliferation by competing with miR-145 in endometrial carcinoma. Oncol Rep 33(4):1745–1752. https://doi.org/10.3892/or.2015.3763

Bailey JA, Gu Z, Clark RA et al (2002) Recent segmental duplications in the human genome. Science 297(5583):1003–1007. https://doi.org/10.1126/science.1072047

Bailey JA, Liu G, Eichler EE (2003) An Alu transposition model for the origin and expansion of human segmental duplications. Am J Hum Genet 73(4):823–834. https://doi.org/10.1086/378594

Barro-Trastoy D, Kohler C (2024) Helitrons: genomic parasites that generate developmental novelties. Trends Genet 40(5):437–448. https://doi.org/10.1016/j.tig.2024.02.002

Barsh GS, Seeburg PH, Gelinas RE (1983) The human growth hormone gene family: structure and evolution of the chromosomal locus. Nucleic Acids Res 11(12):3939–3958. https://doi.org/10.1093/nar/11.12.3939

Batcher K, Varney S, Affolter VK et al (2022) An SNN retrocopy insertion upstream of GPR22 is associated with dark red coat color in Poodles. G3(Bethesda) 12(11). https://doi.org/10.1093/g3journal/jkac227

Batzer MA, Deininger PL (2002) Alu repeats and human genomic diversity. Nat Rev Genet 3(5):370–379. https://doi.org/10.1038/nrg798

Beck CR, Collier P, Macfarlane C et al (2010) LINE-1 retrotransposition activity in human genomes. Cell 141(7):1159–1170. https://doi.org/10.1016/j.cell.2010.05.021

Bergman O, Karry R, Milhem J et al (2018) NDUFV2 pseudogene (NDUFV2P1) contributes to mitochondrial complex I deficits in schizophrenia. Mol Psychiatry. https://doi.org/10.1038/s41380-018-0309-9
Bidaoui G, Assaf A, Marrouche N (2024) Atrial fibrillation in heart failure: novel insights, challenges, and treatment opportunities. Curr Heart Fail Rep 22(1):3. https://doi.org/10.1007/s11897-024-00691-9
Bier A, Oviedo-Landaverde I, Zhao J et al (2009) Connexin43 pseudogene in breast cancer cells offers a novel therapeutic target. Mol Cancer Ther 8(4):786–793. https://doi.org/10.1158/1535-7163.MCT-08-0930
Boeke JD, Fenyo D (2015) Much ado about zero. Cell 163(3):534–535. https://doi.org/10.1016/j.cell.2015.10.033
Boerkoel CF, Takashima H, Garcia CA et al (2002) Charcot-Marie-Tooth disease and related neuropathies: mutation distribution and genotype-phenotype correlation. Ann Neurol 51(2):190–201. https://doi.org/10.1002/ana.10089
Brosius J (2003) The contribution of RNAs and retroposition to evolutionary novelties. Genetica 118(2–3):99–116
Bryzghalov O, Szczesniak MW, Makalowska I (2016) Retroposition as a source of antisense long non-coding RNAs with possible regulatory functions. Acta Biochim Pol 63(4):825–833. https://doi.org/10.18388/abp.2016_1354
Burki F, Kaessmann H (2004) Birth and adaptive evolution of a hominoid gene that supports high neurotransmitter flux. Nat Genet 36(10):1061–1063. https://doi.org/10.1038/ng1431
Buzdin A, Gogvadze E, Kovalskaya E et al (2003) The human genome contains many types of chimeric retrogenes generated through in vivo RNA recombination. Nucleic Acids Res 31(15):4385–4390. https://doi.org/10.1093/nar/gkg496
Campillos M, Doerks T, Shah PK et al (2006) Computational characterization of multiple Gag-like human proteins. Trends Genet 22(11):585–589. https://doi.org/10.1016/j.tig.2006.09.006
Carelli FN, Hayakawa T, Go Y et al (2016) The life history of retrocopies illuminates the evolution of new mammalian genes. Genome Res 26(3):301–314. https://doi.org/10.1101/gr.198473.115
Casola C, Betran E (2017) The genomic impact of gene Retrocopies: what have we learned from comparative genomics, population genomics, and transcriptomic analyses? Genome Biol Evol 9(6):1351–1373. https://doi.org/10.1093/gbe/evx081
Cerbin S, Jiang N (2018) Duplication of host genes by transposable elements. Curr Opin Genet Dev 49:63–69. https://doi.org/10.1016/j.gde.2018.03.005
Chance PF, Bird TD, Matsunami N et al (1992) Trisomy 17p associated with Charcot-Marie-Tooth neuropathy type 1A phenotype: evidence for gene dosage as a mechanism in CMT1A. Neurology 42(12):2295–2299. https://doi.org/10.1212/wnl.42.12.2295
Chen S, Zhang YE, Long M (2010) New genes in Drosophila quickly become essential. Science 330(6011):1682–1685. https://doi.org/10.1126/science.1196380
Chen CL, Tseng YW, Wu JC et al (2015) Suppression of hepatocellular carcinoma by baculovirus-mediated expression of long non-coding RNA PTENP1 and MicroRNA regulation. Biomaterials 44:71–81. https://doi.org/10.1016/j.biomaterials.2014.12.023
Chen X, Zhu H, Wu X et al (2016) Downregulated pseudogene CTNNAP1 promote tumor growth in human cancer by downregulating its cognate gene CTNNA1 expression. Oncotarget 7(34):55518–55528. https://doi.org/10.18632/oncotarget.10833
Chen B, Wang C, Zhang J et al (2018) New insights into long noncoding RNAs and pseudogenes in prognosis of renal cell carcinoma. Cancer Cell Int 18:157. https://doi.org/10.1186/s12935-018-0652-6
Chen X, Wan L, Wang W et al (2020) Re-recognition of pseudogenes: from molecular to clinical applications. Theranostics 10(4):1479–1499. https://doi.org/10.7150/thno.40659
Chiefari E, Iiritano S, Paonessa F et al (2010) Pseudogene-mediated posttranscriptional silencing of HMGA1 can result in insulin resistance and type 2 diabetes. Nat Commun 1:40. https://doi.org/10.1038/ncomms1040

Ciomborowska J, Rosikiewicz W, Szklarczyk D et al (2013) "Orphan" retrogenes in the human genome. Mol Biol Evol 30(2):384–396. https://doi.org/10.1093/molbev/mss235

Ciomborowska-Basheer J, Staszak K, Kubiak MR et al (2021) Not so dead genes-Retrocopies as regulators of their disease-related progenitors and hosts. Cells 10(4). https://doi.org/10.3390/cells10040912

Cordaux R, Batzer MA (2009) The impact of retrotransposons on human genome evolution. Nat Rev Genet 10(10):691–703. https://doi.org/10.1038/nrg2640

Cost GJ, Feng Q, Jacquier A et al (2002) Human L1 element target-primed reverse transcription in vitro. EMBO J 21(21):5899–5910. https://doi.org/10.1093/emboj/cdf592

Costa V, Esposito R, Aprile M et al (2012) Non-coding RNA and pseudogenes in neurodegenerative diseases: "The (un)Usual Suspects". Front Genet 3:231. https://doi.org/10.3389/fgene.2012.00231

de Boer M, van Leeuwen K, Geissler J et al (2014) Primary immunodeficiency caused by an exonized retroposed gene copy inserted in the CYBB gene. Hum Mutat 35(4):486–496. https://doi.org/10.1002/humu.22519

Deininger PL, Batzer MA (1999) Alu repeats and human disease. Mol Genet Metab 67(3):183–193. https://doi.org/10.1006/mgme.1999.2864

Delviks-Frankenberry K, Galli A, Nikolaitchik O et al (2011) Mechanisms and factors that influence high frequency retroviral recombination. Viruses 3(9):1650–1680. https://doi.org/10.3390/v3091650

Denli AM, Narvaiza I, Kerman BE et al (2015) Primate-specific ORF0 contributes to retrotransposon-mediated diversity. Cell 163(3):583–593. https://doi.org/10.1016/j.cell.2015.09.025

Devor EJ (2006) Primate microRNAs miR-220 and miR-492 lie within processed pseudogenes. J Hered 97(2):186–190. https://doi.org/10.1093/jhered/esj022

Dewannieux M, Esnault C, Heidmann T (2003) LINE-mediated retrotransposition of marked Alu sequences. Nat Genet 35(1):41–48. https://doi.org/10.1038/ng1223

DeWitt JJ, Grepo N, Wilkinson B et al (2016) Impact of the autism-associated long noncoding RNA MSNP1AS on neuronal architecture and gene expression in human neural progenitor cells. Genes (Basel) 7(10). https://doi.org/10.3390/genes7100076

Dhellin O, Maestre J, Heidmann T (1997) Functional differences between the human LINE retrotransposon and retroviral reverse transcriptases for in vivo mRNA reverse transcription. EMBO J 16(21):6590–6602. https://doi.org/10.1093/emboj/16.21.6590

Dong Y, Lu X, Song W et al (2011) Structural characterization of helitrons and their stepwise capturing of gene fragments in the maize genome. BMC Genomics 12:609. https://doi.org/10.1186/1471-2164-12-609

Doucet AJ, Droc G, Siol O et al (2015a) U6 snRNA pseudogenes: markers of Retrotransposition dynamics in mammals. Mol Biol Evol 32(7):1815–1832. https://doi.org/10.1093/molbev/msv062

Doucet AJ, Wilusz JE, Miyoshi T et al (2015b) A 3' Poly(A) tract is required for LINE-1 Retrotransposition. Mol Cell 60(5):728–741. https://doi.org/10.1016/j.molcel.2015.10.012

Elliott DJ, Venables JP, Newton CS et al (2000) An evolutionarily conserved germ cell-specific hnRNP is encoded by a retrotransposed gene. Hum Mol Genet 9(14):2117–2124. https://doi.org/10.1093/hmg/9.14.2117

Esnault C, Maestre J, Heidmann T (2000) Human LINE retrotransposons generate processed pseudogenes. Nat Genet 24(4):363–367. https://doi.org/10.1038/74184

Esposito F, De Martino M, Forzati F et al (2014a) HMGA1-pseudogene overexpression contributes to cancer progression. Cell Cycle 13(23):3636–3639. https://doi.org/10.4161/15384101.2014.974440

Esposito F, De Martino M, Petti MG et al (2014b) HMGA1 pseudogenes as candidate proto-oncogenic competitive endogenous RNAs. Oncotarget 5(18):8341–8354. https://doi.org/10.18632/oncotarget.2202

Fablet M, Bueno M, Potrzebowski L et al (2009) Evolutionary origin and functions of retrogene introns. Mol Biol Evol 26(9):2147–2156. https://doi.org/10.1093/molbev/msp125
Feng Q, Moran JV, Kazazian HH Jr et al (1996) Human L1 retrotransposon encodes a conserved endonuclease required for retrotransposition. Cell 87(5):905–916. https://doi.org/10.1016/s0092-8674(00)81997-2
Fueyo R, Judd J, Feschotte C et al (2022) Roles of transposable elements in the regulation of mammalian transcription. Nat Rev Mol Cell Biol 23(7):481–497. https://doi.org/10.1038/s41580-022-00457-y
Garfinkel DJ, Boeke JD, Fink GR (1985) Ty element transposition: reverse transcriptase and virus-like particles. Cell 42(2):507–517. https://doi.org/10.1016/0092-8674(85)90108-4
Gebert D, Ketting RF, Zischler H et al (2015) piRNAs from pig testis provide evidence for a conserved role of the Piwi pathway in post-transcriptional gene regulation in mammals. PLoS One 10(5):e0124860. https://doi.org/10.1371/journal.pone.0124860
Genomes Project C, Abecasis GR, Auton A et al (2012) An integrated map of genetic variation from 1,092 human genomes. Nature 491(7422):56–65. https://doi.org/10.1038/nature11632
Goemann IM, Marczyk VR, Recamonde-Mendoza M et al (2020) Decreased expression of the thyroid hormone-inactivating enzyme type 3 deiodinase is associated with lower survival rates in breast cancer. Sci Rep 10(1):13914. https://doi.org/10.1038/s41598-020-70892-4
Goodier JL, Ostertag EM, Kazazian HH Jr (2000) Transduction of 3′-flanking sequences is common in L1 retrotransposition. Hum Mol Genet 9(4):653–657. https://doi.org/10.1093/hmg/9.4.653
Grabundzija I, Hickman AB, Dyda F (2018) Helraiser intermediates provide insight into the mechanism of eukaryotic replicative transposition. Nat Commun 9(1):1278. https://doi.org/10.1038/s41467-018-03688-w
Guo X, Lin M, Rockowitz S et al (2014) Characterization of human pseudogene-derived non-coding RNAs for functional potential. PLoS One 9(4):e93972. https://doi.org/10.1371/journal.pone.0093972
Hajjar AM, Linial ML (1993) A model system for nonhomologous recombination between retroviral and cellular RNA. J Virol 67(7):3845–3853. https://doi.org/10.1128/JVI.67.7.3845-3853.1993
Hanai A, Ohgi M, Yagi C et al (2016) Class I Arfs (Arf1 and Arf3) and Arf6 are localized to the Flemming body and play important roles in cytokinesis. J Biochem 159(2):201–208. https://doi.org/10.1093/jb/mvv088
Hancks DC, Goodier JL, Mandal PK et al (2011) Retrotransposition of marked SVA elements by human L1s in cultured cells. Hum Mol Genet 20(17):3386–3400. https://doi.org/10.1093/hmg/ddr245
Hawkins PG, Morris KV (2010) Transcriptional regulation of Oct4 by a long non-coding RNA antisense to Oct4-pseudogene 5. Transcription 1(3):165–175. https://doi.org/10.4161/trns.1.3.13332
Herrero MJ, Gitton Y (2018) The untold stories of the speech gene, the FOXP2 cancer gene. Genes Cancer 9(1–2):11–38. https://doi.org/10.18632/genesandcancer.169
Hirt H, Kimelman J, Birnbaum MJ et al (1987) The human growth hormone gene locus: structure, evolution, and allelic variations. DNA 6(1):59–70. https://doi.org/10.1089/dna.1987.6.59
Horton WA, Lunstrum GP (2002) Fibroblast growth factor receptor 3 mutations in achondroplasia and related forms of dwarfism. Rev Endocr Metab Disord 3(4):381–385. https://doi.org/10.1023/a:1020914026829
Hu XY, Ray PN, Worton RG (1991) Mechanisms of tandem duplication in the Duchenne muscular dystrophy gene include both homologous and nonhomologous intrachromosomal recombination. EMBO J 10(9):2471–2477. https://doi.org/10.1002/j.1460-2075.1991.tb07786.x
Hughes SH (2015) Reverse transcription of Retroviruses and LTR Retrotransposons. Microbiol Spectr 3(2):MDNA3-0027-2014. https://doi.org/10.1128/microbiolspec.MDNA3-0027-2014

Jeganathan J, Saraf R, Mahmood F et al (2017) Mitochondrial dysfunction in atrial tissue of patients developing postoperative atrial fibrillation. Ann Thorac Surg 104(5):1547–1555. https://doi.org/10.1016/j.athoracsur.2017.04.060
Johnsson P, Ackley A, Vidarsdottir L et al (2013) A pseudogene long-noncoding-RNA network regulates PTEN transcription and translation in human cells. Nat Struct Mol Biol 20(4):440–446. https://doi.org/10.1038/nsmb.2516
Joly-Lopez Z, Bureau TE (2018) Exaptation of transposable element coding sequences. Curr Opin Genet Dev 49:34–42. https://doi.org/10.1016/j.gde.2018.02.011
Jun J, Ryvkin P, Hemphill E et al (2009) The birth of new genes by RNA- and DNA-mediated duplication during mammalian evolution. J Comput Biol:16(10):1429–1444. https://doi.org/10.1089/cmb.2009.0073
Jurka J (1997) Sequence patterns indicate an enzymatic involvement in integration of mammalian retroposons. Proc Natl Acad Sci USA 94(5):1872–1877. https://doi.org/10.1073/pnas.94.5.1872
Kabza M, Kubiak MR, Danek A et al (2015) Inter-population differences in Retrogene loss and expression in humans. PLoS Genet 11(10):e1005579. https://doi.org/10.1371/journal.pgen.1005579
Kaessmann H (2010) Origins, evolution, and phenotypic impact of new genes. Genome Res 20(10):1313–1326. https://doi.org/10.1101/gr.101386.109
Kaessmann H, Vinckenbosch N, Long M (2009) RNA-based gene duplication: mechanistic and evolutionary insights. Nat Rev Genet 10(1):19–31. https://doi.org/10.1038/nrg2487
Kalyana-Sundaram S, Kumar-Sinha C, Shankar S et al (2012) Expressed pseudogenes in the transcriptional landscape of human cancers. Cell 149(7):1622–1634. https://doi.org/10.1016/j.cell.2012.04.041
Kanavouras K, Mastorodemos V, Borompokas N et al (2007) Properties and molecular evolution of human GLUD2 (neural and testicular tissue-specific) glutamate dehydrogenase. J Neurosci Res 85(5):1101–1109. https://doi.org/10.1002/jnr.21197
Kandouz M, Bier A, Carystinos GD et al (2004) Connexin43 pseudogene is expressed in tumor cells and inhibits growth. Oncogene 23(27):4763–4770. https://doi.org/10.1038/sj.onc.1207506
Kapitonov VV, Jurka J (2001) Rolling-circle transposons in eukaryotes. Proc Natl Acad Sci USA 98(15):8714–8719. https://doi.org/10.1073/pnas.151269298
Kapitonov VV, Jurka J (2004) Harbinger transposons and an ancient HARBI1 gene derived from a transposase. DNA Cell Biol 23(5):311–324. https://doi.org/10.1089/104454904323090949
Kapitonov VV, Jurka J (2007) Helitrons on a roll: eukaryotic rolling-circle transposons. Trends Genet 23(10):521–529. https://doi.org/10.1016/j.tig.2007.08.004
Kazazian HH Jr, Moran JV (1998) The impact of L1 retrotransposons on the human genome. Nat Genet 19(1):19–24. https://doi.org/10.1038/ng0598-19
Kellner M, Makalowski W (2019) Transposable elements significantly contributed to the core promoters in the human genome. Sci China Life Sci 62(4):489–497. https://doi.org/10.1007/s11427-018-9449-0
Kemp JR, Longworth MS (2015) Crossing the LINE toward genomic instability: LINE-1 Retrotransposition in cancer. Front Chem 3:68. https://doi.org/10.3389/fchem.2015.00068
Kerin T, Ramanathan A, Rivas K et al (2012) A noncoding RNA antisense to moesin at 5p14.1 in autism. Sci Transl Med 4(128):128ra140. https://doi.org/10.1126/scitranslmed.3003479
Kim P, Tan H, Liu J et al (2022) FusionGDB 2.0: fusion gene annotation updates aided by deep learning. Nucleic Acids Res 50(D1):D1221–D1230. https://doi.org/10.1093/nar/gkab1056
Kiyosawa H, Chance PF (1996) Primate origin of the CMT1A-REP repeat and analysis of a putative transposon-associated recombinational hotspot. Hum Mol Genet 5(6):745–753. https://doi.org/10.1093/hmg/5.6.745
Kleckner N, Chan RK, Tye BK et al (1975) Mutagenesis by insertion of a drug-resistance element carrying an inverted repetition. J Mol Biol 97(4):561–575. https://doi.org/10.1016/s0022-2836(75)80059-3

Kubiak MR, Makalowska I (2017) Protein-coding genes' Retrocopies and their functions. Viruses 9(4). https://doi.org/10.3390/v9040080
Kubiak MR, Szczesniak MW, Makalowska I (2020) Complex analysis of Retroposed genes' contribution to human genome, proteome and transcriptome. Genes (Basel) 11(5). https://doi.org/10.3390/genes11050542
Kubo S, Seleme MC, Soifer HS et al (2006) L1 retrotransposition in nondividing and primary human somatic cells. Proc Natl Acad Sci USA 103(21):8036–8041. https://doi.org/10.1073/pnas.0601954103
Lehrman MA, Goldstein JL, Russell DW et al (1987a) Duplication of seven exons in LDL receptor gene caused by Alu-Alu recombination in a subject with familial hypercholesterolemia. Cell 48(5):827–835. https://doi.org/10.1016/0092-8674(87)90079-1
Lehrman MA, Russell DW, Goldstein JL et al (1987b) Alu-Alu recombination deletes splice acceptor sites and produces secreted low density lipoprotein receptor in a subject with familial hypercholesterolemia. J Biol Chem 262(7):3354–3361
Li WH, Gojobori T, Nei M (1981) Pseudogenes as a paradigm of neutral evolution. Nature 292(5820):237–239. https://doi.org/10.1038/292237a0
Liao WW, Asri M, Ebler J et al (2023) A draft human pangenome reference. Nature 617(7960):312–324. https://doi.org/10.1038/s41586-023-05896-x
Long M, Langley CH (1993) Natural selection and the origin of jingwei, a chimeric processed functional gene in Drosophila. Science 260(5104):91–95. https://doi.org/10.1126/science.7682012
Luo GX, Taylor J (1990) Template switching by reverse transcriptase during DNA synthesis. J Virol 64(9):4321–4328. https://doi.org/10.1128/JVI.64.9.4321-4328.1990
Luo T, Ou JN, Cao LF et al (2020) The autism-related lncRNA MSNP1AS regulates Moesin protein to influence the RhoA, Rac1, and PI3K/Akt pathways and regulate the structure and survival of neurons. Autism Res 13(12):2073–2082. https://doi.org/10.1002/aur.2413
Lupski JR (1998) Charcot-Marie-Tooth disease: lessons in genetic mechanisms. Mol Med 4(1):3–11
Lupski JR, de Oca-Luna RM, Slaugenhaupt S et al (1991) DNA duplication associated with Charcot-Marie-Tooth disease type 1A. Cell 66(2):219–232. https://doi.org/10.1016/0092-8674(91)90613-4
Lv H, Tong J, Yang J et al (2018) Dysregulated Pseudogene HK2P1 may contribute to Preeclampsia as a competing endogenous RNA for Hexokinase 2 by impairing Decidualization. Hypertension 71(4):648–658. https://doi.org/10.1161/HYPERTENSIONAHA.117.10084
Lynch M, Conery JS (2000) The evolutionary fate and consequences of duplicate genes. Science 290(5494):1151–1155. https://doi.org/10.1126/science.290.5494.1151
Lynch M, Force A (2000) The probability of duplicate gene preservation by subfunctionalization. Genetics 154(1):459–473. https://doi.org/10.1093/genetics/154.1.459
Ma H, Wang M, Zhang YE et al (2023) The power of "controllers": transposon-mediated duplicated genes evolve towards neofunctionalization. J Genet Genomics 50(7):462–472. https://doi.org/10.1016/j.jgg.2023.04.003
Macia A, Widmann TJ, Heras SR et al (2017) Engineered LINE-1 retrotransposition in nondividing human neurons. Genome Res 27(3):335–348. https://doi.org/10.1101/gr.206805.116
Maestre J, Tchénio T, Dhellin O et al (1995) mRNA retroposition in human cells: processed pseudogene formation. EMBO J 14(24):6333–6338
Makalowska I, Kubiak MR (2023) Novel functions of a retroposed gene. Trends Genet 39(6):439–441. https://doi.org/10.1016/j.tig.2023.03.006
Makalowski W, Mitchell GA, Labuda D (1994) Alu sequences in the coding regions of mRNA: a source of protein variability. Trends Genet 10(6):188–193. https://doi.org/10.1016/0168-9525(94)90254-2
Makalowski W, Gotea V, Pande A et al (2019) Transposable elements: classification, identification, and their use as a tool for comparative genomics. Methods Mol Biol 1910:177–207. https://doi.org/10.1007/978-1-4939-9074-0_6

Marcus S, Hellgren D, Lambert B et al (1993) Duplication in the hypoxanthine phosphoribosyltransferase gene caused by Alu-Alu recombination in a patient with Lesch Nyhan syndrome. Hum Genet 90(5):477–482. https://doi.org/10.1007/BF00217444
Martin MD, Brown DN, Ramos KS (2021) Computational modeling of RNase, antisense ORF0 RNA, and intracellular compartmentation and their impact on the life cycle of the line retrotransposon. Comput Struct Biotechnol J 19:5667–5677. https://doi.org/10.1016/j.csbj.2021.10.003
Mathias SL, Scott AF, Kazazian HH Jr et al (1991) Reverse transcriptase encoded by a human transposable element. Science 254(5039):1808–1810. https://doi.org/10.1126/science.1722352
Mc CB (1950) The origin and behavior of mutable loci in maize. Proc Natl Acad Sci USA 36(6):344–355. https://doi.org/10.1073/pnas.36.6.344
Mercuri RLV, Conceicao HB, Guardia GDA et al (2023) Retro-miRs: novel and functional miRNAs originating from mRNA retrotransposition. Mob DNA 14(1):12. https://doi.org/10.1186/s13100-023-00301-w
Milligan MJ, Lipovich L (2014) Pseudogene-derived lncRNAs: emerging regulators of gene expression. Front Genet 5:476. https://doi.org/10.3389/fgene.2014.00476
Milligan MJ, Harvey E, Yu A et al (2016) Global intersection of long non-coding RNAs with processed and unprocessed Pseudogenes in the human genome. Front Genet 7:26. https://doi.org/10.3389/fgene.2016.00026
Moran JV, Holmes SE, Naas TP et al (1996) High frequency retrotransposition in cultured mammalian cells. Cell 87(5):917–927. https://doi.org/10.1016/s0092-8674(00)81998-4
Moran JV, DeBerardinis RJ, Kazazian HH Jr (1999) Exon shuffling by L1 retrotransposition. Science 283(5407):1530–1534. https://doi.org/10.1126/science.283.5407.1530
Mori S, Hayashi M, Inagaki S et al (2016) Identification of multiple forms of RNA transcripts associated with human-specific Retrotransposed gene copies. Genome Biol Evol 8(8):2288–2296. https://doi.org/10.1093/gbe/evw156
Muro EM, Andrade-Navarro MA (2010) Pseudogenes as an alternative source of natural antisense transcripts. BMC Evol Biol 10:338. https://doi.org/10.1186/1471-2148-10-338
Nei M (1969) Gene duplication and nucleotide substitution in evolution. Nature 221(5175):40–42. https://doi.org/10.1038/221040a0
Nestadt G, Wang Y, Grados MA et al (2012) Homeobox genes in obsessive-compulsive disorder. Am J Med Genet B Neuropsychiatr Genet 159B(1):53–60. https://doi.org/10.1002/ajmg.b.32001
Newman S, Hermetz KE, Weckselblatt B et al (2015) Next-generation sequencing of duplication CNVs reveals that most are tandem and some create fusion genes at breakpoints. Am J Hum Genet 96(2):208–220. https://doi.org/10.1016/j.ajhg.2014.12.017
Nigumann P, Redik K, Matlik K et al (2002) Many human genes are transcribed from the antisense promoter of L1 retrotransposon. Genomics 79(5):628–634. https://doi.org/10.1006/geno.2002.6758
Nowak MA, Boerlijst MC, Cooke J et al (1997) Evolution of genetic redundancy. Nature 388(6638):167–171. https://doi.org/10.1038/40618
Nuerzhati Y, Dong R, Song Z et al (2019) Role of the long non-coding RNA-Annexin A2 pseudogene 3/Annexin A2 signaling pathway in biliary atresia-associated hepatic injury. Int J Mol Med 43(2):739–748. https://doi.org/10.3892/ijmm.2018.4023
Nuzziello N, Liguori M (2019) The MicroRNA centrism in the orchestration of neuroinflammation in neurodegenerative diseases. Cells 8(10). https://doi.org/10.3390/cells8101193
Ohno S (1970) Evolution by gene duplication. Springer Verlag, Berlin
Ohshima K, Hattori M, Yada T et al (2003) Whole-genome screening indicates a possible burst of formation of processed pseudogenes and Alu repeats by particular L1 subfamilies in ancestral primates. Genome Biol 4(11):R74. https://doi.org/10.1186/gb-2003-4-11-r74
Okamura K, Nakai K (2008) Retrotransposition as a source of new promoters. Mol Biol Evol 25(6):1231–1238. https://doi.org/10.1093/molbev/msn071

Ostertag EM, Goodier JL, Zhang Y et al (2003) SVA elements are nonautonomous retrotransposons that cause disease in humans. Am J Hum Genet 73(6):1444–1451. https://doi.org/10.1086/380207

Pan CT, Lin YS (2020) MicroRNA retrocopies generated via L1-mediated retrotransposition in placental mammals help to reveal how their parental genes were transcribed. Sci Rep 10(1):20612. https://doi.org/10.1038/s41598-020-77381-8

Pantano L, Jodar M, Bak M et al (2015) The small RNA content of human sperm reveals pseudogene-derived piRNAs complementary to protein-coding genes. RNA 21(6):1085–1095. https://doi.org/10.1261/rna.046482.114

Parker HG, VonHoldt BM, Quignon P et al (2009) An expressed fgf4 retrogene is associated with breed-defining chondrodysplasia in domestic dogs. Science 325(5943):995–998. https://doi.org/10.1126/science.1173275

Pavlicek A, Paces J, Elleder D et al (2002) Processed pseudogenes of human endogenous retroviruses generated by LINEs: their integration, stability, and distribution. Genome Res 12(3):391–399. https://doi.org/10.1101/gr.216902

Perreault J, Noel JF, Briere F et al (2005) Retropseudogenes derived from the human Ro/SS-A autoantigen-associated hY RNAs. Nucleic Acids Res 33(6):2032–2041. https://doi.org/10.1093/nar/gki504

Pickeral OK, Makalowski W, Boguski MS et al (2000) Frequent human genomic DNA transduction driven by LINE-1 retrotransposition. Genome Res 10(4):411–415. https://doi.org/10.1101/gr.10.4.411

Pizarro JG, Cristofari G (2016) Post-transcriptional control of LINE-1 Retrotransposition by cellular host factors in somatic cells. Front Cell Dev Biol 4:14. https://doi.org/10.3389/fcell.2016.00014

Plaitakis A, Spanaki C, Mastorodemos V et al (2003) Study of structure-function relationships in human glutamate dehydrogenases reveals novel molecular mechanisms for the regulation of the nerve tissue-specific (GLUD2) isoenzyme. Neurochem Int 43(4–5):401–410. https://doi.org/10.1016/s0197-0186(03)00028-7

Poliseno L, Salmena L, Zhang J et al (2010) A coding-independent function of gene and pseudogene mRNAs regulates tumour biology. Nature 465(7301):1033–1038. https://doi.org/10.1038/nature09144

Poliseno L, Haimovic A, Christos PJ et al (2011) Deletion of PTENP1 pseudogene in human melanoma. J Invest Dermatol 131(12):2497–2500. https://doi.org/10.1038/jid.2011.232

Pousi B, Hautala T, Heikkinen J et al (1994) Alu-Alu recombination results in a duplication of seven exons in the lysyl hydroxylase gene in a patient with the type VI variant of Ehlers-Danlos syndrome. Am J Hum Genet 55(5):899–906

Prendergast GC (2001) Actin' up: RhoB in cancer and apoptosis. Nat Rev Cancer 1(2):162–168. https://doi.org/10.1038/35101096

Pritham EJ, Feschotte C (2007) Massive amplification of rolling-circle transposons in the lineage of the bat Myotis lucifugus. Proc Natl Acad Sci USA 104(6):1895–1900. https://doi.org/10.1073/pnas.0609601104

Puget N, Sinilnikova OM, Stoppa-Lyonnet D et al (1999) An Alu-mediated 6-kb duplication in the BRCA1 gene: a new founder mutation? Am J Hum Genet 64(1):300–302. https://doi.org/10.1086/302211

Raiz J, Damert A, Chira S et al (2012) The non-autonomous retrotransposon SVA is trans-mobilized by the human LINE-1 protein machinery. Nucleic Acids Res 40(4):1666–1683. https://doi.org/10.1093/nar/gkr863

Rastogi S, Liberles DA (2005) Subfunctionalization of duplicated genes as a transition state to neofunctionalization. BMC Evol Biol 5:28. https://doi.org/10.1186/1471-2148-5-28

Rayan NA, Del Rosario RCH, Prabhakar S (2016) Massive contribution of transposable elements to mammalian regulatory sequences. Semin Cell Dev Biol 57:51–56. https://doi.org/10.1016/j.semcdb.2016.05.004

Richardson SR, Salvador-Palomeque C, Faulkner GJ (2014) Diversity through duplication: whole-genome sequencing reveals novel gene retrocopies in the human population. BioEssays 36(5):475–481. https://doi.org/10.1002/bies.201300181

Rodic N, Burns KH (2013) Long interspersed element-1 (LINE-1): passenger or driver in human neoplasms? PLoS Genet 9(3):e1003402. https://doi.org/10.1371/journal.pgen.1003402

Rohozinski J, Edwards CL, Anderson ML (2012) Does expression of the retrogene UTP14c in the ovary pre-dispose women to ovarian cancer? Med Hypotheses 78(4):446–449. https://doi.org/10.1016/j.mehy.2011.12.016

Rosikiewicz W, Kabza M, Kosinski JG et al (2017) RetrogeneDB-a database of plant and animal retrocopies. Database (Oxford) 2017. https://doi.org/10.1093/database/bax038

Rosso L, Marques AC, Reichert AS et al (2008) Mitochondrial targeting adaptation of the hominoid-specific glutamate dehydrogenase driven by positive Darwinian selection. PLoS Genet 4(8):e1000150. https://doi.org/10.1371/journal.pgen.1000150

Rozmahel R, Heng HH, Duncan AM et al (1997) Amplification of CFTR exon 9 sequences to multiple locations in the human genome. Genomics 45(3):554–561. https://doi.org/10.1006/geno.1997.4968

Rutnam ZJ, Du WW, Yang W et al (2014) The pseudogene TUSC2P promotes TUSC2 function by binding multiple microRNAs. Nat Commun 5:2914. https://doi.org/10.1038/ncomms3914

Sayah DM, Sokolskaja E, Berthoux L et al (2004) Cyclophilin A retrotransposition into TRIM5 explains owl monkey resistance to HIV-1. Nature 430(6999):569–573. https://doi.org/10.1038/nature02777

Schmitz J, Brosius J (2011) Exonization of transposed elements: a challenge and opportunity for evolution. Biochimie 93(11):1928–1934. https://doi.org/10.1016/j.biochi.2011.07.014

Sen SK, Huang CT, Han K et al (2007) Endonuclease-independent insertion provides an alternative pathway for L1 retrotransposition in the human genome. Nucleic Acids Res 35(11):3741–3751. https://doi.org/10.1093/nar/gkm317

Shen F, Cai WS, Feng Z et al (2015) MiR-492 contributes to cell proliferation and cell cycle of human breast cancer cells by suppressing SOX7 expression. Tumour Biol 36(3):1913–1921. https://doi.org/10.1007/s13277-014-2794-z

Shen J, Si J, Wang Q et al (2023) Current status and future perspectives in dysregulated miR-492. Gene 877:147518. https://doi.org/10.1016/j.gene.2023.147518

Sistrom M, Andrews H, Edwards DL (2024) Comparative genomics of Japanese encephalitis virus shows low rates of recombination and a small subset of codon positions under episodic diversifying selection. PLoS Negl Trop Dis 18(1):e0011459. https://doi.org/10.1371/journal.pntd.0011459

Speek M (2001) Antisense promoter of human L1 retrotransposon drives transcription of adjacent cellular genes. Mol Cell Biol 21(6):1973–1985. https://doi.org/10.1128/MCB.21.6.1973-1985.2001

Stambolic V, Suzuki A, de la Pompa JL et al (1998) Negative regulation of PKB/Akt-dependent cell survival by the tumor suppressor PTEN. Cell 95(1):29–39. https://doi.org/10.1016/s0092-8674(00)81780-8

Stankiewicz P, Lupski JR (2002) Genome architecture, rearrangements and genomic disorders. Trends Genet 18(2):74–82. https://doi.org/10.1016/s0168-9525(02)02592-1

Staszak K, Makalowska I (2021) Cancer, Retrogenes, and evolution. Life (Basel) 11(1). https://doi.org/10.3390/life11010072

Sulak M, Fong L, Mika K et al (2016) TP53 copy number expansion is associated with the evolution of increased body size and an enhanced DNA damage response in elephants. elife 5. https://doi.org/10.7554/eLife.11994

Swergold GD (1990) Identification, characterization, and cell specificity of a human LINE-1 promoter. Mol Cell Biol 10(12):6718–6729. https://doi.org/10.1128/mcb.10.12.6718-6729.1990

Taatjes DJ, Roth J (2024) In focus in HCB. Histochem Cell Biol 162(6):443–445. https://doi.org/10.1007/s00418-024-02325-6

Tan S, Cardoso-Moreira M, Shi W et al (2016) LTR-mediated retroposition as a mechanism of RNA-based duplication in metazoans. Genome Res 26(12):1663–1675. https://doi.org/10.1101/gr.204925.116

Tang J, Yu Y, Yang W (2017) Long noncoding RNA and its contribution to autism spectrum disorders. CNS Neurosci Ther 23(8):645–656. https://doi.org/10.1111/cns.12710

Thomas J, Pritham EJ (2015) Helitrons, the eukaryotic rolling-circle transposable elements. Microbiol Spectr 3(4). https://doi.org/10.1128/microbiolspec.MDNA3-0049-2014

Thomas J, Schaack S, Pritham EJ (2010) Pervasive horizontal transfer of rolling-circle transposons among animals. Genome Biol Evol 2:656–664. https://doi.org/10.1093/gbe/evq050

Thomas J, Phillips CD, Baker RJ et al (2014) Rolling-circle transposons catalyze genomic innovation in a mammalian lineage. Genome Biol Evol 6(10):2595–2610. https://doi.org/10.1093/gbe/evu204

Troskie RL, Faulkner GJ, Cheetham SW (2021) Processed pseudogenes: a substrate for evolutionary innovation: Retrotransposition contributes to genome evolution by propagating pseudogene sequences with rich regulatory potential throughout the genome. BioEssays 43(11):e2100186. https://doi.org/10.1002/bies.202100186

Tsujikawa M, Kurahashi H, Tanaka T et al (1999) Identification of the gene responsible for gelatinous drop-like corneal dystrophy. Nat Genet 21(4):420–423. https://doi.org/10.1038/7759

Uechi T, Maeda N, Tanaka T et al (2002) Functional second genes generated by retrotransposition of the X-linked ribosomal protein genes. Nucleic Acids Res 30(24):5369–5375. https://doi.org/10.1093/nar/gkf696

Vitiello M, Evangelista M, Zhang Y et al (2020) PTENP1 is a ceRNA for PTEN: it's CRISPR clear. J Hematol Oncol 13(1):73. https://doi.org/10.1186/s13045-020-00894-2

Vollger MR, Guitart X, Dishuck PC et al (2022) Segmental duplications and their variation in a complete human genome. Science 376(6588):eabj6965. https://doi.org/10.1126/science.abj6965

von Frowein J, Pagel P, Kappler R et al (2011) MicroRNA-492 is processed from the keratin 19 gene and up-regulated in metastatic hepatoblastoma. Hepatology 53(3):833–842. https://doi.org/10.1002/hep.24125

Wang H, Xing J, Grover D et al (2005) SVA elements: a hominid-specific retroposon family. J Mol Biol 354(4):994–1007. https://doi.org/10.1016/j.jmb.2005.09.085

Wang L, Guo ZY, Zhang R et al (2013) Pseudogene OCT4-pg4 functions as a natural micro RNA sponge to regulate OCT4 expression by competing for miR-145 in hepatocellular carcinoma. Carcinogenesis 34(8):1773–1781. https://doi.org/10.1093/carcin/bgt139

Wang Y, Wang Y, Liu Q et al (2014) Comparative RNA-seq analysis reveals potential mechanisms mediating the conversion to androgen independence in an LNCaP progression cell model. Cancer Lett 342(1):130–138. https://doi.org/10.1016/j.canlet.2013.08.044

Wang QS, Shi LL, Sun F et al (2019) High expression of ANXA2 Pseudogene ANXA2P2 promotes an aggressive phenotype in hepatocellular carcinoma. Dis Markers 2019:9267046. https://doi.org/10.1155/2019/9267046

Watanabe T, Totoki Y, Toyoda A et al (2008) Endogenous siRNAs from naturally formed dsRNAs regulate transcripts in mouse oocytes. Nature 453(7194):539–543. https://doi.org/10.1038/nature06908

Wells JN, Feschotte C (2020) A field guide to eukaryotic transposable elements. Annu Rev Genet 54:539–561. https://doi.org/10.1146/annurev-genet-040620-022145

Welsh CF, Moss J, Vaughan M (1994) ADP-ribosylation factors: a family of approximately 20-kDa guanine nucleotide-binding proteins that activate cholera toxin. Mol Cell Biochem 138(1–2):157–166. https://doi.org/10.1007/BF00928458

Whitman SP, Strout MP, Marcucci G et al (2001) The partial nontandem duplication of the MLL (ALL1) gene is a novel rearrangement that generates three distinct fusion transcripts in B-cell acute lymphoblastic leukemia. Cancer Res 61(1):59–63

Wilhelm M, Wilhelm FX (2001) Reverse transcription of retroviruses and LTR retrotransposons. Cell Mol Life Sci 58(9):1246–1262. https://doi.org/10.1007/PL00000937

Witherspoon DJ, Watkins WS, Zhang Y et al (2009) Alu repeats increase local recombination rates. BMC Genomics 10:530. https://doi.org/10.1186/1471-2164-10-530

Wonderlich ER, Leonard JA, Kulpa DA et al (2011) ADP ribosylation factor 1 activity is required to recruit AP-1 to the major histocompatibility complex class I (MHC-I) cytoplasmic tail and disrupt MHC-I trafficking in HIV-1-infected primary T cells. J Virol 85(23):12216–12226. https://doi.org/10.1128/JVI.00056-11

Wu N, Li J, Chen X et al (2020) Identification of long non-coding RNA and circular RNA expression profiles in atrial fibrillation. Heart Lung Circ 29(7):e157–e167. https://doi.org/10.1016/j.hlc.2019.10.018

Xia S, Chen J, Arsala D et al (2025) Functional innovation through new genes as a general evolutionary process. Nat Genet 57(2):295–309. https://doi.org/10.1038/s41588-024-02059-0

Xing J, Wang H, Belancio VP et al (2006) Emergence of primate genes by retrotransposon-mediated sequence transduction. Proc Natl Acad Sci USA 103(47):17608–17613. https://doi.org/10.1073/pnas.0603224103

Xu T, Li D, He Y et al (2018) The expression level of CSDAP1 in lung cancer and its clinical significance. Oncol Lett 16(4):4361–4366. https://doi.org/10.3892/ol.2018.9195

Yang N, Kazazian HH Jr (2006) L1 retrotransposition is suppressed by endogenously encoded small interfering RNAs in human cultured cells. Nat Struct Mol Biol 13(9):763–771. https://doi.org/10.1038/nsmb1141

Yang TP, Stout JT, Konecki DS et al (1988) Spontaneous reversion of novel Lesch-Nyhan mutation by HPRT gene rearrangement. Somat Cell Mol Genet 14(3):293–303. https://doi.org/10.1007/BF01534590

Yeowell HN, Walker LC (2000) Mutations in the lysyl hydroxylase 1 gene that result in enzyme deficiency and the clinical phenotype of Ehlers-Danlos syndrome type VI. Mol Genet Metab 71(1–2):212–224. https://doi.org/10.1006/mgme.2000.3076

Yeowell HN, Walker LC, Murad S et al (1997) A common duplication in the lysyl hydroxylase gene of patients with Ehlers Danlos syndrome type VI results in preferential stimulation of lysyl hydroxylase activity and mRNA by hydralazine. Arch Biochem Biophys 347(1):126–131. https://doi.org/10.1006/abbi.1997.0319

Yin L, Yao J, Deng G et al (2020) Identification of candidate lncRNAs and circRNAs regulating WNT3/beta-catenin signaling in essential hypertension. Aging (Albany NY) 12(9):8261–8288. https://doi.org/10.18632/aging.103137

Yu G, Yao W, Gumireddy K et al (2014) Pseudogene PTENP1 functions as a competing endogenous RNA to suppress clear-cell renal cell carcinoma progression. Mol Cancer Ther 13(12):3086–3097. https://doi.org/10.1158/1535-7163.MCT-14-0245

Zhang W, Tautz D (2022) Tracing the origin and evolutionary fate of recent gene Retrocopies in natural populations of the house mouse. Mol Biol Evol 39(2). https://doi.org/10.1093/molbev/msab360

Zhang Z, Harrison PM, Liu Y et al (2003) Millions of years of evolution preserved: a comprehensive catalog of the processed pseudogenes in the human genome. Genome Res 13(12):2541–2558. https://doi.org/10.1101/gr.1429003

Zhang YW, Liu S, Zhang X et al (2009) A functional mouse retroposed gene Rps23r1 reduces Alzheimer's beta-amyloid levels and tau phosphorylation. Neuron 64(3):328–340. https://doi.org/10.1016/j.neuron.2009.08.036

Zhou W, Yang Y, Wang W et al (2024) Pseudogene OCT4-pg5 upregulates OCT4B expression to promote bladder cancer progression by competing with miR-145-5p. Cell Cycle 23(6):645–661. https://doi.org/10.1080/15384101.2024.2353554

Zhu Z, Tan S, Zhang Y et al (2016) LINE-1-like retrotransposons contribute to RNA-based gene duplication in dicots. Sci Rep 6:24755. https://doi.org/10.1038/srep24755

Chapter 3
Transposable Elements and Their Roles in Genome Architecture

Sampriti Majumder, Snata Pandey, and Sugopa Sengupta

Abstract Three-dimensional organization of genomes in every organism plays an essential role in regulating gene expression. Within the nucleus, chromatin is arranged in a hierarchical manner-forming compartments, topologically associating domains and chromatin loop structures—all of which contribute to gene expression control. Transposable elements, which are quite abundantly distributed within the eukaryotic genomes, seem to play decisive roles in optimizing the genome architecture. They often provide binding sites for various nuclear architectural proteins, which in turn influence chromosomal folding. Transposition events function at the level of chromatin compaction, loop formation, organization into topologically associating domains, and sequestration into chromosomal territories and often aid in the establishment and maintenance of centromeres. At the same time, transpositions in the context of the 3D genome can dynamically reorganize chromatin interactions and contribute to spatio-temporal regulation of gene expression patterns in response to developmental stages, environmental cues, or species-specific needs, ultimately driving evolution. Transposon-mediated alteration of chromatin structure has also been implicated in multiple diseases, including developmental abnormalities and cancers, thereby justifying the need for more extensive research in this field.

Keywords 3D genome architecture · Chromosome territories · Chromosomal compartments · Topologically Associating Domains · Chromatin loops · Transposable elements · Centromere establishment · Heterochromatinization · Neocentromere formation

---

Sampriti Majumder and Snata Pandey contributed equally with all other contributors.

S. Majumder · S. Pandey · S. Sengupta (✉)
Department of Life Sciences, Presidency University Kolkata, Kolkata, India
e-mail: sugopa.dbs@presiuniv.ac.in

L. M. Vaschetto (ed.), *Transposable Elements, Transcriptomics, and Diseases*, Epigenetics and Human Health 14, https://doi.org/10.1007/978-3-032-04125-8_3

Abbreviations

AGR	Adaptive genomic region
CBS	CTCF-binding site
CENP-A/B	Centromere protein-A/B
CRE	Cis-regulatory elements
CTCF	CCCTC-binding factor
CTs	Chromosome territories
ERV	Endogenous retrovirus
HERV-H	Human endogenous retrovirus subfamily H
HOR	Higher-order repeats
HP1	Heterochromatin protein 1
HP1α	Heterochromatin protein 1α
IC	Interchromatin compartment
LINE	Long interspersed nuclear element
LTR	Long terminal repeat
MERV-L	Murine endogenous retrovirus-L
MIR	Mammalian-wide interspersed repeats
Oct4	Octamer-binding transcription factor 4 (Oct4)
PLEs	Penelope-like elements
PML	Promyelocytic leukemia
PSC	Pluripotent stem cells
RNAP	RNA polymerase
Setdb1	SET-domain-bifurcated histone-lysine methyltransferase 1
SINE	Short interspersed nuclear element
SMC	Structural maintenance of chromosomes
SMC3	Structural maintenance of chromosomes protein 3
SSL	Supersized chromatin loops
SVA	SINE-VNTR-Alu retrotransposons
TAD	Topologically associating domain
TE	Transposable element
TFBS	Transcription factor binding site
TIR	Terminal inverted repeats
WAPL	Wings apart-like protein
ZNF143	Zinc finger protein 143

3.1 Introduction

To interpret and understand how a genome functions, complete knowledge about its structural organization is of utmost importance (Li et al. 2022). Although genome sequencing projects have yielded a wealth of information on the primary sequence of the genome, our knowledge is quite limited regarding how the individual portions of the genome are organized within the three-dimensional space. Understanding the basic principles of nuclear architecture and how the nuclear organization changes dynamically during cell growth, differentiation, and senescence is critical to unravel genome functionality (Ren 2016). Chromosomes are one of the most complex molecular entities that remain confined within the densely crowded environment inside the cell nucleus. However, they are organized non-randomly and maintain a strict hierarchy of organization at different levels. Several extrinsic and intrinsic constraints are imposed to control the conformation of chromosomes. A compartmentalized nuclear architecture exists within the cell nucleus during interphase, as defined by distinct CTs and an interchromatin compartment (IC), harboring macromolecular complexes involved in DNA replication, transcription, splicing, and DNA repair (Cremer and Cremer 2001; Kamat et al. 2023). The chromatin fiber is intricately folded in three dimensions within the chromosome territories, and several models have been proposed to explain the higher order of chromatin organization within the CTs. However, more compelling experimental evidence is required to get the final, most reliable model for defining the functional nuclear architecture (Rowley and Corces 2018; Dekker et al. 2023). Major improvements in bioimaging and next-generation sequencing (NGS) technologies have led to rapid revolutions in the study of 3D genome organization. The chromosome conformation capture (3C) approach, a very popular technique for studying spatial organization of chromosomes, enables determination of the crosslinking frequency between any pair of loci within the genome that are spatially close enough (in the range of 10–100 nm) at any given point of time (Dekker et al. 2023). The 3C method provides high-resolution data but is limited to targeted regions within the genome. High-throughput chromosome conformation capture (Hi-C) is an advanced version of the 3C technique that captures chromatin interactions across the entire genome. Briefly, the process involves crosslinking DNA to preserve spatial interactions followed by fragmentation and ligation of interacting loci. Subsequent sequencing of the ligated fragments, followed by extensive data analysis, aids in detailed mapping of genome-wide chromatin interactions (Lajoie et al. 2015).

Based on computational analysis of Hi-C data from genomes of multiple model organisms, large, kilobase/megabase-sized local chromatin interaction domains, known as topologically associating domains (TADs), were identified as the basic structural unit of the hierarchically folded 3D genome (Cremer and Cremer 2001; Kamat et al. 2023). These self-interacting genomic regions constrain the spread of heterochromatin and are stable across different cell types and highly conserved across species, suggesting that they are an integral component of any genome. An alternative model for 3D genome architecture suggests the formation of small compartmental domains based on the transcriptional status or chromatin state. The 3D genome architecture is then established because of interactions between these small compartmental domains and the chromatin loops that are formed by the coordinated action of different chromatin architectural proteins, primarily CTCF protein and cohesin motor protein in mammals (Rowley and Corces 2018).

Transposable elements (TEs) comprise a significant portion of the repetitive fraction of every eukaryotic genome (Kazazian 2004; Feschotte and Pritham 2007). They are specialized segments of DNA that can either replicate themselves in different locations or shift their positions within the genome. They can be further classified into two main classes based on their mechanism of transposition (Wicker et al. 2008; Chakrabarty et al. 2023). *Class I transposons or retrotransposons* move via a copy-paste mechanism, using an RNA intermediate that gets reverse-transcribed into DNA to be inserted into a new location within the genome. This class includes both LTR retrotransposons such as endogenous retroviruses and non-LTR retrotransposons such as long interspersed nuclear elements (LINEs), Penelope-like elements, and short interspersed nuclear elements (SINEs). *Class II transposons (DNA transposons)* move via a "cut-and-paste" mechanism, where the transposon is excised from one location and inserted into another (Wicker et al. 2008; Chakrabarty et al. 2023). As TEs comprise a significant portion of any organism's genome, they contribute to genome size alteration and genetic diversity (Chénais et al. 2012; Wells and Feschotte 2020). Although regulated transposition of TEs occurs through specific mechanisms that control their insertion sites, their movements often lead to insertional mutations, deletions, or chromosomal rearrangements, thereby influencing genome structure and function (Chakrabarty et al. 2023). They further play a role in the compact packaging of DNA and have the potential to facilitate regulatory function. TEs have also been reported to impact multiple levels of 3D genomic architecture, leading to altered functionality of the genome. They often contribute binding sites for major nuclear architectural proteins such as CTCF, which in turn help in shaping the genomic architecture by creating loops, domains, borders of the compartments, and RNA-DNA interactions (Zhang et al. 2019; Choudhary et al. 2020, 2023). In fact, a significant portion (15–23%) of CTCF-binding sites that serve as chromatin loop anchors within the mouse and human genome are contributed by TEs (Choudhary et al. 2020). Although TEs are usually silenced by host DNA and histone methylation machineries, these TE-derived CTCF-binding sites are hypomethylated to ensure that they remain functional as loop anchors. Two candidate TEs, namely the L1M3f subfamily of LINEs and the LTR41 subfamily of endogenous-retrovirus-derived LTR, were deleted using CRISPR-Cas9 technology (Choudhary et al. 2020). L1M3f deletion resulted in collapsing of chromatin domains and loop elimination, thereby altering the 3D genome architecture. The homozygous TE knockout cell lines showed a significant reduction of focal enrichment in the Hi-C contact map when compared to the wild type. Furthermore, the cross-domain interactions significantly increased in the knockout cell lines. Deletion of LTR41 on the other hand, resulted in reinstallation of ancestral CTCF-binding sites, leading to a slight shift in the loop anchor but preserving the local chromatin domains. Thus, TEs appear to strengthen the conserved regulatory 3D genome architecture by providing redundant loop anchors (Choudhary et al. 2020). Retrotransposons seem to be the major player in controlling mammalian genome architecture, as the DNA transposons have been rendered inactive during evolution (Feschotte and Pritham 2007; Pace and Feschotte 2007). Exceptionally in bats, *helitrons*, *hAT* and *piggyBac* families of DNA transposons are still found to be active (Feschotte and Pritham 2007). In a recent study investigating the 3D genome architecture of the mature sperm genome from *Xenopus tropicalis*, supersized (multi-megabase) chromatin loops (SSLs) were observed that are enriched with *helitrons* but lack CTCF, RNA polymerase II, and modified histones (Shi et al. 2023). These SSLs seem to form clusters within the mature sperm nucleus, probably via phase separation (Shi et al. 2023). In this chapter, we elaborate and explore the different ways by which transposons contribute toward shaping genome architecture in different eukaryotic lineages.

3.2 Organization of Interphase Chromosomes into Highly Dynamic, Three-Dimensional Structures

The highly variable appearance of the interphase chromosomes during microscopic observation may often suggest a general lack of organization within the nucleus. However, detailed studies using a variety of improved imaging and sequencing techniques have revealed several organizational principles of chromosomes (Kempfer and Pombo 2020; Jerkovic´ and Cavalli 2021). The nucleus seems to be spatially and functionally separated into two major compartments, depending on accessibility of DNA, chromatin modifications, and gene expression levels (Cremer and Cremer 2001). One of the compartments (Compartment A) comprises an open euchromatin region, usually gene rich, is transcriptionally active and enriched in RNA polymerase II and other components of the transcription and RNA processing machinery. The second compartment (Compartment B) usually includes gene-poor, transcriptionally inactive heterochromatinized segments of the genome. However, this compartmentalization is dynamic and varies between cell types. It also alters during development, differentiation, tumorigenesis, or upon exposure to diverse conditions, including viral infection. At the sub-compartmental level, we find TADs exhibiting a higher frequency of physical interactions within domains in comparison to interactions with other neighboring regions of the genome. Unlike compartments, TADs are relatively constant between different cell types and are also found to be conserved across eukaryotic species. However, more recent research shows inter- and intra-species as well as cell-specific variations in TADs. A comprehensive summary of the hypothesized hierarchy of 3D genome architecture of an interphase nucleus is presented in Fig. 3.1.

The intricate mechanisms that control nuclear architecture at the level of whole chromosomes remain poorly understood. There are multiple *cis*-regulatory DNA sequence elements as well as *trans*-regulatory proteins (such as CTCF, cohesin, transcription factors, and heterochromatin binding proteins) and RNAs with a profound impact on 3D genome architecture (Wang et al. 2016; Dsilva and Galande 2024). Additionally, multiple environmental factors like temperature shifts, light, nutrient availability, osmotic stress, and hypoxia have been shown to alter 3D genome architecture by causing relocalization of architectural proteins. This in turn leads to changes in chromatin interactions and gene expression (Pudelko and Cabianca 2024).

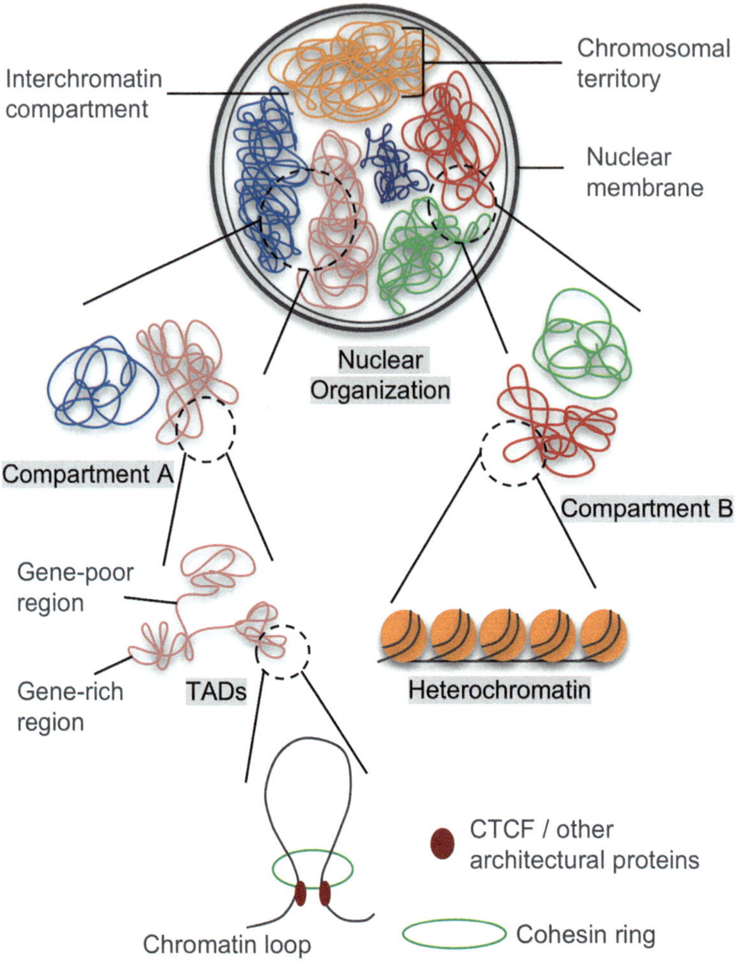

Fig. 3.1 The dynamic three-dimensional genome architecture in interphase. In the nuclear space, the chromosomes are localized into chromosome territories. Further, the chromosomes are compartmentalized into gene-rich, transcriptionally active compartment A and gene-poor, heterochromatinized compartment B. At a finer level, chromosomes are organized into topologically associating domains (TADs) and active extrusion processes form chromatin loops

3.2.1 Cis-Regulatory Elements (CREs) Controlling Genome Architecture

Noncoding DNA constitutes more than 90% of the mammalian genome and includes a variety of sequence classes, such as satellite DNA, long interspersed repeat elements, and shorter DNA sequence motifs. *Cis*-regulatory elements (CREs) are the specific DNA sequence elements, usually noncoding regions of the genome, that can influence the spatial organization of the genome and regulate gene expression (Dsilva and Galande 2024). CREs include *promoters*, which are binding sites for RNA polymerases and other transcription factors and play a critical role in transcription initiation of downstream genes. *Enhancers* are conventionally considered to be CREs that facilitate the transcription process and may be located far away from the promoters. Enhancers interact with promoters via DNA looping, thereby bringing the transcription machinery in close contact. Promoter-enhancer interactions occur frequently within TADs, which are the fundamental units of the 3D genome architecture (Dsilva and Galande 2024). Since enhancers from one chromosome can influence gene expression on another chromosome through chromatin looping or nuclear reorganization, certain enhancers have also been considered to be *trans*-acting elements (D'Aiuto et al. 2006; Blick et al. 2016). *Silencer elements* are antagonistic to enhancers, as they work to repress the transcription of target genes by inhibiting enhancer activity or by blocking the binding of transcription factors. Similarly, *insulator* sequences usually define the boundaries of TADs and block the interaction between promoters and enhancers, thereby inhibiting transcription of target genes. Additionally, *locus control regions (LCRs)* can enhance the expression of linked genes over large distances in a tissue-specific manner. *Boundary elements* prevent the spread of chromatin states and gene regulatory activities from one domain to another, thereby helping in the separation of different regulatory regions within the genome (Dsilva and Galande 2024). As elaborated later in this chapter, transposon-derived sequences can function as CREs such as enhancers, promoters as well as silencers (Guio and González 2019). Transposons also engage in long-range intra-chromosomal interactions guided by the separation of chromosomes into A and B compartments as well as TADs (Raviram et al. 2018). *Repetitive DNA elements,* including transposons, are known to exert profound impacts on the 3D folding of chromosomes during interphase. For example, Long Interspersed Nuclear Elements (LINE-1 or L1) are a type of repetitive DNA that are enriched on the human inactive X chromosome and play a major role in partitioning of the chromosome into two "megadomains" (Bailey et al. 2000; Chow et al. 2010; Bonora et al. 2018). *Short tandem repeat (STR) regions* can help define the boundaries of chromatin domains, ensuring that different regions of the genome are properly compartmentalized (Sun et al. 2018). Alteration in the length of STRs leads to altered transcription status of the target genes via modulating the binding of transcription factors to these (STRs) (Horton et al. 2023). STRs also play a crucial role in the formation, maintenance, and evolution of centromeres. They aggregate into higher-order repeats (HORs) and form the basis of alpha-satellite DNA, which is the primary component of human centromeres (Logsdon et al. 2024). Repetitive DNA sequences along with associated proteins also help in the formation of liquid condensates that cluster the telomeres from many chromosomes, creating structures known as ALT-associated PML bodies (APBs) (Zhang et al. 2020). APBs have been shown to facilitate the alternative lengthening of telomeres (ALT) pathway in cancer (Zhang et al. 2020). Several families of retrotransposons and small RNA repeats have also been reported to provide some of the most important sequence determinants of 3D genome folding (Gunsalus et al. 2023). For example, the telomerase enzyme responsible for maintaining the ends of linear chromosomes in eukaryotes has been evolutionarily linked to Penelope-like elements (PLEs). A subset of PLEs lacking the endonuclease domain is found to be located at telomeres in diverse eukaryotic species. They are considered to be the missing link between ancestral PLEs and modern-day telomerases (Gladyshev and Arkhipova 2007).

3.2.2 Trans-regulatory Factors Controlling Genome Architecture

Phase separation, a process that enables self-organization of chromatin domains, has been implicated to shape genome organization across evolution (Stam et al. 2019; Feric and Misteli 2021; Cai and Wang 2023). Within euchromatin and heterochromatin regions, the chromatin fiber seems to exhibit self-interaction preferentially, thereby leading to organization of the DNA in distinct phases (Misteli 2020). In the phase separation model, chromatin domains are thought to arise due to cooperative interactions between biomolecules harboring multiple interaction domains. Proteins such as RNA polymerase II, multiple transcription factors, and chromatin regulators interact with each other as well as with DNA to form supramolecular assemblies, resulting in chromatin bodies resembling liquid-like droplets. These nuclear bodies can pull interacting chromatin together while pushing away the non-interacting chromatin to control genome organization. Proteins driving phase separation are highly modular with multiple interacting domains, including dimerization domains as well as domains for binding to DNA/RNA/proteins (Feric and Misteli 2021). Moreover, these phase-separating proteins possess intrinsically disordered regions, which add flexibility to enable them to exhibit multivalent interactions and form biomolecular condensates (Cai and Wang 2023). The long chromosomal DNA fiber also contributes to multivalency, particularly through its repetitive DNA sequences (Tang 2017). Multivalent interactions between heterochromatin proteins and modified histones of the nucleosomes and other components tend to form liquid-like droplets, which in turn drive the heterochromatin formation (Stam et al. 2019; Feric and Misteli 2021). Some of the key proteins that have been implicated in driving heterochromatin phase separation are heterochromatin protein 1α (HP1α) and Polycomb protein chromobox 2 (CBX2) (Feric and Misteli 2021). HP1α, a dimeric protein, can interact with trimethylated histone H3 (H3K9me3), associated with compact transcriptionally inactive heterochromatin, and thus form HP1α condensate within the nucleus. CBX2 can form liquid-like droplets using multivalent interactions mediated by its intrinsically disordered regions (Feric and Misteli 2021).

Multiple types of *RNAs, including coding and noncoding RNAs,* can also influence chromatin structure and nuclear organization (Nozawa and Gilbert 2019; Dekker et al. 2023). The noncoding RNA (ncRNA), as for example, XIST, induces silencing and condensation of the X chromosome (Valledor et al. 2023); the repeat-containing ncRNA TERRA promotes telomere condensation (Cusanelli and

Chartrand 2015), and the ncRNA MALAT helps in nuclear speckle formation (Ghosh et al. 2023). High-resolution maps of chromatin-associated RNAs (caRNA) and their associated genomic sequences were determined in genome-edited human cells, wherein specific TAD boundaries were removed (Calandrelli et al. 2021). RNA-DNA association maps were compared before and after deleting a TAD boundary. The results showed that a TAD boundary can insulate both crossover DNA-DNA and crossover RNA-DNA contacts, thereby linking genome architecture to the spatial distribution of RNA in the nucleus (Calandrelli et al. 2021). caRNAs associated with genomic sequences located between anchors of a chromatin loop suppressed this loop's strength, thereby suggesting a role for caRNAs in modulating 3D genome organization (Calandrelli et al. 2021; Dekker et al. 2023).

Cohesin and condensin complexes are well-characterized molecular machines that fold chromatin through ATP-dependent loop extrusion process. *The cohesin complex* is primarily involved in sister chromatid cohesion, forming a ring-like structure encircling sister chromatids. It ensures that the sister chromatids stay together until they are separated during cell division. In addition to its role in sister chromatid cohesion, it functions in gene regulation by contributing toward the formation of TADs and chromatin loops (Rowley and Corces 2018). Depletion of cohesin leads to elimination of chromatin loops (Wutz et al. 2017). Depletion of cohesin-releasing factor WAPL leads to an increase in the amount as well as retention time of cohesin on chromatin. This results in extended chromatin loops and global compaction of the genome (Wutz et al. 2017; Rowley and Corces 2018).

The condensin complex, an important player in chromosome condensation and segregation during mitosis and meiosis, forms ring-like structures in order to compact chromatin into highly condensed mitotic chromosomes. Condensin also extrudes loops of DNA, a process driven by ATP hydrolysis. The process of loop extrusion and its contribution to chromosome folding, nuclear organization, and gene expression are being extensively studied in multiple mammalian systems (Rowley and Corces 2018). It helps in organizing the genome during cell division to ensure that chromosomes are properly aligned and segregated. During evolution, the interphase nuclear architecture underwent switching between two major types across eukaryotic lineages, depending on the presence of subunit(s) of the condensin II complex. The two types are mainly Type 1 or Rab1-like nuclear architecture, comprising features like centromeric clustering, telomeric clustering, and a telomere-to-centromere axis while Type 2 or territorial nuclear architecture comprising of chromosome territories only. Condensin II is shown to be crucial for forming chromatin loops and establishing chromosome territories prior to interphase as part of the Type 2 nuclear architecture.

CCCTC-binding factor *CTCF* plays an important role in setting up the 3D genome architecture, as it helps in anchoring chromatin loops and insulating TAD boundaries along with cohesin and other insulator proteins (Wu et al. 2020). Studies from multiple eukaryotic species have illustrated the importance of CTCF in 3D genome remodeling, with differential CTCF binding leading to altered 3D genome folding. Specifically, CTCF binding seems to have evolved with a selective constraint to preserve existing domain boundaries and loop anchors, along with a stronger drive to evolve novel local interactions like intra-domain chromatin loops to enable phenotypic evolution (Wu et al. 2020). Like CTCF, other architectural proteins such as CP190, BEAF-32, and Chromator in *Drosophila* have also been reported to bind to the TAD borders to mediate the insulating activity (Stam et al. 2019).

3.2.3 Environmental Stimuli Influencing 3D Genome Architecture

Environmental factors can have a profound influence on TADs and genome organization, affecting gene expression and chromatin dynamics. Although individual environmental cues have been shown to elicit unique, often species- or cell-type-specific responses, nonetheless there are similarities in the way the chromatin architecture gets altered in response to diverse environmental stimuli (Pudelko and Cabianca 2024). *Chromatin compaction* is altered by multiple environmental stimuli such as nutrient availability, cold stress, and osmotic stress in different organisms, including various human cell types (Pudelko and Cabianca 2024). For example, the germline chromatin in *C. elegans* larvae gets compacted during starvation in order to repress global gene transcription (Belew et al. 2021). The energy-sensing AMP-activated kinase initiates the process, while condensin II and topoisomerase II are major players involved in this genome compaction process (Belew et al. 2021). Cold stress leads to an increase in intracellular Ca^{2+} levels in mammalian cells, which in turn contributes to chromatin compaction (Martin et al. 2007) Ca^{2+} ions can effectively neutralize the electrostatic repulsion between the negatively charged sugar-phosphate backbone of the DNA strands, thereby facilitating compaction (Martin et al. 2007). *Chromatin accessibility* plays a crucial role in gene regulation by controlling how easily transcription factors and other regulatory proteins can bind to DNA. Heat shock factor 1a plays a crucial role in plant heat stress responses by influencing the chromatin accessibility, particularly in establishing transient enhancer–promoter interactions to drive the expression of heat-responsive genes in tomato (Huang et al. 2023). Hypoxia (Batie et al. 2022) and HIV-1 infection in humans have also been reported to alter chromatin accessibility (Shah et al. 2022; Pudelko and Cabianca 2024). *Chromatin looping* is influenced by dietary, heat stress, cold stress, and hyperosmotic stress and also during various viral and bacterial exposures in different human cell types (Pudelko and Cabianca 2024). For example, during microbial infection in human neutrophils, the cohesin complex is rapidly loaded on the genome. De novo chromatin loops are formed to reposition the enhancer into a single loop domain to facilitate enhancer–promoter communication. This aids in inflammatory gene response (Denholtz et al. 2020). *TAD* sizes and numbers are regulated by heat stress (Liang et al. 2021), hyperosmotic stress (Amat et al. 2019) and viral infection (Wang et al. 2023). *Compartmentalization within the nucleus (A and B compartments)* has also been

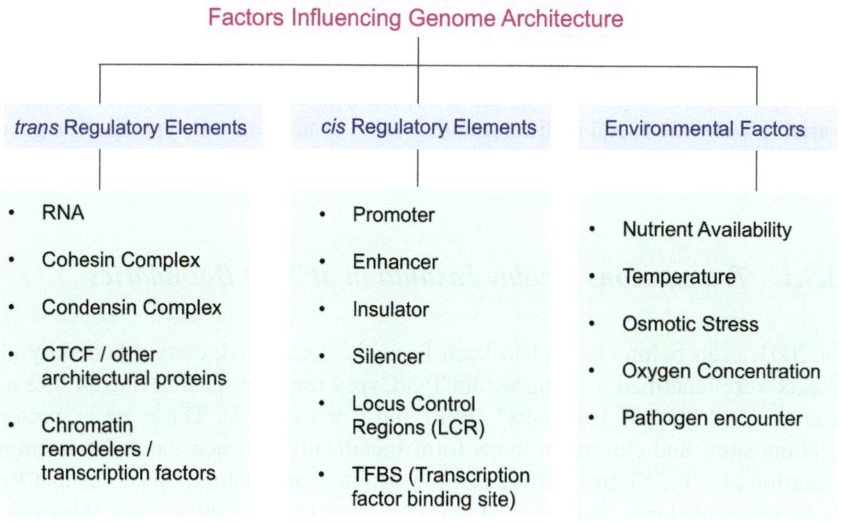

Fig. 3.2 Factors affecting genome architecture. Various *cis*-regulatory elements, often donated by repeat elements including transposons, influence genome architecture and associated gene expression. Additionally, various *trans*-regulatory proteins affect the genome architecture, which often bind sequences donated by transposons. Finally, environmental stimuli have been shown to be important in controlling the dynamics of genome architecture

reported to be altered by heat stress (Liang et al. 2021; Huang et al. 2023), cold stress (Zhang et al. 2023), hyperosmotic stress (Amat et al. 2019), and pathogen encounters (Wang et al. 2023; Pudelko and Cabianca 2024). Figure 3.2 summarizes all the aforementioned factors that influence genome architecture.

With this brief overview on 3D genome architecture, we now delve into discussing specific aspects of TE function in regulation of genome architecture in further details.

3.3 Transposons Help in Defining Topologically Associating Domains

Extensive research over the past decade has revealed how transposons are indispensable in causing optimum genome folding to produce TADs. This ensures correct chromatin interactions and regulates gene expression. To recall, TADs, discovered in 2012, occur as organizational, approximately megabase-sized, conserved yet variable chromosomal regions. They occur within chromosomal compartments. They have a high frequency of self-interaction and are contained within insulating boundaries. Such boundaries usually contain CCCTC sequences, housekeeping genes, SINE elements, and binding sites for other architectural proteins like ZNF143 and SMC3 that separate these domains from the adjacent chromatin (Dixon et al. 2012; Choudhary et al. 2020, 2023). Besides cohesin motors, there may be other loop-extruding factors that form the dynamic chromatin loops within TADs (Fudenberg et al. 2016). TADs are classified into two types—the constitutive TADs are repeat-containing stable domains typically found in heterochromatin, while variable TADs are found in differentiating cells, containing cell-type-specific genes, and transposons can be implicated in both.

3.3.1 Transposons Enable Insulation at TAD Boundaries

In 2003, even before TADs had been formally discovered, certain boundary elements were identified residing within Ty3/Gypsy retrotransposons in fruit flies and were termed "Gypsy insulators" (Byrd and Corces 2003). These act as protein-binding sites, and chromatin loops form specifically between them. Interestingly, insertion of a Ty3/Gypsy retrotransposon within a chromatin loop causes that loop to be subdivided into two smaller loops (Byrd and Corces 2003). Thus, it is evident that transposons donate "insulator" sequences for chromatin loop formation. Insulators are known to contain epigenetic and expression-related TAD features within the TADs, avoiding spillovers and unwanted gene expression (Ichiyanagi et al. 2021). The most common insulators in TADs are CCCTC sequences, which bind CTCF proteins. Alternatively, a mutation at the CTCF-binding site (CBS), preventing CTCF binding, can cause loss of a chromatin loop, ultimately leading to dysregulation of gene expression (Lawson et al. 2023).

A recapitulation of the canonical *CBSs* may be important to understand the origin of insulation demarcating TADs from the surrounding genetic matter in mammals. The human genome has almost 10,000 loops. Each loop is approximately 200,000 base pairs long. When studied at the kilobase resolution, it was found that there are convergently oriented CBSs at the bases of the loops in most cases, thereby making CTCF the predominant factor contributing toward loop formation (Rao et al. 2014). Topology of the chromatin loops within TADs gets altered by changing the orientation of the CBSs (Guo et al. 2015; Nanni et al. 2020). It is established now that CBSs expand throughout the eukaryotic genome via repeat elements. Studies on six mammals, including mice, rats, opossums, dogs, macaques, and humans, showed that retrotransposons are in fact responsible for the distribution of CBSs across genomes in a conserved yet species-specific manner (Schmidt et al. 2012). However, this distribution seems to have been evolving for several million of years, resulting in evolutionarily favorable changes in gene regulation. Only those CBSs that are inserted near TADs govern genome architecture (Kentepozidou et al. 2020). Both conservation and evolution-driven changes balance the clusters of CBSs near TADs (Kentepozidou et al. 2020). In mice, for instance, it was found that SINE-derived CBSs are abundant while LINE-derivations are under-represented (Kentepozidou et al. 2020). CTCF proteins bind to these cognate binding sites throughout the genome forming insulated TADs, facilitating short- and long-range promoter-enhancer interactions, thereby linking 3D genome architecture to gene expression (Ong and Corces 2014). Along with cohesin and other transcription factors, CTCF proteins thus either seal off the boundaries of individual TADs and anchor them to the nuclear matrix within the higher-order chromosomal territories or may even function as transcriptional activator or repressor depending on the position of CBSs.

Interestingly, CTCF may not be the only factor controlling genome architecture, as retrotransposons lacking CTCF-binding motifs have been predicted to contribute toward genome folding by causing abrupt shifts in the GC content of the genome upon insertion (Gunsalus et al. 2023). *Drosophila* is thought to serve as a good model to understand CTCF-independent spatial genome organization (Hug and Vaquerizas 2018). In the human genome, CTCF-independent mammalian-wide interspersed repeats (MIRs), a class of tRNA-derived SINEs, were seen to act as chromatin barriers for highly conserved boundaries separating active and inactive chromatin zones in T-cells (Wang et al. 2015). These sites use histone modifiers and RNA polymerase instead of CTCF proteins (Wang et al. 2015). CTCF-independent boundaries are generally transcriptionally highly active, and its significance in the context of development is discussed later in this chapter.

Interestingly, in the fungal phytopathogen *Verticillium* sp., TADs found in certain genomic regions called "adaptive genomic regions or AGRs" are not insulated properly (Torres et al. 2024). These AGRs are rich in transposons, which are active and also contain genes that mediate infection in host plants. These AGRs also exhibit long-range interactions that seem to cause spatial segregation of AGRs. Thus, the 3D genome organization of these AGRs is distinct and differs from the rest of the core genome, which in turn have impacts on transcriptional regulation and evolution (Torres et al. 2024).

3.3.2 Widespread Consequences of Transposon-Mediated Regulation at TADs

Specific three-dimensional interactions within chromosomes, between chromosomes, and among chromosomes and nuclear structural components are indispensable for regulation of gene expression. TEs have been implicated to play a decisive role in creating such optimal intra-chromosomal and inter-chromosomal interactions (Lawson et al. 2023). Transcripts from activated TEs define boundary and regulatory positions for TADs. They also act as "loop anchors" to stabilize the formed domains against the nuclear periphery (Choudhary et al. 2023). This ensures optimum regulation-related interactions (short- as well as long-range) while maintaining the spatial constraints within or among individual sub-TADs or TADs (Lawson et al. 2023). Transposon-derived ***cis**-regulatory elements* control TADs in multiple ways, as summarized in Fig. 3.3. As discussed in a previous section, regulatory elements like *promoters* and *enhancers* can interact regardless of the distance between the two elements in the linear two-dimensional DNA, owing to the

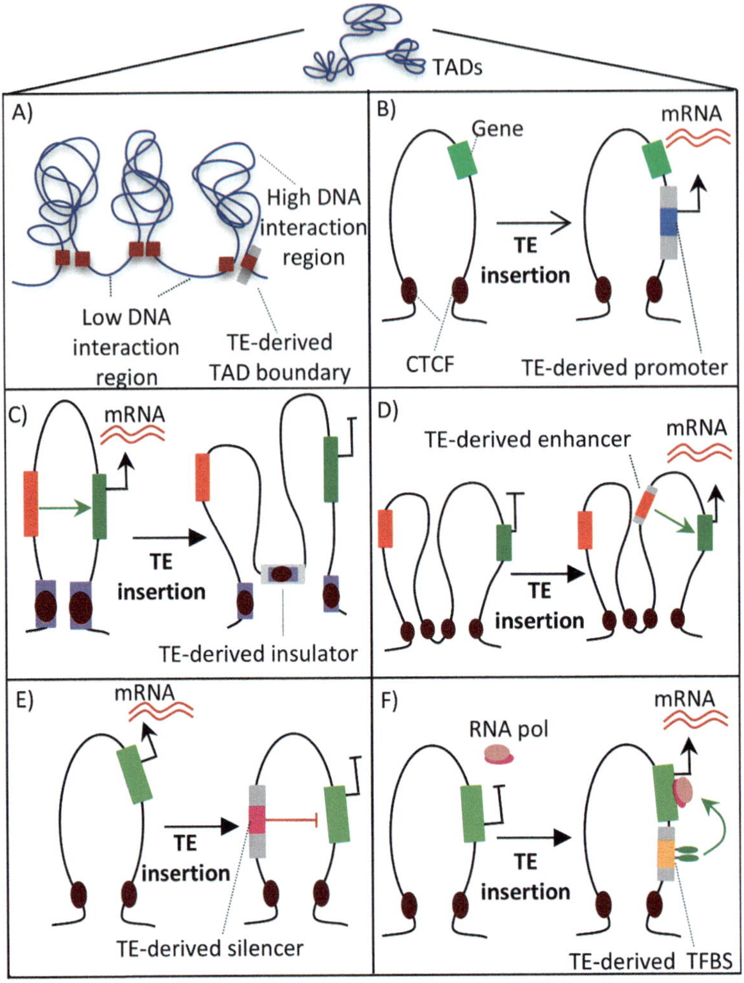

Fig. 3.3 Transposon-derived cis-regulatory elements influence genome architecture. (**a**) TEs such as ERVs and TIR elements provide boundary sequences that define topologically associating domains (TADs). DNA interactions are more frequent within TADs compared to outside TADs. (**b**) TEs (gray box) such as MERVs, HERVs or MIRs donate promoter-like elements (blue box with black arrow) that can initiate gene expression (green box) within TADs. (**c**) TEs (gray box), such as Ty3/Gypsy and Alu elements, provide insulator sequences (brown circles in purple box), which redefine TADs and act as boundary elements. They control interactions between the enhancer (red box) and gene promoter (green box) to regulate gene expression by insulating one regulatory element from the effect of another. (**d**) TEs (gray box), such as MERV, HERV, Alu, and MER elements, donate enhancer sequences (red box) that can upregulate gene expression within TADs. (**e**) Certain Young L1 elements introduce silencer sequences (pink boxes) that can downregulate gene expression within TADs. (**f**) TE-derived transcription factor binding sites (TFBS; yellow box) can bind their cognate transcription factors (TFs; green oval structures) and thereby promote gene expression within TADs. See text for further details

modified genome architecture. This interaction is necessary for facilitating gene expression. Transposons can themselves contribute such promoter-enhancer sites. For example, some Alu elements possess enhancer functions (Su et al. 2014). MIR and LINE-2-derived elements show promoter-like and enhancer-like long-range activity (Cao et al. 2019). A recent study on *Drosophila pseudoobscura* discussed the presence of two kinds of enhancers—housekeeping enhancers for housekeeping genes and specific enhancers for stage- or tissue-specific genes. Housekeeping enhancers often co-localize with the housekeeping genes, unlike the specific enhancers, which seem to frequently co-localize with specific chromatin loop anchors and borders of TADs (Ali et al. 2024). Some of these specific enhancers are contributed by the TEs harboring extensive epigenetic modifications (Ali et al. 2024). This study further reported that on autosomes, evolutionary recent TEs have been generally silenced, but ancient TEs serve as specific enhancers or borders of TADs. However, even the young TEs present in X chromosomes are found to be active and serving as TAD borders, bringing distant chromosomal regions closer on the X chromosome owing to recently acquired epigenetic changes for dosage compensation (Ali et al. 2024).

Transposons also function as *loop anchors*. Deletions of specific transposons-anchoring loops cause the structural loop to fall apart (Choudhary et al. 2020, 2023). For example, in GM12878 B lymphocyte cells, the L1MC1 element anchors a promoter-enhancer loop in humans, deletion of which causes loop collapse and reduced gene expression from that TAD (Choudhary et al. 2023). TE-derived *insulators* control gene repression by bringing about conformational change in the chromatin loops (Comet et al. 2011) and help in localizing chromosomes into distinct nuclear compartments (Pirrotta and Li 2012). TEs also provide regions of scaffolding or matrix attachment, termed "*scaffold/matrix-attachment regions (S/MARs)*"—that enable the formation of chromatin loops and TADs, regulating gene expression and genome evolution (Pathak et al. 2023).

Approximately 20% of *transcription factor binding sites* (TFBSs) in active chromatin of humans are donated by TEs, particularly LTRs, for binding proteins such as octamer-binding transcription factor 4 (Oct4) and tumor suppressor protein P53 (Sundaram et al. 2014; Cao et al. 2019). Specific TEs contribute TFBS for specific transcription factors (Sundaram et al. 2014). Finally, LINE-1 elements have *silencer* functions that can operate across long distances, regulating several gene promoters (Sundaram and Wysocka 2020). TADs occur as the sum of all these CREs provided by various evolutionarily selected TEs. Table 3.1 provides a comprehensive list of *CREs* that have been derived from different families of transposons.

3.3.3 Role of TEs in the Evolution of TADs

Ancient transposition activities are observed throughout the genome, but speciation is usually associated with species-specific evolution of 3D chromatin structure. Establishment of these species-specific TAD structures is made possible by new

Table 3.1 List of transposon families and subfamilies that contribute *cis*-regulatory elements (CREs) and regulate 3D-genome architecture

Sl. No.	Details of TE	CRE Type	Host species	Reference
a. LTR retrotransposons				
1.	Ty3/Gypsy retrotransposons	TAD insulator	*Drosophila* sp.	Byrd and Corces (2003)
2.	Ty3/Gypsy Dictyostelium intermediate repeat sequence (DIRS) Large retrotransposon derivative (LARD) Copia elements	TAD insulator	*Gossypium rotundifolium, Gossypium arboreum* and *G. raimondii*	Wang et al. (2021)
3.	MERV-L	TAD boundaries	*Mus musculus*	Sundaram and Wysocka (2020)
4.	HERV-H	TAD boundaries	*Homo sapiens,* other hominids	Zhang et al. (2019)
5.	ERV-K family elements	Human-specific TAD boundaries	*Homo sapiens*	Okhovat et al. (2023)
6.	LTR13	Loop anchor	*Mus musculus*	Choudhary et al. (2023)
7.	LTR10	p53-binding site	Primates	Wang et al. (2007)
8.	(a) MERVL (b) HERV-K and HERV-H	Alternative promoters, enhancers, and boundary elements (context-dependent)	(a) *Mus musculus* (b) *Homo sapiens*	Sundaram and Wysocka (2020)
b. Non-LTR SINE retrotransposons				
9.	Alu/B1, B2 elements	TAD insulator	*Mus musculus*	Dixon et al. (2012)
10.	Alu elements	TAD insulator	*Homo sapiens*	Dixon et al. (2012)
11.	Mammalian-wide interspersed repeats (MIRs)	TAD insulator	*Homo sapiens*	Wang et al. (2015)
12.	Alu elements	Enhancer	*Homo sapiens*	Su et al. (2014)
13.	MIR, MIRb, MIRc, MIR3, and MIR1_Amn; L2b and L2c	Enhancer-like repeats and promoter-like repeats	Primates	Cao et al. (2019)
c. Non-LTR LINE retrotransposons				
14.	L1MC1	Loop anchor	*Homo sapiens*	Choudhary et al. (2023)
15.	L1MC3, L1MB8 and L1MA7	Matrix-attachment region (MAR)	*Homo sapiens*	Rollini et al. (1999)

(continued)

Table 3.1 (continued)

Sl. No.	Details of TE	CRE Type	Host species	Reference
16.	L3	BHLHE40 TF-binding site	*Homo sapiens* and *Mus musculus*	Sundaram et al. (2014)
17.	Young L1 elements	Silencer	*Homo sapiens*	Sundaram and Wysocka (2020)
d. DNA transposons				
18.	Terminal inverted repeat (TIR)	TAD boundary	*Gossypium rotundifolium, G. arboreum* and *G. raimondii*	Wang et al. (2021)
19.	MER elements	Enhancers	*Homo sapiens*	Raviram et al. (2018)
20.	MER57A1	Loop anchor	*Mus musculus*	Choudhary et al. (2023)
21.	MER91B	CTCF-binding site	*Homo sapiens* and *Mus musculus*	Sundaram et al. (2014)
22.	MER74A	OCT4-binding site	*Homo sapiens*	Kunarso et al. (2010)
23.	MER61	p53-binding site	Primates	Wang et al. (2007)

transposition activities, which are simply the evolutionarily favorable ones. For instance, in genomes of *Drosophila*, mice, and humans, it was hypothesized that repetitive SINEs localized in distant regions of the genome or upstream of orthologous genes showed higher frequencies of contact owing to their sequence homology (Cournac et al. 2016). On one hand, these SINEs help in maintaining the structure of the 3D genome conserved across homologous species (Cournac et al. 2016). On the other hand, these elements also help in fine-tuning the nuclear architecture through evolution (Cournac et al. 2016). Likewise, a study suggests that new species, generated by more recent transposition events, have specific transposon insertions at the TAD boundaries (Kruse et al. 2019).

The initial thought that TAD boundaries are highly conserved across species and within an individual's cell types is thus now refutable. For instance, more conserved TAD boundaries exhibit stronger insulation and possess more CBSs overlapping with ancient MIRs in comparison to species-specific boundaries. On the other hand, human-specific or recent boundaries were found to be rich in LTR elements belonging to the evolutionarily recent ERV-K family (Okhovat et al. 2023). Further, not all transposons have equal contributions to lineage-specific or cell-type-specific 3D genome architecture. Specific subfamilies like B2/B3 SINEs in mice, SINEC_Cf in dogs, and *hAT-Charlie* in humans seem to be actively controlling chromatin organization in these organisms (Choudhary et al. 2023). Thus, species-specific chromatin loops can be formed by retrotransposon insertions (Lawson et al. 2023). Studies on plant species also exhibit relationships between TAD formation, TE transcriptions, lineage-specific chromatin architecture development, and gene expression (Wang et al. 2021; Yu et al. 2024; Xu et al. 2024; He et al. 2024). Some examples of such species-specific TE contributions have also been mentioned in Table 3.1.

It is now established that human-specific TE-dependent additions or deletions in TAD-associated regulatory sequences have created a human-specific 3D genome regulatory network (Glinsky 2018). Among humans, there may be instances of polymorphic retrotransposons, i.e., TE variants, which may bring about differences in chromatin accessibility across humans owing to differential integration (Ferraj et al. 2023). Thus, insertions at the intraspecific level can also bring about phenotypic polymorphism and even cause differences between cell types of any single individual, bringing in the aspect of cell-type-specific TAD boundaries and interactions (Phillips-Cremins et al. 2013; Lawson et al. 2023).

Optimum and favorable activity of transposons ensures a feasible number and positions of chromatin loops and boundaries of TADs. However, the entire system of architecture falls prey to diseases via disrupted insertions or sequence mutations of these TADs. Such diseases include limb and digit malformations and even cancers (Lupiáñez et al. 2015). On a related note, in case of diseases caused by expansions in short tandem repeats (STRs, repeat elements which are different from TE), it has already been shown that the repeats are usually located at the boundaries of TADs, leading to topological changes and altered gene activities (Sun et al. 2018). It has recently been hypothesized that aberrant LINE-1 transpositions, among other mechanisms, may also cause chromatin structural variations, bringing oncogenic promoters and enhancers close, leading to tumorigenesis (Lee et al. 2024). Hypomethylated Alu elements in cancer genomes bring about changes in chromosomal clustering in colon cancer (Jordà et al. 2017).

R-loop, a structure made of a DNA-RNA hybrid and a non-template, single-stranded DNA, closely interacts with 5-hydroxymethylcytosine (5hmC) to cause epigenetic modulation of DNA, particularly affecting gene expression at enhancers. TAR DNA-binding protein 43 (TDP-43) plays a vital role in maintaining the R-loop-5hmC interaction. However, TDP-43 depletion associated with different neurological disorders dysregulates such crosstalk, facilitates TE transcription, and prevents optimum promoter-enhancer interaction (Hou et al. 2024). The potential and frequency of association of transposons with disease-associated genome topology changes must be further investigated to gain valuable insights regarding the role of transposon-mediated altered genome architecture in development and progression of various diseases.

3.4 Transposable Elements Control Heterochromatin Architecture

Heterochromatin is integral in shaping the three-dimensional genome. Among other repetitive elements, LTR and non-LTR TEs often remain compacted within the late-replicating heterochromatin in the centromeric, pericentromeric, and telomeric regions of chromosomes with their expressions being epigenetically regulated (Haws et al.

2022). Fluorescence in situ hybridization studies conducted in 1995 demonstrated for the first time in *Drosophila* that nine TEs are localized in heterochromatic regions in a highly conserved manner (Pimpinelli et al. 1995). In 2010, the role of TE in mammalian heterochromatin formation was elucidated (Zamudio and Bourc'His 2010).

3.4.1 Transposons Help in the Establishment of the Structure of Heterochromatin

Hi-C studies have revealed that the genome is organized into A and B compartments with distinct chromatin environments. Heterochromatin is usually associated with the B chromosomal compartment. It has been established how TEs such as LINE-1 and B1 SINE in mice and Alu elements in humans cluster together and aid in compartmentalization (Lu et al. 2021). TEs also play a decisive role in heterochromatin establishment and compartmentalization via phase separation (Strom et al. 2017).

Initially, TEs are recognized by the transcription factors KRAB-containing zinc finger proteins (KRAB-ZFPs) and their cofactor KAP1. These factors recruit proteins and enzymes needed for epigenetic modification, specifically, methylation of histones (Trono 2015). The di- or tri-methylation of the ninth lysine residue of histone 3 (H3K9me2/3) is a repressive epigenetic modification. These repressive marks bind heterochromatin protein 1 (HP1), one of the previously mentioned *trans*-regulatory factors controlling genome architecture. HP1-bound to H3K9me3 allows cohesin-mediated chromatin loop formation but prevents CTCF binding (Spracklin et al. 2023; Lawson et al. 2023). Proper chromatin interaction along with optimum molecular crowding results in compaction and formation of constitutive heterochromatin (Oh et al. 2018; Gualdrini et al. 2022). Further, methylated H3K9 in different nucleosomes are "bridged" by the HP1 that binds them, bringing about phase separation in interphase nuclei and stabilizing the spatial structure in the B compartment (Singh and Newman 2020). Depletion of DNA methyltransferase activity was reported to alter compartmental structure and increase CTCF binding, validating the significance of TEs in the process (Spracklin et al. 2023; Lawson et al. 2023). Further, TAD cliques, which are series of non-overlapping TADs, are thought to stabilize heterochromatin near the nuclear lamina (Paulsen et al. 2019)). Finally, heterochromatin preserves the correct three-dimensional chromatin architecture, particularly at the centromere and telomere, and thereby ensures normal regulation of gene expression (Oh et al. 2018).

3.4.2 Significance of Correct 3D Heterochromatin in Regulating Genome Function

During embryogenesis and development, global epigenetic changes occur in the heterochromatin. This causes widespread expression of TE, which helps in generating the three-dimensional genome (Rang et al. 2023). Once established, the final three-dimensional structure of heterochromatin in mice has been shown to prevent unwanted chromosomal translocations or recombination events (Mitrentsi et al. 2022). This barrier function is important because repetitive sequences in heterochromatin often act as obstacles to the recombination process. Further, heterochromatin within genome architecture prevents deleterious repeat expansions. Kmt1e/Setdb1 are enzymes that epigenetically silence retrotransposons at heterochromatin. When these enzymes were depleted, it led to unwanted ERV2 activation at an ERV2-enriched heterochromatin site in mouse neurons, causing stress response and neuroinflammation (Chandrasekaran et al. 2021).

Mismanaged heterochromatin has been associated with aging, senescence, progeria-like symptoms, and cancers. Cell senescence occurs in multicellular organisms and is important in physiological functions, including development, repair, aging, and warding off cancer. During senescence, various pro-proliferation genes get suppressed by senescence-associated repressive heterochromatinization (Narita et al. 2003). In senescence and cancer, Alu, SVA, and LINE-1 elements, as well as centromeric and pericentromeric heterochromatin, have been shown to become more active. Retro-transposition events rise late in senescence, bringing about changes in heterochromatin (De Cecco et al. 2013). During senescence, facultative heterochromatin bears the H3K27me3 mark and switches from B to A compartment. However, constitutive heterochromatin bears the H3K9me3 mark and exhibits greater self-interactions within their original compartments, greater chromatin accessibility, and creation of new chromatin loops. Expression changes in LTRs are suggested to be associated with such heterochromatin remodeling and gene activation (Zhang et al. 2021). In Hutchinson-Gilford progeria, the protein progerin causes loss of heterochromatin in G1-arrested cells in addition to DNA damage. Removal of progerin has been shown to revive normal heterochromatin (Chojnowski et al. 2020). LINE-1 RNA accumulates in different human progeroid syndromes, leading to reduction of the enzymatic function of a histone-lysine N-methyltransferase, triggering heterochromatin loss and progeroid symptoms in vitro (Della Valle et al. 2022). The reversal of epigenetic heterochromatin repression and downregulation of senescence-associated genes is possible by depleting the LINE-1 RNA levels in fibroblasts from patients (Della Valle et al. 2022). With prolonged RNA depletion, it was also possible to increase lifespans of progeroid mice and revive their normal histology (Della Valle et al. 2022).

3.5 Role of Transposons in Centromere Establishment, Maintenance, and Stabilization

The centromere is a specialized chromosomal region that is required for proper chromosomal segregation during cell division. It acts as an attachment site for kinetochore formation, a protein complex that binds chromosomes to spindle fibers, allowing duplicated chromosomes to be pulled apart during mitosis and meiosis

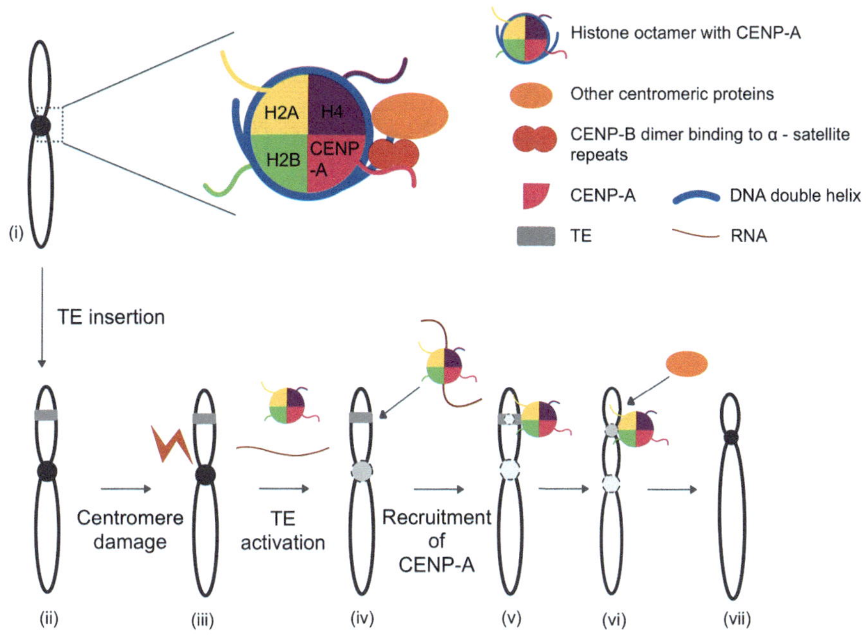

Fig. 3.4 TE-mediated neocentromere formation. (i) A chromosome with a native centromere (black circle) is shown. Centromeric proteins such as CENP-A (pink), CENP-B (red), and other centromeric proteins (orange) localize to this site, ensuring proper kinetochore assembly and chromosome segregation during cell division, (ii) A transposable element (TE) insertion (gray rectangle) occurs in one arm of the chromosome, (iii) The native centromere is damaged, leading to activation of TE transcription, (iv) Transcripts (brown wavy lines) from the TE contribute to the recruitment of CENP-A at a new, alternate location within the same chromosome, (v) CENP-A deposition facilitates the recruitment of other centromeric proteins, (vi) These centromeric proteins attach at the new site, forming a functional centromere, and (vii) A neocentromere is thus established at an alternate site, compensating for the loss of the original centromere

(Lawrimore and Bloom 2022). Thus, a properly functional centromere is needed for the stable inheritance of chromosomes across cycles of cell division. Disarrangement and interference in centromeres can lead to chromosome loss, breakage, or structural damage. In most eukaryotes, centromeres are made up of repetitive DNA sequences that consist of satellite DNA and TEs. TEs play a crucial role in the maintenance of centromere structure, recruitment of centromere-associated proteins, and regulating centromere function as depicted in Figs. 3.4 and 3.5.

Centromeres and their surrounding pericentromere remain in a heterochromatinized state. While the core centromere region harboring microtubular attachment sites primarily consists of satellite DNA, the pericentromeric region mainly consists of TEs, LINEs, and SINEs (Hartley and O'Neill 2019). Human centromeres consist of alpha satellites, an AT-rich 171-bp DNA monomer arranged in tandem, giving rise to HORs. The number and order of repeats are often chromosome-specific (Sullivan et al. 2017). However, species-specific satellite DNA repeats have been

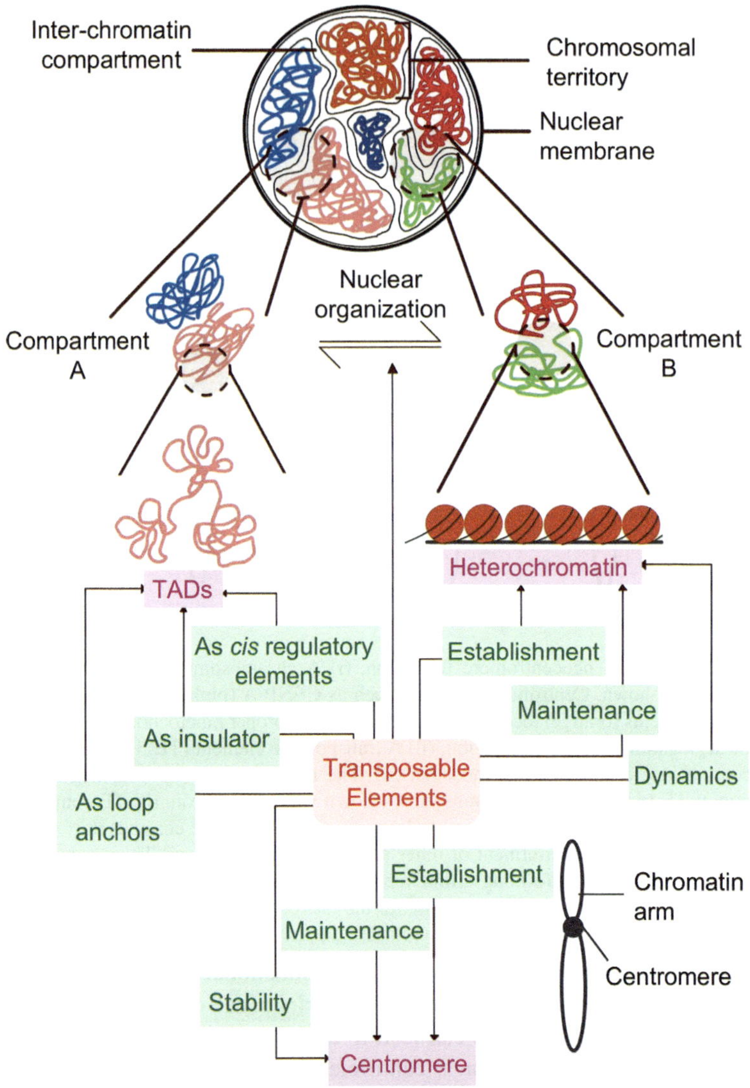

Fig. 3.5 Role of Transposons in modulation of the dynamics of genome architecture. Transposons regulate the changes in chromosome compartmentalization and associated gene expression. Transposons organize topologically associating domains (TADs) by providing insulator or boundary sequences and by acting as loop anchors, and further regulate gene expression from TADs by donating different cis-regulatory elements. Transposons also help in the establishment, maintenance, and dynamics of centromeres and heterochromatin by providing specific sites for binding by trans-regulatory factors

observed as well (Cheng et al. 2002; Klein and O'Neill 2018). It has now been hypothesized that the development of such satellite arrays originated as transposon insertions, with relevant evidence from multiple species, including *Arabidopsis* and *Drosophila* (Hartley and O'Neill 2019). The inserted TE subsequently acquired different mutations and, along with recombinational events, resulted in the formation of the centromeric satellite repeats (Hartley and O'Neill 2019). Within human chromosomal HORs, TE insertions are often found to be present. However, these TEs remain epigenetically silenced to maintain healthy centromere function (Hartley and O'Neill 2019).

Transcription from centromeric DNA is very important, as transcripts from centromeric DNA bind different proteins that form and maintain centromere structure and function (Hartley and O'Neill 2019). One of the most critical proteins is centromere protein-A (CENP-A), which binds RNA as part of a complex. CENP-A, a variant of histone protein H3, is considered a marker of centromere identity (Klein and O'Neill 2018). Transposons contribute specific DNA motifs that favor the recruitment of CENP-A and other centromere-specific proteins (Klein and O'Neill 2018). In fact, CENP-A and other centromeric proteins recognize transposon-rich DNA sequences as potential centromeric sites (Klein and O'Neill 2018). In a recent study from *Arabidopsis*, it was shown that insertion of *ATHILA* retrotransposons in the centromeric region silences transcription from centromeric repeats but produces small RNAs that promote chromosome cohesion and proper segregation during mitosis. Hence, the retrotransposon makes centromeres dependent on it for their proper functioning (Shimada et al. 2024).

Another centromeric protein, CENP-B, shares a high degree of protein sequence identity with *Tigger* elements belonging to the *Tc1/Mariner* family of DNA transposons. It binds a motif (CENP-B box) on the satellite DNA that resembles the terminal inverted repeats (TIR) targeted by Tigger transposase during transposition (Klein and O'Neill 2018). Due to its similarity to a transposase, CENP-B has been hypothesized to facilitate homologous recombination within the satellite DNA arrays by promoting endonucleolytic cleavage (Klein and O'Neill 2018). CENP-B-binding motifs serve as binding sites that aid in positioning CENP-A at centromeres, which, in turn, helps establish the centromere as a distinct chromatin domain (Klein and O'Neill 2018). Finally, CENP-B-mediated DNA looping causes compaction of chromosomes via the centromeres, preserving centromere positioning and DNA integrity during anaphase (Chardon et al. 2022).

Neocentromeres are new centromeres formed on novel chromosome domains following an accidental loss or damage to the original centromere. This in turn might play a role in preventing the loss of acentric chromosomes. Following any damage to the native centromere, transcription occurs from otherwise silenced transposons such as LINE/L1 in humans. This enables loading of CENP-A nucleosomes and formation of a neocentromere elsewhere on the chromosome (Chueh et al. 2005, 2009) (Fig. 3.4). Such transposons are probably newer insertions that are transcriptionally active, as they have not yet been suppressed by the host genome maintenance pathways (Chueh et al. 2005, 2009; Hartley and O'Neill 2019). At neocentromeres, CENP-A acts to cause kinetochore assembly in the absence of the satellite repeats, thus saving the acentric chromosomal fragment from getting lost during chromosome segregation (De Rop et al. 2012). Thus, TE-mediated formation of neocentromeres may change the position of the centromere, thereby leading to evolution of chromosome structure (Marshall et al. 2008).

3.6 Transposable Elements Regulate Genome Architecture During Embryogenesis, Development, and Cellular Differentiation

3.6.1 Role in the Establishment of the Genome Architecture During Early Embryogenesis

As an embryo develops, the establishment and optimum functioning of the 3D genome architecture depend on multiple factors, including replication, transcription, cell cycle length, arrangement of housekeeping genes, and epigenetic changes (Hug et al. 2017; Ke et al. 2017; Hug and Vaquerizas 2018). In addition, TEs also seem to regulate embryonic stem cells through embryogenesis, potency changes, and differentiation (He et al. 2021).

After fertilization, the initial development of the zygote often relies on maternal transcripts until subsequent transcriptional activation of the zygotic genome. Until then, the genome has no significant loops or TADs or short/long-range interactions and is quite flexible, unbiased, random, and unpredictable with regard to its 3D architecture (Hug and Vaquerizas 2018). RNA polymerase is globally recruited and chromatin accessibility increases during zygotic genome activation at the 14th nuclear division (with a minor wave during the eighth nuclear division) in case of *Drosophila* and during the late two-celled stage in mice. Increased chromatin accessibility by RNA polymerase and other transcription factors in turn allows TAD insulator proteins like CTCF and loop-extruding factors like cohesin to bind the chromatin and initiate the formation of TADs (Hug et al. 2017; Hug and Vaquerizas 2018). LINE-1 elements in mice have shown significant involvement in the context of chromatin accessibility for 3D genome packaging. During early mouse development, early suppression of LINE-1 reduces chromatin accessibility. On the other hand, if it is activated for longer than intended, chromatin compaction is increased, negatively impacting the development rate owing to changes in chromatin structure (Jachowicz et al. 2017). LINE-1 transcripts also act as a scaffold for proteins, ultimately helping the mouse embryo to exit the two-celled stage (Percharde et al. 2018). Different TEs have been reported to become operative at different developmental stages in a step-wise fashion. It has been suggested that both evolutionary pressure and the pressure from the host genome affect the development-related transcription from different TEs (Oomen and Torres-Padilla 2024).

3.6.2 Transposons Help in the Establishment of TAD Boundaries

TAD boundaries, once established, are essentially maintained to ensure proper development. Deletion of specific TAD boundaries causes very specific developmental abnormalities, while deletions near some boundaries, like in *Smad3/Smad6*, even cause embryonic lethality (Rajderkar et al. 2023). For CTCF-dependent TAD boundaries, CTCF is indispensable during early embryogenesis. In the preimplantation stage, maternal CTCF mRNA provides the protein, and its absence can cause abnormalities in the blastocyst development (Moore et al. 2012). CTCF deletion in mouse embryonic stem cells (ESCs) resulted in developmental abnormalities and death at embryonic days 4.5–5.5 (Moore et al. 2012). CTCF binding has been related to the methylation status of the DNA. Thus, enormous levels of epigenetic erasure and reprinting that occur during initial zygotic development might affect CTCF binding and DNA topology (Wang et al. 2012). Furthermore, owing to extensive epigenetic erasure during embryogenesis, a lot of retrotransposons become activated. Downregulation of these transposons is enabled by di- or trimethylation of H3K9 and is critical for ensuring proper development. Several methyltransferases have been implicated for this methylation and retrotransposon suppression. *Set-domain-bifurcated 1* (*Setdb1/Eset/Kmt1e*) plays a crucial role in maintaining TAD structures in the neurons (Jiang et al. 2017). They are crucial in development by controlling CTCF binding and unwanted epigenetic patterns (Jiang et al. 2017). At the growth hormone gene locus of mouse embryos, the B2 SINE element acts as an insulator element, keeping the repression marker H3K9me3 histone restricted from invading the gene locus, thereby allowing expression of the growth hormone gene. This in turn facilitates development of the pituitary gland (Lunyak et al. 2007).

CTCF-independent TAD boundaries are generally transcriptionally highly active, and RNAP may function like CTCF to pause cohesin function at these boundaries (Bonev et al. 2017). In *Drosophila*, transcription and chromatin architecture development seem to occur simultaneously, while transcription, including of several retrotransposons, usually precedes structure establishment in mice (Hug et al. 2017; Ke et al. 2017). In fact, transcription enables formation of genome architecture in its wake (Hug et al. 2017; Bonev et al. 2017). Some TAD boundaries lacking either CTCF or RNAP are thought to depend on other factors (Bonev et al. 2017). Within the nucleus, dynamic higher-order compartments are created during development on the basis of transcriptional activity and optimum contacts between chromatin (Hug and Vaquerizas 2018).

3.6.3 Transposons Help in the Establishment of Chromosomal Compartments

LINE-1 poses an interesting example of a retrotransposon transcript auto-regulating its own DNA sequence folding and compartmentalization, as seen in mice and humans (Lu et al. 2021). DNA sequences with LINE-1 repeats and B1/Alu repeats tend to autonomously self-cluster in the nucleus during embryogenesis in an evolutionarily conserved manner. They form distinct nuclear domains, which correspond respectively to the B and A compartments of chromatin. Transcripts of LINE-1 repeats facilitate optimum homotypic repeat contacts and chromatin compartmentalization, particularly for heterochromatin, in a *cis*-acting manner (Lu et al. 2021). LINE-1 transcripts bind to DNA in ESCs extensively, without which the heterochromatin dislocates from the nuclear periphery and gets derepressed (Lu et al. 2020). It has been suggested that LINE-1 transcripts aid in heterochromatin formation by promoting the recruitment of chromatin crosslinker protein HP1α. HP1α stabilizes DNA binding of heterochromatin-associated proteins, or it helps in anchoring LINE-1-rich chromatin to nuclear lamina (Lu et al. 2021). The repeats are thought to first initiate A and B compartmentalization, with B1-Alu-rich sequences and LINE-1-rich sequences undergoing homotypic contacts to cause autonomous genome folding (Lu et al. 2021). This is followed by LINE-1-transcript-mediated phase separation within the so-formed compartments into euchromatin and heterochromatin (Lu et al. 2021). Finally, the chromatin regions are stabilized by anchoring to the nuclear periphery (Lu et al. 2021).

It can also be noted here that, considering the role of histone proteins in higher-order chromatin structure, a recent study has shown how six different human histone 1 (H1) protein variants are differentially localized into the A and B compartments (Salinas-Pena et al. 2024). Certain H1 protein variants are associated with specific TE. For example, H1X is found to be localized more in the A compartment in recently transposed SVA elements, wherein it helps in TE repression (Salinas-Pena et al. 2024).

3.6.4 Transposons Have Been Implicated in the Establishment of Cell Fate and Cellular Differentiation

Owing to their impact in controlling genome architecture, TE families like ERV and LINE-1 have been implicated in regulating cellular potency and fate determination from a mature oocyte through a blastocyst (Peaston et al. 2004; Macfarlan et al. 2012). In mice, MERV-L, which occurs at TAD boundaries, produces transcripts that suppress genes of totipotency, making the cells pluripotent (Macfarlan et al. 2012). In humans, human endogenous retrovirus subfamily H (HERV-H) has been reported to be crucial for pluripotency maintenance and prevention of differentiation (Santoni et al. 2012).

A study characterizing four neural cell types from mid-gestational human cortex found that cell type-specific super-promoters contain several ERVL-MaLR ERVs (Song et al. 2020). Targeted inhibition of such retrotransposons resulted in reduced expression from the corresponding super-promoters (Song et al. 2020). In a recent study on neurogenesis in mice, epigenetic regulation of SINEs has been demonstrated to be essential for proper brain development (Sun et al. 2024). The study elucidated how DNA and histone methylation of specific SINEs associated with neural precursors prevent SINE transcription. It was further shown that loss of methylation results in increased accessibility of these SINEs, generation of new chromatin loops, and altered mitotic gene activation (Sun et al. 2024).

The genome architecture undergoes changes during cellular differentiation. During cardiomyocyte differentiation, the HERV-H subfamily of retrotransposons became transcriptionally active and was seen to donate TAD boundaries in pluripotent stem cells (PSCs) in humans and other primates (Zhang et al. 2019). They also act as enhancers, leading to optimum transcription necessary for differentiation (Zhang et al. 2019). It was suggested that, similar to early development, high rates of transcription by RNAP II determine TAD boundaries during differentiation as well via regulating cohesin positioning (Zhang et al. 2019). Some of the constituent TADs are reported to be gained or lost during adipose stem cell differentiation. All these structural changes in the 3D genome are associated with the regulatory expression of TEs during differentiation (Paulsen et al. 2019; Zhang et al. 2019; Lawson et al. 2023).

3.7 Conclusions

Throughout this chapter, the significance of TEs in organizing the genome architecture has been discussed. Figure 3.5 shows a summary of these multi-faceted roles of TEs in regulating 3D genome organization. The concept of a three-dimensional genome architecture is still quite vague and remains an underexplored domain of research. However, it is quite evident that TE effectively contribute in every step of the way toward establishing and maintaining genome architecture. At the very beginning of life, transposon activities during fertilization, embryogenesis, cellular potency, differentiation, until birth, in the context of shaping and modifying the 3D genome are exciting avenues of further research. TE, despite their established characteristics, have also proved to be functional for genome architecture and corresponding gene expression throughout life and during senescence. Disruption of their normal activities has been implicated in several abnormalities. With regard to the mechanisms of establishment of chromatin loops, direct evidence for the extrusion process in the context of CTCF loops in vivo is lacking in mammalian cells. In vitro experiments with naked DNA and working with similar Structural Maintenance of Chromosomes (SMC) protein complexes in bacterial systems have provided some valuable insights. For example, in *Staphylococcus aureus*, SMC complexes load onto the chromosome and translocate away from their loading positions, thereby generating DNA loops to establish long-range DNA contacts. This in turn promotes chromatin compaction and simultaneously facilitates chromosome segregation (Izquierdo-Martinez et al. 2025). Further work with mammalian cohesin on actual chromatin will be an important step forward in accepting the chromatin extrusion model for 3D genome architecture in mammalian cells.

It is undebatable that transposons define TADs, ensuring normal chromosomal spatial arrangements, gene interaction, and expression, and so transposons are not just selfish or useless genetic components. Their favorable expansion has been evolutionarily selected for specific purposes. However, despite advancements in Hi-C technology and computational skills, the entire narrative is still obscure. First, what regulates expansions and locations of genomic transposons through development and evolution? What is the degree of uniformity or functionally significant variety of transposition in the context of chromatin structure? How are transposons relevant in operating the switch between expression and repression at TADs depending on environmental factors? How can this knowledge aid the treatment of diseases? Further exploration might provide valuable insights in addressing these questions.

TEs indeed are indispensable for generating heterochromatin and dividing the nucleus into chromosome compartments. However, many questions in this field remain to be addressed. What is the detailed model and precise role of extensively distributed TEs in heterochromatin compartmentalization? How do the epigenetics of TEs at three-dimensionally organized heterochromatin and their related functions alter across cell types or species? How can this knowledge of TE regulating heterochromatin homeostasis and expression be translated into the therapeutics of aging-related disorders?

Transposons are present across all domains of life, including bacteria, archaea, and eukaryotes, and are expected to have similar impacts on the genome architecture in prokaryotes as well. Although most of our initial efforts have been directed toward constructing the three-dimensional structures of chromosomes of eukaryotic species based on 3C/Hi-C data, these powerful tools are now being applied successfully for analyzing several prokaryote genomes as well (Liu et al. 2024). Preliminary 3D-genomic analysis on five bacterial model species (*Caulobacter crescentus*, *Escherichia coli*, *Bacillus subtilis*, *Vibrio cholerae,* and *Mycoplasma pneumoniae*) has provided some knowledge about the spatial organization pattern of bacterial chromosomes and revealed the important roles of nucleoid-associated proteins in regulation of genome architecture. TAD-like structures in bacteria are called chromosomal interaction domains (CIDs), and a transcription factor Rok is considered a functional analog of CTCF in prokaryotes (Dugar et al. 2022). However, compared with eukaryotes, the research on the three-dimensional genome of prokaryotes is still insufficient. Understanding the impact of TEs on the bacterial three-dimensional genomes from a systematic perspective will be the research focus in the future.

Acknowledgments We are highly indebted to all the researchers whose pathbreaking research work has been compiled in this book chapter. Their relentless pursuit of knowledge has greatly enriched our understanding about how transposons may contribute toward dynamic three-dimensional genome architecture. This chapter would not have been possible without their invaluable contributions. We also apologize in advance to those researchers whose work could not be mentioned here due to space constraints. We do hold their contributions in the highest regard.

Funding The authors received no financial support for the research, authorship, and/or publication of this chapter.

Compliance with Ethical Standards

Conflict of Interest The authors declare that they have no competing interests as defined by Springer or other interests that might be perceived to influence the discussion reported in this chapter.

Ethical Approval This chapter is a review of previously published accounts; as such, no animal or human studies were performed.

References

Ali M, Younas L, Liu J et al (2024) Development and evolution of Drosophila chromatin landscape in a 3D genome context. Nat Commun 15:9452. https://doi.org/10.1038/s41467-024-53892-0

Amat R, Böttcher R, Le Dily F et al (2019) Rapid reversible changes in compartments and local chromatin organization revealed by hyperosmotic shock. Genome Res 29. https://doi.org/10.1101/gr.238527.118

Bailey JA, Carrel L, Chakravarti A, Eichler EE (2000) Molecular evidence for a relationship between LINE-1 elements and X chromosome inactivation: the Lyon repeat hypothesis. Proc Natl Acad Sci U S A 97. https://doi.org/10.1073/pnas.97.12.6634

Batie M, Frost J, Shakir D, Rocha S (2022) Regulation of chromatin accessibility by hypoxia and HIF. Biochem J 479. https://doi.org/10.1042/BCJ20220008

Belew MD, Chien E, Wong M, Michael WM (2021) A global chromatin compaction pathway that represses germline gene expression during starvation. J Cell Biol 220. https://doi.org/10.1083/jcb.202009197

Blick AJ, Mayer-Hirshfeld I, Malibiran BR et al (2016) The capacity to act in trans varies among drosophila enhancers. Genetics 203. https://doi.org/10.1534/genetics.115.185645

Bonev B, Mendelson Cohen N, Szabo Q et al (2017) Multiscale 3D genome rewiring during mouse neural development. Cell 171:557–572.e24. https://doi.org/10.1016/j.cell.2017.09.043

Bonora G, Deng X, Fang H et al (2018) Orientation-dependent Dxz4 contacts shape the 3D structure of the inactive X chromosome. Nat Commun 9. https://doi.org/10.1038/s41467-018-03694-y

Byrd K, Corces VG (2003) Visualization of chromatin domains created by the gypsy insulator of Drosophila. J Cell Biol 162:565–574. https://doi.org/10.1083/jcb.200305013

Cai L, Wang GG (2023) Through the lens of phase separation: intrinsically unstructured protein and chromatin looping. Nucleus 14

Calandrelli R, Wen X, Charles Richard JL et al (2021) Three-dimensional organization of chromatin associated RNAs and their role in chromatin architecture in human cells. bioRxiv

Cao Y, Chen G, Wu G et al (2019) Widespread roles of enhancer-like transposable elements in cell identity and long-range genomic interactions. Genome Res 29:40–52. https://doi.org/10.1101/gr.235747.118

Chakrabarty P, Sen R, Sengupta S (2023) From parasites to partners: exploring the intricacies of host-transposon dynamics and coevolution. Funct Integr Genomics 23:278

Chandrasekaran S, Espeso-Gil S, Loh YHE et al (2021) Neuron-specific chromosomal megadomain organization is adaptive to recent retrotransposon expansions. Nat Commun 12. https://doi.org/10.1038/s41467-021-26862-z

Chardon F, Japaridze A, Witt H et al (2022) CENP-B-mediated DNA loops regulate activity and stability of human centromeres. Mol Cell 82(9):1751–1767.e8. https://doi.org/10.1016/j.molcel.2022.02.032

Chénais B, Caruso A, Hiard S, Casse N (2012) The impact of transposable elements on eukaryotic genomes: from genome size increase to genetic adaptation to stressful environments. Gene 509:7

Cheng Z, Dong F, Langdon T et al (2002) Functional rice centromeres are marked by a satellite repeat and a centromere-specific retrotransposon. Plant Cell 14(8):1691–1704. https://doi.org/10.1105/tpc.003079
Chojnowski A, Ong PF, Foo MXR et al (2020) Heterochromatin loss as a determinant of progerin-induced DNA damage in Hutchinson–Gilford Progeria. Aging Cell 19. https://doi.org/10.1111/acel.13108
Choudhary MNK, Friedman RZ, Wang JT et al (2020) Co-opted transposons help perpetuate conserved higher-order chromosomal structures. Genome Biol 21. https://doi.org/10.1186/s13059-019-1916-8
Choudhary MNK, Quaid K, Xing X et al (2023) Widespread contribution of transposable elements to the rewiring of mammalian 3D genomes. Nat Commun 14. https://doi.org/10.1038/s41467-023-36364-9
Chow JC, Ciaudo C, Fazzari MJ et al (2010) LINE-1 activity in facultative heterochromatin formation during X chromosome inactivation. Cell 141. https://doi.org/10.1016/j.cell.2010.04.042
Chueh AC, Wong LH, Wong N, Choo KHA (2005) Variable and hierarchical size distribution of L1-retroelement-enriched CENP-A clusters within a functional human neocentromere. Hum Mol Genet 14. https://doi.org/10.1093/hmg/ddi008
Chueh AC, Northrop EL, Brettingham-Moore KH et al (2009) LINE retrotransposon RNA is an essential structural and functional epigenetic component of a core neocentromeric chromatin. PLoS Genet 5. https://doi.org/10.1371/journal.pgen.1000354
Comet I, Schuettengruber B, Sexton T, Cavalli G (2011) A chromatin insulator driving three-dimensional Polycomb response element (PRE) contacts and Polycomb association with the chromatin fiber. Proc Natl Acad Sci U S A 108:2294–2299. https://doi.org/10.1073/pnas.1002059108
Cournac A, Koszul R, Mozziconacci J (2016) The 3D folding of metazoan genomes correlates with the association of similar repetitive elements. Nucleic Acids Res 44:245–255. https://doi.org/10.1093/nar/gkv1292
Cremer T, Cremer C (2001) Chromosome territories, nuclear architecture and gene regulation in mammalian cells. Nat Rev Genet 2:292
Cusanelli E, Chartrand P (2015) Telomeric repeat-containing RNA TERRA: a noncoding RNA connecting telomere biology to genome integrity. Front Genet 6. https://doi.org/10.3389/fgene.2015.00143
D'Aiuto L, De Marco R, Edward N et al (2006) Evidence of the capability of the CMV enhancer to activate in trans gene expression in mammalian cells. DNA Cell Biol 25. https://doi.org/10.1089/dna.2006.25.171
De Cecco M, Criscione SW, Peckham EJ et al (2013) Genomes of replicatively senescent cells undergo global epigenetic changes leading to gene silencing and activation of transposable elements. Aging Cell 12:247–256. https://doi.org/10.1111/acel.12047
De Rop V, Padeganeh A, Maddox PS (2012) CENP-A: the key player behind centromere identity, propagation, and kinetochore assembly. Chromosome 121(6):527–538. https://doi.org/10.1007/s00412-012-0386-5
Dekker J, Alber F, Aufmkolk S et al (2023) Spatial and temporal organization of the genome: current state and future aims of the 4D nucleome project. Mol Cell 83:2624
Della Valle F, Reddy P, Yamamoto M et al (2022) LINE-1 RNA causes heterochromatin erosion and is a target for amelioration of senescent phenotypes in progeroid syndromes. Sci Transl Med 14(657):eabl6057. https://doi.org/10.1126/scitranslmed.abl6057
Denholtz M, Zhu Y, He Z et al (2020) Upon microbial challenge, human neutrophils undergo rapid changes in nuclear architecture and chromatin folding to orchestrate an immediate inflammatory gene program. Genes Dev 34. https://doi.org/10.1101/gad.333708.119
Dixon JR, Selvaraj S, Yue F et al (2012) Topological domains in mammalian genomes identified by analysis of chromatin interactions. Nature 485:376–380. https://doi.org/10.1038/nature11082
Dsilva GJ, Galande S (2024) From sequence to consequence: deciphering the complex cis-regulatory landscape. J Biosci 49

Dugar G, Hofmann A, Heermann DW, Hamoen LW (2022) A chromosomal loop anchor mediates bacterial genome organization. Nat Genet 54(2):194–201. https://doi.org/10.1038/s41588-021-00988-8
Feric M, Misteli T (2021) Phase separation in genome organization across evolution. Trends Cell Biol 31:671
Ferraj A, Audano PA, Balachandran P et al (2023) Resolution of structural variation in diverse mouse genomes reveals chromatin remodeling due to transposable elements. Cell Genom 3. https://doi.org/10.1016/j.xgen.2023.100291
Feschotte C, Pritham EJ (2007) DNA transposons and the evolution of eukaryotic genomes. Annu Rev Genet 41:331
Fudenberg G, Imakaev M, Lu C et al (2016) Formation of chromosomal domains by loop extrusion. Cell Rep 15:2038–2049. https://doi.org/10.1016/j.celrep.2016.04.085
Ghosh A, Pandey SP, Joshi DC et al (2023) Identification of G-quadruplex structures in MALAT1 lncRNA that interact with nucleolin and nucleophosmin. Nucleic Acids Res 51. https://doi.org/10.1093/nar/gkad639
Gladyshev EA, Arkhipova IR (2007) Telomere-associated endonuclease-deficient Penelope-like retroelements in diverse eukaryotes. Proc Natl Acad Sci U S A 104. https://doi.org/10.1073/pnas.0702741104
Glinsky GV (2018) Contribution of transposable elements and distal enhancers to evolution of human-specific features of interphase chromatin architecture in embryonic stem cells. Chromosome Res 26(1–2):61–84. https://doi.org/10.1007/s10577-018-9571-6. Erratum in: Chromosome Res. 26(1–2):85–92. doi: 10.1007/s10577-018-9574-3
Gualdrini F, Polletti S, Simonatto M et al (2022) H3K9 trimethylation in active chromatin restricts the usage of functional CTCF sites in SINE B2 repeats. Genes Dev 36:414–432. https://doi.org/10.1101/gad.349282.121
Guio L, González J (2019) New insights on the evolution of genome content: population dynamics of transposable elements in flies and humans. Methods Mol Biol 1910:505-530
Gunsalus LM, Keiser MJ, Pollard KS (2023) In silico discovery of repetitive elements as key sequence determinants of 3D genome folding. Cell Genom 3. https://doi.org/10.1016/j.xgen.2023.100410
Guo Y, Xu Q, Canzio D et al (2015) CRISPR inversion of CTCF sites alters genome topology and enhancer/promoter function. Cell 162:900–910. https://doi.org/10.1016/j.cell.2015.07.038
Hartley G, O'Neill RJ (2019) Centromere repeats: Hidden gems of the genome. Genes 10(3). https://doi.org/10.3390/genes10030223
Haws SA, Simandi Z, Barnett RJ, Phillips-Cremins JE (2022) 3D genome, on repeat: higher-order folding principles of the heterochromatinized repetitive genome. Cell 185:2690–2707
He J, Babarinde IA, Sun L et al (2021) Identifying transposable element expression dynamics and heterogeneity during development at the single-cell level with a processing pipeline scTE. Nat Commun 12. https://doi.org/10.1038/s41467-021-21808-x
He X, Dias Lopes C, Pereyra-Bistrain LI et al (2024) Genetic–epigenetic interplay in the determination of plant 3D genome organization. Nucleic Acids Res. https://doi.org/10.1093/nar/gkae690
Horton CA, Alexandari AM, Hayes MGB et al (2023) Short tandem repeats bind transcription factors to tune eukaryotic gene expression. Science:381. https://doi.org/10.1126/science.add1250
Hou Y, Li Y, Xiang JF et al (2024) TDP-43 chronic deficiency leads to dysregulation of transposable elements and gene expression by affecting R-loop and 5hmC crosstalk. Cell Rep 43. https://doi.org/10.1016/j.celrep.2023.113662
Huang Y, An J, Sircar S et al (2023) HSFA1a modulates plant heat stress responses and alters the 3D chromatin organization of enhancer-promoter interactions. Nat Commun 14. https://doi.org/10.1038/s41467-023-36227-3
Hug CB, Vaquerizas JM (2018) The birth of the 3D genome during early embryonic development. Trends Genet 34:903–914

Hug CB, Grimaldi AG, Kruse K, Vaquerizas JM (2017) Chromatin architecture emerges during zygotic genome activation independent of transcription. Cell 169:216–228.e19. https://doi.org/10.1016/j.cell.2017.03.024
Ichiyanagi T, Katoh H, Mori Y et al (2021) B2 SINE copies serve as a transposable boundary of DNA methylation and histone modifications in the mouse. Mol Biol Evol 38:2380–2395. https://doi.org/10.1093/molbev/msab033
Izquierdo-Martinez A, Schäper S, Brito AD et al (2025) Chromosome segregation dynamics during the cell cycle of *Staphylococcus aureus*. bioRxiv 19:2025.02.18.638847
Jachowicz JW, Bing X, Pontabry J et al (2017) LINE-1 activation after fertilization regulates global chromatin accessibility in the early mouse embryo. Nat Genet 49:1502–1510. https://doi.org/10.1038/ng.3945
Jerkovic´ I, Cavalli G (2021) Understanding 3D genome organization by multidisciplinary methods. Nat Rev Mol Cell Biol 22:511
Jiang Y, Loh YHE, Rajarajan P et al (2017) The methyltransferase SETDB1 regulates a large neuron-specific topological chromatin domain. Nat Genet 49:1239–1250. https://doi.org/10.1038/ng.3906
Jordà M, Díez-Villanueva A, Mallona I et al (2017) The epigenetic landscape of Alu repeats delineates the structural and functional genomic architecture of colon cancer cells. Genome Res 27:118–132. https://doi.org/10.1101/gr.207522.116
Kamat K, Lao Z, Qi Y et al (2023) Compartmentalization with nuclear landmarks yields random, yet precise, genome organization. Biophys J 122. https://doi.org/10.1016/j.bpj.2023.03.003
Kazazian HH (2004) Mobile elements: drivers of genome evolution. Science 303:1626–1632. https://doi.org/10.1126/science.1089670
Ke Y, Xu Y, Chen X et al (2017) 3D chromatin structures of mature gametes and structural reprogramming during mammalian embryogenesis. Cell 170:367–381.e20. https://doi.org/10.1016/j.cell.2017.06.029
Kempfer R, Pombo A (2020) Methods for mapping 3D chromosome architecture. Nat Rev Genet 21:207
Kentepozidou E, Aitken SJ, Feig C et al (2020) Clustered CTCF binding is an evolutionary mechanism to maintain topologically associating domains. Genome Biol 21. https://doi.org/10.1186/s13059-019-1894-x
Klein SJ, O'Neill RJ (2018) Transposable elements: genome innovation, chromosome diversity, and centromere conflict. Chromosome Res 26(1–2). https://doi.org/10.1007/s10577-017-9569-5
Kruse K, Díaz N, Enriquez-Gasca R et al (2019) Transposable elements drive reorganisation of 3D chromatin during early embryogenesis. https://doi.org/10.1101/523712
Kunarso G, Chia NY, Jeyakani J et al (2010) Transposable elements have rewired the core regulatory network of human embryonic stem cells. Nat Genet 42(7):631–634. https://doi.org/10.1038/ng.600
Lajoie BR, Dekker J, Kaplan N (2015) The hitchhiker's guide to hi-C analysis: practical guidelines. Methods 72. https://doi.org/10.1016/j.ymeth.2014.10.031
Lawrimore J, Bloom K (2022) Shaping centromeres to resist mitotic spindle forces. J Cell Sci 135. https://doi.org/10.1242/jcs.259532
Lawson HA, Liang Y, Wang T (2023) Transposable elements in mammalian chromatin organization. Nat Rev Genet 24:712–723
Lee M, Ahmad SF, Xu J (2024) Regulation and function of transposable elements in cancer genomes. Cell Mol Life Sci 81:157
Li D, He M, Tang Q et al (2022) Comparative 3D genome architecture in vertebrates. BMC Biol 20. https://doi.org/10.1186/s12915-022-01301-7
Liang Z, Zhang Q, Ji C et al (2021) Reorganization of the 3D chromatin architecture of rice genomes during heat stress. BMC Biol 19. https://doi.org/10.1186/s12915-021-00996-4
Liu T, Qiu QT, Hua KJ, Ma BG (2024) Chromosome structure modeling tools and their evaluation in bacteria. Brief Bioinform 25

Logsdon GA, Rozanski AN, Ryabov F et al (2024) The variation and evolution of complete human centromeres. Nature 629. https://doi.org/10.1038/s41586-024-07278-3
Lu JY, Shao W, Chang L et al (2020) Genomic repeats categorize genes with distinct functions for orchestrated regulation. Cell Rep 30:3296–3311.e5. https://doi.org/10.1016/j.celrep.2020.02.048
Lu JY, Chang L, Li T et al (2021) Homotypic clustering of L1 and B1/Alu repeats compartmentalizes the 3D genome. Cell Res 31:613–630. https://doi.org/10.1038/s41422-020-00466-6
Lunyak V, Prefontaine G, Nunez E, Cramer T et al (2007) Developmentally regulated activation of a SINE B2 repeat as a domain boundary in organogenesis. Science 317:248–251. https://doi.org/10.1126/science.1140637
Lupiáñez DG, Kraft K, Heinrich V et al (2015) Disruptions of topological chromatin domains cause pathogenic rewiring of gene-enhancer interactions. Cell 161:1012–1025. https://doi.org/10.1016/j.cell.2015.04.004
Macfarlan TS, Gifford WD, Driscoll S et al (2012) Embryonic stem cell potency fluctuates with endogenous retrovirus activity. Nature 487:57–63. https://doi.org/10.1038/nature11244
Marshall OJ, Chueh AC, Wong LH, Choo KHA (2008) Neocentromeres: new insights into centromere structure, disease development, and karyotype evolution. Am J Hum Genet 82(2):261–282. https://doi.org/10.1016/j.ajhg.2007.11.009
Martin RM, Görisch SM, Leonhardt H, Cardoso MC (2007) An unexpected link between energy metabolism, calcium, chromatin condensation and cell cycle. Cell Cycle 6
Misteli T (2020) The self-organizing genome: principles of genome architecture and function. Cell 183
Mitrentsi I, Lou J, Kerjouan A et al (2022) Heterochromatic repeat clustering imposes a physical barrier on homologous recombination to prevent chromosomal translocations. Mol Cell 82:2132–2147.e6. https://doi.org/10.1016/j.molcel.2022.03.033
Moore JM, Rabaia NA, Smith LE et al (2012) Loss of maternal CTCF is associated with peri-implantation lethality of CtCf null embryos. PLoS One 7. https://doi.org/10.1371/journal.pone.0034915
Nanni L, Ceri S, Logie C (2020) Spatial patterns of CTCF sites define the anatomy of TADs and their boundaries. Genome Biol 21. https://doi.org/10.1186/s13059-020-02108-x
Narita M, Nunez S, Heard E et al (2003) Rb-Mediated heterochromatin formation and silencing of E2F target genes during cellular senescence. Cell 113(6):703–716
Nozawa RS, Gilbert N (2019) RNA: nuclear glue for folding the genome. Trends Cell Biol 29. https://doi.org/10.1016/j.tcb.2018.12.003
Oh I, Choi S, Jung Y, Kim JS (2018) Entropic effect of macromolecular crowding enhances binding between nucleosome clutches in heterochromatin, but not in euchromatin. Sci Rep 8. https://doi.org/10.1038/s41598-018-23753-0
Okhovat M, VanCampen J, Nevonen KA et al (2023) TAD evolutionary and functional characterization reveals diversity in mammalian TAD boundary properties and function. Nat Commun 14. https://doi.org/10.1038/s41467-023-43841-8
Ong CT, Corces VG (2014) CTCF: an architectural protein bridging genome topology and function. Nat Rev Genet 15:234–246
Oomen ME, Torres-Padilla ME (2024) Jump-starting life: balancing transposable element co-option and genome integrity in the developing mammalian embryo. EMBO Rep 25:1721–1733
Pace JK, Feschotte C (2007) The evolutionary history of human DNA transposons: evidence for intense activity in the primate lineage. Genome Res 17. https://doi.org/10.1101/gr.5826307
Pathak RU, Phanindhar K, Mishra RK (2023) Transposable elements as scaffold/matrix attachment regions: shaping organization and functions in genomes. Front Mol Biosci 10. https://doi.org/10.3389/fmolb.2023.1326933
Paulsen J, Liyakat Ali TM, Nekrasov M et al (2019) Long-range interactions between topologically associating domains shape the four-dimensional genome during differentiation. Nat Genet 51:835–843. https://doi.org/10.1038/s41588-019-0392-0

Anne E. Peaston, Alexei V. Evsikov, Joel H. Graber, et al (2004) Retrotransposons regulate host genes in mouse oocytes and preimplantation embryos
Percharde M, Lin CJ, Yin Y et al (2018) A LINE1-Nucleolin partnership regulates early development and ESC identity. Cell 174:391–405.e19. https://doi.org/10.1016/j.cell.2018.05.043
Phillips-Cremins JE, Sauria MEG, Sanyal A et al (2013) Architectural protein subclasses shape 3D organization of genomes during lineage commitment. Cell 153:1281–1295. https://doi.org/10.1016/j.cell.2013.04.053
Pimpinelli S, Berloco M, Fanti L (1995) TE in heterochromatin of drosophila. Proc Natl Acad Sci U S A 92:3804–3808
Pirrotta V, Li HB (2012) A view of nuclear Polycomb bodies. Curr Opin Genet Dev 22:101–109
Pudelko L, Cabianca DS (2024) The influencers' era: how the environment shapes chromatin in 3D. Curr Opin Genet Dev 85:102173
Rajderkar S, Barozzi I, Zhu Y et al (2023) Topologically associating domain boundaries are required for normal genome function. Commun Biol 6. https://doi.org/10.1038/s42003-023-04819-w
Rang FJ, Kind J, Guerreiro I (2023) The role of heterochromatin in 3D genome organization during preimplantation development. Cell Rep 42:112248
Rao SSP, Huntley MH, Durand NC et al (2014) A 3D map of the human genome at kilobase resolution reveals principles of chromatin looping. Cell 159:1665–1680. https://doi.org/10.1016/j.cell.2014.11.021
Raviram R, Rocha PP, Luo VM et al (2018) Analysis of 3D genomic interactions identifies candidate host genes that transposable elements potentially regulate. Genome Biol 19. https://doi.org/10.1186/s13059-018-1598-7
Ren B (2016) The 3D organization of the mammalian genome. FASEB J 30. https://doi.org/10.1096/fasebj.30.1_supplement.99.1
Rollini P, Namciu SJ, Marsden MD, Fournier REK (1999) Identification and characterization of nuclear matrix-attachment regions in the human serpin gene cluster at 14q32.1. Nucleic Acids Research 27(19):3779–3791. https://doi.org/10.1093/nar/27.19.3779
Rowley MJ, Corces VG (2018) Organizational principles of 3D genome architecture. Nat Rev Genet 19. https://doi.org/10.1038/s41576-018-0060-8
Salinas-Pena M, Serna-Pujol N, Jordan A (2024) Genomic profiling of six human somatic histone H1 variants denotes that H1X accumulates at recently incorporated transposable elements. Nucleic Acids Res 52:1793–1813. https://doi.org/10.1093/nar/gkae014
Santoni FA, Guerra J, Luban J (2012) HERV-H RNA is abundant in human embryonic stem cells and a precise marker for pluripotency
Schmidt D, Schwalie PC, Wilson MD et al (2012) Waves of retrotransposon expansion remodel genome organization and CTCF binding in multiple mammalian lineages. Cell 148. https://doi.org/10.1016/j.cell.2011.11.058
Shah R, Gallardo CM, Jung YH et al (2022) Activation of HIV-1 proviruses increases downstream chromatin accessibility. iScience 25. https://doi.org/10.1016/j.isci.2022.105490
Shi Z, Xu J, Niu L et al (2023) Evolutionarily distinct and sperm-specific supersized chromatin loops are marked by Helitron transposons in xenopus tropicalis. Cell Rep 42:112151. https://doi.org/10.1016/j.celrep.2023.112151
Shimada A, Cahn J, Ernst E et al (2024) Retrotransposon addiction promotes centromere function via epigenetically activated small RNAs. Nat Plants 10:1304–1316. https://doi.org/10.1038/s41477-024-01773-1
Singh PB, Newman AG (2020) On the relations of phase separation and Hi-C maps to epigenetics. R Soc Open Sci 7
Song M, Pebworth MP, Yang X et al (2020) Cell-type-specific 3D epigenomes in the developing human cortex. Nature 587:644–649. https://doi.org/10.1038/s41586-020-2825-4
Spracklin G, Abdennur N, Imakaev M et al (2023) Diverse silent chromatin states modulate genome compartmentalization and loop extrusion barriers. Nat Struct Mol Biol 30:38–51. https://doi.org/10.1038/s41594-022-00892-7

Stam M, Tark-Dame M, Fransz P (2019) 3D genome organization: a role for phase separation and loop extrusion? Curr Opin Plant Biol 48:36
Strom AR, Emelyanov AV, Mir M et al (2017) Phase separation drives heterochromatin domain formation. Nature 547:241–245. https://doi.org/10.1038/nature22989
Su M, Han D, Boyd-Kirkup J et al (2014) Evolution of Alu elements toward enhancers. Cell Rep 7:376–385. https://doi.org/10.1016/J.CELREP.2014.03.011
Sullivan LL, Chew K, Sullivan BA (2017) α satellite DNA variation and function of the human centromere. Nucleus 8(4):331–339. https://doi.org/10.1080/19491034.2017.1308989
Sun JH, Zhou L, Emerson DJ et al (2018) Disease-associated short tandem repeats co-localize with chromatin domain boundaries. Cell 175:224–238.e15. https://doi.org/10.1016/j.cell.2018.08.005
Sun D, Zhu Y, Peng W et al (2024) SETDB1 regulates short interspersed nuclear elements and chromatin loop organization in mouse neural precursor cells. Genome Biol 25. https://doi.org/10.1186/s13059-024-03327-2
Sundaram V, Wysocka J (2020) Transposable elements as a potent source of diverse cis-regulatory sequences in mammalian genomes. Philos Trans R Soc B Biol Sci 375:20190347
Sundaram V, Cheng Y, Ma Z et al (2014) Widespread contribution of transposable elements to the innovation of gene regulatory networks. Genome Res 24:1963–1976. https://doi.org/10.1101/gr.168872.113
Tang SJ (2017) Potential role of phase separation of repetitive DNA in chromosomal organization. Genes (Basel) 8. https://doi.org/10.3390/genes8100279
Torres DE, Kramer HM, Tracanna V et al (2024) Implications of the three-dimensional chromatin organization for genome evolution in a fungal plant pathogen. Nat Commun 15. https://doi.org/10.1038/s41467-024-45884-x
Trono D (2015) Transposable elements, polydactyl proteins, and the genesis of human-specific transcription networks. Cold Spring Harbor Symposia on Quantitative Biology 80:281–288. Cold Spring Harbor Laboratory Press, Cold Spring Harbor, NY, USA. https://doi.org/10.1101/sqb.2015.80.027573
Valledor M, Byron M, Dumas B et al (2023) Early chromosome condensation by XIST builds A-repeat RNA density that facilitates gene silencing. Cell Rep 42. https://doi.org/10.1016/j.celrep.2023.112686
Wang H, Maurano MT, Qu H et al (2012) Widespread plasticity in CTCF occupancy linked to DNA methylation. Genome Res 22:1680–1688. https://doi.org/10.1101/gr.136101.111
Wang J, Vicente-García C, Seruggia D et al (2015) MIR retrotransposon sequences provide insulators to the human genome. Proc Natl Acad Sci U S A 112:E4428–E4437. https://doi.org/10.1073/pnas.1507253112
Wang S, Su JH, Beliveau BJ et al (2016) Spatial organization of chromatin domains and compartments in single chromosomes. Science 1979:353. https://doi.org/10.1126/science.aaf8084
Wang T, Zeng J, Lowe CB et al (2007) Species-specific endogenous retroviruses shape the transcriptional network of the human tumor suppressor protein p53. Proc Natl Acad Sci U S A 104(47):18613–18618. https://doi.org/10.1073/pnas.0703637104
Wang M, Li J, Wang P et al (2021) Comparative genome analyses highlight transposon-mediated genome expansion and the evolutionary architecture of 3D genomic folding in cotton. Mol Biol Evol 38:3621–3636. https://doi.org/10.1093/molbev/msab128
Wang C, Liu X, Liang J et al (2023) A DNA tumor virus globally reprograms host 3D genome architecture to achieve immortal growth. Nat Commun 14. https://doi.org/10.1038/s41467-023-37347-6
Wells JN, Feschotte C (2020) A field guide to eukaryotic transposable elements. Annu Rev Genet 54:539
Wicker T, Sabot F, Hua-Van A et al (2008) A universal classification of eukaryotic transposable elements implemented in Repbase. Nat Rev Genet 9:414
Wu Q, Liu P, Wang L (2020) Many facades of CTCF unified by its coding for three-dimensional genome architecture. J Genet Genomics 47:407

Wutz G, Várnai C, Nagasaka K et al (2017) Topologically associating domains and chromatin loops depend on cohesin and are regulated by CTCF, WAPL, and PDS5 proteins. EMBO J 36. https://doi.org/10.15252/embj.201798004

Xu Z, Wei H, Li M et al (2024) Impact of chromosomal fusion and transposable elements on the genomic evolution and genetic diversity of ilex species. Plants 13. https://doi.org/10.3390/plants13182649

Yu Z, Li J, Wang H et al (2024) Transposable elements in Rosaceae: insights into genome evolution, expression dynamics, and syntenic gene regulation. Hortic Res 11. https://doi.org/10.1093/hr/uhae118

Zamudio N, Bourc'His D (2010) Transposable elements in the mammalian germline: a comfortable niche or a deadly trap. Heredity (Edinb) 105:92–104

Zhang Y, Li T, Preissl S et al (2019) Transcriptionally active HERV-H retrotransposons demarcate topologically associating domains in human pluripotent stem cells. Nat Genet 51:1380–1388. https://doi.org/10.1038/s41588-019-0479-7

Zhang H, Zhao R, Tones J et al (2020) Nuclear body phase separation drives telomere clustering in ALT cancer cells. Mol Biol Cell 31. https://doi.org/10.1091/mbc.E19-10-0589

Zhang X, Liu X, Du Z et al (2021) The loss of heterochromatin is associated with multiscale three-dimensional genome reorganization and aberrant transcription during cellular senescence. Genome Res 31:1121–1135. https://doi.org/10.1101/gr.275235.121

Zhang X, Yu G, Dai Y et al (2023) High-resolution Hi-C maps highlight multiscale chromatin architecture reorganization during cold stress in Brachypodium distachyon. BMC Plant Biol 23. https://doi.org/10.1186/s12870-023-04269-w

Chapter 4
Epigenetic Control of Transposable Elements

Zhengyan Liang, Chengyu Jia, Xinyi Yao, and Yin Zhang

Abstract This chapter reviews the epigenetic control of transposable elements (TEs) and their roles in various biological processes. We first demonstrated how epigenetic modifications influence the transposition activity and transcriptional expression of transposons, focusing on key regulatory mechanisms such as DNA methylation and histone modifications—including methylation, demethylation, and acetylation—and their effects on transposon silencing and activation. We then describe the contributions of transposons and transposon-derived genes to genome architecture and epigenetic regulation, including their roles in genome size evolution, chromatin accessibility, DNA and histone methylation/demethylation, histone acetylation/deacetylation, and the regulation of noncoding RNA (ncRNA) expression. Finally, we highlight the significance of epigenetic regulation of TEs in embryonic development, cellular reprogramming, tumorigenesis, and age-related diseases, underscoring the central role of coordinated epigenetic control of transposons in determining cell fate and disease outcomes.

Z. Liang
Guangdong Provincial Key Laboratory of Malignant Tumor Epigenetics and Gene Regulation, Sun Yat-sen Memorial Hospital, Sun Yat-sen University, Guangzhou, Guangdong, China

Guangdong Pharmaceutical University, Guangzhou, Guangdong, China

C. Jia · X. Yao
Guangdong Provincial Key Laboratory of Malignant Tumor Epigenetics and Gene Regulation, Sun Yat-sen Memorial Hospital, Sun Yat-sen University, Guangzhou, Guangdong, China

Y. Zhang (✉)
Guangdong Provincial Key Laboratory of Malignant Tumor Epigenetics and Gene Regulation, Sun Yat-sen Memorial Hospital, Sun Yat-sen University, Guangzhou, Guangdong, China

Department of Cellular and Molecular Diagnostics Center, Sun Yat-Sen Memorial Hospital, Sun Yat-Sen University, Guangzhou, China
e-mail: zhangy525@mail.sysu.edu.cn

L. M. Vaschetto (ed.), *Transposable Elements, Transcriptomics, and Diseases*, Epigenetics and Human Health 14, https://doi.org/10.1007/978-3-032-04125-8_4

Keywords Transposable Elements · Epigenetic Regulation · DNA Methylation · Histone Modification · Chromatin Accessibility · Aging · Embryonic Development · Cellular Reprogramming

Abbreviations

Alu	A type of SINE element
ERVs	Endogenous retroviruses
ESCs	Embryonic stem cells
HMT	Histone Methyltransferase
KDM	Lysine-specific demethylase
LINE-1	Long interspersed nuclear element-1
LSD1	Lysine-specific demethylase 1
LTR	Long terminal repeat
ncRNAs	Noncoding RNAs
ORF	Open reading frame
PRC2	Polycomb repressive complex 2
RdDM	RNA-directed DNA methylation
TAD	Topologically associating domain
TEs	Transposable elements
TIR	Terminal inverted repeat ZGA: zygotic genome activation

4.1 Introduction

4.1.1 Definition and Basic Characteristics of Transposable Elements

Transposable elements (TEs) are a class of DNA sequences that can change their positions within the genome, with this process of movement referred to as transposition. These elements are ubiquitous across all forms of life, ranging from bacteria to humans. Their abundance varies significantly, accounting for 4% to 85% of genomes. For instance, it is estimated that TEs constitute approximately 44% of the human genome. (J. He et al. 2019).TEs were first discovered in maize by Barbara McClintock (Feschotte 2023). Her groundbreaking research revealed the dynamic nature of genes, challenging the conventional belief that genes have fixed positions. She was awarded the Nobel Prize in 1983 for this work.

The basic characteristics of TEs are as follows: autonomy and non-autonomy. Autonomous TEs, like the *Activator* (Ac) element in maize, can transpose independently. Non - autonomous TEs, like the *Dissociation* (Ds) element, rely on other elements for transposase. Diverse transposition mechanisms: Class I involves retrotransposons (copy - and - paste via RNA intermediates). Class II involves DNA transposons (cut - and - paste).

TEs can cause mutations by inserting into genes, disrupting repair processes, or inducing recombination. But they also play a key role in genome evolution, new gene formation, and the evolution of regulatory networks (He and Chen 2021).

4.1.2 The Epigenetic Regulation of Transposable Elements

Epigenetic regulation is central to suppressing the activity of TEs, playing a crucial role in maintaining genome stability and cellular function. In normal tissues, TEs are usually tightly repressed through mechanisms like DNA methylation, histone modifications (e.g., H3K9me3), and ncRNAs (e.g., piRNAs) (He and Chen 2021).

DNA methylation involves adding methyl groups (5mC) to transposon promoters or regulatory regions (e.g., CpG islands), directly inhibiting transcription. For example, in the human cancer genome, *LINE-1* elements are activated by hypomethylation in their 5' UTR regions, which serves as a hallmark of TEs activation in cancer (Jing et al. 2024). In plants, DNA methylation acts synergistically with H3K9me2 to suppress TE activity (Cui et al. 2013). Histone modifications regulate TE silencing or activation by altering chromatin states. Repressive modifications mark heterochromatic regions, recruit HP1 proteins to silence TE chromatin, and inhibit transcription. In mouse embryonic stem cells, H3K9me3 works with DNA methylation to regulate endogenous retroviruses (ERVs) (He et al. 2019). H3K27me3 in *Arabidopsis thaliana* is crucial for silencing some TEs and dynamically interplays with DNA methylation (Hure et al. 2025). NcRNAs also play a role in controlling TEs' activities. In germline cells, piRNAs form complexes with PIWI proteins to cleave transposon RNAs or recruit DNA methyltransferases (Li et al. 2025). In plants, small interfering RNAs (siRNAs) guide the RNA-directed DNA methylation (RdDM) pathway to silence TEs.

Chromatin remodeling complexes are also involved, For example, the polycomb repressive complex 2 (PRC2) deposits H3K27me3 to inhibit TE transcription. In mouse neural progenitor cells, SETDB1-mediated H3K9me3 works with DNA methylation to suppress SINE activity and influence cell proliferation (Sun et al. 2024).

4.1.3 Types of Transposable Elements and Their Epigenetic Regulatory Patterns

Epigenetic regulation of TEs shows remarkable heterogeneity, reflecting their transposition mechanisms and chromatin contexts. TEs are diverse and can be classified based on whether they require a reverse transcription step: Class I (retrotransposons) and Class II (DNA transposons). Retrotransposons are further divided into two categories based on the presence of long terminal repeats (LTRs)sequences at their ends: LTR retrotransposons and non-LTR retrotransposons. LTR retrotransposons have Pol and Gag proteins closely related to retroviral proteins, but lack envelope proteins needed to exit cells. Retrotransposons undergo replicative transposition, increasing their numbers with each transposition and potentially causing genome expansion. In humans, the non-LTR retrotransposon family*LINE-1* makes up 17% of the human genome. Type II TEs, or DNA transposons, integrate into the genome without reverse transcription. Transposases encoded by transposons recognize terminal inverted repeats (TIRs) sequences at the transposon's ends, excise it from the donor site, and insert it into a new target site. The gap left at the donor site can be repaired in a "cut-and-paste" manner without replacing the element or filled in through gap repair with a transposon copy. Helitrons, newly discovered DNA transposons, replicate via a unique rolling circle mechanism. Autonomous helitrons contain DNA helicase proteins and a replicase protein similar to replication protein A (RPA).

TEs can also be classified as autonomous or non-autonomous. Both retrotransposon and DNA transposon families have autonomous and non-autonomous types. Non-autonomous elements are often mutated remnants of autonomous members and sometimes have only limited sequence similarity to autonomous ones. Non-autonomous DNA transposons are typically flanked by a pair of TIRs surrounding non-transposon DNA, often copied from unrelated protein-coding genes. Some non-autonomous elements are called miniature inverted repeat transposable elements (MITEs), composed of two TIRs joined tip to tip. Non-autonomous elements can greatly outnumber autonomous ones in the genome. For example, the human genome has an estimated 1.2 million copies of Alu palindromic transposons, a non-autonomous short interspersed nuclear element (SINE).

Transposable elements (TEs) contribute to epigenetic regulation in the genome through several mechanisms. They can supply cis-regulatory regions, such as promoters and enhancer elements. These elements serve as binding sites for transcription factors, enabling them to modulate gene expression (Gebrie 2023). Furthermore, ncRNAs like microRNAs (miRNAs) and lncRNAs can be derived from TEs. These ncRNAs play a crucial role in regulating gene expression. TEs also regulate gene expression by altering chromatin states. Their activity varies across tissues, endowing TE-derived regulatory sites with tissue-specific characteristics and making them key players in tissue-specific gene regulation. For example, in the human genome, genes linked to TEs show greater expression variation in active chromatin regions than those without TEs. These studies show that TEs are crucial for gene regulation, as they can generate novel regulatory elements and functional RNA sequences.

4.2 Mechanisms of Epigenetic Regulation of Transposable Elements Activity

As TEs are ubiquitous in the genome and have a significant impact on genomic stability and gene expression, the activity of TEs is strictly regulated. Epigenetic regulation plays a crucial role in controlling TE activities through various mechanisms, including DNA methylation, histone modification, chromatin accessibility regulation, and small RNA-mediated control.

4.2.1 The Role of DNA Methylation in Transposon Silencing

DNA methylation is a key epigenetic modification for transposon silencing. In mammals and plants, the symmetrical methylation of cytosine residues at CpG sites is crucial for suppressing transposon transcription. This methylation is perpetuated during DNA replication by copying the methylation pattern onto the new DNA strand, ensuring the inheritance of transposon silencing. For example, in mice, the DNA methyltransferase DNMT1 maintains DNA methylation. In Dnmt1-deficient embryos, IAP retrotransposon transcription increases, leading to embryonic death around days 9–10. However, this doesn't happen in undifferentiated embryonic stem cells (Walsh et al. 1998). This shows that DNA methylation is crucial for transposon-specific epigenetic silencing during certain mammalian development stages (Woodcock et al. 1997). The DNMT3 methyltransferase is responsible for performing new (mainly CpG) DNA methylation. Its non-catalytic homolog, DNMT3L, is required for the methylation and silencing of IAP retrotransposons in pre-meiotic male germ cells (Bourc'his and Bestor 2004).

In *Arabidopsis thaliana*, RdDM has been extensively studied. The RNAi gene family in *Arabidopsis* has undergone duplication and diversification, contributing to post-transcriptional regulation and siRNA-mediated chromatin modification. Four distinct Dicer proteins with specific yet overlapping functions are present in this species. Additionally, two classes of siRNAs with different sizes have been identified in Arabidopsis. The smaller siRNAs are involved in post-transcriptional regulation (RNAi and miRNA) (Bühler et al. 2006). Larger siRNAs mainly arise from transposons and tandem repeats and are involved in RNA-dependent DNA methylation (Xie et al. 2004; Bourc'his and Bestor 2004). The DCL3 gene generates the larger siRNA species from transposons, and AGO4, one of the ten Argonaute proteins, binds a subset of these siRNAs. The siRNA-loaded AGO4 complex appears to have both catalytic and non-catalytic roles in silencing various transposons by targeting asymmetric DNA methylation. The formation of heterochromatin targeted by siRNAs raises a question: how do transposons silenced at the chromatin level produce transcripts that are processed into siRNAs? This is achieved by the plant-specific RNA polymerase IV (Pol IV). The absence of Pol IV leads to the reactivation of transcription of several silenced transposons and isolated single LTRs (Kanno et al. 2005b). Pol IV exists in two forms, Pol IVa and Pol IVb, distinguished by their unique largest subunits. Both have specific roles in RNA-mediated DNA methylation: Pol IVa is responsible for generating siRNAs, while Pol IVb is involved in transmitting siRNA signals to produce silencing outcomes. Pol IVb can be directed to the correct location by its longer carboxyl (C)-terminal domain compared to Pol IVa, which serves as a platform for different protein factors and may interact with AGO4 in a specific nuclear body in an RNA-dependent manner. The "trigger" transcripts generated by Pol IV (from full-length or cryptic transposon elements) may be cotranscriptionally targeted during transcription, similar to the process in *S. pombe*. Pol IV may also act directly on the RNA template (rather than DNA) to produce double-stranded RNA. Through mutation screening of genes involved in RNA-mediated DNA methylation, Pol IV and a SWI/SNF chromatin remodeling gene, DRD1, were identified (Kanno et al. 2005a). DRD1 may cooperate with Pol IV to transcribe heterochromatic DNA (and possibly produce double-stranded RNA) and then target these regions for DNA methylation. While symmetric DNA methylation is heritable, asymmetric DNA methylation must be replaced through targeting mechanisms after each cell division, likely involving siRNAs. In *Arabidopsis*, DRM2, the DNMT3 homolog, works with another methyltransferase, CMT3, to impose asymmetric DNA methylation and is essential for RNA-dependent DNA methylation. Asymmetric methylation of transposons requires DRM2 and CMT3, but only when MET1 (the Dnmt1 homolog) is still present are transposons and transgenes selectively reactivated. This suggests that asymmetric DNA methylation is much less important than symmetric CpG methylation in maintaining transposon silencing. In *Arabidopsis*, only a subset of transposons is regulated by RNAi when MET1 and DDM1 are present, further highlighting the importance of chromatin remodeling and CpG methylation in maintaining transposon silencing.

4.2.2 The Role of Histone Modification in Transposable Element Silencing and Activation

4.2.2.1 Histone Methylation and Demethylation

Histone methylation and demethylation, mediated by histone methyltransferases or demethylases, respectively, play critical roles in the silencing and activation of TEs by modulating chromatin states that either repress or permit TE transcription and mobility. Histone methylation is classified into repressive marks (e.g., H3K9me3, H3K27me3) and active marks (e.g., H3K4me3).

H3K9me3, a mark of constitutive heterochromatin, is associated with TE silencing. In mammalian genomes, SETDB1-mediated H3K9me3 cooperates with DNA methylation to suppress the chromatin activity of certain SINE sites. When SETDB1 is knocked out, these SINE sites become more chromatin-accessible, allowing CTCF binding and leading to chromatin loop redistribution, which affects the cell cycle and proliferation (Zhao et al. 2023). SETDB1 also teams up with KAP1 and KRAB zinc-finger proteins to form repressive chromatin structures, maintaining H3K9me3 and inhibiting TEs like endogenous retroviruses.

The dynamic changes in H3K9me3 levels are crucial for controllingTE activity and genome stability. During specific environmental conditions or cellular processes, such as plant germ cell development, H3K9me3 can be removed by demethylases, activating TEs and enhancing genetic diversity and adaptive evolution. For instance, *Arabidopsis thaliana* Ddm1 mutants show TE reactivation, highlighting DDM1's key role in maintaining TE silencing (Gendrel et al. 2002).

In *Arabidopsis thaliana*, the SUVH4/5/6 methyltransferases contribute to TE silencing. They introduce H3K9me3 into TE chromatin by recognizing TE DNA features or interacting with RNA interference pathways, thus repressing TE transcription. The triple mutants of *Arabidopsis* suvh4/5/6 show a significant increase in TE transcripts, highlighting the role of SUVH proteins in TE silencing (J. Zhang et al. 2023).

H3K4me3 is typically linked to actively transcribed genes but can also contribute to TE silencing. In mammals, TRIM28 proteins bind the histone tails with H3K4me3 modifications, recruiting corepressors to form transcription repression complexes, thereby silencing TEs (Hu et al. 2023). However, in mouse embryonic stem cells, about 19% of various pluripotency factor binding sites are located within TEs. These TEs usually contain multiple transcription factor binding sites, which may expand their regulatory capacity and transcriptional activity, influencing gene expression in multiple cell types, tissues, and developmental stages (Hu et al. 2023). Additionally, H3K4me1 modifications are associated with the activation of TEs in specific cell types.

In addition to H3K9me3 and H3K4me3, other histone methylation modifications also participate in the silencing and activation of transposable elements. H3K36me3 is generally associated with active transcription. Nevertheless, when present in the internal regions of certain transposable elements, it may influence their transcriptional elongation, thereby indirectly affecting their silencing or activation state (Yao et al. 2025). Moreover, H3K27me3, a repressive modification mediated by polycomb group proteins, has a role in silencing some transposable elements. In mammals, the PRC2 complex can silence certain transposable elements by catalyzing the H3K27me3 modification (Guo et al. 2021). In *Arabidopsis thaliana*, many transposable elements can be targeted and silenced solely by H3K27me3(Hure et al. 2025).

H3K9me3 and H3K4me1 do not function independently in TE regulation but instead coordinate to ensure precise control. In human embryos, H3K9me3 in ERV regions represses transposition, maintaining genome stability, while H3K4me1 can turn some areas into enhancers, guiding development-related gene expression. The chromatin state here is shaped by various histone modifications, balancing genome stability and TE activity (Yu et al. 2022). When SVA elements lose H3K9me3, their enhancer activity is activated by H3K4me1, boosting zygotic genome activation (ZGA) gene transcription. Histone-modifying enzymes also matter in TE regulation. In T-cell differentiation, MLL4 sets up H3K4me1 patterns, managing chromatin interactions and coordinating early-stage gene expression. The dynamic H3K4me1 distribution is key for enhancer activity and TE function shift (Zhen et al. 2024).

In mouse embryonic stem cells (ESCs), TEs like ERVs have both activating (H3K4me1, H3K27ac) and repressive (H3K9me3, H4K20me3) histone modifications, forming "multiply marked transposons" (He et al. 2019). The dynamic balance of these modifications, mediated by epigenetic regulators like NOCR2 and RNF2, ensures TEs' dual role in genome stability and cell reprogramming.

Histone demethylases (KDMs) play a key role in the silencing or activation of TEs by dynamically regulating histone methylation levels. Demethylases affect TE transcriptional activity and transposition by removing these marks, altering chromatin structure, or recruiting other epigenetic factors. For instance, removing repressive marks like H3K9me3 can activate TEs by relaxing heterochromatin, while removing active marks like H3K4me3 may suppress TE expression (Xiang et al. 2019). Moreover, histone demethylases often synergize with other epigenetic mechanisms, such as DNA methylation and noncoding RNAs, forming a multi-layered regulatory network. This ensures the dynamic balance of TEs in both physiological and pathological conditions (Sepulveda et al. 2024).

Histone demethylase KDM4B activates *LINE-1* transcription by removing repressive H3K9me3 maker and disrupting heterochromatin. In tumor cells, high KDM4B levels boost *LINE-1* transposition, causing DNA double-strand breaks and genomic instability (Xiang et al. 2019). KDM4B inhibition reduces *LINE-1*-induced DNA damage, showing potential in cancer therapy. PHF8 (KDM7B) suppresses retrotransposons by maintaining H3K9me3. When PHF8 is absent, H3K9me3 methyltransferase SETDB1 degrades, leading to retrotransposon transcription of double-stranded RNA (dsRNA). This triggers antiviral immune responses, raises tumor immunogenicity, and increases ICB sensitivity (Liu et al. 2023).

Active methyl marks (e.g., H3K4me3) are enriched in promoters of actively transcribed genes, promoting open chromatin and gene expression. In TE regulation, removing these active marks can suppress TE activity. For instance, in rice, the H3K4me3 demethylase JMJ703 removes H3K4me3 to inhibit non-LTR retrotransposon reactivation. JMJ703 loss leads to TE abnormal expression and genomic instability, highlighting the importance of dynamic H3K4me3 regulation for TE silencing in plants (Cui et al. 2013).

Moreover, removing active methyl marks can recruit repressive complexes. After removing active marks, some demethylases can recruit repressive methyltransferases to form complexes that enhance silencing. For example, KDM5B (JARID1B) removes H3K4me3 and recruits the H3K9me3 methyltransferase SETDB1 to TE regions, forming an epigenetic silencing complex that inhibits TE transcription (Qian et al. 2025). This collaboration is crucial in cancer but may cause immune evasion if hyperactive.

The "dual-track regulation" of H3K27me3 and DNA methylation in *Arabidopsis* involves dynamic transitions between these two epigenetic marks for some TEs, a process termed "bifrons" phenomenon. In certain ecotypes, the silencing of specific TEs relies on H3K27me3 rather than DNA methylation. This transition is determined by both the sequence features of TEs and trans-acting factors. Research indicates that about 15% of TEs are marked solely by H3K27me3, with longer TEs (> 3.5 kb) tending to depend more on DNA methylation, revealing a functional division of labor and complementarity between the two mechanisms in TE silencing (Hure et al. 2025). The OGT-TET interaction maintains DNA methylation homeostasis. OGT, an O-linked N-acetylglucosamine transferase, inhibits TET enzyme activity to prevent DNA demethylation, thus preserving the silencing of TEs in heterochromatin. When the OGT gene is knocked out, the TET catalytic product 5hmC accumulates significantly across the genome, including in heterochromatin, leading to TE activation. Acting as an "epigenetic brake," OGT restricts TET-mediated oxidative demethylation and works together with histone demethylases to maintain genomic stability (Sepulveda et al. 2024).

4.2.2.2 The Role of Histone Acetylation in the Regulation of Transposable Elements

Histone acetylation plays a crucial role in regulating TEs by modulating their transcriptional activity and chromatin state. Generally, histone acetylation promotes the transcriptional activation of TEs and supports their enhancer-like regulatory functions, whereas histone deacetylation represses TE activity. For example, H3K27ac, a marker of active enhancers, is usually linked to gene activation. In TE regulation, it shows a complex dual-faced nature: it can open chromatin to promote TE transcription but also inhibit TE activity by working with other epigenetic mechanisms.

Since histone is positively charged and histone acetylation can neutralize and weaken nucleosome-DNA interactions, histone acetylation usually enhances chromatin accessibility. This open state enables transcription factors and RNA polymerase to bind to transposon promoter/enhancer regions, promoting their transcription. For example, in tumor cells, abnormal H3K27ac upregulation is closely linked to *LINE-1* activation and genomic instability.

Besides regulating the expression of TEs, histone acetylation is also crucial for genomic stability through spatiotemporal-specific dynamic changes. In normal physiology, the balance between H3K27ac-mediated activation and silencing prevents TE-induced DNA damage. During early embryonic ZGA, H3K27ac broadly enriches at *LINE-1* regions to initiate transcription but is promptly erased to curb excessive TE activity, preserving genomic integrity (Wu et al. 2023). Dysregulation of this mechanism has a double-edged effect in disease and evolution. In cancer, elevated H3K27ac activates *LINE-1* elements, causing genome rearrangements and mutations and increasing tumor heterogeneity. However, TE-derived dsRNA can trigger antiviral immune responses, offering new immunotherapeutic targets (Yu et al. 2022). Evolutionarily, H3K27ac-mediated spatiotemporal TE activation may drive genomic diversity, promoting species adaptability. For example, domesticated TE sequences can give rise to novel functional genes or regulatory elements, highlighting the deeper role of epigenetic regulation in shaping biological complexity and environmental adaptation.

Moreover, the histone acetylation at TEs is highly dynamic and interacts with other epigenetic mechanisms. For example, it can medicate antagonistic regulation with DNA methylation. H3K27ac and DNA methylation (5mC) have a dynamic balance in transposon regulation. In *Arabidopsis*, some transposons are silenced by H3K27me3, not DNA methylation. But long ones (>3.5 kb) mainly depend on DNA methylation. This "dual-track" mechanism, via the antagonism of H3K27ac and DNA methylation, ensures transposon dynamic balance in different development stages or under environmental stress (Hure et al. 2025). It is also reported that histone acetylation coordinates with ncRNAs. In mammals, H3K27ac-marked enhancers produce enhancer RNAs (eRNAs), these eRNAs indirectly regulate nearby transposon transcription by recruiting chromatin remodelers or transcription complexes. In pig liver studies, super-enhancers marked by H3K27ac form eRNA-based co-regulatory networks with gene expression, influencing metabolic phenotypes, with some mechanisms possibly extending to transposon regulation (Ling et al. 2024).

In human early embryos, H3K27ac signals show an atypical broad-peak pattern before ZGA, and then shift to a sharp-peak pattern. H3K27ac is significantly enriched on human early embryonic retrotransposons (e.g., LINE-1), and its dynamics are closely related to their active transcription. P300 histone acetyltransferase knockdown reduces H3K27ac signals and markedly suppresses transposon expression, indicating H3K27ac's important role in transposon activation (Wu et al. 2023).

Cheng Daojun's team at Southwest University found that during insect metamorphosis, the transcription factor Schlank recruits deacetylase Rpd3 and methyltransferase PRC2, turning H3K27ac into the repressive H3K27me3. This silences juvenile hormone-related genes (e.g., Hairy) and suppresses transposons. It reveals a fine-tuned control of transposons and genome stability by histone modifications in the developmental context (Yuan et al. 2024).

4.2.3 Chromatin Accessibility and Transposable Elements

Chromatin accessibility refers to the extent of chromatin openness and the availability of DNA for binding with proteins, as indicated by the looseness or compactness of the chromatin structure (Klemm et al. 2019). In open chromatin regions, DNA is less tightly associated with proteins like histones. This makes DNA more accessible to transcription factors and RNA polymerase, facilitating gene transcription. Conversely, in closed chromatin regions, DNA is tightly packed within protein complexes, making it inaccessible to these proteins and inhibiting gene expression.

Chromatin accessibility and TEs activity have a complex bidirectional regulatory relationship. Open chromatin provides a physical environment for transposon insertion and transcription. Meanwhile, transposon activities can remodel local chromatin structure and influence 3D genome organization. The following sections explore their interaction mechanisms and related research progress.

4.2.3.1 The Impact of Chromatin Accessibility on Transposable Element Activity

TEs often insert into open chromatin regions, such as near gene promoters or enhancers. This open structure facilitates transposase access and cutting, promoting transposition (Hong et al. 2024). Techniques like ATAC-seq show TE insertion sites often overlap with these open regions. Here, TE promoters and enhancers are more accessible to transcription factors (TFs) and RNA polymerase II (PolII), activating TE transcription. For example, Liu Nian's team found active *LINE-1* retrotransposons recruit PolII to form chimeric transcripts and chromatin topologically associating domain (TAD) boundaries, relying on open chromatin (Hong et al. 2024). Also, environmental stresses (e.g., DNA damage) can alter chromatin accessibility to activate TEs. For instance, Zhu's team at Shenzhen University discovered that histone H1 deimination and acetylation weaken its DNA binding, loosening chromatin. This helps recruit DNA repair factors and may indirectly activate TEs (Tian et al. 2025).

4.2.3.2 Remodeling of Chromatin Structure by Transposable Elements

TEs play crucial roles in chromatin structure and gene regulation. Research by Liu Nian's team shows that active *LINE-1* TEs recruit Pol II and form chimeric transcripts, creating CTCF-independent TAD boundaries. This is cell- and species-specific and linked to *LINE-1* transcriptional activity (Hong et al. 2024). SAFB, a nuclear matrix protein, binds *LINE-1* RNA to inhibit PolII aggregation and regulate the formation of TAD boundaries. Some TEs carry regulatory elements like enhancers or insulators. For example, certain LTR TEs can act as enhancers upon insertion, altering local chromatin accessibility and activating nearby gene expression. Fudan University's ATAC-MS technology has uncovered dynamic binding between TE insertion sites and transcription factor complexes.Furthermore,TE insertions can also cause local changes in DNA methylation or histone modifications. In rice, the chromatin remodeler OsINO80 maintains TE silencing by promoting H3K27me3 and H3K9me2 deposition. OsINO80 knockout reduces H3K9me2 in heterochromatin, activates TEs, and triggers transposition (Du et al. 2024).

4.2.3.3 Significance in Disease and Evolution

Aberrant transposon activation is linked to multiple diseases and serves as a key evolutionary driver. In cancer, abnormal chromatin accessibility (e.g., genome-wide hypomethylation) can lead to transposon derepression. For instance,Insertions near oncogenes or tumor-suppressor genes can disrupt local chromatin. For example, HERV-K-derived sequencescan activate oncogenes like MYC by forming novel enhancer-like elements, promoting tumorigenesis (Rodriguez-Martin et al. 2020). In neurodegenerative diseases, age-related hypomethylation in neurons may activate LINE-1 transposons. Their retrotransposition causes DNA double-strand breaks and disrupts tau protein homeostasis, playing a role in Alzheimer's disease pathology (Guo et al. 2018). Evolutionarily, transposons contribute to genomic innovation by providing regulatory elements. Studies indicate that about 40% of mammalian enhancers originate from transposon sequences. For instance, the MER41 transposon's STAT1 binding site participates in interferon signaling evolution (Du et al. 2024). Primate-specific SVA transposons regulate chromatin accessibility and drive 3D genomic reorganization of vitamin D-related genes, showing species-specific adaptation. Overall, transposons are both a potential threat to genomic stability and a key medium for evolutionary innovation.

4.2.4 Small RNA Regulation of Transposable Elements

TEs, the "jumping genes" in the genome, are central to maintaining genomic stability. The epigenetic silencing mediated by siRNA is a key way to repress TEs in plants and animals.

4.2.4.1 siRNA Regulation of Transposons

SiRNA biogenesis starts with Dicer-like (DCL) enzymes processing dsRNA. In plants, transposon-derived long RNAs can form dsRNA through two pathways. In the RdDM pathway, RNA polymerase Pol IV makes ssRNA from transposon transcripts, which is converted to dsRNA by RDR2 and then cut by DCL3 to form 24nt siRNAs for DNA methylation guidance (Matzke and Mosher 2014). In the non-canonical RdDM pathway, activated transposon or invasive TE transcripts are converted to dsRNA by RDR6 and cut by DCL4 to form 21–22 nt siRNAs, triggering RNAi and de novo DNA methylation. Notably, plant transposon RNAs often have non-optimal codons, causing ribosome stalling. This stalling shortens the RNA, creating substrates easily recognizable by RDR6.

RDR6 plays a core role in secondary siRNA amplification. When transposon RNA is ribosome-stalled or nuclease-cleaved, RDR6 uses truncated RNA as a template to synthesize complementary strands, forming dsRNA precursors for DCL-mediated secondary siRNA generation. This boosts silencing signal propagation (Kim et al. 2021). Also, SGS3's prion-like domain (PrLD) mediates liquid-liquid phase separation (LLPS), co-localizing with RDR6 to form cytoplasmic siRNA bodies. This phase-separated structure enhances siRNA production efficiency and specificity for targeting transposon transcripts.

SiRNA forms the RNA-induced silencing complex (RISC) by loading onto Argonaute (AGO) proteins and binds to transposon RNA via sequence complementarity. In Arabidopsis, AGO4 and AGO6 preferentially bind 24 nt siRNAs and guide DNA methyltransferases (e.g., DRM2) to target sites, while AGO1 is involved in post-transcriptional silencing mediated by 21–22 nt siRNAs (Carbonell 2017).

Phased small interfering RNAs (phasiRNAs) are a distinct class of secondary siRNAs in plants, requiring RDR6 and specific trigger RNAs (such as miRNA cleavage products) for their biogenesis. For example, silencing of the AtREP transposon in *Arabidopsis* involves a miRNA-triggered phasiRNA cascade. Multiple rounds of RDR6-mediated amplification generate a siRNA array covering the entire length of the transposon, achieving efficient silencing (Nuthikattu et al. 2013).

4.2.4.2 piRNA Regulation of Transposons

Piwi-interacting RNAs (piRNAs) are small noncoding RNAs that silence transposons in animals. Their biogenesis has two phases: primary piRNA generation. These come from long single-stranded RNA precursors transcribed from transposon-rich regions and intergenic noncoding RNAs. In fruit flies, piRNA sequences originate from piRNA clusters in the genome, which can be uni- or bidirectionally transcribed. Cytoplasmic nucleases like Zucchini (Zuc) cut these precursors into primary piRNAs. The Rhino-Deadlock-Cutoff (RDC) complex in fruit flies also helps by binding to piRNA precursors, promoting their transcription and nuclear export, providing more material for processing (Huang et al. 2014)Secondary piRNA amplification via the "Ping-Pong" cycle: This cycle involves PIWI proteins (e.g., Aub, Ago3) with endonuclease activity and RNA helicases like Vasa. PIWI-piRNA complexes cleave transposon-complementary RNA strands, creating new secondary piRNA precursors (Zhou et al. 2024a). RNA helicases facilitate dsRNA unwinding, enhancing PIWI-piRNA binding and cleavage. The cycle starts with PIWI-piRNA complexes cleaving transposon RNA to make secondary piRNA precursors. These are then loaded onto another PIWI protein to form new complexes, which repeat the process. This self-amplifying positive feedback loop quickly increases piRNA numbers, improving transposon silencing and safeguarding genomic stability.

PiRNA-mediated transposon silencing occurs at both the transcriptional and post-transcriptional levels, and this precise regulation is crucial for maintaining genomic stability. At the post-transcriptional level, piRNAs bind to Ago proteins, directly cleaving transposon RNAs or recruiting factors to inhibit their translation, thus blocking transposon expression and spread. At the transcriptional level, piRNAs work with PIWI proteins to guide epigenetic modifications like chromatin remodeling and DNA methylation, silencing transposon transcription. This dual mechanism prevents abnormal transposon activation and protects the genome from instability. The precise regulation of piRNAs also offers cells a flexible response to transposon threats, ensuring genomic integrity and stability.

In *Drosophila*, the PIWI-piRNA complex anchors transposon mRNA via sequence complementarity. Aub, a PIWI family member with endonuclease activity, directly cleaves transposon RNA complementary to piRNA, reducing transposon protein expression and providing a defense against transposon activation (Li et al. 2025). In mice, MIWI2 binds to the 3'UTR of target RNA, blocking ribosome-mRNA interaction and translation. This suppresses transposon protein production, offering an example of precise translational control and fine-tuned gene expression (Carmell et al. 2007).

In the nucleus, the PIWI-piRNA complex plays a significant regulatory role. It can recruit histone-modifying enzymes, such as the H3K9me3 methyltransferase SETDB1, to methylate histones near transposons. This promotes heterochromatin formation and a more compact chromatin structure, physically limiting transposon transcription (Claro-Linares and Rojas-Ríos 2025). In mammalian germ cells, the complex also guides the DNA methyltransferase DNMT3 to transposon regions to establish DNA methylation. This creates an epigenetic basis for long-term transposon silencing. This transgenerational silencing potential is crucial for maintaining the stability of the germ cell genome (Greenberg and Bourc'his 2019).

The piRNA system doesn't function randomly but shows significant spatiotemporal specificity, which is closely related to its key functions. In germ cells, which bear the responsibility of transmitting genetic information, the piRNA system is highly active during gametogenesis. For example, in mouse spermatogonia, the MIWI2-piRNA complex precisely silences *LINE-1* retrotransposons. This prevents chromosomal catastrophes like recombination or breakage caused by transposon activation during meiosis, safeguarding healthy sperm production and highlighting the piRNA system's irreplaceable role in protecting germ cell genomes (Wang et al. 2023b). In early embryos, maternally derived piRNAs play a crucial role. They protect the zygotic genome from transposon activation, paving the way for normal embryonic development. For instance, early Drosophila embryonic development is greatly influenced by maternal piRNAs (Wang et al. 2023b). As the gonads mature, piRNA diversity dramatically increases, enabling precise matching and inhibition of specific transposon subfamilies. In zebrafish ovaries, the piRNA system efficiently silences transposon subfamilies like CR1 elements. This dynamic regulation ensures genomic stability across different developmental stages, underpinning the individual's lifelong health and development (Li et al. 2025). The association between piRNA pathways and TE activity is discussed in Chap. 6.

4.2.4.3 Other Noncoding RNAs Regulatory Mechanisms

Noncoding RNAs have diverse roles in transposon regulation. Alongside canonical siRNA and piRNAs, lncRNAs, endogenous siRNAs (endo-siRNAs), miRNAs, and circular RNAs (circRNAs) contribute to controlling transposon activity through unique mechanisms, creating a multi-layered regulatory network.

lncRNAs can act as molecular scaffolds to recruit epigenetic modification complexes to transposon regions. For example, XIST lncRNA binds to the polycomb repressive complex PRC2, targeting H3K27me3 to transposon-rich regions on the X chromosome (e.g., LINE-1 clusters), inducing heterochromatin formation and suppressing transposon activity (Slotkin and Martienssen 2007). In humans, lncRNAs transcribed from the antisense strand of *LINE-1* (e.g., LINE-1-AS) form dsRNA with *LINE-1* mRNA, triggering RNAi. They also enhance local DNA methylation by binding to the DNA methyltransferase DNMT3B (Denli et al. 2015).

MiRNAs can indirectly regulate transposon activity by targeting mRNAs of transposon-suppressing factors. For instance, the miRNA-Ago2 axis involves miR-34 family members (e.g., miR-34c) directly targeting Ago2 mRNA, reducing Ago2 levels and weakening siRNA-mediated transposon silencing. This negative feedback loop may be critical in cancer-associated transposon activation (Tomasetti et al. 2024). Additionally, miR-29b suppresses the histone deacetylase HDAC4, increasing transposon-region histone acetylation (H3K27ac), enhancing chromatin accessibility and transposon transcription (Ha and Yu 2021).

In early mammalian embryos, DNA methylation pattern remodeling is crucial for zygotic genome reprogramming, and miRNAs play a role in this process. By targeting the 3′UTR of DNMTs, miRNAs regulate DNMT expression, impacting DNA methylation. For example, the miR-29 family targets the 3′UTR of DNMT3A and DNMT3B, showing a negative correlation with their expression (Sun et al. 2023). This miRNA regulation of DNMTs is vital in embryonic development. It temporarily reduces DNA methylation in transposon regions, permitting transient transposon activation and aiding genome reprogramming. This process is essential for normal early embryonic development and cell fate determination.

4.3 The Role of Transposable Elements in the Genome and Epigenetic Regulation

4.3.1 Transposons Regulate the Evolutionary Changes in Genome Size

4.3.1.1 Transposable Element Types and Genomic Amplification Mechanisms

DNA transposons move via a "cut-and-paste" mechanism. Transposase binds to the transposon's terminal inverted repeat (ITR) sequences, excises the transposon from its original site, and inserts it into a new genomic location. While this mechanism does not contribute to the expansion of repetitive sequences in the genome, it is a potent source of genetic variation. For example, the Mutator transposon family in maize increases genomic instability through continuous excision and insertion (Bourque et al. 2018). Tc1-like transposons, a diverse and widespread DNA transposon superfamily, have abundant transposase genes. They induce genomic or gene recombination and variation, accelerating biodiversity and evolutionary rates (Slotkin and Martienssen 2007).

Retrotransposons (e.g., LINEs, SINEs, and LTRs) amplify via a "copy-and-paste" mechanism. Take *LINE-1* as an example: it is first transcribed into RNA, then reverse-transcribed into cDNA by reverse transcriptase, and finally inserted into a new genomic site. Examples are the P elements in fruit flies and human Alu sequences, typical retrotransposons. Alu sequences, a type of SINE, are widespread in the human genome and contribute significantly to genomic expansion through retrotransposition (Lee et al. 2024). Accounting for about 10% of the human genome, Alu elements are widely distributed and tend to insert into gene-rich regions, thereby impacting genomic structure and function in multiple ways (Chuong et al. 2017). On the one hand, Alu insertion can create new CpG sites for DNA methylation, potentially activate nearby gene transcription as an enhancer, or regulate gene expression by affecting alternative splicing or translation initiation rates. On the other hand, Alu retrotransposition and recombination can cause genomic variations, such as chromosomal structural changes and gene dosage alterations, and lead to insertional mutations. For instance, inserting into a gene can disrupt transcription, interfere with splicing, or alter regulatory elements, subsequently affecting protein function.

4.3.1.2 Transposon-Driven Dynamic Genomic Changes

In the maize genome, approximately 80% of sequences are derived from transposons, with LTR retrotransposons being the primary contributors. These elements have amplified rapidly through a "copy-and-paste" mechanism, making the maize genome much larger than that of its close relative sorghum. This explosive amplification mainly occurs in intergenic regions, which are subject to relaxed purifying selection. Transposons can also drive genome contraction through non-allelic recombination or DNA deletion mediated by transposase activity. Non-allelic recombination refers to recombination between transposons at different loci on homologous chromosomes, leading to DNA fragment deletion or inversion. For example, in yeast, non-allelic recombination mediated by the Ty1 retrotransposon has been observed to cause genomic rearrangements (Qi et al. 2023).

4.3.1.3 Species Differences and Evolutionary Adaptability

Transposons contribute significantly to genome size differences. In wheat, 80%–90% of the genome is transposon-derived, with LTR retrotransposons like Gypsy and Copia proliferating via "copy-and-paste," leading to a large genome (e.g., 16 Gb in bread wheat) (Ragupathy et al. 2010). Corn (2.3Gb) has 70% LTR retrotransposons, while sorghum (730 Mb) has a smaller genome due to more efficient transposon silencing (Soderlund et al. 2009). Corn's recent LTR retrotransposon burst within three million years is the main cause of its genome expansion. In humans, about 45% of the genome is transposon-derived, but only 0.1% consists of active retrotransposons like *LINE-1* and Alu. *LINE-1* elements amplify autonomously, while Alu elements rely on LINE-1's enzymatic machinery for non-autonomous amplification, together contributing around 30% of human genomic sequences (Li et al. 2024). In fruit flies, the horizontal transfer of P elements in the mid-twentieth century caused rapid genome expansion in wild populations. However, the genome stabilized due to strong piRNA-mediated repression (Kofler et al. 2015).

During evolution, plants and animals have developed different genomic trajectories due to their survival strategy differences. Plants, with sufficient energy from photosynthesis, have a low metabolic cost for genome expansion. This resource redundancy allows transposon-driven "genomic obesity," where transposon activity isn't strongly suppressed but instead enhances environmental adaptability by creating new regulatory elements. In polyploid plants (e.g., hexaploid wheat), chromosome duplication reduces the harmful effects of transposon insertion on single-copy genes while retaining the advantages of promoting genetic innovation, leading to a co-evolutionary pattern of genome expansion and adaptability enhancement (Liu et al. 2025). In contrast, animals' mobility-related physiological constraints have shaped a selective pressure for genome streamlining. A large genome can significantly impact DNA replication in terminally differentiated cells like neurons. Also, animal embryonic development relies on the precise spatiotemporal expression of regulatory networks such as Hox gene clusters, which can be disrupted by transposon insertion (Oomen et al. 2025). Under these dual selective pressures, animals have evolved stricter transposon silencing mechanisms, leading to a more compact genome that balances developmental precision and cell division efficiency (Tan et al. 2021).

4.3.2 Transposon Regulation of Chromatin Accessibility

4.3.2.1 Direct Alterations to Chromatin Structure by Transposon Insertion

Transposon insertion can directly alter local chromatin structure. When inserted into a genome, a transposon's regulatory elements (like promoters or enhancers) can affect chromatin openness by modifying transcription factor binding sites. For example, LTR retrotransposons have LTR regions with multiple transcription factor binding sites. Upon insertion, these can attract transcription factors, changing chromatin accessibility and impacting nearby gene expression (Maston et al. 2006).

4.3.2.2 Transposable Elements as Recruitment Platforms for Regulatory Factors

Many TEs have specific sequences or structures that serve as platforms for recruiting regulatory factors. Conserved sequences or motifs on TEs can be recognized and bound by transcription factors, localizing them to specific chromatin regions. For instance, some LTR retrotransposons have promoter sequences in their LTR regions that recruit RNA polymerase II and other transcription factors. This initiates the transcription of transposons and nearby genes, opening up otherwise closed chromatin (Hong et al. 2024). Additionally, TEs can interact with epigenetic regulators to modify chromatin states. Some can bind histone-modifying enzymes, recruiting them to specific chromatin regions and altering histone modifications like acetylation or methylation. These changes regulate chromatin accessibility (Yang et al. 2023).

4.3.2.3 Long-Range Regulation of Chromatin Accessibility Mediated by Transposable Elements

Transposons can regulate chromatin accessibility over long distances through several mechanisms. First, their insertion sites and interactions between transposon family members can affect chromatin's 3D structure. For example, transposons often cluster in specific chromatin regions. These clusters can alter chromatin topology by looping or folding, impacting distant accessibility (Lawson et al. 2023). Additionally, noncoding RNAs from transposons can mediate long-range chromatin regulation by binding to chromatin and recruiting remodeling complexes or other regulatory factors (Gan et al. 2024). Third, long-range regulation can occur through the spread of epigenetic modifications. Changes in histone modifications caused by transposon insertion can propagate along the chromatin, altering the accessibility of regions distant from the insertion site (García-Vílchez and Guallar 2025).

4.3.3 Transposon Regulation of DNA Methylation

Transposons carry specific sequences, such as CpG-rich regions or inverted repeats, that can directly recruit DNA methyltransferases (DNMTs) and trigger DNA methylation. In mammals, the 5'UTR of *LINE-1* transposons has CpG islands that bind DNMT3A/3B, inducing local methylation to silence transposons and regulate nearby gene expression (Auclair et al. 2014). In *Arabidopsis*, some LTR transposons form dsRNA via inverted repeats, activating the RdDM pathway and causing methylation of themselves and surrounding regions to maintain transposon silencing (Deniz et al. 2019).

Transposon-encoded proteins may directly interact with DNMTs or demethylases like TET, thereby influencing DNA methylation dynamics. The interactions between transposon-encoded proteins and DNMTs can alter the catalytic activity of DNMTs, affect their degradation to change cellular levels and distribution, and guide DNMTs to specific genomic regions to modulate methylation. For instance, transposon proteins can sometimes inhibit DNMT activity, reducing methylation in specific genomic areas (Bhat et al. 2022). Proteins encoded by transposons may also influence the degradation of DNMTs, thereby altering their intracellular levels and distribution, and consequently affecting the methylation patterns of the genome (de Mendoza et al. 2018). Research indicates that in mouse embryonic stem cells, while TET enzymes can demethylate *LINE-1* elements, potentially activating these transposons, they are counteracted by robust repressive mechanisms. Host-encoded repressors, such as SIN3A to maintain *LINE-1* silencing. This might involve modulating TET enzyme activity or its DNA binding, which is crucial for genomic stability and normal cell function.

4.3.4 Transposon Regulation of Histone Methylation and Demethylation

Histone methylation involves adding methyl groups to specific amino acid residues (mostly lysine and arginine) on histones, primarily at the N-terminal tails of H3 and H4, via histone methyltransferases (HMTs). Lysine can be mono-, di-, or tri-methylated (me1, me2, and me3), while arginine can be mono-methylated (me1) or asymmetrically/symmetrically di-methylated (me2a/me2s) (Tee and Reinberg 2014). Histone demethylation is the reverse process, mediated by histone demethylases (HDMs), which remove methyl groups from histone tails. HDMs are mainly divided into two classes: lysine-specific demethylases (LSDs) and JumonjiC (JmjC)-domain-containing demethylases (JMJs) (Black et al. 2012). LSD1 primarily catalyzes demethylation of mono-methylated and di-methylated H3K4 and H3K9, whereas JmjC family members can remove tri-methyl modifications from lysine residues. These modifications play a crucial role in gene expression regulation by altering chromatin structure and recruiting transcription factors. For instance, H3K4me3 and H3K36me3 are linked to active gene transcription, while H3K9me3 and H3K27me3 are associated with gene silencing (Wang et al. 2023a).

Transposons can regulate histone methylation in various ways, impacting gene expression and genomic stability. Some transposon transcriptional or translational products can interact with histone methyltransferases or demethylases, altering their activity and changing histone methylation. For instance, specific proteins derived from transposon can bind to SUVH-family methyltransferases and affect histone H3K9 methylation. In *Arabidopsis*, when LTR retrotransposons are activated, their transcripts are processed into siRNAs, which then influence histone methylation through gene regulatory networks (Hung et al. 2023). Transposons can recruit histone demethylases, thereby influencing the expression of adjacent genes. For instance, in hepatocellular carcinoma, liver-specific transposon elements recruit the histone demethylase KDM1A. This action negatively impacts the methylation of histone H3 lysine 4 (H3K4) at liver transposon elements, leading to the epigenetic silencing of HNF4A expression and subsequently promoting hepatocellular carcinoma cell proliferation (Jing et al. 2024). Transposons can recruit histone demethylases to alter the histone methylation status of chromatin regions associated with neighboring genes, thereby silencing or activating gene expression. This regulatory mechanism offers a new dimension for gene expression regulation and provides a crucial perspective for understanding the role of transposons in disease development.

4.3.5 Transposons Regulate Histone Acetylation/Deacetylation

As dynamic genomic elements, transposon activities are precisely controlled by epigenetic modifications, particularly histone acetylation/deacetylation. Histone acetylation (e.g., H3K9ac, H3K27ac), by neutralizing the positive charge of histone tails' lysine residues, reduces histone-DNA affinity, opens chromatin, and promotes transcriptional activation. In contrast, histone deacetylation (e.g., via HDACs) condenses chromatin and silences genes. Studying the interplay between transposons and histone acetylation/deacetylation is vital for understanding genome dynamic regulation. Transposon activities can influence histone modification patterns, while histone modifications can regulate transposon transposition and genomic distribution. This interaction affects genome stability and has broad implications for development and diseases. For example, abnormal transposon activities may cause genomic instability and trigger cancers and neurodegenerative diseases (Huda and Jordan 2009). During normal development, transposon and histone modification interplay may regulate gene spatiotemporal expression. Exploring this can enhance our grasp of genomic epigenetic regulation and boost disease treatment and genome stability maintenance.

4.3.5.1 Recruitment of Chromatin-Modifying Enzymes

Transposons can regulate histone acetylation levels and affect gene expression by interacting with histone acetyltransferases or deacetylases. For example, in *Arabidopsis*, proteins derived from PIF/Harbinger transposons with SANT domains can form complexes with HDA6, regulating histone deacetylation and gene expression at specific loci (Zhou et al. 2024b). This interaction enhances plant thermotolerance by enabling transposons to recruit specific chromatin-modifying enzymes to particular genomic regions. Once there, these enzymes can alter the chromatin state and influence transcriptional activity. HDA6, a histone deacetylase, reduces histone acetylation in corresponding regions. This typically correlates with gene silencing, as deacetylated histones lead to more compact chromatin, hindering transcription factor binding and reducing gene expression (Zhou et al. 2021).

4.3.5.2 Change the High Structure of Chromatin

Histone acetylation can regulate the openness of chromatin, thereby influencing the binding of TFs and other regulatory proteins to the promoter regions of target genes or other regulatory elements. Transposons can alter the tightness of chromatin by affecting the level of histone acetylation, thus regulating gene expression. When transposons increase the level of histone acetylation, the chromatin structure becomes loose, which makes it easier for regulatory proteins such as TFs to bind to the promoter regions of genes, thereby promoting gene transcription and expression. Conversely, if transposons lead to histone deacetylation, the chromatin will become more compact, making it difficult for TFs to bind and inhibiting gene expression (Fueyo et al. 2022). In human brain progenitor cells, primate-specific endogenous retroviruses (ERVs) can serve as docking sites for the co-repressor protein KAP1 (TRIM28), generating local heterochromatin and thus controlling the expression of genes that are crucial for brain development. KAP1 binds to ERVs and recruits histone deacetylases (HDACs), reducing the level of histone acetylation, making the chromatin structure more compact, and inhibiting the transcription of nearby genes.

4.3.5.3 Synergistic Effects with Other Epigenetic Modifications

Transposons also interact with other epigenetic modifications, such as DNA methylation and histone methylation, through histone acetylation to intricately regulate gene expression. This synergistic effect holds significant importance in maintaining genomic stability and normal cellular functions. For instance, DNA methylation typically occurs in the CpG island and is associated with gene silencing, while histone deacetylation is often accompanied by gene silencing as well. Transposons can recruit DNA methyltransferases and histone deacetylases to specific gene loci, simultaneously carrying out DNA methylation and histone deacetylation, thereby more effectively suppressing gene expression. Moreover, under certain circumstances, histone methylation and acetylation can also influence each other, jointly regulating the chromatin structure and gene expression. For example, certain histone methylation modifications may affect the activity or binding sites of histone acetyltransferases, and vice versa. This complex network of interactions enables transposons to flexibly regulate gene expression through multiple epigenetic mechanisms in response to various intra- and extracellular environmental signals, thus maintaining genomic stability and normal physiological functions of cells.

4.3.6 Transposons Regulate the Expression of Noncoding RNAs

Noncoding RNA plays important regulatory role at the levels of gene transcription, post-transcription, and translation. Based on whether its length exceeds 200 nucleotides, it can be classified into long ncRNA (over 200 nt in length) and short ncRNA (under 200 nt in length). The latter includes miRNAs, siRNAs, piRNAs, etc. These ncRNA molecules participate in gene expression and regulation at different levels and in different ways. Recent studies have further revealed the relationship between transposons and their related noncoding regulatory RNAs. In eukaryotes, small RNAs such as siRNA, miRNA, and piRNA are mostly derived from transposons. In plants, these transposon-derived small RNAs can regulate gene expression by modulating DNA methylation. In animals, transposon-derived piRNAs can lead to mRNA degradation in spermatocytes, thereby affecting sperm maturation in mice. Transposon expression can also be silenced due to the piRNA-mediated ping-pong cycle. HARDING et al. performed high-throughput sequencing of small RNAs from mid-gastrula stage embryos of Xenopus tropicalis at three developmental time points, identifying a novel class of ncRNAs derived from intronic transposon remnants, which they termed siteRNAs (Gardner et al. 2012).

4.3.6.1 Regulation of miRNAs

MiRNAs are endogenous, single-stranded noncoding RNAs approximately 20-24 nucleotides in length, functioning as critical regulators within biological gene networks. They exert inhibitory effects on mRNA expression at both transcriptional and post-transcriptional levels through partial or complete complementary base pairing with target mRNAs. In animals, imperfect base pairing between miRNAs and mRNAs typically represses translation, whereas in plants, extensive complementarity triggers endonucleolytic cleavage of target transcripts (Macrae et al. 2006). Accumulating evidence indicates that a substantial fraction of miRNAs originates from transposable element (TE) sequences. In mammals, numerous miRNAs are derived from TEs or their derivatives, which contribute to the formation of miRNA hairpin structures. Approximately 20% of annotated human miRNAs are associated with MITE transposons (Murchison et al. 2007). miRNAs containing inverted TE sequences primarily arise from SINE and LINE elements; those partially overlapping with non-inverted TEs are often derived from SINEs, LINEs, and DNA transposons, while miRNAs entirely derived from TEs predominantly originate from DNA transposons and SINEs.

In plants, transposons serve as a significant source of miRNAs. LI et al. demonstrated that certain plant miRNAs have evolved from TEs, which can form hairpin structures within noncoding regions of transcripts, ultimately giving rise to mature miRNAs. In the rapidly growing rhizomes, stems, and leaves of bamboo, 2297 miRNAs were identified, with 29.1% originating from MITE-related sequences—the hAT transposon superfamily making the most substantial contribution—further confirming MITEs as a primary source of miRNA biogenesis.

Building upon research into miRNA origins and their gene regulatory roles, it can be hypothesized that TEs in certain organisms modulate the expression of both themselves and other genes via TE-derived miRNAs. For example, in wheat, miR820—originating from TEs—targets transcripts of the DRM2 gene; repression of this gene reduces DNA methylation, leading to increased expression of corresponding TE transcripts, exemplifying how TE-derived miRNAs regulate plant DNA methylation (Nosaka et al. 2013). In the human genome, target genes of miRNAs derived from TE transcripts have been identified as TE sequences embedded within 3′ untranslated regions (UTRs). These miRNAs regulate gene expression by binding to TE sequences in 3′ UTRs; for instance, functional miRNA target sites associated with the AGO2 protein are derived from L2 transposon sequences within 3′ UTRs (Petri et al. 2019)

4.3.6.2 Regulation of piRNAs

As mentioned above, piRNAs are a class of small noncoding RNAs primarily derived from transposon sequences that play a crucial role in silencing TEs to maintain genome integrity, especially in the germline. Active transcription of transposons and piRNA clusters provides the precursor RNAs essential for piRNA biogenesis. The prevailing model of piRNA biogenesis posits that small piRNA precursors are generated via endonucleolytic cleavage of long precursor transcripts, which is followed by processing by the endonuclease Zucchini (ZUC). The 5′ ends of processed piRNAs associate with PIWI proteins to form functional complexes. Based on their biogenesis pathways, piRNAs are classified into primary and secondary piRNAs. Primary piRNAs are transcribed from single-stranded precursors, known as piRNA clusters, by RNA polymerase II. These precursors are transported to the cytoplasm by MAEL (maelstrom), a conserved RNA-binding protein containing an HMG-box domain. In the cytoplasm, interaction between MAEL and the endonuclease PLD6 generates the 5′ ends of primary piRNAs (Chen et al. 2015). piRNA precursors then bind to PIWI proteins for stabilization, after which the PIWI-piRNA complex undergoes 3′ end processing and modification to generate mature primary piRNAs. Primary piRNAs guide PIWI proteins to cleave complementary TE transcripts, producing secondary piRNAs with a 10-nucleotide offset from the primary piRNA—a process known as the "ping-pong cycle." Secondary piRNAs, in turn, can initiate the production of new primary piRNAs, through deep sequencing and bioinformatics analysis of mouse type A spermatogonia, pachytene spermatocytes, and round spermatids. GAN (Choi et al. 2024) demonstrated that TEs are a major source of piRNAs in the first two cell types, with abundant piRNAs derived from the three major retrotransposon classes: LTRs, LINEs, and SINEs. Similar findings have been reported in humans and Drosophila. piRNAs silence TEs at both transcriptional and post-transcriptional levels. The ping-pong cycle represents a post-transcriptional TE silencing mechanism wherein piRNAs direct PIWI proteins to cleave TE transcripts, generating new piRNAs in the opposite orientation. These newly generated piRNAs associate with AGO3 and undergo modifications involving ZUC and other nucleases to form AUG-piRNA complexes. Collectively, during spermatogenesis, TE-derived piRNAs regulate sperm development by inducing the degradation of target mRNAs.

4.3.6.3 Regulation of siRNAs

siRNA, a double-stranded RNA molecule with a length ranging from 20 to 25 nucleotides, plays a pivotal role in regulating gene expression and maintaining genomic stability. It exerts three primary functions: first, it impedes mRNA stability by cleaving mRNA transcripts through the RNA interference pathway; second, it participates in DNA methylation via the RdDM process; and third, it is involved in the plant's response to heat stress. In both plants and animals, TEs serve as the predominant sources of siRNA. For instance, endo-siRNA (endogenous short interfering RNA), a type of siRNA, has been identified in non-gonadal cells. In the non-gonadal cells of Drosophila, TEs have been demonstrated to be one of the crucial sources of endo-siRNA (Biryukova and Ye 2015). Notably, a multitude of TE-associated siRNAs exert, to varying degrees, an influence on gene expression in both animals and plants.

In plants, TEs and the siRNAs derived from them can engage in the DNA methylation process via the plant's RdDM pathway, dsRNA p t- the siRNAs play a role in gene silencing by directing the cytosine methylation of complementary DNA sequences. The regulation of the FLOWERING LOCUS C (FLC) gene serves as a prime example. The FLC gene acts as a major repressor of flowering and is negatively regulated by the vernalization mechanism (Shao et al. 2025). This phenotypic mechanism is associated with the insertion of a 1224-bp non-autonomous Mutator-like transposon into the first intron of the FLC gene. Such transposon insertion suppresses the expression of FLC alleles and also subjects the FLC gene to siRNA-mediated repressive chromatin modifications generated by homologous transposons located at other genomic sites. Moreover, during plant gamete production and embryonic development, both RNA interference and RdDM regulate genes in relation to transposons. In plants, suppressing transposon activity in gametes represents a vital defense strategy to prevent mutations and ensure the stability of the host genome. Therefore, TE-associated siRNAs assume a crucial role in this process by directing DNA methylation to silence TEs.

4.4 The Role of Epigenetic Control of Transposable Elements in Cell Fate Determination

4.4.1 Embryonic Development

4.4.1.1 The Expression Pattern of Transposable Elements in Early Embryonic Development

In the intricate regulatory network at the beginning of life, TEs, as movable DNA sequences in the genome, are gradually revealing their crucial regulatory functions from the inherent perception of being "genomic dark matter." These genetic elements, once regarded as "molecular parasites," deeply participate in the process of early embryonic development through complex and precise expression regulatory mechanisms, playing an indispensable role in determining the fate of embryonic development, maintaining totipotency, and cell lineage differentiation.

During the 2-cell stage of mouse embryonic development, MERVL (murine endogenous retrovirus-L) shows characteristic high expression, and its expression level is closely related to the totipotent state of the embryo, thus being an important molecular marker for evaluating embryonic totipotency (Sakashita et al. 2023). After fertilization, the paternal genome rapidly initiates active demethylation, and the maternal genome undergoes passive demethylation, and the peak expression of MERVL highly coincides with this epigenetic reprogramming process. Mechanistic studies have shown that MERVL participates in the dynamic regulation of DNA methylation modification in cooperation with the TET3 protein. TET3 catalyzes the oxidation of 5-methylcytosine (5mC) into 5-hydroxymethylcytosine (5hmC), initiating a demethylation cascade reaction, and MERVL may ensure the accuracy of epigenetic reprogramming by regulating the activity distribution of TET3. It is worth noting that an oxidative stress environment can significantly inhibit the expression of MERVL, leading to the developmental arrest of in vitro fertilized (IVF) mouse embryos, accompanied by increased DNA damage and apoptosis, and the blastocyst formation rate sharply drops from the normal level of 60% to about 20%, highlighting its necessity for normal embryonic development.

LINE-1 plays a unique regulatory function at key nodes of cell fate specification (Li et al. 2024). Manipulatethe expression of *LINE-1* in Mouse ESCs reveals that L1 selectively regulates the expression of specific genes in the two-cell phase. The RNA transcripts of *LINE-1* acts as a molecular scaffold for DNA organization in the three-dimensional space of the nucleus. By interacting with specific DNA sequences and structural proteins, it mediates the migration of chromosome 19, which is rich in genes, to the gene-silencing region, reshaping the spatial conformation of chromatin. At the same time, *LINE-1* RNA can also cooperate with the chromatin remodeling complex to dynamically regulate the openness of chromatin, thereby precisely regulating gene transcription activity and ensuring the orderly progress of the embryonic development process.

4.4.1.2 Transposable Elements in the Maintenance of Totipotency and Cell Fate Determination

In ESCs, the transcriptional activity of transposable elements is precisely regulated by the KRAB-ZFP/KAP1 system (Tie and Rowe 2017). As a core member of the genomic defense system, this complex recognizes the DNA sequences of transposable elements and recruits heterochromatin-related proteins, leading to the compaction of the chromatin structure and achieving transcriptional silencing of TEs. However, ESCs in different developmental states exhibit distinct epigenetic modification characteristics of transposable elements: in naive ESCs, the HERVH elements are in a transcriptionally repressed state due to the high enrichment of H3K27me3 (Theunissen et al. 2016); the SVA elements show a decrease in the level of H3K27me3 modification and a significant increase in H3K4me3 modification, demonstrating higher transcriptional activity. These dynamic changes in epigenetic markers are not isolated events but work in concert with the core transcription factor network to maintain the totipotent state of ESCs. Primate-specific transposable elements and their derived regulatory sequences play the role of a "molecular switch" during embryonic development through their interaction network with transcription factors (Pontis et al. 2022). The transcription factor binding sites hidden in the DNA sequences of transposable elements can be specifically recognized by transcription factors at different differentiation stages. Taking the key event of gastrulation in embryonic development as an example, the KRAB-ZFP protein encoded by the activated KZFP gene binds to TEs and re-establishes the inhibition of TEs by recruiting the KAP1 complex. This regulatory mechanism not only restricts the abnormal activation of TEs but also endows transcriptional regulation with significant germ layer specificity. For instance, during mesoderm differentiation, a specific

KZFP-TE interaction pattern activates genes related to the development of tissues such as the heart and muscles; during the differentiation process of the neural ectoderm, another set of regulatory combinations drives the directional differentiation of neural stem cells.

During the ZGA stage of mouse embryos, the MERVL elements reach a transcriptional peak, which is highly synchronized with the establishment of embryonic totipotency. As a "pioneer regulatory factor" of ZGA, MERVL transcripts cooperate with key transcription factors (such as Dux) to activate downstream totipotency gene networks and regulate core pathways such as the cell cycle process and metabolic reprogramming (Sakashita et al. 2023). At the same time, MERVL is also involved in the dynamic remodeling of the three-dimensional chromatin structure. By interacting with structural proteins such as CTCF, it regulates the formation and maintenance of topologically associating domains (TADs), ensuring the spatiotemporal specificity of gene expression. Gene knockout experiments have confirmed that the lack of MERVL expression will lead to embryonic development arrest and ultimately death, fully revealing its central position in cell fate determination.

Transposable elements deeply participate in various key links of early embryonic development through multi-level and multi-dimensional regulatory mechanisms. The synergistic relationship formed between these elements and the host genome during long-term evolution makes them an indispensable part of the precise regulatory network of embryonic development and provides a new research perspective and theoretical basis for a deeper understanding of the mechanisms of the origin and development of life.

4.4.2 Epigenetic Regulation of Cell Reprogramming by Transposable Elements

In the crucial process of reprogramming somatic cells into induced pluripotent stem cells (iPSCs), alterations in DNA methylation and histone modifications serve as the core driving force for the activation of TEs. Taking the *LINE-1* retrotransposon as an example, studies have revealed that, compared with parental fibroblasts, during the formation of iPSCs, the CpG methylation level in the promoter region of LINE-1 significantly decreases (Gerdes et al. 2022). This reduction in methylation levels directly leads to a substantial increase in the transcriptional activity of *LINE-1*, and its retrotransposition efficiency is approximately 10 times higher than that in parental cells. Bisulfite sequencing technology has further confirmed that there is a direct correlation between the hypomethylated state of the LINE-1 promoter in iPSCs and transcriptional activation, which clearly indicates that DNA demethylation is an essential key mechanism for the activation of TEs during somatic cell reprogramming.

During the process of embryonic development, genomic reprogramming relieves the repression of TEs, and a similar phenomenon occurs during the reprogramming of somatic cells into iPSCs. Numerous studies have shown that retrotransposons such as *LINE-1* are significantly activated in both mouse and human somatic cell reprogramming experiments. Moreover, this activation is not an isolated case, and numerous other TE families also undergo the activation process synchronously. However, what function the overall derepression of TEs plays in the reprogramming of pluripotent cells remains a scientific question that needs to be answered.

The functional studies of epigenetic modification enzymes in the process of reprogramming and in ESCs have provided important clues for deciphering the mechanisms of TE derepression. Numerous experiments have demonstrated that the decrease in the levels of epigenetic factors exhibits distinctly different biological effects in somatic cell reprogramming and ESC maintenance: during the reprogramming process, reducing the levels of epigenetic factors often promotes cell fate transition; conversely, knocking down the same factors in ESCs leads to an imbalance in the cell state and a tendency toward differentiation. Taking KDM1A (also known as LSD1) as an example, a decrease in its expression level can significantly enhance the efficiency of somatic cell reprogramming. However, when KDM1A is knocked down in ESCs, it promotes the differentiation of cells toward the extraembryonic endoderm by regulating the CoREST-HDAC signaling axis (Karakaidos et al. 2019). Additionally, the H3K9me3 modification is a crucial obstructive factor in somatic cell reprogramming, resulting in an "pre-iPSC" state with incomplete reprogramming (Becker et al. 2016). This series of findings reveals a deep correlation between epigenetic remodeling and TE derepression during cell fate transition.

It is worth noting that the expression of MuERV-L endogenous retrovirus can serve as a marker to identify a special subpopulation of ESCs. These cells have a gene expression pattern highly similar to that of the 2-cell stage of developing embryos and possess certain totipotency characteristics (Macfarlan et al. 2012). This process is closely related to the chromatin assembly factor 1 (CAF-1) complex, which is composed of Chaf1a, Chaf1b, and Rbbp4 subunits and mainly participates in histone H3/H4 deposition and heterochromatin assembly. Studies have found that the loss of CAF-1 activity leads to a significant increase in the number of 2C-like cells. This phenomenon may be related to the dynamic changes of heterochromatin during the early embryonic development process, and embryos with Chaf1a mutations are unable to complete this crucial recombination process.

Currently, there is still controversy in the academic community: is the derepression of TEs a necessary biological event in somatic cell reprogramming or a concomitant phenomenon of the DNA demethylation process? Existing studies have confirmed that DNA demethylation is a key event in the early stage of reprogramming—a necessary condition for mesenchymal-epithelial transition (MET), and there is a significant temporal and spatial overlap between TE activation and the DNA demethylation process.

4.4.3 Epigenetic Regulation of Transposable Elements and Tumorigenesis

In the late 1980s, people recognized the structural variations derived from *LINE-1* associated with hemophilia A, which drew attention to the adverse effects of *LINE-1* transposition in diseases. Subsequent studies have revealed that LINE-1 can act as a driver mutation in colorectal cancer. In this context, it inserts into and disrupts tumor-suppressor genes such as adenomatous polyposis coli (APC), thus triggering an investigation into the oncogenic potential of TE transposition. In various cancer types, including colorectal cancer (Miki et al. 1992), esophageal cancer, hepatocellular carcinoma, and other cancers, *LINE-1* exhibits increased retrotransposition in the early stages of tumorigenesis (Nam et al. 2023). Somatic insertional mutations of *LINE-1* occur in the early stages of gastrointestinal tumor development, even preceding dysplasia. Different levels and locations of retrotransposition reflect clonal expansion in cancer (Burns 2017). It is worth noting that driver mutations caused by TE transposition have been found only in limited scenarios in cancer, which may be due to insufficient understanding and technical difficulties. Technological advancements in recent years have spurred a research upsurge on the functional impacts of transposons on gene regulatory networks during evolution and development, as well as their roles in diseases such as cancer. The genetic and epigenetic evolution of cancer involves the utilization of transposons, with transposons directly participating in cancer-specific gene activities. The functional link between TEs and cancer is discussed in detail in Chap. 10.

4.4.3.1 Abnormal Activation of Transposable Elements in Tumorigenesis

LINE-1 belongs to the non-LTR retrotransposons. In the human body, it is the only known element with the ability of autonomous retrotransposition. Its full length is approximately 6 kb, and it contains two crucial open reading frames, namely ORF1 and ORF2 (Smith et al. 2020). ORF1 is responsible for encoding an RNA-binding protein, which plays a key role in the formation of ribonucleoprotein complexes (RNPs), while ORF2 has endonuclease and reverse transcriptase activities. It can accurately cleave the target DNA and reverse transcribe RNA into cDNA. The transcription process of *LINE-1* is strictly regulated by its 5′-end CpG-rich bidirectional promoter. In a normal cellular environment, this promoter is usually silenced due to DNA methylation modification, resulting in the inhibition of *LINE-1* transcription. However, during the development of cancer, global hypomethylation occurs in many tumor types, such as colon cancer and lung cancer. This leads to the release of the inhibitory state of the *LINE-1* promoter, thereby activating its expression (Xiao-Jie et al. 2016).

From the perspective of epigenetic regulation, DNA methylation is one of the main mechanisms for inhibiting the activity of TEs. In the cancer state, the activity decline or mutation of DNA methyltransferases (DNMTs) triggers widespread genomic hypomethylation. This hypomethylation state is particularly evident in the *LINE-1* promoter region and is significantly correlated with the RNA and protein expression levels of LINE-1 (Xiao-Jie et al. 2016). At the same time, KRAB domain zinc-finger proteins (KZFPs) and PIWI-interacting RNAs (piRNAs) can originally inhibit the activity of TEs through a series of complex mechanisms, such as recruiting related proteins to form heterochromatin, thus suppressing the transcription of TEs. However, in tumor cells, these key inhibitory pathways are often non-functional , causing **LINE-1** to lose effective regulation.

The expression of *LINE-1* also has the characteristics of tissue specificity and tumor heterogeneity. Taking pancreatic ductal adenocarcinoma (PDAC) as an example, high expression of ORF1p occurs in up to 90% of cases (Rodić et al. 2015); but in glioblastoma and low-grade B-cell lymphoma, ORF1p is almost not expressed (Banaz-Yaşar et al. 2010). Even for tumors of the same tissue type, such as lung adenocarcinoma, due to differences in individual genetic backgrounds (such as different copy numbers of active *LINE-1*) or the influence of the tumor microenvironment (such as p53 mutations, etc.), the expression of LINE-1 varies. Such differences in expression may lead to significantly different biological behaviors of tumor cells, thereby affecting the development process of tumors and the prognosis of patients.

LINE-1 mainly completes the transposition process through the complex and exquisite mechanism of "target-primed reverse transcription (TPRT)." Firstly, in the nucleus, RNA polymerase II (RNApolII) transcribes using the DNA sequence of *LINE-1* as a template to generate full-length LINE-1 RNA. Subsequently, these RNAs are transported to the cytoplasm, where they are translated on cytoplasmic ribosomes to form ORF1p and ORF2p. These two proteins will further assemble to form an RNP complex (Yang et al. 2023). Next, the RNP complex re-enters the nucleus by virtue of its special localization signals. In the nucleus, the endonuclease activity of ORF2p is activated. It can recognize specific target sites and cleave the DNA at this site, generating a 3'-OH end. This end serves as the starting point for the subsequent reverse transcription reaction. The reverse transcriptase of ORF2p uses *LINE-1* RNA as a template to synthesize cDNA starting from the 3'-OH end. Finally, with the help of the DNA repair mechanism of the host cell, the newly synthesized cDNA is integrated into the genome. It is worth noting that the newly inserted *LINE-1* usually appears in a 5′-truncated form and is accompanied by the phenomenon of target site duplication (TSD).

During this process, *LINE-1* proteins exhibit cis-preference, that is, they tend to bind to their own RNA for subsequent transposition. However, non-autonomous SINE elements such as Alu have unique competitive strategies. They can rely on their highly structured RNA to compete with *LINE-1* RNA for ORF2p, thus achieving their own retrotransposition. Nevertheless, in the environment of somatic cells (such as tumor cells), LINE-1 still plays a dominant role in driving transposition activities. This might be because in tumor cells, the expression level of LINE-1 is relatively high, and the quantity and activity of its proteins are sufficient to maintain its transposition advantage. Meanwhile, it also reflects the influence of the complex molecular regulatory network in tumor cells on the activities of TEs.

The aberrant activation of *LINE-1* exerts multifaceted oncogenic effects during tumor initiation and progression. A prototypical example of its role in driving mutations involves direct disruption of tumor-suppressor genes through insertional mutagenesis. As mentioned above, in colorectal cancer, *LINE-1* insertions have been shown to truncate exons of the APC gene, thereby abrogating its tumor-suppressive function and triggering aberrant activation of the β-catenin signaling pathway (Rodriguez-Martin et al. 2020). Dysregulation of this pathway promotes uncontrolled cell proliferation, aberrant differentiation, and ultimately, tumorigenesis.

Even when *LINE-1* inserts into noncoding regions, it profoundly impacts genomic stability. By introducing reverse transcriptase activity or inducing DNA double-strand breaks, it promotes chromosomal abnormalities such as translocations, deletions, or amplifications. In hepatocellular carcinoma, *LINE-1* insertions have been observed to activate the DNA damage response (DDR) (Li et al. 2024). When combined with TP53 mutations, this DDR activation accelerates the rate of genomic alterations, driving increased malignancy in tumor cells.

Additionally, the bidirectional activity of the *LINE-1* promoter introduces further complexities, generating chimeric transcripts that induce epigenetic reprogramming. In the blast crisis of chronic myeloid leukemia (CML) and esophageal cancer, for instance, the LINE-1 antisense promoter (ASP) drives aberrant splicing of the MET gene, resulting in the production of truncated tyrosine kinase receptors (Cervantes-Ayalc et al. 2020). Such aberrant receptors not only fail to function properly but also constitutively activate downstream signaling pathways, promoting abnormal cell proliferation and facilitating tumor progression. Our analysis of *LINE-1* ASP-associated genes across pan-cancer also reveals that these genes are significantly dysregulated in multiple cancer types, which is closely related to the tumor microenvironment, progression, and patient prognosis (Xu et al. 2023).

4.4.3.2 The Role of Transposable Element-Related Molecular Events in Cancer Progression

Tumor progression is a highly complex process encompassing critical phases such as clonal evolution, invasion, and metastasis. TEs, through diverse mechanisms, are profoundly implicated in every facet of this progression—driving sustained genomic diversification, facilitating tumor cell immune evasion, and remodeling the tumor microenvironment—thereby exerting profound effects on tumor development.

During carcinogenesis, approximately 72 germline *LINE-1* loci—such as 22q12 and 6p23.1—emerge as "hotspot" regions that frequently contribute to somatic insertions, with their 3′ transduction events accounting for 25% of all insertions. The primacy of these loci as major mutational sources may arise from their intrinsic hyperactive retrotranscriptase activity, enabling more frequent initiation of transposition, or from their enhanced ability to evade host silencing mechanisms, allowing unimpeded transpositional activity (E. Lee et al. 2012). From spatiotemporal perspectives, studies in pancreatic cancer illustrate distinct *LINE-1* insertion profiles between primary tumors and metastatic lesions (Han et al. 2023). In the early stages of primary tumor formation, specific *LINE-1* source elements are activated by microenvironmental stresses—such as hypoxia and nutrient deprivation—leading to insertional mutations. During metastasis, metastatic foci retain insertions that promote cell dissemination through clonal selection, underscoring the dynamic role of LINE-1 insertions in tumor progression and their impact on tumor biological behavior.

When *LINE-1* inserts into noncoding genomic regions, it can disrupt the normal regulation of gene expression by interfering with critical regulatory elements. Enhancers, insulators, and noncoding RNAs (such as lncRNAs) play pivotal roles in gene expression control, and LINE-1 insertions may undermine their functional integrity. For example, in breast cancer studies, *LINE-1* insertions within the regulatory region of the TPFI2 gene have been shown to suppress its expression through epigenetic silencing mechanisms, thereby abrogating its metastasis-suppressive function and rendering tumor cells more prone to invasion and metastasis (Cruickshanks et al. 2013). The chimeric RNAs generated by *LINE-1* ASPs exhibit unique structural and functional properties. These transcripts may incorporate ORF0 from the *LINE-1* antisense strand, which, upon fusion with downstream exons, can encode truncated proteins (Mendez-Dorantes and Burns 2023). In Hodgkin lymphoma, endogenous retrovirus (ERV) promoters drive aberrant transcription of CSF1R, producing protein products that promote macrophage-dependent immune suppression (Stacey and Sagulenko 2010). This mechanism enables tumor cells to evade immune system surveillance, creating a permissive microenvironment for tumor progression.

The Rec and Np9 proteins encoded by the env gene of the endogenous retrovirus HERV-K (particularly the HML-2 subgroup) exhibit the functional capacity to suppress the interferon signaling pathway, which plays a pivotal role in the body's antiviral defense mechanisms and immune surveillance. Following their suppression of this pathway, these proteins induce a promotive effect on cell migration, thereby facilitating the invasive and metastatic potential of tumor cells (X. Zhou et al. 2021). Additionally, the envelope protein of HERV-H demonstrates analogous immunosuppressive functionality, enabling tumor cells to evade immune system surveillance and thus fostering an environment conducive to unchecked proliferation and disseminated spread within the host organism (Dervan et al. 2021).

DsRNA transcribed from TEs activates the MDA5/MAVS signaling cascade, thereby inducing a type I interferon response—a primordial immune defense mechanism evolved to combat viral infections (Gazquez-Gutierrez et al. 2021). However, an intriguing observation has emerged in oncology: in the treatment of acute myeloid leukemia, DNA methylation inhibitors such as 5-azacytidine derepress endogenous retroviruses , triggering an antiviral response that paradoxically enhances tumor sensitivity to immunotherapy. This phenomenon unveils unanticipated intersections between retroelement reactivation, innate immunity, and cancer therapeutics, presenting novel therapeutic avenues that leverage these ancient viral relics to augment antitumor immune surveillance (Noronha et al. 2024).

4.4.4 *Epigenetic Regulation of Aging-Related Diseases by Transposable Elements*

In recent years, the reactivation of retrotransposable elements (RTEs) has been implicated in aging phenotypes, the pathogenesis of age-related diseases, and the acceleration of senescence; however, the causal nature of these associations remains undefined. Existing research demonstrates that aberrant RTE activation elicits multiple aging-related processes, including inflammation, alterations in heterochromatin structure and organization, and DNA instability (Tsai et al. 2024; Zhu et al. 2021). Notably, exposure to challenging and stressful environments has also been linked to enhanced RTE transcriptional activity. Environmental stressors such as chemical agents, radiation, and metabolic stress (Yushkova and Moskalev 2023) are recognized as critical factors that accelerate aging and elevate the risk of age-related diseases. The roles of TEs in aging are discussed in detail in Chap. 11.

4.4.4.1 Reactivation of Transposable Elements in Cellular Senescence

Throughout the lifespan of an organism, cellular capacity to maintain homeostasis of the epigenome and transcriptome diminishes progressively with age. This process is accompanied by marked dysregulation of epigenetic mechanisms that normally silence RTEs, rendering such dysregulation a central driver of RTE reactivation during cellular senescence.

Mechanistically, aging exerts profound effects on epigenetic pathways responsible for transcriptional silencing of intergenic regions enriched in RTEs and maintenance of chromatin homeostasis. A defining feature of senescent cells is the pronounced and locus-specific loss of H3K9me3, a constitutive heterochromatin mark, which disrupts the stability of chromatin architecture. Concomitantly, reduced DNA methylation at CpG islands—a highly conserved biological hallmark of eukaryotic aging (Kane and Sinclair 2019)—is particularly pronounced in RTE-enriched genomic regions, directly leading to substantial upregulation of RTE RNA expression. Studies in aging monozygotic twins have demonstrated significant hypomethylation of CpG islands at RTE loci, a trend exacerbated in individuals with unhealthy lifestyles and chronic exposure to adverse environmental factors (e.g., air pollution, prolonged psychological stress) (Della Valle et al. 2025). These findings underscore the synergistic interplay between environmental insults and epigenetic alterations in driving RTE activation.

Beyond epigenetic alterations during natural aging, genetic intervention studies have further corroborated the critical regulatory role of epigenetic pathways in controlling RTE activation. By leveraging gene editing techniques to disrupt key epigenetic mechanisms responsible for RTE silencing—such as impairing the deacetylation activity of silent information regulator (SIRT) proteins (Kumar and Lombard 2015), disrupting methylation dynamics mediated by H3K9 methyltransferases, or inhibiting the catalytic function of DNA methyltransferases (Yan et al.

2023)—researchers have observed accelerated cellular senescence accompanied by robust RTE reactivation in both murine and human cell models. This reciprocal validation not only underscores the role of intact epigenetic machinery in maintaining RTE quiescence but also establishes a direct causal link between epigenetic dysregulation and RTE activation. Upon reactivation, RTEs rapidly accumulate their transcribed RNAs and reverse-transcribed cDNAs, molecules harboring canonical retroviral features—most notably, pathogen-associated molecular patterns (PAMPs) (Della Valle et al. 2025). These PAMPs are efficiently recognized by pattern-recognition receptors (PRRs), the frontline sentinels of the innate immune system specialized in detecting single- and double-stranded RNAs (ssRNA/dsRNA) as well as ssDNA/dsDNA. Recognition of RTE-derived nucleic acids triggers the immediate activation of intracellular immune signaling programs, instigating a cascade of complex immune responses and functional reprogramming within affected cells.

4.4.4.2 The Relationship Between Transposable Elements and Aging-Related Diseases

The reactivation of TEs exhibits intricate and multifaceted associations with the initiation and progression of multiple aging-related diseases, whereby their mechanistic impacts primarily occur through two key pathways: induction of inflammatory responses and disruption of epigenetic homeostasis. These processes profoundly impact normal cellular physiology, thereby driving disease progression through cumulative functional impairments.

Upon reactivation, RTE-derived RNAs and cDNAs—once recognized by PRRs—swiftly activate innate immune cascades (Lagisquet et al. 2021), a process mediated by specialized cellular nucleic acid sensors: RIG-I and MDA5 detect viral-like RNAs in the cytoplasm and mitochondrial membranes (Rehwinkel and Gack 2020); cGAS identifies cytosolic double-stranded DNA (dsDNA) (Zhang et al. 2020); and Toll-like receptors TLR3, TLR7, and TLR9 facilitate nucleic acid recognition within endosomal vesicles (Vijay 2018), a function conserved even in the tumor microenvironment. Activation of these sensors triggers the IKK phosphorylation pathway, culminating in robust induction of innate immunity and massive secretion of type I/II interferons. This inflammatory axis profoundly influences aging-related pathologies: in cardiovascular diseases, RTE-derived nucleic acids in vascular endothelial cells instigate immune responses that promote inflammatory cell infiltration into the vascular wall, accelerating atherosclerotic plaque formation and vascular stenosis (Pedro-Botet et al. 2020); in systemic lupus erythematosus, aberrant RTE-derived nucleic acids chronically stimulate the immune system, driving uncontrolled autoantibody production and multi-organ damage (Tsokos 2020); and in Alzheimer's disease, RTE-mediated inflammation disrupts neuronal function, facilitating the formation of neurofibrillary tangles and senile plaques—hallmark pathological features—and exacerbating disease progression (Kinney et al. 2018).

In both cellular and tissue biopsy samples from elderly individuals and SIRT6-knockout progeria mouse models, *LINE-1* expression is markedly upregulated (Van Meter et al. 2014), accompanied by substantial accumulation of *LINE-1* cDNA that activates the interferon signaling pathway, thereby instigating robust inflammatory responses. Beyond autocrine effects, the influence of RTE expression extends to neighboring cells: virus-like particles released by expressed endogenous retroviruses (ERVs) act as pro-inflammatory paracrine signals (Zhao et al. 2024), stimulating adjacent cells to propagate an inflammatory cascade that amplifies both the scope and intensity of inflammation.

During the normal aging trajectory, a broad spectrum of RTEs and other repetitive genomic constituents undergo widespread transcriptional activation, whose products persistently disrupt the homeostatic equilibrium of the epigenome. In embryonic development, RNA transcripts derived from satellite DNA sequences have been shown to impair the enzymatic activity of SUV39H1/2 histone methyltransferases (Vojvoda Zeljko et al. 2021), leading to a genome-wide reduction in H3K9me3 (trimethylation of histone H3 at lysine 9) levels that disrupts normal embryonic cell differentiation and developmental programs. Emerging evidence highlights intricate mechanistic connections between environmental stressors, RTE activation, and aging-associated pathologies (Della Valle et al. 2025). Phenotypic disparities in aging onset and progression among genetically identical individuals residing in distinct environments underscore the pivotal role of environmental effects mediated through the epigenome in modulating senescence processes.

Key disruptors of epigenomic homeostasis include environmental stressors such as chemical pollutants (e.g., polycyclic aromatic hydrocarbons, heavy metals), ionizing radiation, reactive oxygen species accumulation, and dietary stressors (Bachmann et al. 2020). These stressors exert dual impacts on epigenetic regulation: they not only alter DNA methylation landscapes but also interfere with post-translational modifications of histones. Mechanistically, they disrupt the stoichiometry of metabolic cofactors essential for chromatin-modifying enzymes—for instance, perturbing the intracellular concentrations of S-adenosylmethionine required for histone methylation or acetyl-CoA for histone acetylation—and disrupt cellular redox homeostasis, such as by interfering with the NADH/NAD^+ balance that modulates the activity of SIRTuin (SIRTs) enzymes governing chromatin structure. Such multifaceted perturbations to RTE-regulatory epigenetic mechanisms ultimately culminate in the release of long interspersed nuclear elements (LINEs) and endogenous retroviruses (ERVs) from their transcriptionally silenced states.

Substantial research has established that factors including chemical pollutants, ionizing radiation, and metabolic perturbations induce epigenetic alterations and upregulate RTE expression, with these changes and associated phenotypes closely linked to accelerated aging. Experimental evidence from both in vitro human cell systems and in vivo murine models demonstrates that exposure to genotoxic contaminants such as benzo(a)pyrene or 1,3-butadiene (Lewis et al. 2019) elicits demethylation of RTEs, leading to enhanced transcriptional output and active transposition events. Epidemiological investigations in populations exposed to low-dose arsenic (Hossain et al. 2017) have identified significant hypomethylation of *LINE-1* elements, particularly in these cohorts. In murine models of metabolic stress induced by high-fat/high-sugar diets, notable upregulation of RTE expression has been observed (Moreno-Fernández et al. 2018). Concordantly, studies in 488 obese women have demonstrated a positive correlation between *LINE-1* demethylation levels and increased body mass index (BMI) (Maugeri et al. 2021).

Extreme climatic conditions likewise disrupt the regulatory equilibrium of retrotransposons, precipitating multiple diseases and accelerating senescence trajectories. Drosophila populations isolated from divergent ecosystems exhibit heightened RTE reactivation under harsh environmental conditions (Tahami et al. 2024). Thermal stressors also interfere with RTE expression in mammals: heat shock induces endogenous retrovirus expression by inhibiting HSP90-dependent recruitment of the KAP1–SETDB1 silencing complex (Hummel et al. 2017), a mechanism phenotypically analogous to the effects of declining HSP90 expression during physiological aging. Noteworthy is the observation that long-lived microbat species harbor lower *LINE-1* copy numbers compared to their short-lived counterparts (Huang et al. 2019). Pathogen exposure represents another critical stressor intimately linked to RTE reactivation. Upon cellular infection by DNA or RNA viruses (Macchietto et al. 2020), significant alterations occur in the expression and activity of epigenetic enzymes normally involved in RTE silencing, concurrently with suppression of piwi-interacting RNAs (piRNAs)—small noncoding RNAs that clear RTE transcripts via RNA interference. Collectively, these findings establish that RTE-derived nucleic acids, akin to epigenomic aberrations, play a significant and direct role in inducing genomic instability and driving cellular trajectories toward dysfunctional states, ultimately culminating in shortened healthspan and accelerated aging processes.

4.5 Summary and Prospects

This chapter has comprehensively described the epigenetic regulation of TEs, elucidating their pivotal roles in multiple biological processes. As DNA sequences capable of altering their genomic locations, TEs ubiquitously inhabit the genomes of diverse organisms, exerting profound impacts on genomic structure, function, and evolution. Epigenetic regulatory mechanisms—including DNA methylation, histone modifications, chromatin accessibility control, and small RNA-mediated regulation—orchestrate TE activity with precision, ensuring genomic stability and maintaining normal cellular functions.

DNA methylation assumes a central role in repressing transposon transcription, with RdDM representing a plant-specific pathway of notable importance. In the realm of histone modifications, various methylation, demethylation, and acetylation events modulate chromatin states, critically influencing TE silencing or activation. Synergistic interactions among these modifications further underscore their collective regulatory potency. Chromatin accessibility and transposon activity exhibit bidirectional interplay: open chromatin provides a permissive environment for transposon insertion and transcription, while TE mobilization reshapes chromatin architecture. Small RNAs, including siRNAs, piRNAs, and other noncoding RNAs, employ intricate biogenesis and target recognition mechanisms to effectively govern transposon activity.

TEs contribute in various ways to genomic and epigenetic landscapes. They modulate genome size evolution through diverse transposition mechanisms, influencing both genomic expansion and contraction. Additionally, TEs regulate chromatin accessibility by directly altering chromatin structure or serving as recruitment platforms for regulatory factors. These elements are also implicated in the regulation of DNA methylation, histone methylation/demethylation, histone acetylation/deacetylation, and noncoding RNA expression, highlighting their versatile regulatory capacities.

During embryonic development, TE expression patterns are intimately linked to the maintenance of embryonic pluripotency and cellular fate specification. In somatic cell reprogramming to induce pluripotent stem cells, TE activation coincides with epigenetic remodeling, underscoring their functional integration in cellular reprogramming. In oncogenesis, aberrant activation of TEs such as *LINE-1* promotes tumor progression by inducing driver mutations, genomic instability, and epigenetic reprogramming. Moreover, TE reactivation in aging-related diseases triggers inflammatory responses via nucleic acid byproducts, disrupting epigenetic homeostasis and exacerbating pathological processes. .

4.6 Challenges and Future Directions

Despite substantial advances, numerous frontiers in TE epigenetic regulation remain uncharted. Technological Limitations and Innovations: Current methodologies face constraints in detecting low-frequency transposition events, deciphering complex epigenetic modification networks, and conducting single-cell analyses. Development of more sensitive tools—such as high-resolution single-cell epigenomic sequencing—will enable precise tracking of TE dynamics and regulatory mechanisms at the single-cell level, facilitating insights into cellular heterogeneity and context-dependent TE functions.

Acknowledgments The language of the manuscript was refined using the artificial intelligence tool Kimi (version 1.0.3). Every section was carefully reviewed, revised, and approved by authors to ensure that the text is appropriate in terms of content and scientific correctness and maintains the integrity of the original content.

Compliance with Ethical Standards

Competing Interests The authors declare that they have no competing interests.

Ethical Approval This chapter contains no studies with human/animal participants performed by the authors.

References

Auclair G, Guibert S, Bender A et al (2014) Ontogeny of CpG Island methylation and specificity of DNMT3 methyltransferases during embryonic development in the mouse. Genome Biol 15(12):545. https://doi.org/10.1186/s13059-014-0545-5

Bachmann MC, Bellalta S, Basoalto R et al (2020) The challenge by multiple environmental and biological factors induce inflammation in aging: their role in the promotion of chronic disease. Front Immunol 11:570083. https://doi.org/10.3389/fimmu.2020.570083

Banaz-Yaşar F, Steffen G, Hauschild J et al (2010) LINE-1 retrotransposition events affect endothelial proliferation and migration. Histochem Cell Biol 134(6):581–589. https://doi.org/10.1007/s00418-010-0758-y

Becker JS, Nicetto D, Zaret KS (2016) H3K9me3-dependent heterochromatin: barrier to cell fate changes. Trends Genet 32(1):29–41. https://doi.org/10.1016/j.tig.2015.11.001

Bhat A, Ghatage T, Bhan S et al (2022) Role of transposable elements in genome stability: implications for health and disease. Int J Mol Sci 23(14). https://doi.org/10.3390/ijms23147802

Biryukova I, Ye T (2015) Endogenous siRNAs and piRNAs derived from transposable elements and genes in the malaria vector mosquito Anopheles gambiae. BMC Genomics 16(1):278. https://doi.org/10.1186/s12864-015-1436-1

Black JC, Van Rechem C, Whetstine JR (2012) Histone lysine methylation dynamics: establishment, regulation, and biological impact. Mol Cell 48(4):491–507. https://doi.org/10.1016/j.molcel.2012.11.006

Bourc'his D, Bestor TH (2004) Meiotic catastrophe and retrotransposon reactivation in male germ cells lacking Dnmt3L. Nature 431(7004):96–99. https://doi.org/10.1038/nature02886

Bourque G, Burns KH, Gehring M et al (2018) Ten things you should know about transposable elements. Genome Biol 19(1):199. https://doi.org/10.1186/s13059-018-1577-z

Bühler M, Verdel A, Moazed D (2006) Tethering RITS to a nascent transcript initiates RNAi- and heterochromatin-dependent gene silencing. Cell 125(5):873–886. https://doi.org/10.1016/j.cell.2006.04.025

Burns KH (2017) Transposable elements in cancer. Nat Rev Cancer 17(7):415–424. https://doi.org/10.1038/nrc.2017.35

Carbonell A (2017) Plant ARGONAUTEs: features, functions, and unknowns. Methods Mol Biol 1640:1–21. https://doi.org/10.1007/978-1-4939-7165-7_1

Carmell MA, Girard A, van de Kant HJ et al (2007) MIWI2 is essential for spermatogenesis and repression of transposons in the mouse male germline. Dev Cell 12(4):503–514. https://doi.org/10.1016/j.devcel.2007.03.001

Cervantes-Ayalc A, Ruiz Esparza-Garrido R, Velázquez-Flores M (2020) Long interspersed nuclear elements 1 (LINE1): the chimeric transcript L1-MET and its involvement in cancer. Cancer Genet 241:1–11. https://doi.org/10.1016/j.cancergen.2019.11.004

Chen KM, Campbell E, Pandey RR et al (2015) Metazoan maelstrom is an RNA-binding protein that has evolved from an ancient nuclease active in protists. RNA 21(5):833–839. https://doi.org/10.1261/rna.049437.114

Choi H, Zhou L, Zhao Y et al (2024) RNA helicase D1PAS1 resolves R-loops and forms a complex for mouse pachytene piRNA biogenesis required for male fertility. Nucleic Acids Res 52(19):11973–11994. https://doi.org/10.1093/nar/gkae712

Chuong EB, Elde NC, Feschotte C (2017) Regulatory activities of transposable elements: from conflicts to benefits. Nat Rev Genet 18(2):71–86. https://doi.org/10.1038/nrg.2016.139

Claro-Linares F, Rojas-Ríos P (2025) PIWI proteins and piRNAs: key regulators of stem cell biology. Front Cell Dev Biol 13:1540313. https://doi.org/10.3389/fcell.2025.1540313

Cruickshanks HA, Vafadar-Isfahani N, Dunican DS et al (2013) Expression of a large LINE-1-driven antisense RNA is linked to epigenetic silencing of the metastasis suppressor gene TFPI-2 in cancer. Nucleic Acids Res 41(14):6857–6869. https://doi.org/10.1093/nar/gkt438

Cui X, Jin P, Cui X et al (2013) Control of transposon activity by a histone H3K4 demethylase in rice. Proc Natl Acad Sci U S A 110(5):1953–1958. https://doi.org/10.1073/pnas.1217020110

de Mendoza A, Bonnet A, Vargas-Landin DB et al (2018) Recurrent acquisition of cytosine methyltransferases into eukaryotic retrotransposons. Nat Commun 9(1):1341. https://doi.org/10.1038/s41467-018-03724-9
Della Valle F, Reddy P, Aguirre Vazquez A et al (2025) Reactivation of retrotransposable elements is associated with environmental stress and ageing. Nat Rev Genet. https://doi.org/10.1038/s41576-025-00829-y
Deniz Ö, Frost JM, Branco MR (2019) Regulation of transposable elements by DNA modifications. Nat Rev Genet 20(7):417–431. https://doi.org/10.1038/s41576-019-0106-6
Denli AM, Narvaiza I, Kerman BE et al (2015) Primate-specific ORF0 contributes to retrotransposon-mediated diversity. Cell 163(3):583–593. https://doi.org/10.1016/j.cell.2015.09.025
Dervan E, Bhattacharyya DD, McAuliffe JD et al (2021) Ancient adversary - HERV-K (HML-2) in cancer. Front Oncol 11:658489. https://doi.org/10.3389/fonc.2021.658489
Du K, Wu J, Wang J et al (2024) The chromatin remodeling factor OsINO80 promotes H3K27me3 and H3K9me2 deposition and maintains TE silencing in rice. Nat Commun 15(1):10919. https://doi.org/10.1038/s41467-024-55387-4
Feschotte C (2023) Transposable elements: McClintock's legacy revisited. Nat Rev Genet 24(11):797–800. https://doi.org/10.1038/s41576-023-00652-3
Fueyo R, Judd J, Feschotte C et al (2022) Roles of transposable elements in the regulation of mammalian transcription. Nat Rev Mol Cell Biol 23(7):481–497. https://doi.org/10.1038/s41580-022-00457-y
Gan Y, Wang L, Liu G et al (2024) Transposable elements contribute to the regulation of long noncoding RNAs in Drosophila melanogaster. Insects 15(12). https://doi.org/10.3390/insects15120950
García-Vílchez R, Guallar D (2025) Interplay of transposable elements and ageing: epigenetic regulation and potential epitranscriptomic influence. Curr Opin Genet Dev 92:102331. https://doi.org/10.1016/j.gde.2025.102331
Gardner EJ, Nizami ZF, Talbot CC Jr et al (2012) Stable intronic sequence RNA (sisRNA), a new class of noncoding RNA from the oocyte nucleus of Xenopus tropicalis. Genes Dev 26(22):2550–2559. https://doi.org/10.1101/gad.202184.112
Gazquez-Gutierrez A, Witteveldt J, Heras SR et al (2021) Sensing of transposable elements by the antiviral innate immune system. RNA 27(7):735–752. https://doi.org/10.1261/rna.078721.121
Gebrie A (2023) Transposable elements as essential elements in the control of gene expression. Mob DNA 14(1):9. https://doi.org/10.1186/s13100-023-00297-3
Gendrel AV, Lippman Z, Yordan C et al (2002) Dependence of heterochromatic histone H3 methylation patterns on the Arabidopsis gene DDM1. Science 297(5588):1871–1873. https://doi.org/10.1126/science.1074950
Gerdes P, Lim SM, Ewing AD et al (2022) Retrotransposon instability dominates the acquired mutation landscape of mouse induced pluripotent stem cells. Nat Commun 13(1):7470. https://doi.org/10.1038/s41467-022-35180-x
Greenberg MVC, Bourc'his D (2019) The diverse roles of DNA methylation in mammalian development and disease. Nat Rev Mol Cell Biol 20(10):590–607. https://doi.org/10.1038/s41580-019-0159-6
Guo C, Jeong HH, Hsieh YC et al (2018) Tau activates transposable elements in Alzheimer's disease. Cell Rep 23(10):2874–2880. https://doi.org/10.1016/j.celrep.2018.05.004
Guo Y, Zhao S, Wang GG (2021) Polycomb gene silencing mechanisms: PRC2 chromatin targeting, H3K27me3 'readout', and phase separation-based compaction. Trends Genet 37(6):547–565. https://doi.org/10.1016/j.tig.2020.12.006
Ha ZL, Yu ZY (2021) Downregulation of miR-29b-3p aggravates podocyte injury by targeting HDAC4 in LPS-induced acute kidney injury. Kaohsiung J Med Sci 37(12):1069–1076. https://doi.org/10.1002/kjm2.12431

Han M, Perkins MH, Novaes LS et al (2023) Advances in transposable elements: from mechanisms to applications in mammalian genomics. Front Genet 14:1290146. https://doi.org/10.3389/fgene.2023.1290146
He JP, Chen JK (2021) Epigenetic control of transposable elements and cell fate decision. Yi Chuan 43(9):822–834. https://doi.org/10.16288/j.yczz.21-113
He J, Fu X, Zhang M et al (2019) Transposable elements are regulated by context-specific patterns of chromatin marks in mouse embryonic stem cells. Nat Commun 10(1):34. https://doi.org/10.1038/s41467-018-08006-y
Hong Y, Bie L, Zhang T et al (2024) SAFB restricts contact domain boundaries associated with L1 chimeric transcription. Mol Cell 84(9):1637–1650.e10. https://doi.org/10.1016/j.molcel.2024.03.021
Hossain K, Suzuki T, Hasibuzzaman MM et al (2017) Chronic exposure to arsenic, LINE-1 hypomethylation, and blood pressure: a cross-sectional study in Bangladesh. Environ Health 16(1):20. https://doi.org/10.1186/s12940-017-0231-7
Hu S, Song A, Peng L et al (2023) H3K4me2/3 modulate the stability of RNA polymerase II pausing. Cell Res 33(5):403–406. https://doi.org/10.1038/s41422-023-00794-3
Huang H, Li Y, Szulwach KE et al (2014) AGO3 slicer activity regulates mitochondria-nuage localization of Armitage and piRNA amplification. J Cell Biol 206(2):217–230. https://doi.org/10.1083/jcb.201401002
Huang Z, Whelan CV, Foley NM et al (2019) Longitudinal comparative transcriptomics reveals unique mechanisms underlying extended healthspan in bats. Nat Ecol Evol 3(7):1110–1120. https://doi.org/10.1038/s41559-019-0913-3
Huda A, Jordan IK (2009) Epigenetic regulation of mammalian genomes by transposable elements. Ann N Y Acad Sci 1178:276–284. https://doi.org/10.1111/j.1749-6632.2009.05007.x
Hummel B, Hansen EC, Yoveva A et al (2017) The evolutionary capacitor HSP90 buffers the regulatory effects of mammalian endogenous retroviruses. Nat Struct Mol Biol 24(3):234–242. https://doi.org/10.1038/nsmb.3368
Hung FY, Feng YR, Hsin KT et al (2023) Arabidopsis histone H3 lysine 9 methyltransferases KYP/SUVH5/6 are involved in leaf development by interacting with AS1-AS2 to repress KNAT1 and KNAT2. Commun Biol 6(1):219. https://doi.org/10.1038/s42003-023-04607-6
Hure V, Piron-Prunier F, Yehouessi T et al (2025) Alternative silencing states of transposable elements in Arabidopsis associated with H3K27me3. Genome Biol 26(1):11. https://doi.org/10.1186/s13059-024-03466-6
Jing T, Wei D, Xu X et al (2024) Transposable elements-mediated recruitment of KDM1A epigenetically silences HNF4A expression to promote hepatocellular carcinoma. Nat Commun 15(1):5631. https://doi.org/10.1038/s41467-024-49926-2
Kane AE, Sinclair DA (2019) Epigenetic changes during aging and their reprogramming potential. Crit Rev Biochem Mol Biol 54(1):61–83. https://doi.org/10.1080/10409238.2019.1570075
Kanno T, Aufsatz W, Jaligot E et al (2005a) A SNF2-like protein facilitates dynamic control of DNA methylation. EMBO Rep 6(7):649–655. https://doi.org/10.1038/sj.embor.7400446
Kanno T, Huettel B, Mette MF et al (2005b) Atypical RNA polymerase subunits required for RNA-directed DNA methylation. Nat Genet 37(7):761–765. https://doi.org/10.1038/ng1580
Karakaidos P, Verigos J, Magklara A (2019) LSD1/KDM1A, a gate-keeper of cancer stemness and a promising therapeutic target. Cancers (Basel) 11(12). https://doi.org/10.3390/cancers11121821
Kim EY, Wang L, Lei Z et al (2021) Ribosome stalling and SGS3 phase separation prime the epigenetic silencing of transposons. Nat Plants 7(3):303–309. https://doi.org/10.1038/s41477-021-00867-4
Kinney JW, Bemiller SM, Murtishaw AS et al (2018) Inflammation as a central mechanism in Alzheimer's disease. Alzheimers Dement (N Y) 4:575–590. https://doi.org/10.1016/j.trci.2018.06.014
Klemm SL, Shipony Z, Greenleaf WJ (2019) Chromatin accessibility and the regulatory epigenome. Nat Rev Genet 20(4):207–220. https://doi.org/10.1038/s41576-018-0089-8

Kofler R, Hill T, Nolte V et al (2015) The recent invasion of natural Drosophila simulans populations by the P-element. Proc Natl Acad Sci U S A 112(21):6659–6663. https://doi.org/10.1073/pnas.1500758112

Kumar S, Lombard DB (2015) Mitochondrial sirtuins and their relationships with metabolic disease and cancer. Antioxid Redox Signal 22(12):1060–1077. https://doi.org/10.1089/ars.2014.6213

Lagisquet J, Zuber K, Gramberg T (2021) Recognize yourself-innate sensing of non-LTR retrotransposons. Viruses 13(1). https://doi.org/10.3390/v13010094

Lawson HA, Liang Y, Wang T (2023) Transposable elements in mammalian chromatin organization. Nat Rev Genet 24(10):712–723. https://doi.org/10.1038/s41576-023-00609-6

Lee E, Iskow R, Yang L et al (2012) Landscape of somatic retrotransposition in human cancers. Science 337(6097):967–971. https://doi.org/10.1126/science.1222077

Lee K, Ku J, Ku D et al (2024) Inverted Alu repeats: friends or foes in the human transcriptome. Exp Mol Med 56(6):1250–1262. https://doi.org/10.1038/s12276-024-01177-3

Lewis L, Chappell GA, Kobets T et al (2019) Sex-specific differences in genotoxic and epigenetic effects of 1,3-butadiene among mouse tissues. Arch Toxicol 93(3):791–800. https://doi.org/10.1007/s00204-018-2374-x

Li X, Bie L, Wang Y et al (2024) LINE-1 transcription activates long-range gene expression. Nat Genet 56(7):1494–1502. https://doi.org/10.1038/s41588-024-01789-5

Li Z, Xu Q, Zhong J et al (2025) Structural insights into RNA cleavage by PIWI Argonaute. Nature 639(8053):250–259. https://doi.org/10.1038/s41586-024-08438-1

Ling Z, Li J, Jiang T et al (2024) Omics-based construction of regulatory variants can be applied to help decipher pig liver-related traits. Commun Biol 7(1):381. https://doi.org/10.1038/s42003-024-06050-7

Liu Y, Hu L, Wu Z et al (2023) Loss of PHF8 induces a viral mimicry response by activating endogenous retrotransposons. Nat Commun 14(1):4225. https://doi.org/10.1038/s41467-023-39943-y

Liu S, Li K, Dai X et al (2025) A telomere-to-telomere genome assembly coupled with multi-omic data provides insights into the evolution of hexaploid bread wheat. Nat Genet 57(4):1008–1020. https://doi.org/10.1038/s41588-025-02137-x

Macchietto MG, Langlois RA, Shen SS (2020) Virus-induced transposable element expression up-regulation in human and mouse host cells. Life Sci Alliance 3(2). https://doi.org/10.26508/lsa.201900536

Macfarlan TS, Gifford WD, Driscoll S et al (2012) Embryonic stem cell potency fluctuates with endogenous retrovirus activity. Nature 487(7405):57–63. https://doi.org/10.1038/nature11244

Macrae IJ, Zhou K, Li F et al (2006) Structural basis for double-stranded RNA processing by Dicer. Science 311(5758):195–198. https://doi.org/10.1126/science.1121638

Maston GA, Evans SK, Green MR (2006) Transcriptional regulatory elements in the human genome. Annu Rev Genomics Hum Genet 7:29–59. https://doi.org/10.1146/annurev.genom.7.080505.115623

Matzke MA, Mosher RA (2014) RNA-directed DNA methylation: an epigenetic pathway of increasing complexity. Nat Rev Genet 15(6):394–408. https://doi.org/10.1038/nrg3683

Maugeri A, Barchitta M, Magnano San Lio R et al (2021) The relationship between body mass index, obesity, and LINE-1 methylation: a cross-sectional study on women from southern Italy. Dis Markers 2021:9910878. https://doi.org/10.1155/2021/9910878

Mendez-Dorantes C, Burns KH (2023) LINE-1 retrotransposition and its deregulation in cancers: implications for therapeutic opportunities. Genes Dev 37(21–24):948–967. https://doi.org/10.1101/gad.351051.123

Miki Y, Nishisho I, Horii A et al (1992) Disruption of the APC gene by a retrotransposal insertion of L1 sequence in a colon cancer. Cancer Res 52(3):643–645

Moreno-Fernández S, Garcés-Rimón M, Vera G et al (2018) High fat/high glucose diet induces metabolic syndrome in an experimental rat model. Nutrients 10(10). https://doi.org/10.3390/nu10101502

Murchison EP, Stein P, Xuan Z et al (2007) Critical roles for dicer in the female germline. Genes Dev 21(6):682–693. https://doi.org/10.1101/gad.1521307

Nam CH, Youk J, Kim JY et al (2023) Widespread somatic L1 retrotransposition in normal colorectal epithelium. Nature 617(7961):540–547. https://doi.org/10.1038/s41586-023-06046-z

Noronha N, Durette C, Cahuzac M et al (2024) Autophagy degrades immunogenic endogenous retroelements induced by 5-azacytidine in acute myeloid leukemia. Leukemia 38(5):1019–1031. https://doi.org/10.1038/s41375-024-02250-6

Nosaka M, Ono A, Ishiwata A et al (2013) Expression of the rice microRNA miR820 is associated with epigenetic modifications at its own locus. Genes Genet Syst 88(2):105–112. https://doi.org/10.1266/ggs.88.105

Nuthikattu S, McCue AD, Panda K et al (2013) The initiation of epigenetic silencing of active transposable elements is triggered by RDR6 and 21-22 nucleotide small interfering RNAs. Plant Physiol 162(1):116–131. https://doi.org/10.1104/pp.113.216481

Oomen ME, Rodriguez-Terrones D, Kurome M et al (2025) An atlas of transcription initiation reveals regulatory principles of gene and transposable element expression in early mammalian development. Cell 188(4):1156–1174.e20. https://doi.org/10.1016/j.cell.2024.12.013

Pedro-Botet J, Climent E, Benaiges D (2020) Atherosclerosis and inflammation. New therapeutic approaches. Med Clin (Barc) 155(6):256–262. https://doi.org/10.1016/j.medcli.2020.04.024

Petri R, Brattås PL, Sharma Y et al (2019) LINE-2 transposable elements are a source of functional human microRNAs and target sites. PLoS Genet 15(3):e1008036. https://doi.org/10.1371/journal.pgen.1008036

Pontis J, Pulver C, Playfoot CJ et al (2022) Primate-specific transposable elements shape transcriptional networks during human development. Nat Commun 13(1):7178. https://doi.org/10.1038/s41467-022-34800-w

Qi L, Sui Y, Tang XX et al (2023) Shuffling the yeast genome using CRISPR/Cas9-generated DSBs that target the transposable Ty1 elements. PLoS Genet 19(1):e1010590. https://doi.org/10.1371/journal.pgen.1010590

Qian W, Jiang P, Niu M et al (2025) Selective identification of epigenetic regulators at methylated genomic sites by SelectID. Nat Commun 16(1):3709. https://doi.org/10.1038/s41467-025-59002-y

Ragupathy R, Banks T, Cloutier S (2010) Molecular characterization of the Sasanda LTR copia retrotransposon family uncovers their recent amplification in Triticum aestivum (L.) genome. Mol Genet Genomics 283(3):255–271. https://doi.org/10.1007/s00438-009-0509-8

Rehwinkel J, Gack MU (2020) RIG-I-like receptors: their regulation and roles in RNA sensing. Nat Rev Immunol 20(9):537–551. https://doi.org/10.1038/s41577-020-0288-3

Rodić N, Steranka JP, Makohon-Moore A et al (2015) Retrotransposon insertions in the clonal evolution of pancreatic ductal adenocarcinoma. Nat Med 21(9):1060–1064. https://doi.org/10.1038/nm.3919

Rodriguez-Martin B, Alvarez EG, Baez-Ortega A et al (2020) Pan-cancer analysis of whole genomes identifies driver rearrangements promoted by LINE-1 retrotransposition. Nat Genet 52(3):306–319. https://doi.org/10.1038/s41588-019-0562-0

Sakashita A, Kitano T, Ishizu H et al (2023) Transcription of MERVL retrotransposons is required for preimplantation embryo development. Nat Genet 55(3):484–495. https://doi.org/10.1038/s41588-023-01324-y

Sepulveda H, Li X, Yue X et al (2024) OGT prevents DNA demethylation and suppresses the expression of transposable elements in heterochromatin by restraining TET activity genome-wide. bioRxiv. https://doi.org/10.1101/2024.01.31.578097

Shao Y, Ma J, Zhang S et al (2025) NERD-dependent m(6)A modification of the nascent FLC transcript regulates flowering time in Arabidopsis. Nat Plants 11(3):468–482. https://doi.org/10.1038/s41477-025-01945-7

Slotkin RK, Martienssen R (2007) Transposable elements and the epigenetic regulation of the genome. Nat Rev Genet 8(4):272–285. https://doi.org/10.1038/nrg2072

Smith CG, Moser T, Mouliere F et al (2020) Comprehensive characterization of cell-free tumor DNA in plasma and urine of patients with renal tumors. Genome Med 12(1):23. https://doi.org/10.1186/s13073-020-00723-8

Soderlund C, Descour A, Kudrna D et al (2009) Sequencing, mapping, and analysis of 27,455 maize full-length cDNAs. PLoS Genet 5(11):e1000740. https://doi.org/10.1371/journal.pgen.1000740

Stacey KJ, Sagulenko V (2010) A clear link between endogenous retroviral LTR activity and Hodgkin's lymphoma. Cell Res 20(8):869–871. https://doi.org/10.1038/cr.2010.96

Sun H, Chen Z, Jiang J et al (2023) Analyses of regulation between miRNA and DNA methyltransferase 1 related genes in sea cucumber Apostichopusjaponicus. Fish Shellfish Immunol 143:109169. https://doi.org/10.1016/j.fsi.2023.109169

Sun D, Zhu Y, Peng W et al (2024) SETDB1 regulates short interspersed nuclear elements and chromatin loop organization in mouse neural precursor cells. Genome Biol 25(1):175. https://doi.org/10.1186/s13059-024-03327-2

Tahami MS, Vargas-Chavez C, Poikela N et al (2024) Transposable elements in Drosophila montana from harsh cold environments. Mob DNA 15(1):18. https://doi.org/10.1186/s13100-024-00328-7

Tan S, Ma H, Wang J et al (2021) DNA transposons mediate duplications via transposition-independent and -dependent mechanisms in metazoans. Nat Commun 12(1):4280. https://doi.org/10.1038/s41467-021-24585-9

Tee WW, Reinberg D (2014) Chromatin features and the epigenetic regulation of pluripotency states in ESCs. Development 141(12):2376–2390. https://doi.org/10.1242/dev.096982

Theunissen TW, Friedli M, He Y et al (2016) Molecular criteria for defining the naive human pluripotent state. Cell Stem Cell 19(4):502–515. https://doi.org/10.1016/j.stem.2016.06.011

Tian Y, Feng T, Zhang J et al (2025) Histone H1 deamidation facilitates chromatin relaxation for DNA repair. Nature. https://doi.org/10.1038/s41586-025-08835-0

Tie CH, Rowe HM (2017) Epigenetic control of retrotransposons in adult tissues: implications for immune regulation. Curr Opin Virol 25:28–33. https://doi.org/10.1016/j.coviro.2017.06.007

Tomasetti M, Monaco F, Rubini C et al (2024) AGO2-RIP-Seq reveals miR-34/miR-449 cluster targetome in sinonasal cancers. PLoS One 19(1):e0295997. https://doi.org/10.1371/journal.pone.0295997

Tsai YT, Seymen N, Thompson IR et al (2024) Expression of most retrotransposons in human blood correlates with biological aging. Elife 13. https://doi.org/10.7554/eLife.96575

Tsokos GC (2020) Autoimmunity and organ damage in systemic lupus erythematosus. Nat Immunol 21(6):605–614. https://doi.org/10.1038/s41590-020-0677-6

Van Meter M, Kashyap M, Rezazadeh S et al (2014) SIRT6 represses LINE1 retrotransposons by ribosylating KAP1 but this repression fails with stress and age. Nat Commun 5:5011. https://doi.org/10.1038/ncomms6011

Vijay K (2018) Toll-like receptors in immunity and inflammatory diseases: past, present, and future. Int Immunopharmacol 59:391–412. https://doi.org/10.1016/j.intimp.2018.03.002

Vojvoda Zeljko T, Ugarković Đ, Pezer Ž (2021) Differential enrichment of H3K9me3 at annotated satellite DNA repeats in human cell lines and during fetal development in mouse. Epigenetics Chromatin 14(1):47. https://doi.org/10.1186/s13072-021-00423-6

Walsh CP, Chaillet JR, Bestor TH (1998) Transcription of IAP endogenous retroviruses is constrained by cytosine methylation. Nat Genet 20(2):116–117. https://doi.org/10.1038/2413

Wang H, Fan Z, Shliaha PV et al (2023a) H3K4me3 regulates RNA polymerase II promoter-proximal pause-release. Nature 615(7951):339–348. https://doi.org/10.1038/s41586-023-05780-8

Wang X, Ramat A, Simonelig M et al (2023b) Emerging roles and functional mechanisms of PIWI-interacting RNAs. Nat Rev Mol Cell Biol 24(2):123–141. https://doi.org/10.1038/s41580-022-00528-0

Woodcock DM, Lawler CB, Linsenmeyer ME et al (1997) Asymmetric methylation in the hypermethylated CpG promoter region of the human L1 retrotransposon. J Biol Chem 272(12):7810–7816. https://doi.org/10.1074/jbc.272.12.7810

Wu K, Fan D, Zhao H et al (2023) Dynamics of histone acetylation during human early embryogenesis. Cell Discov 9(1):29. https://doi.org/10.1038/s41421-022-00514-y
Xiang Y, Yan K, Zheng Q et al (2019) Histone demethylase KDM4B promotes DNA damage by activating long interspersed nuclear Element-1. Cancer Res 79(1):86–98. https://doi.org/10.1158/0008-5472.Can-18-1310
Xiao-Jie L, Hui-Ying X, Qi X et al (2016) LINE-1 in cancer: multifaceted functions and potential clinical implications. Genet Med 18(5):431–439. https://doi.org/10.1038/gim.2015.119
Xie Z, Johansen LK, Gustafson AM et al (2004) Genetic and functional diversification of small RNA pathways in plants. PLoS Biol 2(5):E104. https://doi.org/10.1371/journal.pbio.0020104
Xu B, Li X, Zhang S et al (2023) Pan cancer characterization of genes whose expression has been associated with LINE-1 antisense promoter activity. Mob DNA 14(1):13. https://doi.org/10.1186/s13100-023-00300-x
Yan B, Yuan Q, Guryanova OA (2023) Epigenetic mechanisms in hematologic aging and premalignant conditions. Epigenomes 7(4). https://doi.org/10.3390/epigenomes7040032
Yang Z, Su W, Wei X et al (2023) HIF-1α drives resistance to ferroptosis in solid tumors by promoting lactate production and activating SLC1A1. Cell Rep 42(8):112945. https://doi.org/10.1016/j.celrep.2023.112945
Yao Y, Zhou J, Wang J et al (2025) H3K36 methylation stamps transcription resistive to preserve development in plants. Nat Plants. https://doi.org/10.1038/s41477-025-01962-6
Yu H, Chen M, Hu Y et al (2022) Dynamic reprogramming of H3K9me3 at hominoid-specific retrotransposons during human preimplantation development. Cell Stem Cell 29(7):1031–1050.e12. https://doi.org/10.1016/j.stem.2022.06.006
Yuan D, Zhang X, Yang Y et al (2024) Schlank orchestrates insect developmental transition by switching H3K27 acetylation to trimethylation in the prothoracic gland. Proc Natl Acad Sci U S A 121(35):e2401861121. https://doi.org/10.1073/pnas.2401861121
Yushkova E, Moskalev A (2023) Transposable elements and their role in aging. Ageing Res Rev 86:101881. https://doi.org/10.1016/j.arr.2023.101881
Zhang X, Bai XC, Chen ZJ (2020) Structures and mechanisms in the cGAS-STING innate immunity pathway. Immunity 53(1):43–53. https://doi.org/10.1016/j.immuni.2020.05.013
Zhang J, Yuan J, Lin J et al (2023) Molecular basis of locus-specific H3K9 methylation catalyzed by SUVH6 in plants. Proc Natl Acad Sci U S A 120(1):e2208525120. https://doi.org/10.1073/pnas.2211155120
Zhao S, Lu J, Pan B et al (2023) TNRC18 engages H3K9me3 to mediate silencing of endogenous retrotransposons. Nature 623(7987):633–642. https://doi.org/10.1038/s41586-023-06688-z
Zhao XR, Zong JB, Liu YX et al (2024) Endogenous retroviruses unveiled: a comprehensive review of inflammatory signaling/senescence-related pathways and therapeutic strategies. Aging Dis 16(2):738–756. https://doi.org/10.14336/ad.2024.0123-1
Zhen S, Jia Y, Zhao Y et al (2024) NEAT1_1 confers gefitinib resistance in lung adenocarcinoma through promoting AKR1C1-mediated ferroptosis defence. Cell Death Dis 10(1):131. https://doi.org/10.1038/s41420-024-01892-w
Zhou X, He J, Velanis CN et al (2021) A domesticated Harbinger transposase forms a complex with HDA6 and promotes histone H3 deacetylation at genes but not TEs in Arabidopsis. J Integr Plant Biol 63(8):1462–1474. https://doi.org/10.1111/jipb.13108
Zhou F, Guo H, Xia Y et al (2024a) The changing treatment landscape of EGFR-mutant non-small-cell lung cancer. Nat Rev Clin Oncol. https://doi.org/10.1038/s41571-024-00971-2
Zhou X, Fan Y, Zhu X et al (2024b) SANT proteins modulate gene expression by coordinating histone H3KAc and Khib levels and regulate plant heat tolerance. Plant Physiol 196(2):902–915. https://doi.org/10.1093/plphys/kiae348
Zhu X, Chen Z, Shen W et al (2021) Inflammation, epigenetics, and metabolism converge to cell senescence and ageing: the regulation and intervention. Signal Transduct Target Ther 6(1):245. https://doi.org/10.1038/s41392-021-00646-9

Chapter 5
Transposable Element-Derived Noncoding RNAs

Francesco Panariello, Alen Stambolliu, Michele Panepuccia, Filippo Vittorio Burattin, and Beatrice Bodega

Abstract The human genome is composed of almost 98% of noncoding DNA, much of which gives rise to functional noncoding RNAs (ncRNAs). A significant but often overlooked source of ncRNAs is transposable elements (TEs)—mobile genetic sequences capable of integrating into various genomic regions. TEs contribute extensively to the ncRNA landscape, yet their transcriptional activity is often entangled with that of host genes, complicating functional interpretation. They have been shown to be co-opted by cells for various functions, such as cryptic cis-regulatory sequences, and for generating transcripts (TE transcripts) that could impact cellular functions and disease processes. In contexts like stemness, cancer, and immunity, TEs are crucial in maintaining cell identity through transcriptional regulation. However, TE-derived ncRNAs remain insufficiently characterized, partly due to reliance on outdated genomic tools and databases. This chapter addresses this gap by providing a comprehensive overview of TE-derived ncRNAs based on the most current genomic resources. We first examine how TEs, or portions of them, are incorporated into annotated ncRNA transcripts based on the most recent human genome reference, highlighting recent technological advances that enhance this analysis. The discussion then narrows to specific examples of TE insertions within two prominent ncRNA classes: microRNAs (miRNAs) and long noncoding

F. Panariello · F. V. Burattin
INGM, Istituto Nazionale di Genetica Molecolare "Romeo ed Enrica Invernizzi", Milan, Italy
e-mail: panariello@ingm.org; burattin@ingm.org

A. Stambolliu · M. Panepuccia
INGM, Istituto Nazionale di Genetica Molecolare "Romeo ed Enrica Invernizzi", Milan, Italy

SEMM, European School of Molecular Medicine, Milan, Italy
e-mail: stambolliu@ingm.org; panepuccia@ingm.org

B. Bodega (✉)
INGM, Istituto Nazionale di Genetica Molecolare "Romeo ed Enrica Invernizzi", Milan, Italy

Department of Clinical Sciences and Community Health, University of Milan, Milan, Italy
e-mail: bodega@ingm.org

L. M. Vaschetto (ed.), *Transposable Elements, Transcriptomics, and Diseases*, Epigenetics and Human Health 14, https://doi.org/10.1007/978-3-032-04125-8_5

RNAs (lncRNAs). Finally, we explore emerging evidence linking TE-containing ncRNAs to various disease contexts, offering insights into their potential biological and clinical relevance.

Keywords Transposable elements · Noncoding RNA · NGS · Genomics · Health and disease

Abbreviations

ncRNA	Noncoding RNA
miRNA	MicroRNAs
lncRNA	Long noncoding RNA
TE	Transposable element
RE	Retrotransposable element
LINEs	Long interspersed nuclear elements
SINEs	Short interspersed nuclear elements
LTRs	Long terminal repeats
NMD	Nonsesnse-mediated decay
NGS	Next-generation sequencing
snRNA	Small nuclear RNA
rRNA	Ribosomal RNA
ORF	Open reading frame
hPSC	Human pluripotent stem cells
hESCs	Human embryonic stem cells
hiPSC	Human induced pluripotent stem cells
XCI	X chromosome inactivation
XIC	X-inactivation center
HCC	Hepatocellular carcinoma
CPCs	Cardiac precursor cells
CMs	Cardiomyocytes
SMCs	Smooth muscle cells
SMD	Staufen1 mediated decay
SBS	Stau1-binding sites
UTRs	Untranslated regions
eRNAs	Enhancer RNAs
uaRNAs	Upstream antisense RNAs
PROMPTs	Promoter upstream transcripts
RIC-seq	In Situ Conformation Sequencing
IRAlus	Inverted repeats Alu elements
Pol II	RNA polymerase II
Pol III	RNA polymerase III
Alu-RA	Alu right arm
Alu-LA	Alu left arm

Cryo-EM	Cryogenic electron microscopy
circRNAs	Circular RNAs
ecircRNAs	Exon-derived circular RNAs
pri-miRNAs	Primary transcripts of micro RNAs
pre-miRNAs	Precursor of micro RNAs
mRNA	Messenger RNA
MDTEs	Micro RNAs derived from transposable elements
MITEs	Miniature inverted-repeat transposable elements
CRC	Colorectal cancer
ASCVD	Atherosclerotic cardiovascular diseases
CAD	Coronary artery disease
PAD	Peripheral artery disease
ESCC	Esophageal squamous cell carcinoma
ASP	Antisense promoter
BET	Bromodomain and extraterminal proteins
TNBC	Triple-negative breast cancer
TILs	Tumor-infiltrating lymphocytes
NSCLC	Non-small cell lung cancer
HLA-I	Human leukocytes antigen I
UCA1	Urothelial cancer-associated 1
hTSCs	Human trophoblast stem cells
EO-PE	Early onset preeclampsia
siRNAs	Small interfering RNAs
ASOs	Antisense oligonucleotides

5.1 Introduction

5.1.1 Background

Protein-coding sequences, primarily located within exons, represent a small portion of the human genome (1.5%), while a substantial portion is made up of non-protein-coding DNA, which was once labeled as "junk" because it was initially believed to have no biological function (Palazzo and Gregory 2014). Although Barbara McClintock had already proposed a functional role for these elements as early as the mid-twentieth century, it was the advancement of next-generation sequencing (NGS) technologies and the emergence of large-scale projects such as ENCODE (Luo et al. 2020) and FANTOM (Kawaji et al. 2017) that later demonstrated that much of this non-protein-coding DNA is transcribed into biologically active RNA molecules, collectively known as noncoding RNAs (ncRNAs) (Pennisi 2012; De Hoon et al. 2022). Afterwards, advances in sequencing technologies have led to the identification and classification of tens of thousands of ncRNAs (Lagarde et al. 2017; Marx 2023; Mattick et al. 2023). These molecules act as key regulators of gene expression in crucial biological processes through different mechanisms: chromatin modification, molecular sponges, degradation of coding transcripts, enhancers, and so on (Ranzani et al. 2015; Zhang et al. 2019a; Mattick et al. 2023; Ferrer and Dimitrova 2024; Oo et al. 2025).

Recent studies have highlighted the close relationship between ncRNAs and transposable elements (TEs), showing that a large proportion of ncRNAs are derived from these elements (Kapusta et al. 2013; Carlevaro-Fita et al. 2019; Park et al. 2022). Nonetheless, relatively little is understood about ncRNAs derived from TEs, which contribute significantly to the overall ncRNA pool (Kapusta et al. 2013). TEs are mobile genetic elements that can insert into various regions of the genome and are classified into two main types based on their transposition mechanism (Wicker et al. 2007; Wells and Feschotte 2020). Retrotransposons (RE—Class 1, 43%), which include long interspersed nuclear elements (LINEs), short interspersed nuclear elements (SINEs), and long terminal repeats (LTRs), move through the genome via a 'copy-and-paste' mechanism, utilizing an RNA intermediate that is reverse-transcribed into DNA before being reinserted elsewhere (Boeke et al. 1985). DNA transposons (Class 2, 3%), in contrast, utilize a 'cut-and-paste' mechanism, directly excising and reinserting DNA sequences (Yuan et al. 2011a). TEs are subdivided into superfamilies, families, and subfamilies, based on their sequence homology, which reflects their evolutionary history (Koning et al. 2011). Retrotransposons are further classified as autonomous or non-autonomous based on their ability to encode the proteins required for their own mobilization. Autonomous elements contain the necessary genetic information to produce these proteins, whereas non-autonomous elements rely on the machinery encoded by the autonomous ones. So-called "full-length" elements, which are capable of mobilization, whether autonomous or non-autonomous, are typically transcribed using an internal promoter that is carried along with the element during transposition and allows their further propagation into the genome (Richardson et al. 2015). The same occurs for full-length DNA transposons. However, the transposition process can be subjected to 5′ truncation, ectopic recombination, or host-induced repressive mechanisms (e.g., RNA interference), resulting in the loss of their promoter region (Feschotte and Pritham 2007; Chuong et al. 2017). As a result, many TEs are now remnants of once-active copies, losing their ability to transpose. They can be found as short fossils, often inserted in the sequence of genes or their vicinity, with their transcription often closely linked with gene expression. Therefore, besides their transcription as "full-length" units, pervasive transcription includes many TEs in the sequence of their host genes, a process known as co-transcription (Lanciano and Cristofari 2020), generating chimeric transcripts. The tools to study TEs have, hence, to be tailored to distinguish the two phenomena.

Given that ncRNAs are increasingly recognized for their potential as diagnostic biomarkers for various diseases (Yang et al. 2023; Bakinowska et al. 2024; Eldakhakhny et al. 2024; Chang et al. 2024), it is paradigmatic to deepen our knowledge on how ncRNA and, especially, TE-derived ncRNAs are involved in fundamental biological processes, both in health and disease, to find novel potential markers and therapeutic targets in complex and multifactorial pathological conditions.

5.1.2 *Aim of This Chapter*

This chapter aims to provide an overview of TE-derived ncRNAs involved in health and disease, as evidenced by the most current state of the art. We will first leverage the latest genomic resources to characterize how full-length TEs (or portions of them) are included in annotated noncoding transcripts. We will also tackle which technological improvements have been made to deepen our knowledge in this matter.

Then, in the following paragraphs, we will focus on some examples of TEs inserted in two of the most represented classes of noncoding RNAs, specifically micro-RNAs (miRNAs) and long ncRNAs (lncRNAs). We will conclude this chapter by giving some insights into TE-containing ncRNAs involved in several pathogenic contexts.

5.2 Almost 60% of Noncoding RNAs Contain Transposable Elements

Although several studies have investigated the contribution of TE insertions to long noncoding RNAs using earlier GENCODE releases (Kapusta et al. 2013; Carlevaro-Fita et al. 2019; Park et al. 2022), in this chapter, we aim to extend the analysis to all noncoding RNA species as annotated in the latest GENCODE version (v47, released in October 2024) (Mudge et al. 2025). To achieve this goal, we have intersected the information about transcripts with the RepeatMasker annotation v4.0.3 (Smit et al. 2013) using the *intersect* module from Bedtools (Quinlan and Hall 2010; Quinlan 2014). We have defined as TE-containing those transcripts that showed a sequence overlap of at least 20 bp with a genomic instance annotated as TE in the RepeatMasker, as we did in Marasca et al. (2022). Among the 387,944 transcripts annotated in the GENCODE database, it was found that 56.75% (220,174) contained at least one Transposable element, with the SINE class being the most abundant (20.57%), followed by LINE (15.26%), LTR (13.38%), and DNA transposons (6.08%—Fig. 5.1a). We then looked at their distribution across different transcript types. To highlight the difference between coding and noncoding transcripts, we narrowed down the GENCODE definitions to three categories of transcripts:

(a) *Protein coding*
(b) *ncRNAs*, in which we considered canonical noncoding species (e.g., lncRNAs, snRNAs, miRNAs, rRNAs, etc.), protein-coding transcripts lacking an open reading frame (ORF) and those with a retained intron
(c) *Other*, which contained pseudogenes, artifacts, and transcripts predicted to be subjected to nonsesnse-mediated decay (NMD)

Protein-coding transcripts were mostly found to be deprived of TEs in their sequence. On the other hand, the number of TE-containing transcripts annotated as ncRNAs was more than double their protein-coding counterpart (Fig. 5.1b). This

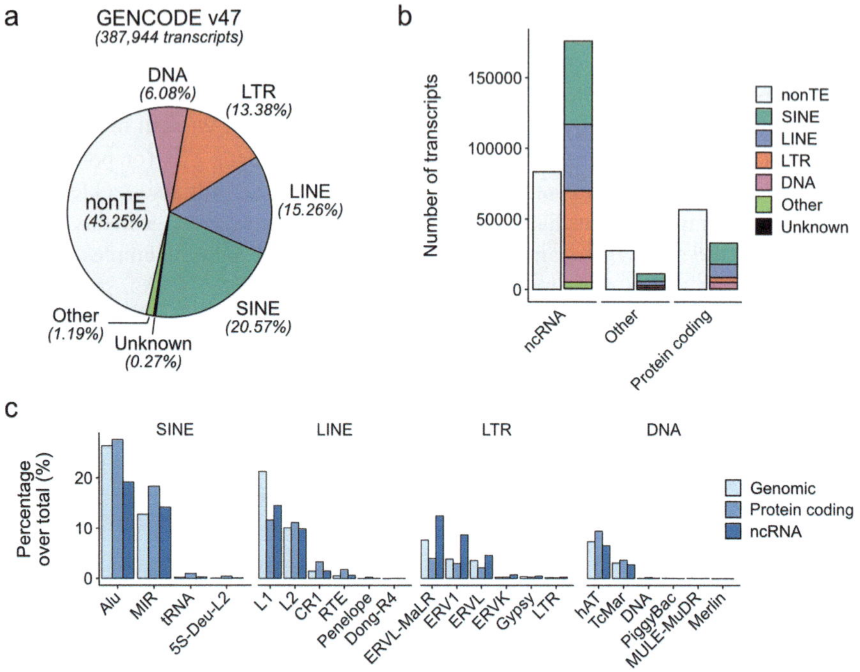

Fig. 5.1 (**a**) Pie chart displaying the percentage of transcripts from GENCODE v47 annotation that contain a TE, colored by class. Transcripts that do not contain any TE instance are reported as nonTE (gray slice). (**b**) Bar plot showing the number of TE-containing and TE-deprived (nonTE) transcripts, colored by class and divided by the transcript type. (**c**) Distribution of TE families (expressed in percentage) across the entire genome (Genomic), protein-coding or noncoding transcripts (ncRNA)

relationship is in line with previous studies (Park et al. 2022), and it suggests an insertional preference of TEs into noncoding regions of the genome. A huge determinant of TEs' localization stands in the process of natural selection they are subjected to (Dazeniere et al. 2022). Indeed, TEs are intrinsically mutagenic elements, and they can be selected against in order to preserve host fitness (Campos-Sánchez et al. 2016). The underrepresentation of TEs within protein transcripts might be indicative of the deleterious effects that they could have on gene expression, whether at the transcript or at the protein level (Maksakova et al. 2006).

One reason TE insertions are subjected to negative selection is their tendency to mediate ectopic recombination—that is, homologous recombination between similar elements located at different genomic positions, including on separate chromosomes—which can lead to large-scale chromosomal rearrangements (Balachandran et al. 2022). Hence, we compared the enrichment of the most represented TE families along the entire genome with their representation in protein-coding and noncoding transcripts (Fig. 5.1c). *Alu* repeats, members of the SINE class, are very abundant along the genome (~22%); their prevalence becomes even higher when looking at the number of their insertions in protein-coding transcripts (~28%). Also, *MIR*, another SINE element, displays the same pattern, following *Alu* as the most enriched TE family in protein-coding transcripts (~18%). On the other hand, the LINE member *L1* and LTR families (*ERV1*, *ERVL-MaLR*, *ERVL*) exhibit a higher enrichment in noncoding transcripts compared to coding ones. This evidence is in line with studies that describe how longer homologous regions are more prone to recombination events leading to large-scale chromosomal rearrangements (Langley et al. 1988; Han et al. 2008).

5.3 Emerging Technologies for the Detection of TE-Containing ncRNAs: From Long-Read Sequencing to De Novo Assembly

ncRNAs exhibit highly tissue-specific expression patterns, which can vary significantly across different cell types and biological contexts (Jiang et al. 2016; Seifuddin et al. 2020), as well as physiological or pathological conditions (Jiang et al. 2016; Navandar et al. 2024). This specificity plays a key role in gene regulation, cellular differentiation, and other essential processes (Wagner et al. 2023; Xie et al. 2024). However, due to this variability, many ncRNAs remain underrepresented or even missing in reference annotations such as GENCODE, which tend to favor genes with broad or high expression. As a result, transcripts with low (Xue et al. 2008) or highly context-dependent expression—common features of ncRNAs—are often overlooked. For example, many lncRNAs are expressed only in specific tissues or under certain conditions, making comprehensive cataloging difficult (Bjorklund et al. 2022). This complexity increases further when transposable elements (TEs) are co-opted into ncRNA sequences, contributing to regulatory or functional domains. TE-derived regulatory elements may also be active only in certain tissues or disease states, meaning their roles—and the full regulatory potential of the ncRNAs they affect—can be missed in datasets focused solely on well-characterized, broadly expressed genes (Marasca et al. 2022; Burattin et al. 2024). This phenomenon may be overlooked if databases primarily focus on well-characterized genes without incorporating the diversity and functional potential of TEs.

NGS technologies have revolutionized the study of ncRNAs, enabling the discovery of numerous context-specific ncRNAs that were previously uncharacterized (Jarroux et al. 2017; Mihaljevic et al. 2024). De novo assembly approaches (Raghavan et al. 2022) have played a crucial role in this advancement by allowing the identification of novel ncRNAs across a wide array of tissues and conditions (Ranzani et al. 2015; Yamada et al. 2018; Zhang et al. 2023a). These methods have revealed tissue-specific lncRNAs, contributing significantly to our understanding of gene regulation in diverse biological contexts and processes, such as cellular differentiation (Fatica and Bozzoni 2014), stress responses (Nickerson and Momen-Heravi 2024), and disease mechanisms (De Goede et al. 2021). The continuous refinement of NGS techniques and bioinformatics tools is poised to uncover even more previously overlooked ncRNAs (Li and Liang 2022; Huang et al. 2024), offering new avenues for understanding cellular regulation and disease pathogenesis. Nevertheless, these approaches prevalently rely on the usage of sequencing methods based on short reads, which pose significant limitations, especially in the field of TEs. These repetitive regions, indeed, frequently produce multiple alignments and are therefore excluded by standard analysis pipelines (Polimeni et al. 2024). From this perspective, long-read sequencing technologies, such as those provided by PacBio and Oxford Nanopore, offer significant advantages for de novo assembly of transcriptomes compared to short-read approaches (Pardo-Palacios et al. 2024). Besides their ability to uniquely map chimeric transcripts, another benefit is the generation of reads that span entire isoforms. For instance, studies have leveraged long-read sequencing to discover novel genes and isoforms related to pathologies affecting immune cells (Inamo et al. 2024) and clear cell renal cell carcinoma (Lee et al. 2024a). Hence, coupling de novo assembly with long-read sequencing has led to the identification of genomic variation in Drosophila (Rech et al. 2022), as well as regulatory TE-containing ncRNAs in plants (Berthelier et al. 2023) and during human embryo development (Wang et al. 2024).

5.4 TE-Derived LncRNAs in Physiology

As highlighted by Kapusta et al. (2013) and confirmed in this chapter, more than two-thirds of mature lncRNAs include a TE sequence, thus suggesting that apart from being the main source of lncRNAs, TEs may also constitute the functional portion by which lncRNAs exert their functions. TEs add an extra layer of complexity to gene regulation by providing crucial functional domains that ncRNAs can exploit to fulfill diverse roles. These domains enable ncRNAs to participate in a wide range of cellular processes, such as chromatin remodeling, transcriptional regulation, RNA stability, and signaling pathways (Kornienko et al. 2013; De Majo and Calore 2018; Marasca et al. 2022). By integrating TEs into their structure, ncRNAs gain the ability to interact with various cellular components and adapt to evolving biological demands, thereby contributing to both normal development and disease pathogenesis (Payer et al. 2019).

5.4.1 LTR7-HERVH

To start with, it is well documented how a specific class of TEs, the LTR retrotransposons, are particularly proficient in lncRNAs generation. LTRs often exhibit diverse and intricate motifs, yet they share a conserved structural organization consisting of three key regions: the U3 region, followed by the R and U5 regions. These regions collectively house critical functional elements, namely the enhancer and promoter, a TATA box, and a polyadenylation signal, respectively (Zhang et al. 2014a). Over ages, LTRs acquired so much sequence variability, developing specific functional binding sites that confer cell- and developmental-specific expression patterns (Carter et al. 2022). To sum up, LTRs alone contain all the functional domains to initiate RNA transcription. Human pluripotent stem cells (hPSC), such as hESCs and hiPSCs for instance, display high levels of LTR-derived lncRNAs that regulate cell identity and pluripotency (Sexton et al. 2022). Among such lncRNAs, those displaying the highest expression originate from an endogenous retrovirus type I (ERV1) of long terminal repeat retroelements called LTR7-HERVH (Lu et al. 2014; Zhang et al. 2019b). *Wang* et al. discovered, in 2014, that such elements are under the transcriptional control of a cocktail of naïve pluripotency transcription factors such as OCT4, NANOG, KLF4, and LBP9. Transcription of these elements commonly initiates at 5′ LTR7-HERVH boundaries, to then splice within nearby genes, generating chimeric transcripts (Wang et al. 2014). Given their similar transcription initiation mechanism, such RNA species share an LTR7-HERVH conserved core domain, predicted to be bound by pluripotency factors (e.g., NANOG) and pluripotency-associated histone modifiers (e.g., SET1A, SETDB1). The authors further demonstrated that LBP9-driven HERVH transcription is a primate-specific network, and its downregulation abolishes self-renewal, leading to loss of pluripotency and upregulation of multi-lineage differentiation markers (Wang et al. 2014).

5.4.2 *XIST*

Another compelling example of how transposable elements contribute to the emergence of lncRNAs is found in the regulation of X chromosome inactivation (XCI), a process controlled by the X-inactivation center (XIC) (Kay 1998). The XIC encodes several cis-acting lncRNAs, among which Xist stands out as the master regulator of XCI, promoting chromosome-wide silencing through chromatin condensation and widespread H3K27me3 deposition (Schwämmle and Schulz 2023). Comparative genomic analyses suggest that the X-inactivation center (XIC) originated through the transformation of a protein-coding–rich genomic region into a noncoding domain, likely via a combination of pseudogenization and insertions of mobile genetic elements (Duret et al. 2006). Xist, the master regulator of X chromosome inactivation, is composed of ten unique exons arranged in different isoforms. Interestingly, four of these exons—as well as the Xist promoter—share sequence homology with exons of the chicken Lnx3 gene (Elisaphenko et al. 2008). The remaining six exons include simple tandem repeats and show similarity to various transposable element (TE) families. According to curated RefSeq annotations, all human XIST isoforms include one exon that is entirely embedded within an L1MC3 element, a LINE1 subfamily.

The choice of which X chromosome will be inactivated is regulated by Tsix (Lee et al. 1999; Maclary et al. 2014), an antisense transcript of *Xist*. Tsix is a large, mono-exonic transcript whose transcription start site (TSS) lies within a MER58B element, a DNA transposon. The Tsix sequence itself is a composite of multiple transposable elements and repeated sequences. It represses Xist expression through several cis-acting mechanisms that modify chromatin accessibility or alter the methylation status of the Xist promoter. Of note, the antisense region complementary to the 3′ end of Xist contains multiple retrotransposon fragments, including those from the L2b and MIRc families. Despite the fact that Tsix is almost entirely composed of TE-derived sequences, their possible roles in Tsix RNA folding or in the recruitment of regulatory factors remain unknown.

5.4.3 lncMER52A

lncMER52A (Wu et al. 2020) exemplifies how transposable elements can give rise to functional noncoding RNAs that are specifically expressed in pathological contexts, such as hepatocellular carcinoma (HCC). This four-exon, 828-nucleotide lncRNA is transcribed from within an LTR MER52A retrotransposon and terminates within an MLT1A0 element, which provides a transcriptional termination signal. lncMER52A is exclusively expressed in HCC, where it localizes to the cytoplasm and stabilizes p120-catenin by preventing its ubiquitin-mediated degradation. Its expression is regulated by chromatin accessibility and the recruitment of the YY1 transcription factor to its LTR-derived promoter.

5.4.4 CARMEN

Plaisance et al. (2023) have dissected the role of a specific retrotransposon-containing isoform of the noncoding CARMEN gene in directing lineage specification in cardiac precursor cells (CPCs). CARMEN, a lncRNA locus known to regulate differentiation toward both cardiomyocytes (CMs) and smooth muscle cells (SMCs), expresses multiple alternatively spliced isoforms during development (Plaisance et al. 2023). Through isoform usage analysis during CPC differentiation, the authors identified a TE-containing variant, CARMEN-201, as a key driver in skewing cell fate toward the SMC lineage. This isoform includes an exon that originates from an exonized MIRc element, a SINE retrotransposon. The inclusion of this exon appears to be facilitated by a nearby AluSx3 element, which contributes a functional splice donor site. Importantly, the MIRc-derived exon is conserved among primates but absent in other species, suggesting a relatively recent evolutionary innovation with functional impact in humans. The spliced MIRc sequence enables CARMEN-201 to recruit the RE1-Silencing Transcription Factor (REST) (Thiel et al. 2015), which is then redirected to repress CM-specific gene expression—thereby promoting the SMC program. This finding exemplifies how TE-derived sequences can modulate isoform structure and functional output of lncRNAs in a lineage-specific manner, with potential implications for cardiovascular disease and regenerative medicine.

5.4.5 LINE1 ncRNAs

We have recently reported how alternative splicing of LINE1 into noncoding transcripts represents a key regulatory mechanism in T cells, influencing their functional state across physiological conditions—such as quiescence and activation—and pathological contexts like exhaustion and age-related immunosenescence (Marasca et al. 2022). In human CD4+ naïve T cells, the chromatin is decorated by LINE1 RNAs. By applying custom transcriptome assembly approaches, as well as expression- and epigenetic-based filtering procedures, we finally reveal that intronically evolutionary old LINE1 elements (i.e., L1Ms) are spliced into chromatin-associated non-canonical transcript variants of T-cell activation genes. Such novel alternative LINE1-containing isoforms (henceforth LINE1-transcripts) are in complex with Kap1 and nucleolin, with the latter known to be a LINE1-binding protein (Percharde et al. 2018). Mechanistically, the complex acts in cis, hampering the deposition of H3K36me3 and, therefore, inhibiting transcription elongation and stalling gene expression. Upon T cell activation, LINE1 transcripts are downregulated via splicing suppression of L1M elements mediated by PTBP1 under the control of the mTORC pathway, resulting in increased abundance of the corresponding protein-coding isoform. In addition, DNA-RNA-FISH experiments pinpointed how the function of LINE1 transcripts may not be restricted in cis but may be spread via trans interaction. The relevance of LINE1 transcripts in the establishment of a dysfunctional T cell state will be discussed in the next paragraph.

Two years later, we further demonstrated that LINE1 transcripts expression progressively increases during aging, with neonatal naïve T cells being poor in LINE1 transcripts, whose level peaks in elderly individuals (Burattin et al. 2024). LINE1 absence correlates with increased protein synthesis. In detail, LINE1 transcripts downregulation induces upregulation of translation- and, to a lesser extent, cell cycle-related genes. Among translation-associated genes, several translation initiating, elongating, and terminating factors were found, albeit most of them were constituted by a core of ribosomal-associated proteins. Nonetheless, further mechanistic explanations and validations are needed to fully reveal whether LINE1 transcripts directly regulate such genes in trans or indirectly, for instance, by sequestering specific regulators of these gene sets.

In the neurological field, different studies have revealed that LINE1 RNAs play a critical regulatory role in neuronal development by acting as lncRNAs (Percharde et al. 2018; Della Valle et al. 2020; Mangoni et al. 2023). These RNAs influence corticogenesis by modulating the balance between neural progenitor proliferation and differentiation and guiding the migration of post-mitotic neurons. They achieve this by binding to chromatin and interacting with Polycomb Repressive Complex 2 (PRC2) components, such as Ezh2 and Suz12, to regulate the deposition of repressive histone marks (H3K27me3) on genes essential for neuronal differentiation. Silencing LINE1 RNAs disrupts these processes, impairing transcriptional programs and chromatin landscapes, necessary for brain development. This regulatory mechanism likely evolved to preserve neural progenitor pools while enabling the progressive differentiation required for cortical complexity, emphasizing the importance of L1s in brain development and evolution.

5.4.6 Half-STAU1-Binding Site RNAs

A more direct example of translation control by RE-containing lncRNAs emerged from the study of Staufen 1 (STAU1)-mediated mRNA decay (SMD) pathway (Gong et al. 2011). This intracellular pathway degrades translationally active mRNAs owning STAU1-binding sites (SBS) in their 3′-untranslated regions (3′ UTRs). Previous research identified the SBS within ADP-ribosylation factor 1 (ARF1) mRNA as a 19-base-pair stem-loop structure with a 100-nucleotide apex (Kim et al. 2007). However, similar structures were not found in the 3′ UTRs of other SMD targets by the authors. In fact, their analysis reveals that SBSs can also be formed through imperfect base-pairing between single Alu elements (Batzer and Deininger 2002) in the 3′ UTR of an SMD target and Alu elements embedded in cytoplasmic, polyadenylated long noncoding RNAs, which they refer to as half-STAU1-binding site RNAs (1/2-sbsRNAs). These lncRNAs (378 candidates) contain a single Alu element as their target mRNA and can selectively downregulate subsets of SMD targets. At the same time, multiple 1/2-sbsRNAs can act on the same target mRNA. This finding highlights previously unrecognized roles for noncoding RNAs and Alu elements in the transactivation of RNA translation regulation. Interestingly, not all mRNAs with Alu elements in their 3′ UTR are targeted for SMD, even in the presence of lncRNAs with complementary sequences. This implies that additional factors influence SMD selectivity, such as adequate homology between the paired Alu elements or the involvement of additional RNA-binding proteins. The introduction of this new class of lncRNAs underscores an innovative cellular strategy for recruiting proteins to mRNAs to regulate their turnover.

5.4.7 Enhancer RNAs and Upstream Antisense RNAs

Alu elements also play a central role in positively regulating gene expression via Alu-mediated duplex formation between two classes of lncRNA: enhancer RNAs (eRNAs) and upstream antisense RNAs (uaRNA), also known as promoter upstream transcripts (PROMPTs) (Liang et al. 2023; Wen and Zhong 2023). Liang et al. performed RNA in situ conformation sequencing (RIC-seq) (Cai et al. 2020) in six different cell lines and found that 37.9% of enhancer–promoter RNA interaction sites involve Alu sequences. Alu can also form complementary RNA duplexes with non-Alu regions. Disrupting Alu elements impairs enhancer–promoter looping, while introducing Alu sequences or tethering them to unregulated promoter RNAs can establish new functional loops. This study highlights Alu elements as key regulators of enhancer–promoter connectivity and provides a framework to link noncoding variants to their molecular functions. Alu enrichment at RIC-seq chimeric reads from enhancers and promoters was revealed by motif analysis. In transcribed regions, Alu elements were highly enriched near interacting enhancer–promoter fragments compared to shuffled controls. Around enhancer–promoter RNA interaction junctions (i.e., EPRI), 61% of the enriched Alu elements were from the AluS subfamily, highlighting a high degree of co-option and specialization of such elements in mediating long-range promoter-enhancer interactions.

5.4.8 *hLincRNA-p21*

Chillon and Pyle demonstrated that pairing between two Alu elements within the same lncRNA has implications in its localization and function (Chillon and Pyle 2016; Lee et al. 2024b). Sequencing of human lincRNA-p21 (hLincRNA-p21) revealed a single exon containing inverted-repeat Alu elements (IRAlus). Furthermore, the authors discovered a second and longer isoform and validated both of them by northern blotting and RNA-seq data from ENCODE project. These transcript variants shared sense and antisense Alu elements that independently form conserved secondary structures in primates. These IRAlus structures are crucial for the co-localization of hLincRNA-p21 with paraspeckles into the nucleus. The disruption of the secondary structure of the IRAlus by mutating exclusively the sense or antisense Alu element (dS-Alu and dA-Alu, respectively), or both Alu elements (dIR-Alu), decreased hLincRNA-p21 nuclear localization. Nonetheless, authors commented that other factors may be implicated in this phenomenon, as they observed a significant—yet not total—nuclear localization reduction of hLincRNA-p21 upon Alu deletions. To conclude, hLincRNA-p21 have implication in cellular stress response, discussed in the next paragraph.

5.4.9 *Alu ncRNAs*

It has been recently demonstrated that human Alu ncRNAs act as mammalian transcriptional repressors through cryogenic electron microscopy (Tlučková et al. 2025). RNA polymerase II (Pol II), a 12-subunit enzyme essential for generating messenger RNA, is a key target for transcriptional regulation via various accessory factors, including proteins and RNAs (Malik and Roeder 2023). While protein regulators are well-characterized, RNA-based mechanisms remain less understood. The authors showed that Alu ncRNA is transcribed by RNA polymerase III (Pol III) with an average length of 300 nucleotides. Under normal conditions, Alu RNA levels are low; however, during stress scenarios (e.g., heat shock), they behave as natural transcription inhibitors, repressing Pol II transcription (Liu et al. 1995). Alu RNA binds Pol II directly through its two halves, the left arm (Alu-LA) and right arm (Alu-RA), with high affinity. This results in Alu-derived RNAs accumulating at promoters of repressed genes. However, only Alu-RA inhibits transcription in the presence of TFIIF, a heterodimeric general transcription factor. Cryo-EM studies revealed Alu RNA binding within Pol II's DNA-binding cleft, although the resulting RNA structure could not be resolved. As such, the exact structural interaction as well as the complete overview of the selective repression mechanism involving TFIIF remains unclear. Another group showed that the function of these noncoding free RNAs transcribed from genomic Alu elements is not nuclear-restricted (Magliacane Trotta et al. 2024). By exosome-mediated transfer, Alu RNAs can be shuttled to nearby cells, displaying a particular proficiency in promoting cancer progression. To note, the function and phenotypic effect of Alu RNA seems to be dose dependent. Low level of Alu RNA increases cell proliferation, while higher levels lead to cell death.

5.4.10 NORAD

Expanding beyond the regulatory roles of Alu RNAs in stress responses and cancer, other studies have unveiled how Alu motifs contribute to immune responses and antiviral defenses (Lee et al. 2024b), showcasing an intricate interplay between noncoding RNAs and intracellular pathways, such as in the case of NORAD. A novel role for the lncRNA NORAD in regulating the innate immune response in human cells has been identified (Argoetti et al. 2025). NORAD facilitates unphosphorylated STAT3 nuclear translocation via Importin-β1, where STAT3 binds the promoters of interferon-stimulated genes (ISGs) to suppress transcription. During viral infections, NORAD levels decrease as it is recruited to stress granules and cleaved by RNaseL, rapidly inhibiting viral replication without requiring gene expression changes. This downregulation releases STAT3-mediated inhibition, activating ISGs and enhancing the STAT1-driven immune response. NORAD's role bridges intrinsic defense mechanisms with STAT1-mediated pathways, offering a rapid, noncanonical activation that bypasses interferon-dependent processes often targeted by viruses.

Interestingly, NORAD-STAT3 interaction is mediated by a structured region downstream of an Alu element unique to hominoids, suggesting evolutionary adaptation. The authors speculated that during primate evolution, an Alu inserted in the 3′ region of NORAD, inducing a change in the sequence that ultimately reflects a structural change responsible for a gain of function. RNA structure prediction by RNAfold (Lorenz et al. 2011) of the STAT3-binding region, highlighted that sequences containing Alu in primates fold into stem-loop-like structures. This work provides evidence that transposable elements insertional activities can confer novel regulatory functions by inducing lncRNA structural changes.

5.4.11 Circular RNAs

Lastly, Alu elements play a pivotal role in the formation of circular RNAs (circRNAs) (Liang et al. 2021; Zhou et al. 2020), contributing to their biogenesis and stability (Jeck et al. 2013). It is important to mention that some circRNAs, despite being in principle characterized as a new class of lncRNAs, can be translated into proteins in particular biological contexts and if presenting specific RNA modifications (Yang et al. 2017; Legnini et al. 2017; Pamudurti et al. 2017). Circular RNAs are composed of non-linear exon arrangements known as "backsplices" and were once thought to result from splicing errors and lacked recognized functions. High-throughput RNA sequencing of ribosome-depleted RNA treated with RNase R identified over 25,000 distinct circular RNA species in human fibroblasts, with many validated as exon-derived circRNAs (ecircRNAs) (Jeck et al. 2013). These ecircRNAs, often more stable and abundant than their linear mRNA counterparts, were identified in 14.4% of actively transcribed genes, with conservation observed in murine orthologues. Notably, bioinformatic analysis revealed that the long introns flanking circularized exons frequently contain complementary Alu repeats, implicating these transposable elements as critical drivers of circRNA formation (Zhang et al. 2014b; Ivanov et al. 2015). Circularized exons exhibit a notable enrichment of Alu repeats in their flanking introns, with a two-fold higher occurrence compared to non-circularized exons. Flanking Alu elements are frequently found in complementary (inverted) orientations, a configuration closely linked to RNA circularization (Wilusz 2015). This highlights a nonrandom, conserved process of RNA splicing where Alu elements influence the generation of abundant and stable ecircRNAs, potentially regulating gene expression through mechanisms such as acting as competing endogenous RNAs (Mitra et al. 2018).

In summary, TEs represent a cornerstone in the origin, evolution, and functionality of lncRNAs, shaping their regulatory landscapes, cell specificity, and biological roles. Recent studies reinforce and expand upon the seminal work of Kapusta et al. (2013), revealing that beyond contributing to the majority of mature lncRNA sequences, TEs profoundly influence their structure, splicing patterns, nuclear localization, and functional versatility. Through mechanisms such as exonization, promoter activity, and enhancer modulation, TEs endow lncRNAs with the capacity to modulate gene expression dynamically across diverse cellular contexts, including pluripotency, differentiation, immune responses, and aging (Table 5.1).

5.5 TE-Derived MiRNAs in Physiology

MiRNAs are small, noncoding RNA molecules approximately 22 nucleotides long that regulate post-transcriptionally gene expression (Shang et al. 2023). They are derived from longer primary transcripts (pri-miRNAs) containing hairpin structures (Shang et al. 2023). Many conserved miRNAs are processed via the canonical biogenesis pathway (Bartel 2004). In the nucleus, the microprocessor complex, composed of Drosha and its cofactor DGCR8, excises the hairpin structure from pri-miRNAs to generate precursor miRNAs (pre-miRNAs) of about 55–70 nucleotides in length (Lee et al. 2003; Wang et al. 2007). These pre-miRNAs are then exported to the cytoplasm, where the RNase III enzyme Dicer cleaves them to produce an miRNA duplex (Gebert et al. 2018). One strand of this duplex (the guide

Table 5.1 Representative examples of TE-derived ncRNA and their biological function

Noncoding RNA type	TE type	Biological context	Function	References
LTR7-HERVH	LTR/ERV1	hPSC	Regulation of cell identity and pluripotency	Wang et al. (2014) Zhang et al. (2019a, b) Sexton et al. (2022)
XIST	LINE/L1 Others	Female cells	X chromosome inactivation	Elisaphenko et al. (2008)
Tsix	LINE/L2 SINE/MIR	Female cells	XIST downregulation	Maclary et al. (2014)
lncMER52A	LTR/ERV1 LTR/ERVL-MaLR	Hepatocellular carcinoma	Inhibition of p120-catenin degradation	Wu et al. (2020)
lncRNA CARMEN	SINE/MIR	Cardiac precursor cells	Differentiation to SMC	Plaisance et al. (2023)
lncRNAs	LINE/L1	Naïve T cells	Transcriptional inhibition of T-cell activation genes	Marasca et al. (2022)
lncRNAs	LINE/L1	Naïve T cells from elders	Inhibition of translation/protein synthesis	Burattin et al. (2024)
lncRNAs	LINE/L1	Neural progenitors	Inhibition of genes related to neuronal differentiation	Mangoni et al. (2023)
Half-STAU1-binding site RNAs	SINE/Alu	//	Degradation of SBS-containing mRNA	Gong et al. (2011)
eRNAs and uaRNA	SINE/Alu	//	Regulation of enhancer–promoter looping	Liang et al. (2023)
hLincRNA-p21	SINE/Alu	//	Co-localization of hLincRNA-p21 with paraspeckles	Chillon and Pyle (2016)
Alu	SINE/Alu	Cellular stress	Transcriptional repression via PolII inhibition	Tlučková et al. (2025)
NORAD	SINE/Alu	Innate immune cells	Inhibition of ISGs transcription via STAT3	Argoetti et al. (2025)
ecircRNAs	SINE/Alu	//	Drive ecircRNA formation	Jeck et al. (2013)

strand) is loaded into an Argonaute (AGO) protein to form the miRNA-induced silencing complex (miRISC), which mediates the silencing of its targets (Shang et al. 2023). The miRISC binds complementary sequences in the target mRNAs, primarily within the 3′ UTR. The "seed" region, comprising nucleotides 2–8 of the miRNA, is crucial for target recognition, with additional specificity provided by the supplemental region (nucleotides 13–16) (Schirle et al. 2014). miRISC mediates gene silencing through two principal mechanisms: translational repression and mRNA degradation. Translational repression involves the inhibition of initiation, potentially via the release of eukaryotic initiation factors such as eIF4A-I and eIF4A-II (Fukao et al. 2014). mRNA degradation is mediated by GW182 proteins, which interact with polyadenylate-binding protein (PABPC) and recruit deadenylation complexes such as PAN2–PAN3 and CCR4–NOT (Chen et al. 2009; Braun et al. 2011; Fabian et al. 2011). This is followed by decapping through the DCP1–DCP2 complex and exonucleolytic degradation in the 5′–3′ direction (Braun et al. 2012). miRNAs exhibit remarkable versatility, with one miRNA capable of regulating multiple mRNAs and a single mRNA often regulated by several miRNAs (Gebert et al. 2018). These interactions create intricate regulatory networks essential for maintaining cellular homeostasis and responding to environmental cues.

Most miRNAs originate from noncoding regions or introns, although some of them overlap with exons of protein-coding genes. miRNAs can arise from various genomic sources, including TEs (MDTEs) (Roberts et al. 2014; Qin et al. 2015; Park et al. 2022). MDTEs either completely or partially overlap with TE sequences, exhibiting unique relationships with their TE of origin (Wei et al. 2016). MDTEs can be classified into three distinct types based on their structural overlap with TEs: Type I MDTEs originate from pre-miRNA that includes inverted TE sequences. Type II MDTEs partially overlap with a TE sequence, while the other portion is unique and complementary to the TE. Type III MDTEs completely originate within a TE sequence (Qin et al. 2015). Usually MDTEs originate when TEs insert into neighboring genomic positions, forming hairpin structures that act as miRNA precursors (Smalheiser and Torvik 2005). Alternatively, the formation and expansion of some MDTE families is associated with segmental duplications or other evolutionary events that shaped the architecture of the genome (Yuan et al. 2011b). These duplications often involve repetitive elements and promote the genesis of novel miRNAs. Indeed, large-scale duplication events may have driven the diversification and expansion of specific miRNA families (Yuan et al. 2011b).

To study MDTEs in the human genome, a study from 2022 analyzed 2883 mature miRNAs with known chromosomal locations from the miRBase database (Kozomara et al. 2019), cross-referencing them with human genome repeat sequences (Park et al. 2022). This analysis identified 474 MDTEs, either fully or partially overlapping with TE sequences. Among these, multi-copy MDTEs (i.e., miRNAs with identical sequences originating from distinct pre-miRNAs) were traced to the same TE family, albeit with variations in subfamilies. When multi-copy MDTEs were excluded, 405 unique MDTEs were identified, with 352 fully overlapping TEs and the remaining partially overlapping. MDTEs constitute approximately 15% of the 2652 annotated human miRNAs, emphasizing their significant contribution to the miRNA repertoire (Wei et al. 2016). Among the four major TE subclasses, DNA transposons emerged as the primary contributors, accounting for 144 MDTEs. LINE elements were followed by 116 MDTEs, SINE elements by 90, and LTR transposons by 50. At the subfamily level, TcMar-Mariner was the predominant source within DNA transposons (58%), while L1, MIR, and ERVL-MaLR were the leading contributors for LINE (55%), SINE (60%), and LTR transposons (44%), respectively.

5.5.1 LINE2 miRNAs

A prominent example of MDTEs includes those originating from LINE2 (L2) transposable elements (Smalheiser and Torvik 2005; Piriyapongsa et al. 2007). These L2-derived miRNAs are generated from hairpin structures formed entirely of L2 sequences, which arise at the junction of two adjacent L2 elements arranged in opposite orientations. This specific configuration is crucial for miRNA biogenesis, as it facilitates the processing of precursor miRNAs into mature forms. L2-derived miRNAs display remarkable conservation across species, including humans and mice. This conservation is noteworthy, as TE-derived miRNA genes are generally less conserved compared to other miRNA genes (Piriyapongsa et al. 2007). Similar conservation was also found in MDTEs originating from MIR elements (Piriyapongsa et al. 2007). The functional importance of this conservation lies in its role in regulating gene expression and maintaining genomic stability. Some L2-derived miRNAs exhibit perfect complementarity to specific families of mRNA (Petri et al. 2019). Many of these targets contain fragments of MIR elements or other L2 sequences within the 3′ UTRs of protein-coding genes. This complementarity enables these miRNAs to regulate a diverse array of genes, highlighting their significant role in gene expression. Both L2-miRNAs and their targets are widely expressed across various human tissues, primarily regulating housekeeping genes (Petri et al. 2019). Examples of functionally validated L2-derived miRNAs include miR-28, miR-95, and miR-151 (Shin et al. 2010; Frankel et al. 2014; Ding et al. 2010), which will be further discussed in the final paragraph, as they have been associated with pathological conditions.

5.5.2 hsa-mir-548

Another example of miRNAs derived from TEs is the hsa-mir-548 gene family, which originates from Made1 miniature inverted-repeat transposable elements (MITEs) (Piriyapongsa and Jordan 2007). These MITEs are inserted into transcriptionally active genomic regions and are characterized by a palindromic structure that promotes the formation of imperfect RNA hairpins when transcribed (Piriyapongsa and Jordan 2007). Thus, the biogenesis of has-mir-548 does not follow the canonical pathway but instead is akin to the mechanism for miRNA production described for the miRNA cluster on chromosome 19 (Borchert et al. 2006). The emergence of Made1 elements is a primate-specific evolutionary event, with hsa-mir-548 orthologues identified exclusively in humans (Piriyapongsa and Jordan 2007). This lineage-specific evolution underscores the contribution of recently evolved TEs to the diversification of miRNA genes. Functional studies have identified hsa-mir-548 genes in various biological contexts. Members of the hsa-mir-548 family, such as hsa-miR-548ba, exhibit tissue-specific expression, with the highest levels detected in granulosa cells (Rooda et al. 2021). These miRNAs are implicated in ovarian follicle activation, granulosa cell differentiation, and proliferation (Rooda et al. 2021). Their involvement in regulating these processes suggests a critical role in ovarian function and reproductive health. In addition, elevated expression of hsa-mir-548 has been observed in colorectal cancer (CRC), both in cell lines and in tissue samples (Piriyapongsa and Jordan 2007). This upregulation may suppress genes that normally regulate cellular proliferation, thereby contributing to cancer progression (Piriyapongsa and Jordan 2007). Such findings suggest a potential role for hsa-mir-548 as a biomarker or therapeutic target in CRC.

5.5.3 *miR-1302*

Finally, all members of the miR-1302 family are derived from a single transposon, the MER53 element (Yuan et al. 2010; Ahn et al. 2013). MER53 is a DNA transposable element with a 193-base pair consensus sequence, found specifically in eutherian species (Kapitonov and Jurka 1998). It is characterized by terminal inverted repeats and TA target site duplications, which allow the formation of palindromic structures. miRNAs derived from the MER53 element likely evolved after the divergence of eutherian mammals from marsupials and monotremes, approximately 180 million years ago (Wei et al. 2016). This miRNA gene family has undergone multiple gene duplications, resulting in some duplicated genes acquiring new functions, while others have become pseudogenes or were deleted (Kornienko et al. 2013). Recently, the miR-1302 family has been suggested to play a role in successful implantation and placental development (Cho et al. 2021).

In summary, these examples of miRNAs derived from transposable elements illustrate the intricate relationship between TEs and microRNA evolution. However, compared to the extensive research on non-MDTEs, relatively few studies have explored MDTEs in the context of human physiology and diseases. Therefore, further investigation is required to uncover the functions of miRNAs overlapping with TEs, which could offer valuable insights into disease pathogenesis.

5.6 TE-Derived ncRNAs in Pathology

Having previously highlighted their role as key regulators of physiological processes, it is not surprising that noncoding RNAs—when dysregulated—have also been implicated in the pathogenesis of various human diseases (Nemeth et al. 2024). Indeed, alterations of ncRNAs have been investigated across a broad range of conditions, including cardiovascular diseases, neurodegenerative disorders, immunological pathologies, and cancer (Ziats and Rennert 2013; Lorenzen et al. 2016; Dhawan et al. 2018). The most extensively studied ncRNAs in the context of human disease are miRNAs and lncRNAs. To facilitate their characterization, several databases have been developed to catalog disease-associated ncRNAs, including The Human MicroRNA Disease Database (Huang et al. 2019) for miRNAs and LncRNADisease (Chen et al. 2013; Lin et al. 2024) for lncRNAs.

Although the direct association between TE-derived ncRNAs and human disease remains a relatively novel area of research with limited publications, evidence indicates that many essential ncRNAs previously studied in pathogen-associated disease and cancer harbor TE sequences, which may represent an unexplored player in driving pathogenic genetic mechanisms (Lee et al. 2020; Park et al. 2022). Regarding TE-derived miRNAs, the MDTE database (Wei et al. 2016) was established in 2016 as a resource to study TEs origins, evolution, and functions, enhancing predictions within novel miRNAs and their regulatory roles.

In this paragraph, we will summarize the latest literature focusing only on TE-derived ncRNA in which the TE sequence has been demonstrated to be the functional element associated with the disease. These studies reveal the diverse mechanisms by which TE sequences contribute to disease pathology, such as modulating gene expression, influencing RNA stability, or serving as binding platforms for regulatory proteins (Liang et al. 2024). By directly linking specific TE sequences within ncRNAs to functional outcomes in various diseases, this body of work underscores the critical role of TEs in driving disease mechanisms and their potential as diagnostic and therapeutic targets.

5.6.1 Cardiovascular Diseases

We have already mentioned how CARMEN-201 is a TE-containing alternative isoform of CARMEN, an lncRNA that regulates the specification toward the CM and the SMC fates (Plaisance et al. 2023). Given the aforementioned physiological role, the authors found that the expression of CARMEN-201 is upregulated in response to myocardial infarction in humans, suggesting its potential as a valuable biomarker for evaluating the extent of cardiovascular damage in various pathological situations.

Another study by Vlachogiannis and colleagues (Vlachogiannis et al. 2021) underscores the significant role of Alu sequences within the 3' UTR of the lncRNA NEAT1 in regulating its stability and function, with implications for atherosclerotic cardiovascular disease (ASCVD). NEAT1 expression is markedly elevated in peripheral blood mononuclear cells and atherosclerotic plaques of patients with coronary artery disease (CAD) and peripheral arterial disease (PAD), driven by TNF-α stimulation in endothelial cells, which amplifies the inflammatory response. The regulatory mechanism involves ADAR1-mediated A-to-I RNA editing, targeting adenosines near AU-rich elements within the AluSx3+/AluJo double-stranded RNA complex in NEAT1's 3' UTR. This editing enhances NEAT1 stability by promoting its interaction with the RNA-binding protein AUF1. When ADAR1 is silenced, AUF1 binding is disrupted, reducing NEAT1 levels. A strong correlation between ADAR1 and NEAT1 expression in ASCVD further supports the critical role of RNA editing in maintaining NEAT1 stability and function. These findings highlight how Alu elements and RNA editing shape lncRNA activity, contributing to vascular inflammation and disease progression.

5.6.2 Cancer

Among the aforementioned L2-derived miRNAs, miR-28 has been identified as a regulator of the Nrf2 pathway, which is essential for cellular responses to oxidative stress (Yang et al. 2011). It has been associated with the suppression of oncogenic pathways, indicating its potential as a tumor suppressor in various cancers, including B-cell lymphomas (Schneider et al. 2014). Another example is miR-151, located on chromosome 8q24.3. It is the most frequently amplified microRNA in hepatocellular carcinoma (HCC) and is co-expressed with its host gene, FAK. Chromosomal gains in this region contribute to the upregulation of both miR-151 and FAK, which synergistically promote HCC cell motility and invasion. miR-151 directly suppresses RhoGDIA, leading to the activation of Rho GTPases (Rac1, Cdc42, and Rho), further enhancing cell migration. This regulatory axis functions alongside FAK signaling to drive HCC progression (Ding et al. 2010).

Related to cancer research, a study (Chen et al. 2024) investigated the overexpression of the lncRNA CASC9, which acts as an oncogene in multiple cancer types (Liu et al. 2020; Zhan et al. 2020), focusing on its role in esophageal squamous cell carcinoma (ESCC). The researchers found that CASC9 transcripts composed of LINE1-containing exon combinations (exons 2–6 and 4–6) are highly expressed in ESCC, while absent in normal esophageal cells. The LINE1 plays a pivotal role in CASC9 evolution and activation. The antisense promoter (ASP) of LINE1 provides cis-regulatory elements for CASC9 activation and contributes to the formation of its exons. This activation is driven by aberrant hypomethylation of LINE1 elements, which enables interaction between the ASP and an active enhancer element located downstream of CASC9, confirmed through 3C analysis. These findings reveal that LINE1 ASP activation and enhancer interactions are critical mechanisms underlying CASC9 overexpression in ESCC, highlighting the potential of using CASC9 as a therapeutic target for this malignancy.

Another lncRNA involved in tumorigenesis is lncMER52A, whose biological function has been elucidated in previous paragraphs (Wu et al. 2020). With regard to HCC, it is considered a key player in its progression and correlates with more advanced TNM stages, poorly differentiated tumors, and reduced overall survival in patients. Functional experiments show that lncMER52A promotes HCC cell invasion and metastasis both in vitro and in vivo. Being regulated by the binding of YY1 to the MER52A LTR, lncMER52A stabilizes p120-catenin and induces the activation of the Rho GTPase pathway, including Rac1 and Cdc42. This signaling axis enhances cancer cell motility and invasiveness. These findings establish lncMER52A as an LTR-activated oncogenic lncRNA that promotes HCC progression through the p120-ctn/Rac1/Cdc42 axis. This lncRNA holds promise as both a biomarker and a therapeutic target for HCC.

In the context of CRC, the loss of ARID1A, a key epigenetic regulator commonly mutated in various cancers (Kamori et al. 2022; Zhang et al. 2023b), specifically leads to the de-repression of HERVH (Yu et al. 2022). Normally suppressed through various epigenetic mechanisms, including DNA methylation and histone modifications, ERVs like HERVH are reactivated in the absence of ARID1A, generating ncRNAs or producing chimeric transcripts with protein-coding genes via alternative splicing. The study shows that ARID1B partially compensates for the loss of ARID1A, contributing to the transcriptional activation of HERVH, which is crucial for the survival of CRC cells. This activation impacts the dynamics of BRD4, a key regulator of transcription, and modulates the transcriptional network that drives tumorigenesis. The results suggest that the activation of HERVH, a pluripotency-associated retrovirus, plays a central role in the vulnerabilities of ARID1A-deficient cells, including their sensitivity to BET inhibitors and ARID1B targeting. Furthermore, the study highlights how mutations in tumor suppressor genes like ARID1A can lead to the reactivation of ERVs, contributing to cancer progression by co-opting these elements in the tumorigenic process. This interaction between ARID1A, HERVH, and BRD4 presents potential therapeutic targets, as targeting these pathways could help treat ARID1A-mutated cancers.

Jin and colleagues (Jin et al. 2019) investigated the role of HERVs that were highly expressed in human triple-negative breast cancer (TNBC). Their research led to the identification of TROJAN, a primate-specific lncRNA that contains an LTR70 sequence. The research shows that TROJAN promotes TNBC cell proliferation and invasion, correlating with poor patient outcomes. TROJAN functions by binding to ZMYND8, a metastasis-repressing factor, leading to its degradation through the ubiquitin-proteasome pathway, facilitated by the repulsion of ZNF592. Additionally, TROJAN epigenetically upregulates metastasis-related genes. Clinical samples further confirmed the correlation between TROJAN and ZMYND8. The study also demonstrated that targeting TROJAN with antisense oligonucleotides effectively suppressed TNBC progression in vivo.

The IRAlus-containing transcript hLincRNA-p21, described earlier, has been demonstrated to be involved in tumors (Yang et al. 2014). They associated its function with the Warburg effect, particularly in the context of hypoxia, which has been linked to enhanced glycolysis. The researchers revealed that hLincRNA-p21 is a hypoxia-responsive lncRNA that plays a crucial role in hypoxia-enhanced glycolysis. Under hypoxic conditions, lincRNA-p21 is induced by HIF-1α and interacts with both HIF-1α and VHL, disrupting the VHL-HIF-1α interaction. This disruption impedes VHL-mediated HIF-1α ubiquitination, leading to HIF-1α accumulation. The study elucidates a positive feedback loop between HIF-1α and hLincRNA-p21, which further promotes glycolysis during hypoxia. Additionally, the ability of hLincRNA-p21 to promote tumor growth is validated in mouse xenograft models.

Finally, EVADR has been identified as a novel lncRNA activated by the MER48 ERV element. It is specifically expressed in adenocarcinomas, including colon, rectal, lung, pancreas, and stomach cancers (Gibb et al. 2015). Analysis of RNA-seq data from colorectal tumors and TCGA datasets revealed that EVADR is expressed in 25–53% of these tumors and correlates with decreased patient survival. In nonglandular tumors, EVADR expression was negligible. Mechanistic insights demonstrate that the MER48 ERV element acts as an active promoter driving EVADR transcription, highlighting the tumor-specific regulatory role of retroviral LTRs. Genome-wide analysis revealed that MER48 is associated with nine lncRNAs, but only EVADR is consistently activated in adenocarcinomas. The EVADR locus, absent in New World monkeys and prosimians, is conserved in Old World monkeys and apes, indicating evolutionary importance. Structural analysis further suggests that the EVADR RNA sequence and secondary structure are under strong purifying selection, emphasizing its potential functional relevance. This study implicates retroviral elements in the tissue-specific regulation of lncRNAs and suggests EVADR as a potential biomarker and therapeutic target in glandular-origin cancers.

5.6.3 Cancer Immunology

We have already discussed that noncoding LINE1-containing splicing variants were shown to play a critical role in enforcing the quiescence of naïve CD4+ T cells (Marasca et al. 2022). These LINE1-containing transcripts, derived from genes upregulated during T cell activation, stall gene expression in quiescent T cells and, upon T cell activation, are downregulated, allowing gene expression to proceed. In tumor-infiltrating lymphocytes (TILs), the accumulation of LINE1-containing transcripts at chromatin impairs effector function. Notably, depleting LINE1-containing transcripts restores TIL function, highlighting their role in T cell exhaustion, offering potential therapeutic insights for improving cancer immunotherapy (Marasca et al. 2022).

It is worth noting that recent studies have reported that the inclusion of TEs in canonical transcripts, lncRNAs included, can be the source of tumor-specific neoantigens in non-small cell lung cancer (NSCLC) (Merlotti et al. 2023). TE exonization induces the generation of aberrant Open Reading Frames (ORFs) that derive human leukocyte antigen I (HLA-I)–bound peptides and, thus, can actively induce an antitumoral immunological response.

5.6.4 Lysosomal Storage Disorders

miR-95, which is made by L2, plays a crucial role in regulating sulfate homeostasis and lysosomal function by targeting SUMF1, the key activator of all cellular sulfatases. Its disruption results in a severe blockade of autophagy-mediated degradation, as evidenced by the accumulation of autophagosomes, leading to lysosomal dysfunction (Frankel et al. 2014).

5.6.5 Embryonic Developmental Disorders

A study from 2024 explores the role of the LTR7C/HERVH, a component of the ERV1 family, in human trophoblast development and its implications in placental disorders (Kong et al. 2024). While HERVH is known for its role in pluripotency and early epiblast development, its functions in trophoblast lineage and placentation are less understood. The researchers reveal that LTR7C/HERVH is highly active in human trophoblasts, maintaining an accessible chromatin state. It transcribes urothelial cancer associated 1 (UCA1), a primate-specific lncRNA that plays a dual role. On the one hand, UCA1 promotes the proliferation of human trophoblast stem cells (hTSCs). On the other hand, it impairs trophoblast syncytialization, a process essential for normal placental function. This disruption is associated with heightened interferon signaling. Importantly, UCA1 upregulation is observed in placental samples from patients with early onset preeclampsia (EO-PE). Transcriptomic analysis shows striking similarities between EO-PE placentas and syncytiotrophoblasts derived from UCA1-overexpressing hTSCs, suggesting that dysregulated UCA1 could serve as a biomarker for EO-PE. These findings expand our understanding of HERVH's regulatory roles in early development and placental syndromes, revealing that endogenous retroviral elements influence human placentation beyond acting as enhancers.

Table 5.2 summarizes the most relevant TE-derived ncRNAs associated to different human pathologies, as described in this paragraph.

5.7 Challenges and Future Perspective of TE-Derived ncRNA-Based Targeted Therapies

As discussed in the previous sections, a growing body of evidence demonstrates the critical involvement of TE-derived ncRNA in the pathogenesis of diverse human diseases, including cancer (Schneider et al. 2014; Chen et al. 2024; Yu et al. 2022; Jin et al. 2019; Yang et al. 2014; Gibb et al. 2015) and cardiovascular diseases (Plaisance et al. 2023), highlighting their emerging potential as novel targets for therapeutic intervention and drug development (Jang et al. 2024).

Table 5.2 Representative examples of TE-derived ncRNA associated with human disease

Disease	Noncoding RNA type	TE type	Condition	References
Cardiovascular	lncRNA CARMEN-201	SINE/MIR	Myocardial infarction	Plaisance et al. (2023)
	lncRNA NEAT1	SINE/Alu	Coronary artery disease	Vlachogiannis et al. (2021)
Cancer	miR-28	LINE/L2	B-cell lymphoma	Schneider et al. (2014)
	miR-151	LINE/L2	Hepatocellular carcinoma	Ding et al. (2010)
	lncRNA CASC9	LINE/L1	Esophageal squamous cell carcinoma	Chen et al. (2024)
	lncRNA TROJAN	LTR/HERV	Triple-negative breast cancer	Jin et al. (2019)
	lncRNA LincRNA-p21	SINE/Alu	p53-mediated stress response/tumor microenvironment	Chillon and Pyle (2016) Yang et al. (2014)
	lncRNA lncMER52A	LTR/ERV1	Hepatocellular carcinoma	Wu et al. (2020)
	lncRNA EVADR	LTR/ERV1	Adenocarcinomas	Gibb et al. (2015)
	lncRNA	LTR/HERVH	Colorectal cancer	Yu et al. (2022)
Cancer immunology	lncRNA	LINE/L1	Exhausted tumor-infiltrating lymphocytes	Marasca et al. (2022)
Lysosomal storage disorders	miR-95	LINE/L2	Multiple sulfatase deficiency	Frankel et al. (2014)
Embryonic development	lncRNA UCA1	LTR/HERVH	Human trophoblast development—Placental disorders	Kong et al. (2024)

We will here discuss recent approaches employed to target TE-derived ncRNAs, which are broadly applicable to ncRNAs in general, with particular emphasis on the specific challenges posed by the presence of TE sequences within these transcripts. Although no current technology can specifically distinguish between TE-derived and TE-devoid ncRNAs, the difference lies in the design of the targeting strategies, which must be carefully optimized to mitigate the risk of nonspecific sequence pairing due to the repetitive nature of TE-derived regions.

Current methodologies for targeting TE-derived ncRNAs include small interfering RNAs (siRNAs), antisense oligonucleotides (ASOs), and CRISPR/Cas9 gene-editing systems.

siRNAs operate predominantly in the cytoplasm, where they recruit RNA-induced silencing complexes (RISCs) to bind and degrade specific ncRNAs (Liang et al. 2018; Winkle et al. 2021). This strategy has demonstrated efficacy in various preclinical cancer models, such as targeting LINE1-containing lncRNA CASC9 in esophageal squamous cell carcinoma, where siRNA-mediated knockdown reduced cellular invasiveness and migration (Chen et al. 2024).

ASOs are chemically synthesized single-stranded oligonucleotides that primarily function in the nucleus. Based on their chemical structure and modification, they can act in several cellular compartments and via different mechanisms, such as the induction of RNA degradation through RNase H activation or obstruction of mRNA processing by sterically blocking specific RNA regions (Dhuri et al. 2020). ASO-based therapies have achieved notable clinical success, with 12 ASO drugs approved as of 2024, primarily targeting mRNAs involved in genetic disorders (Vinjamuri et al. 2024). Recent preclinical studies have expanded ASO applications to lncRNAs, demonstrating reduced expression and oncogenic activity of targets such as the previously described HERV-derived lncRNA TROJAN in triple-negative breast cancer (Jin et al. 2019), suggesting the potential for clinical translation in oncology.

CRISPR/Cas9 represents another emerging strategy for targeting TE-derived ncRNAs, particularly those with dual nuclear and cytoplasmic localization or uncharacterized subcellular distribution (Li et al. 2023). Despite promising preclinical data, significant challenges persist regarding in vivo delivery and systemic toxicity.

Overall, despite the technological advancements applied to the investigation of TEs, the use of TE-derived targeted therapies has not yet advanced to clinical trials. Nevertheless, several preclinical studies have yielded promising results, underscoring the potential of these strategies for future therapeutic development.

The clinical translation of TE-derived RNA-based therapeutics faces considerable challenges related to target specificity. These difficulties largely arise from the repetitive nature of TE sequences, which poses a major obstacle in the design of selective and effective drugs. Therefore, a precise understanding of both the sequence and secondary structure of the TE-derived ncRNA to be targeted is essential.

Under this perspective, TE-derived ncRNAs can be broadly categorized into two groups: (i) transcripts entirely composed of TE sequences—including full-length TE transcripts and most TE-derived microRNAs—and (ii) transcripts in which the TE constitutes only a portion of the RNA molecule. The first group, consisting of fully TE-covered transcripts, presents a high risk of off-target effects due to sequence redundancy.

For spliced transcripts that include a TE segment within a broader sequence context, more refined and tailored approaches are required. Given the complexity of such genomic loci—including the potential presence of alternative isoforms lacking the TE—a promising strategy involves designing targeting agents that span the junction between the repetitive TE sequence and adjacent unique, non-repetitive regions. This approach may allow for selective targeting of the TE-derived isoform while minimizing off-target interactions.

To overcome these limitations and enhance the specificity and efficacy of TE-derived RNA-targeted therapies, comprehensive annotation of TE-derived isoforms through transcriptome reconstruction approaches will be essential. Coupled with advanced target design strategies, these efforts are critical for unlocking the therapeutic potential of TE-derived ncRNAs.

5.8 Conclusion

In this chapter, we aimed at reviewing the current knowledge about TEs included in ncRNAs and their role in both physiological and pathological conditions. Exploring the current GENCODE annotation, we could appreciate a preferential insertion of TEs in transcripts that do not encode for proteins. Among the hypotheses that explain this phenomenon, studies propose the specificity of the enzymes that catalyze TE insertion and the intrinsic mutagenicity associated with the transposition process (Sultana et al. 2017; Maksakova et al. 2006; Campos-Sánchez et al. 2016). However, we also point out that this phenomenon differently affects TE families, with shorter elements (i.e., Alu, MIR) being more frequently co-opted by protein-coding transcripts with respect to their longer counterparts (i.e., LINE1, LTRs).

Many studies focus on the molecular function of ncRNAs, especially lncRNA and miRNAs, and their relevance in the pathogenesis of the disease. Yet, there is still a lack of knowledge about the role of TEs overlapping noncoding transcripts, with few exceptions mentioned here.

Despite the technological improvements of the last decade, one of the biggest challenges is to systematically study TE-containing ncRNAs, which are: a) often co-opted by tissue-specific transcripts and b) repetitive and interspersed by nature. For this reason, we here highlight the importance of de novo transcriptome assembly based on the use of long-read sequencing that might improve our current strategies to stratify the involvement of TEs in the onset and progression of pathological conditions.

Acknowledgments We want to thank Dr. Federica Marasca and Dr. Valeria Ranzani for their feedback and suggestions. This chapter was refined with the assistance of the artificial intelligence (AI) tool OpenAI, ChatGPT model GPT-4o, 2025. Every section was carefully reviewed, revised, and approved by authors to ensure that the text is appropriate in terms of content and scientific correctness and maintains the integrity of the original content.

Compliance with Ethical Standards

Conflict of Interest None.

Ethical Approval This chapter contains no studies with human/animal participants performed by the author.

References

Ahn K, Gim J-A, Ha H-S, Han K, Kim H-S (2013) The novel MER transposon-derived miRNAs in human genome. Gene 512:422

Argoetti A, Shalev D, Polyak G, Shima N, Biran H, Lahav T, Hashimshony T, Mandel-Gutfreund Y, Argoetti A, Shalev D, Polyak G, Shima N, Biran H, Lahav T, Hashimshony T, Mandel-Gutfreund Y (2025) lncRNA NORAD modulates STAT3/STAT1 balance and innate immune responses in human cells via interaction with STAT3. Nat Commun 16:571

Bakinowska E, Kielbowski K, Skorka P, Dach A, Olejnik-Wojciechowska J, Szwedkowicz A, Pawlik A (2024) Non-coding RNA as biomarkers and their role in the pathogenesis of gastric cancer-a narrative review. Int J Mol Sci 25:5144

Balachandran P, Walawalkar IA, Flores JI, Dayton JN, Audano PA, Beck CR, Balachandran P, Walawalkar IA, Flores JI, Dayton JN, Audano PA, Beck CR (2022) Transposable element-mediated rearrangements are prevalent in human genomes. Nat Commun 13:7115

Bartel DP (2004) MicroRNAs: genomics, biogenesis, mechanism, and function. Cell 116:281–297

Batzer MA, Deininger PL (2002) Alu repeats and human genomic diversity. Nat Rev Genet 3:370–379

Berthelier J, Furci L, Asai S, Sadykova M, Shimazaki T, Shirasu K, Saze H, Berthelier J, Furci L, Asai S, Sadykova M, Shimazaki T, Shirasu K, Saze H (2023) Long-read direct RNA sequencing reveals epigenetic regulation of chimeric gene-transposon transcripts in Arabidopsis thaliana. Nat Commun 14(1):3248

Bjorklund SS, Aure MR, Hakkinen J, Vallon-Christersson J, Kumar S, Evensen KB, Fleischer T, Tost J, Osbreac SKK, Mathelier A, Bhanot G, Ganesan S, Tekpli X, Kristensen VN (2022) Subtype and cell type specific expression of lncRNAs provide insight into breast cancer. Commun Biol 5:834

Boeke JD, Garfinkel DJ, Styles CA, Fink GR (1985) Ty elements transpose through an RNA intermediate. Cell 40:491–500

Borchert GM, Lanier W, Davidson BL, Borchert GM, Lanier W, Davidson BL (2006) RNA polymerase III transcribes human microRNAs. Nat Struct Mol Biol 13:1097

Braun JE, Huntzinger E, Fauser M, Izaurralde E (2011) GW182 proteins directly recruit cytoplasmic deadenylase complexes to miRNA targets. Mol Cell 44:120–133

Braun JE, Truffault V, Boland A, Huntzinger E, Chang CT, Haas G, Weichenrieder O, Coles M, Izaurralde E (2012) A direct interaction between DCP1 and XRN1 couples mRNA decapping to 5′ exonucleolytic degradation. Nat Struct Mol Biol 19:1324–1331

Burattin F, Vadalà R, Panepuccia M, Ranzani V, Crosti M, Colombo F, Ruberti C, Erba E, Prati D, Nittoli T, Montini G, Ronchi A, Pugni L, Mosca F, Ricciardi S, Abrignani S, Pietrasanta C, Marasca F, Bodega B (2024) LINE1 modulate human T cell function by regulating protein synthesis during the life span. Sci Adv 10:eado2134

Cai Z, Cao C, Ji L, Ye R, Wang D, Xia C, Wang S, Du Z, Hu N, Yu X, Chen J, Wang L, Yang X, He S, Xue Y, Cai Z, Cao C, Ji L, Ye R, Wang D, Xia C, Wang S, Du Z, Hu N, Yu X, Chen J, Wang L, Yang X, He S, Xue Y (2020) RIC-seq for global in situ profiling of RNA–RNA spatial interactions. Nature 582:7812

Campos-Sánchez R, Cremona MA, Pini A, Chiaromonte F, Makova KD (2016) Integration and fixation preferences of human and mouse endogenous retroviruses uncovered with functional data analysis. PLoS Comput Biol 12:e1004956

Carlevaro-Fita J, Polidori T, Das M, Navarro C, Zoller TI, Johnson R (2019) Ancient exapted transposable elements promote nuclear enrichment of human long noncoding RNAs. Genome Res 29:208–222

Carter TA, Singh M, Dumbovic G, Chobirko JD, Rinn JL, Feschotte C (2022) Mosaic cis-regulatory evolution drives transcriptional partitioning of HERVH endogenous retrovirus in the human embryo. elife 11:e76257

Chang J, Zhang L, Li Z, Qian C, Du J (2024) Exosomal non-coding RNAs (ncRNAs) as potential biomarkers in tumor early diagnosis. Biochim Biophys Acta Rev Cancer 1879:189188

Chen C-YA, Zheng D, Xia Z, Shyu A-B, Chen C-YA, Zheng D, Xia Z, Shyu A-B (2009) Ago–TNRC6 triggers microRNA-mediated decay by promoting two deadenylation steps. Nat Struct Mol Biol 16:1160

Chen G, Wang Z, Wang D, Qiu C, Liu M, Chen X, Zhang Q, Yan G, Cui Q (2013) LncRNADisease: a database for long-non-coding RNA-associated diseases. Nucleic Acids Res 41:D983–D986

Chen X, Li J, Guan X, Bai Y, Wang K (2024) Abnormal activation of genomic LINE1 elements caused by DNA demethylation contributes to lncRNA CASC9 overexpression in esophageal squamous cell carcinoma. Heliyon 10:e32857

Chillon I, Pyle AM (2016) Inverted repeat Alu elements in the human lincRNA-p21 adopt a conserved secondary structure that regulates RNA function. Nucleic Acids Res 44:9462–9471

Cho SH, Kim JH, An HJ, Kim YR, Ahn EH, Lee JR, Kim JO, Ko JJ, Kim NK (2021) Genetic polymorphisms in miR-604A>G, miR-938G>A, miR-1302-3C>T and the risk of idiopathic recurrent pregnancy loss. Int J Mol Sci 22:6127

Chuong EB, Elde NC, Feschotte C (2017) Regulatory activities of transposable elements: from conflicts to benefits. Nat Rev Genet 18:71–86

Dazeniere J, Bousios A, Eyre-Walker A (2022) Patterns of selection in the evolution of a transposable element. G3 (Bethesda) 12:jkac056

De Goede OM, Nachun DC, Ferraro NM, Gloudemans MJ, Rao AS, Smail C, Eulalio TY, Aguet F, Ng B, Xu J, Barbeira AN, Castel SE, Kim-Hellmuth S, Park Y, Scott AJ, Strober BJ, Consortium GT, Brown CD, Wen X, Hall IM, Battle A, Lappalainen T, Im HK, Ardlie KG, Mostafavi S, Quertermous T, Kirkegaard K, Montgomery SB (2021) Population-scale tissue transcriptomics maps long non-coding RNAs to complex disease. Cell 184:2633–2648 e19

De Hoon M, Bonetti A, Plessy C, Ando Y, Hon CC, Ishizu Y, Itoh M, Kato S, Lin D, Maekawa S, Murata M, Nishiyori H, Shin JW, Stolte J, Suzuki AM, Tagami M, Takahashi H, Thongjuea S, Forrest ARR, Hayashizaki Y, Kere J, Carninci P (2022) Deep sequencing of short capped RNAs reveals novel families of noncoding RNAs. Genome Res 32:1727–1735

De Majo F, Calore M (2018) Chromatin remodelling and epigenetic state regulation by non-coding RNAs in the diseased heart. Non-coding RNA Research 3(1):20–28. https://doi.org/10.1016/j.ncrna.2018.02.003

Della Valle F, Thimma MP, Caiazzo M, Pulcrano S, Celii M, Adroub SA, Liu P, Alanis-Lobato G, Broccoli V, Orlando V (2020) Transdifferentiation of mouse embryonic fibroblasts into dopaminergic neurons reactivates LINE-1 repetitive elements. Stem Cell Rep 14:60–74

Dhawan A, Scott JG, Harris AL, Buffa FM (2018) Pan-cancer characterisation of microRNA across cancer hallmarks reveals microRNA-mediated downregulation of tumour suppressors. Nat Commun 9:5228

Dhuri K, Bechtold C, Quijano E, Pham H, Gupta A, Vikram A, Bahal R (2020) Antisense oligonucleotides: An emerging area in drug discovery and development. J Clin Med 9:2004

Ding J, Huang S, Wu S, Zhao Y, Liang L, Yan M, Ge C, Yao J, Chen T, Wan D, Wang H, Gu J, Yao M, Li J, Tu H, He X (2010) Gain of miR-151 on chromosome 8q24.3 facilitates tumour cell migration and spreading through downregulating RhoGDIA. Nat Cell Biol 12:390–399

Duret L, Chureau C, Samain S, Weissenbach J, Avner P (2006) The Xist RNA gene evolved in eutherians by pseudogenization of a protein-coding gene. Science 312:1653–1655

Eldakhakhny B, Sutaih AM, Siddiqui MA, Aqeeli YM, Awan AZ, Alsayegh MY, Elsamanoudy SA, Elsamanoudy A (2024) Exploring the role of noncoding RNAs in cancer diagnosis, prognosis, and precision medicine. Noncoding RNA Res 9:1315–1323

Elisaphenko EA, Kolesnikov NN, Shevchenko AI, Rogozin IB, Nesterova TB, Brockdorff N, Zakian SM (2008) A dual origin of the Xist gene from a protein-coding gene and a set of transposable elements. PLoS One 3:e2521

Fabian MR, Cieplak MK, Frank F, Morita M, Green J, Srikumar T, Nagar B, Yamamoto T, Raught B, Duchaine TF, Sonenberg N (2011) miRNA-mediated deadenylation is orchestrated by GW182 through two conserved motifs that interact with CCR4-NOT. Nat Struct Mol Biol 18:1211–1217

Fatica A, Bozzoni I (2014) Long non-coding RNAs: new players in cell differentiation and development. Nat Rev Genet 15:7–21
Ferrer J, Dimitrova N (2024) Transcription regulation by long non-coding RNAs: mechanisms and disease relevance. Nat Rev Mol Cell Biol 25:396–415
Feschotte C, Pritham EJ (2007) DNA transposons and the evolution of eukaryotic genomes. Annu Rev Genet 41:331–368
Frankel LB, Di Malta C, Wen J, Eskelinen E-L, Ballabio A, Lund AH, Frankel LB, Di Malta C, Wen J, Eskelinen E-L, Ballabio A, Lund AH (2014) A non-conserved miRNA regulates lysosomal function and impacts on a human lysosomal storage disorder. Nat Commun 2014(5):5840
Fukao A, Mishima Y, Takizawa N, Oka S, Imataka H, Pelletier J, Sonenberg N, Thoma C, Fujiwara T (2014) MicroRNAs trigger dissociation of eIF4AI and eIF4AII from target mRNAs in humans. Mol Cell 56:79–89
Gebert LFR, Macrae IJ, Gebert LFR, Macrae IJ (2018) Regulation of microRNA function in animals. Nat Rev Mol Cell Biol 20(1):21–37
Gibb EA, Warren RL, Wilson GW, Brown SD, Robertson GA, Morin GB, Holt RA (2015) Activation of an endogenous retrovirus-associated long non-coding RNA in human adenocarcinoma. Genome Med 7:22
Gong C, Maquat LE, Gong C, Maquat LE (2011) lncRNAs transactivate STAU1-mediated mRNA decay by duplexing with 3′ UTRs via Alu elements. Nature 470:7333
Han K, Lee J, Meyer TJ, Remedios P, Goodwin L, Batzer MA, Han K, Lee J, Meyer TJ, Remedios P, Goodwin L, Batzer MA (2008) L1 recombination-associated deletions generate human genomic variation. Proc Natl Acad Sci 105:19366
Huang Z, Shi J, Gao Y, Cui C, Zhang S, Li J, Zhou Y, Cui Q (2019) HMDD v3.0: a database for experimentally supported human microRNA-disease associations. Nucleic Acids Res 47:D1013–D1017
Huang W, Xiong T, Zhao Y, Heng J, Han G, Wang P, Zhao Z, Shi M, Li J, Wang J, Wu Y, Liu F, Xi JJ, Wang Y, Zhang QC (2024) Computational prediction and experimental validation identify functionally conserved lncRNAs from zebrafish to human. Nat Genet 56:124–135
Inamo J, Suzuki A, Ueda MT, Yamaguchi K, Nishida H, Suzuki K, Kaneko Y, Takeuchi T, Hatano H, Ishigaki K, Ishihama Y, Yamamoto K, Kochi Y, Inamo J, Suzuki A, Ueda MT, Yamaguchi K, Nishida H, Suzuki K, Kaneko Y, Takeuchi T, Hatano H, Ishigaki K, Ishihama Y, Yamamoto K, Kochi Y (2024) Long-read sequencing for 29 immune cell subsets reveals disease-linked isoforms. Nat Commun 15:4285
Ivanov A, Memczak S, Wyler E, Torti F, Porath HT, Orejuela MR, Piechotta M, Levanon EY, Landthaler M, Dieterich C, Rajewsky N (2015) Analysis of intron sequences reveals hallmarks of circular RNA biogenesis in animals. Cell Rep 10:170–177
Jang HJ, Shah NM, Maeng JH, Liang Y, Basri NL, Ge J, Qu X, Mahlokozera T, Tzeng SC, Williams RB, Moore MJ, Annamalai D, Chen JY, Lee HJ, Desouza PA, Li D, Xing X, Kim AH, Wang T (2024) Epigenetic therapy potentiates transposable element transcription to create tumor-enriched antigens in glioblastoma cells. Nat Genet 56:1903–1913
Jarroux J, Morillon A, Pinskaya M (2017) History, discovery, and classification of lncRNAs. Adv Exp Med Biol 1008:1–46
Jeck WR, Sorrentino JA, Wang K, Slevin MK, Burd CE, Liu J, Marzluff WF, Sharpless NE (2013) Circular RNAs are abundant, conserved, and associated with ALU repeats. RNA 19:141–157
Jiang C, Li Y, Zhao Z, Lu J, Chen H, Ding N, Wang G, Xu J, Li X (2016) Identifying and functionally characterizing tissue-specific and ubiquitously expressed human lncRNAs. Oncotarget 7:7120
Jin X, Xu XE, Jiang YZ, Liu YR, Sun W, Guo YJ, Ren YX, Zuo WJ, Hu X, Huang SL, Shen HJ, Lan F, He YF, Hu GH, Di GH, He XH, Li DQ, Liu S, Yu KD, Shao ZM (2019) The endogenous retrovirus-derived long noncoding RNA TROJAN promotes triple-negative breast cancer progression via ZMYND8 degradation. Sci Adv 5:eaat9820

Kamori T, Oki E, Shimada Y, Hu Q, Hisamatsu Y, Ando K, Shimokawa M, Wakai T, Oda Y, Mori M (2022) The effects of ARID1A mutations on colorectal cancer and associations with PD-L1 expression by stromal cells. Cancer Rep (Hoboken) 5:e1420

Kapitonov VV, Jurka J (1998) MER53, a non-autonomous DNA transposon associated with a variety of functionally related defense genes in the human genome. DNA Seq 8:277–288

Kapusta A, Kronenberg Z, Lynch VJ, Zhuo X, Ramsay L, Bourque G, Yandell M, Feschotte C (2013) Transposable elements are major contributors to the origin, diversification, and regulation of vertebrate long noncoding RNAs. PLoS Genet 9:e1003470

Kawaji H, Kasukawa T, Forrest A, Carninci P, Hayashizaki Y, Kawaji H, Kasukawa T, Forrest A, Carninci P, Hayashizaki Y (2017) The FANTOM5 collection, a data series underpinning mammalian transcriptome atlases in diverse cell types. Sci Data 4:170113

Kay GF (1998) Xist and X chromosome inactivation. Mol Cell Endocrinol 140:71

Kim YK, Furic L, Parisien M, Major F, Desgroseillers L, Maquat LE (2007) Staufen1 regulates diverse classes of mammalian transcripts. EMBO J 26:2670–2681

Kong X, Li R, Chen M, Zheng R, Wang J, Sun C, Qu Y (2024) Endogenous retrovirus HERVH-derived lncRNA UCA1 controls human trophoblast development. Proc Natl Acad Sci USA 121:e2318176121

Koning APJD, Gu W, Castoe TA, Batzer MA, Pollock DD (2011) Repetitive elements may comprise over two-thirds of the human genome. PLoS Genet 7:e1002384

Kornienko AE, Guenzl PM, Barlow DP, Pauler FM (2013) Gene regulation by the act of long non-coding RNA transcription. Abstract BMC Biology 11(1). https://doi.org/10.1186/1741-7007-11-59

Kozomara A, Birgaoanu M, Griffiths-Jones S (2019) miRBase: from microRNA sequences to function. Nucleic Acids Res 47:D155–D162

Lagarde J, Uszczynska-Ratajczak B, Carbonell S, Perez-Lluch S, Abad A, Davis C, Gingeras TR, Frankish A, Harrow J, Guigo R, Johnson R (2017) High-throughput annotation of full-length long noncoding RNAs with capture long-read sequencing. Nat Genet 49:1731–1740

Lanciano S, Cristofari G (2020) Measuring and interpreting transposable element expression. Nat Rev Genet 21:721

Langley CH, Montgomery E, Hudson R, Kaplan N, Charlesworth B (1988) On the role of unequal exchange in the containment of transposable element copy number. Genet Res 52:223–235

Lee J, Davidow LS, Warshawsky D, Lee J, Davidow LS, Warshawsky D (1999) Tsix, a gene antisense to Xist at the X-inactivation centre. Nat Genet 21:400

Lee Y, Ahn C, Han J, Choi H, Kim J, Yim J, Lee J, Provost P, Radmark O, Kim S, Kim VN (2003) The nuclear RNase III Drosha initiates microRNA processing. Nature 425:415–419

Lee H-E, Huh J-W, Kim H-S (2020) Bioinformatics analysis of evolution and human disease related transposable element-derived microRNAs. Life 10:95

Lee J, Snell EA, Brown J, Booth CE, Banks RE, Turner DJ, Vasudev NS, Lagos D (2024a) Long-read RNA sequencing of archival tissues reveals novel genes and transcripts associated with clear cell renal cell carcinoma recurrence and immune evasion. Genome Res 34:1849–1864

Lee K, Ku J, Ku D, Kim Y (2024b) Inverted Alu repeats: friends or foes in the human transcriptome. Exp Mol Med 56:1250–1262

Legnini I, Timoteo GD, Rossi F, Morlando M, Briganti F, Sthandier O, Fatica A, Santini T, Andronache A, Wade M, Laneve P, Rajewsky N, Bozzoni I (2017) Circ-ZNF609 is a circular RNA that can be translated and functions in Myogenesis. Mol Cell 66:22

Li M, Liang C (2022) LncDC: a machine learning-based tool for long non-coding RNA detection from RNA-Seq data. Sci Rep 12:19083

Li T, Yang Y, Qi H, Cui W, Zhang L, Fu X, He X, Liu M, Li PF, Yu T (2023) CRISPR/Cas9 therapeutics: progress and prospects. Signal Transduct Target Ther 8:36

Liang Y, Chen X, Wu Y, Li J, Zhang S, Wang K, Guan X, Yang K, Bai Y (2018) LncRNA CASC9 promotes esophageal squamous cell carcinoma metastasis through upregulating LAMC2 expression by interacting with the CREB-binding protein. Cell Death Differ 25:1980–1995

Liang Y, Liu N, Yang L, Tang J, Wang Y, Mei M (2021) A brief review of circRNA biogenesis, detection, and function. Curr Genomics 22:485–495

Liang L, Cao C, Ji L, Cai Z, Wang D, Ye R, Chen J, Yu X, Zhou J, Bai Z, Wang R, Yang X, Zhu P, Xue Y, Liang L, Cao C, Ji L, Cai Z, Wang D, Ye R, Chen J, Yu X, Zhou J, Bai Z, Wang R, Yang X, Zhu P, Xue Y (2023) Complementary Alu sequences mediate enhancer–promoter selectivity. Nature 619:7971

Liang Y, Qu X, Shah NM, Wang T, Liang Y, Qu X, Shah NM, Wang T (2024) Towards targeting transposable elements for cancer therapy. Nat Rev Cancer 24:123

Lin X, Lu Y, Zhang C, Cui Q, Tang Y-D, Ji X, Cui C (2024) LncRNADisease v3.0: an updated database of long non-coding RNA-associated diseases. Nucleic Acids Res 52:D1365

Liu W-M, Chu W-M, Choudary PV, Schmid CW (1995) Cell stress and translational inhibitors transiently increase the abundance of mammalian SINE transcripts. Nucleic Acids Res 23:1758

Liu HZ, Shan TD, Han Y, Liu XS (2020) Silencing long non-coding RNA CASC9 inhibits colorectal cancer cell proliferation by acting as a competing endogenous RNA of miR-576-5p to regulate AKT3. Cell Death Discov 6:115

Lorenz R, Bernhart SH, Höner Zu Siederdissen C, Tafer H, Flamm C, Stadler PF, Hofacker IL, Lorenz R, Bernhart SH, Höner Zu Siederdissen C, Tafer H, Flamm C, Stadler PF, Hofacker IL (2011) ViennaRNA Package 2.0. Algorithms Mol Biol 6:26

Lorenzen JM, Thum T, Lorenzen JM, Thum T (2016) Long noncoding RNAs in kidney and cardiovascular diseases. Nat Rev Nephrol 12:360

Lu X, Sachs F, Ramsay L, Jacques P-É, Göke J, Bourque G, Ng H-H, Lu X, Sachs F, Ramsay L, Jacques P-É, Göke J, Bourque G, Ng H-H (2014) The retrovirus HERVH is a long noncoding RNA required for human embryonic stem cell identity. Nat Struct Mol Biol 21:423

Luo Y, Hitz BC, Gabdank I, Hilton JA, Kagda MS, Lam B, Myers Z, Sud P, Jou J, Lin K, Baymuradov UK, Graham K, Litton C, Miyasato SR, Strattan JS, Jolanki O, Lee JW, Tanaka FY, Adenekan P, O'neill E, Cherry JM (2020) New developments on the encyclopedia of DNA elements (ENCODE) data portal. Nucleic Acids Res 48:D882–D889

Maclary E, Buttigieg E, Hinten M, Gayen S, Harris C, Sarkar MK, Purushothaman S, Kalantry S, Maclary E, Buttigieg E, Hinten M, Gayen S, Harris C, Sarkar MK, Purushothaman S, Kalantry S (2014) Differentiation-dependent requirement of Tsix long non-coding RNA in imprinted X-chromosome inactivation. Nat Commun 5:8564

Magliacane Trotta S, Adinolfi A, D'orsi L, Panico S, Mercadante G, Mehlen P, Ambati J, De Falco S, Tarallo V, Magliacane Trotta S, Adinolfi A, D'orsi L, Panico S, Mercadante G, Mehlen P, Ambati J, De Falco S, Tarallo V (2024) Cancer-derived exosomal Alu RNA promotes colorectal cancer progression. Exp Mol Med 56:700

Maksakova IA, Romanish MT, Gagnier L, Dunn CA, Lagemaat LNVD, Mager DL (2006) Retroviral elements and their hosts: insertional mutagenesis in the mouse germ line. PLoS Genet 2:e2

Malik S, Roeder RG (2023) Regulation of the RNA polymerase II pre-initiation complex by its associated coactivators. Nat Rev Genet 24:767–782

Mangoni D, Simi A, Lau P, Armaos A, Ansaloni F, Codino A, Damiani D, Floreani L, Di Carlo V, Vozzi D, Persichetti F, Santoro C, Pandolfini L, Tartaglia GG, Sanges R, Gustincich S (2023) LINE-1 regulates cortical development by acting as long non-coding RNAs. Nat Commun 14:4974

Marasca F, Sinha S, Vadalà R, Polimeni B, Ranzani V, Paraboschi E, Burattin F, Ghilotti M, Crosti M, Negri M, Campagnoli S, Notarbartolo S, Sartore-Bianchi A, Siena S, Prati D, Montini G, Viale G, Torre O, Harari S, Grifantini R, Soldà G, Biffo S, Abrignani S, Bodega B (2022) LINE1 are spliced in non-canonical transcript variants to regulate T cell quiescence and exhaustion. Nat Genet 54:180

Marx V (2023) Method of the year: long-read sequencing. Nat Methods 20:6–11

Mattick JS, Amaral PP, Carninci P, Carpenter S, Chang HY, Chen L-L, Chen R, Dean C, Dinger ME, Fitzgerald KA, Gingeras TR, Guttman M, Hirose T, Huarte M, Johnson R, Kanduri C, Kapranov P, Lawrence JB, Lee JT, Mendell JT, Mercer TR, Moore KJ, Nakagawa S, Rinn JL, Spector DL, Ulitsky I, Wan Y, Wilusz JE, Wu M, Mattick JS, Amaral PP, Carninci P, Carpenter S, Chang HY, Chen L-L, Chen R, Dean C, Dinger ME, Fitzgerald KA, Gingeras TR, Guttman

M, Hirose T, Huarte M, Johnson R, Kanduri C, Kapranov P, Lawrence JB, Lee JT, Mendell JT, Mercer TR, Moore KJ, Nakagawa S, Rinn JL, Spector DL, Ulitsky I, Wan Y, Wilusz JE, Wu M (2023) Long non-coding RNAs: definitions, functions, challenges and recommendations. Nat Rev Mol Cell Biol 24:430

Merlotti A, Sadacca B, Arribas YA, Ngoma M, Burbage M, Goudot C, Houy A, Rocanin-Arjo A, Lalanne A, Seguin-Givelet A, Lefevre M, Heurtebise-Chretien S, Baudon B, Oliveira G, Loew D, Carrascal M, Wu CJ, Lantz O, Stern MH, Girard N, Waterfall JJ, Amigorena S (2023) Noncanonical splicing junctions between exons and transposable elements represent a source of immunogenic recurrent neo-antigens in patients with lung cancer. Sci Immunol 8:eabm6359

Mihaljevic A, Rubin PD, Chouvardas P, Esposito R (2024) Cell type specific long non-coding RNA targets identified by integrative analysis of single-cell and bulk colorectal cancer transcriptomes. Sci Rep 14:10939

Mitra A, Pfeifer K, Park KS (2018) Circular RNAs and competing endogenous RNA (ceRNA) networks. Transl Cancer Res 7:S624–S628

Mudge JM, Carbonell-Sala S, Diekhans M, Martinez JG, Hunt T, Jungreis I, Loveland JE, Arnan C, Barnes I, Bennett R, Berry A, Bignell A, Cerdan-Velez D, Cochran K, Cortes LT, Davidson C, Donaldson S, Dursun C, Fatima R, Hardy M, Hebbar P, Hollis Z, James BT, Jiang Y, Johnson R, Kaur G, Kay M, Mangan RJ, Maquedano M, Gomez LM, Mathlouthi N, Merritt R, Ni P, Palumbo E, Perteghella T, Pozo F, Raj S, Sisu C, Steed E, Sumathipala D, Suner MM, Uszczynska-Ratajczak B, Wass E, Yang YT, Zhang D, Finn RD, Gerstein M, Guigo R, Hubbard TJP, Kellis M, Kundaje A, Paten B, Tress ML, Birney E, Martin FJ, Frankish A (2025) GENCODE 2025: reference gene annotation for human and mouse. Nucleic Acids Res 53:D966–D975

Navandar M, Vennin C, Lutz B, Gerber S, Navandar M, Vennin C, Lutz B, Gerber S (2024) Long non-coding RNAs expression and regulation across different brain regions in primates. Sci Data 11:545

Nemeth K, Bayraktar R, Ferracin M, Calin GA (2024) Non-coding RNAs in disease: from mechanisms to therapeutics. Nat Rev Genet 25:211–232

Nickerson JA, Momen-Heravi F (2024) Long non-coding RNAs: roles in cellular stress responses and epigenetic mechanisms regulating chromatin. Nucleus 15:2350180

Oo JA, Warwick T, Palfi K, Lam F, Mcnicoll F, Prieto-Garcia C, Gunther S, Cao C, Zhou Y, Gavrilov AA, Razin SV, Cabrera-Orefice A, Wittig I, Pullamsetti SS, Kurian L, Gilsbach R, Schulz MH, Dikic I, Muller-Mcnicoll M, Brandes RP, Leisegang MS (2025) Long non-coding RNAs direct the SWI/SNF complex to cell type-specific enhancers. Nat Commun 16:131

Palazzo AF, Gregory TR (2014) The case for junk DNA. PLoS Genet 10:e1004351

Pamudurti NR, Bartok O, Jens M, Ashwal-Fluss R, Stottmeister C, Ruhe L, Hanan M, Wyler E, Perez-Hernandez D, Ramberger E, Shenzis S, Samson M, Dittmar G, Landthaler M, Chekulaeva M, Rajewsky N, Kadener S (2017) Translation of CircRNAs. Mol Cell 66:9–21 e7

Pardo-Palacios FJ, Wang D, Reese F, Diekhans M, Carbonell-Sala S, Williams B, Loveland JE, De Maria M, Adams MS, Balderrama-Gutierrez G, Behera AK, Gonzalez Martinez JM, Hunt T, Lagarde J, Liang CE, Li H, Meade MJ, Moraga Amador DA, Prjibelski AD, Birol I, Bostan H, Brooks AM, Celik MH, Chen Y, Du MRM, Felton C, Goke J, Hafezqorani S, Herwig R, Kawaji H, Lee J, Li JL, Lienhard M, Mikheenko A, Mulligan D, Nip KM, Pertea M, Ritchie ME, Sim AD, Tang AD, Wan YK, Wang C, Wong BY, Yang C, Barnes I, Berry AE, Capella-Gutierrez S, Cousineau A, Dhillon N, Fernandez-Gonzalez JM, Ferrandez-Peral L, Garcia-Reyero N, Gotz S, Hernandez-Ferrer C, Kondratova L, Liu T, Martinez-Martin A, Menor C, Mestre-Tomas J, Mudge JM, Panayotova NG, Paniagua A, Repchevsky D, Ren X, Rouchka E, Saint-John B, Sapena E, Sheynkman L, Smith ML, Suner MM, Takahashi H, Youngworth IA, Carninci P, Denslow ND, Guigo R, Hunter ME, Maehr R, Shen Y, Tilgner HU, Wold BJ, Vollmers C, Frankish A, Au KF, Sheynkman GM, Mortazavi A, Conesa A, Brooks AN (2024) Systematic assessment of long-read RNA-seq methods for transcript identification and quantification. Nat Methods 21:1349–1363

Park EG, Ha H, Lee DH, Kim WR, Lee YJ, Bae WH, Kim HS (2022) Genomic analyses of non-coding RNAs overlapping transposable elements and its implication to human diseases. Int J Mol Sci 23:8950

Payer LM, Burns KH (2019) Transposable elements in human genetic disease. Nature Reviews Genetics 20(12):760–772. https://doi.org/10.1038/s41576-019-0165-8

Pennisi E (2012) Genomics. ENCODE project writes eulogy for junk DNA. Science 337(1159):1161

Percharde M, Lin CJ, Yin Y, Guan J, Peixoto GA, Bulut-Karslioglu A, Biechele S, Huang B, Shen X, Ramalho-Santos M (2018) A LINE1-Nucleolin partnership regulates early development and ESC identity. Cell 174:391–405 e19

Petri R, Brattas PL, Sharma Y, Jonsson ME, Pircs K, Bengzon J, Jakobsson J (2019) LINE-2 transposable elements are a source of functional human microRNAs and target sites. PLoS Genet 15:e1008036

Piriyapongsa J, Jordan IK (2007) A family of human microRNA genes from miniature inverted-repeat transposable elements. PLoS One 2:e203

Piriyapongsa J, Marino-Ramirez L, Jordan IK (2007) Origin and evolution of human microRNAs from transposable elements. Genetics 176:1323–1337

Plaisance I, Chouvardas P, Sun Y, Nemir M, Aghagolzadeh P, Aminfar F, Shen S, Shim WJ, Rochais F, Johnson R, Palpant N, Pedrazzini T (2023) A transposable element into the human long noncoding RNA CARMEN is a switch for cardiac precursor cell specification. Cardiovasc Res 119:1361

Polimeni B, Marasca F, Ranzani V, Bodega B (2024) IRescue: uncertainty-aware quantification of transposable elements expression at single cell level. Nucleic Acids Res 52:e93

Qin S, Jin P, Zhou X, Chen L, Ma F (2015) The role of transposable elements in the origin and evolution of MicroRNAs in human. PLoS One 10:e0131365

Quinlan AR (2014) BEDTools: the Swiss-Army tool for genome feature analysis. Curr Protoc Bioinformatics 47:11.12.1-34

Quinlan AR, Hall IM (2010) BEDTools: a flexible suite of utilities for comparing genomic features. Bioinformatics 26:841–842

Raghavan V, Kraft L, Mesny F, Rigerte L (2022) A simple guide to de novo transcriptome assembly and annotation. Brief Bioinform 23:bbab563

Ranzani V, Rossetti G, Panzeri I, Arrigoni A, Bonnal RJ, Curti S, Gruarin P, Provasi E, Sugliano E, Marconi M, De Francesco R, Geginat J, Bodega B, Abrignani S, Pagani M (2015) The long intergenic noncoding RNA landscape of human lymphocytes highlights the regulation of T cell differentiation by linc-MAF-4. Nat Immunol 16:318–325

Rech GE, Radío S, Guirao-Rico S, Aguilera L, Horvath V, Green L, Lindstadt H, Jamilloux V, Quesneville H, González J, Rech GE, Radío S, Guirao-Rico S, Aguilera L, Horvath V, Green L, Lindstadt H, Jamilloux V, Quesneville H, González J (2022) Population-scale long-read sequencing uncovers transposable elements associated with gene expression variation and adaptive signatures in Drosophila. Nat Commun 13:1948

Richardson SR, Doucet AJ, Kopera HC, Moldovan JB, Garcia-Perez JL, Moran JV (2015) The influence of LINE-1 and SINE retrotransposons on mammalian genomes. Microbiol Spectr 3:MDNA3-0061-2014

Roberts JT, Cardin SE, Borchert GM (2014) Burgeoning evidence indicates that microRNAs were initially formed from transposable element sequences. Mob Genet Elements 4:e29255

Rooda I, Kaselt B, Liivrand M, Smolander O-P, Salumets A, Velthut-Meikas A, Rooda I, Kaselt B, Liivrand M, Smolander O-P, Salumets A, Velthut-Meikas A (2021) Hsa-mir-548 family expression in human reproductive tissues. BMC Genomic Data 22:40

Schirle NT, Sheu-Gruttadauria J, Macrae IJ (2014) Structural basis for microRNA targeting. Science 346:608–613

Schneider C, Setty M, Holmes AB, Maute RL, Leslie CS, Mussolin L, Rosolen A, Dalla-Favera R, Basso K (2014) MicroRNA 28 controls cell proliferation and is down-regulated in B-cell lymphomas. Proc Natl Acad Sci USA 111:8185–8190

Schwämmle T, Schulz EG (2023) Regulatory principles and mechanisms governing the onset of random X-chromosome inactivation. Curr Opin Genet Dev 81:102063
Seifuddin F, Singh K, Suresh A, Judy JT, Chen YC, Chaitankar V, Tunc I, Ruan X, Li P, Chen Y, Cao H, Lee RS, Goes FS, Zandi PP, Jafri MS, Pirooznia M (2020) lncRNAKB, a knowledgebase of tissue-specific functional annotation and trait association of long noncoding RNA. Sci Data 7:326
Sexton CE, Tillett RL, Han MV (2022) The essential but enigmatic regulatory role of HERVH in pluripotency. Trends Genet 38:12
Shang R, Lee S, Senavirathne G, Lai EC (2023) microRNAs in action: biogenesis, function and regulation. Nat Rev Genet 24:816–833
Shin C, Nam J-W, Farh KK-H, Chiang HR, Shkumatava A, Bartel DP (2010) Expanding the MicroRNA targeting code: functional sites with centered pairing. Mol Cell 38:789
Smalheiser NR, Torvik VI (2005) Mammalian microRNAs derived from genomic repeats. Trends Genet 21:322–326
Smit AF, Hubley R, Green P (2013) RepeatMasker Open-4.0 [Online]. Available: http://www.repeatmasker.org [Accessed]
Sultana T, Zamborlini A, Cristofari G, Lesage P, Sultana T, Zamborlini A, Cristofari G, Lesage P (2017) Integration site selection by retroviruses and transposable elements in eukaryotes. Nat Rev Genet 18:292
Thiel G, Ekici M, Rossler OG (2015) RE-1 silencing transcription factor (REST): a regulator of neuronal development and neuronal/endocrine function. Cell Tissue Res 359:99–109
Tlučková K, Kaczmarek B, Salmazo A, Bernecky C, Tlučková K, Kaczmarek B, Salmazo A, Bernecky C (2025) Mechanism of mammalian transcriptional repression by noncoding RNA. Nat Struct Mol Biol 32:607
Vinjamuri BP, Pan J, Peng P (2024) A review on commercial oligonucleotide drug products. J Pharm Sci 113:1749–1768
Vlachogiannis NI, Sachse M, Georgiopoulos G, Zormpas E, Bampatsias D, Delialis D, Bonini F, Galyfos G, Sigala F, Stamatelopoulos K, Gatsiou A, Stellos K (2021) Adenosine-to-inosine Alu RNA editing controls the stability of the pro-inflammatory long noncoding RNA NEAT1 in atherosclerotic cardiovascular disease. J Mol Cell Cardiol 160:111–120
Wagner V, Kern F, Hahn O, Schaum N, Ludwig N, Fehlmann T, Engel A, Henn D, Rishik S, Isakova A, Tan M, Sit R, Neff N, Hart M, Meese E, Quake S, Wyss-Coray T, Keller A, Wagner V, Kern F, Hahn O, Schaum N, Ludwig N, Fehlmann T, Engel A, Henn D, Rishik S, Isakova A, Tan M, Sit R, Neff N, Hart M, Meese E, Quake S, Wyss-Coray T, Keller A (2023) Characterizing expression changes in noncoding RNAs during aging and heterochronic parabiosis across mouse tissues. Nat Biotechnol 42:109
Wang Y, Medvid R, Melton C, Jaenisch R, Blelloch R (2007) DGCR8 is essential for microRNA biogenesis and silencing of embryonic stem cell self-renewal. Nat Genet 39:380–385
Wang J, Xie G, Singh M, Ghanbarian AT, Raskó T, Szvetnik A, Cai H, Besser D, Prigione A, Fuchs NV, Schumann GG, Chen W, Lorincz MC, Ivics Z, Hurst LD, Izsvák Z, Wang J, Xie G, Singh M, Ghanbarian AT, Raskó T, Szvetnik A, Cai H, Besser D, Prigione A, Fuchs NV, Schumann GG, Chen W, Lorincz MC, Ivics Z, Hurst LD, Izsvák Z (2014) Primate-specific endogenous retrovirus-driven transcription defines naive-like stem cells. Nature 516:7531
Wang C, Shi Z, Huang Q, Liu R, Su D, Chang L, Xiao C, Fan X (2024) Single-cell analysis of isoform switching and transposable element expression during preimplantation embryonic development. PLoS Biol 22:e3002505
Wei G, Qin S, Li W, Chen L, Ma F (2016) MDTE DB: a database for microRNAs derived from transposable element. IEEE/ACM Trans Comput Biol Bioinform 13:1155–1160
Wells JN, Feschotte C (2020) A field guide to eukaryotic transposable elements. Annu Rev Genet 54:539
Wen X, Zhong S (2023) Alu transposable elements rewire enhancer-promoter network through RNA pairing. Mol Cell 83:3234–3235

Wicker T, Sabot F, Hua-Van A, Bennetzen JL, Capy P, Chalhoub B, Flavell A, Leroy P, Morgante M, Panaud O, Paux E, Sanmiguel P, Schulman AH, Wicker T, Sabot F, Hua-Van A, Bennetzen JL, Capy P, Chalhoub B, Flavell A, Leroy P, Morgante M, Panaud O, Paux E, Sanmiguel P, Schulman AH (2007) A unified classification system for eukaryotic transposable elements. Nat Rev Genet 8:973

Wilusz JE (2015) Repetitive elements regulate circular RNA biogenesis. Mob Genet Elements 5:1–7

Winkle M, El-Daly SM, Fabbri M, Calin GA (2021) Noncoding RNA therapeutics – challenges and potential solutions. Nat Rev Drug Discov 20:629–651

Wu Y, Zhao Y, Huan L, Zhao J, Zhou Y, Xu L, Hu Z, Liu Y, Chen Z, Wang L, Huang S, He X, Liang L (2020) An LTR retrotransposon-derived long noncoding RNA lncMER52A promotes hepatocellular carcinoma progression by binding p120-catenin. Cancer Res 80:976

Xie B, Wu Y, Wang S, Ruan L, Liu X (2024) Expression profile of long noncoding RNAs and comprehensive analysis of lncRNA-cisTF-DGE regulation in condyloma acuminatum. BMC Med Genet 17:167

Xue C, Li F, Li F (2008) Finding noncoding RNA transcripts from low abundance expressed sequence tags. Cell Res 18:695–700

Yamada A, Yu P, Lin W, Okugawa Y, Boland CR, Goel A, Yamada A, Yu P, Lin W, Okugawa Y, Boland CR, Goel A (2018) A RNA-Sequencing approach for the identification of novel long non-coding RNA biomarkers in colorectal cancer. Sci Rep 8:575

Yang M, Yao Y, Eades G, Zhang Y, Zhou Q (2011) MiR-28 regulates Nrf2 expression through a Keap1-independent mechanism. Breast Cancer Res Treat 129:983–991

Yang F, Zhang H, Mei Y, Wu M (2014) Reciprocal regulation of HIF-1alpha and lincRNA-p21 modulates the Warburg effect. Mol Cell 53:88–100

Yang Y, Fan X, Mao M, Song X, Wu P, Zhang Y, Jin Y, Yang Y, Chen L-L, Wang Y, Wong CC, Xiao X, Wang Z, Yang Y, Fan X, Mao M, Song X, Wu P, Zhang Y, Jin Y, Yang Y, Chen L-L, Wang Y, Wong CC, Xiao X, Wang Z (2017) Extensive translation of circular RNAs driven by N6-methyladenosine. Cell Res 27:626

Yang Z, Tang N, Zheng M, Chang Y (2023) Exosomal long noncoding RNAs serve as biomarkers for liver disease. Turk J Gastroenterol 34:674–680

Yu C, Lei X, Chen F, Mao S, Lv L, Liu H, Hu X, Wang R, Shen L, Zhang N, Meng Y, Shen Y, Chen J, Li P, Huang S, Lin C, Zhang Z, Yuan K, Yu C, Lei X, Chen F, Mao S, Lv L, Liu H, Hu X, Wang R, Shen L, Zhang N, Meng Y, Shen Y, Chen J, Li P, Huang S, Lin C, Zhang Z, Yuan K (2022) ARID1A loss derepresses a group of human endogenous retrovirus-H loci to modulate BRD4-dependent transcription. Nat Commun 13:3501

Yuan Z, Sun X, Jiang D, Ding Y, Lu Z, Gong L, Liu H, Xie J, Yuan Z, Sun X, Jiang D, Ding Y, Lu Z, Gong L, Liu H, Xie J (2010) Origin and evolution of a placental-specific microRNA family in the human genome. BMC Evol Biol 10:346

Yuan Y-W, Wessler SR, Yuan Y-W, Wessler SR (2011a) The catalytic domain of all eukaryotic cut-and-paste transposase superfamilies. Proc Natl Acad Sci USA 108:7884

Yuan Z, Sun X, Liu H, Xie J (2011b) MicroRNA genes derived from repetitive elements and expanded by segmental duplication events in mammalian genomes. PLoS One 6:e17666

Zhan Y, Zhang L, Yu S, Wen J, Liu Y, Zhang X (2020) Long non-coding RNA CASC9 promotes tumor growth and metastasis via modulating FZD6/Wnt/beta-catenin signaling pathway in bladder cancer. J Exp Clin Cancer Res 39:136

Zhang L, Yan L, Jiang J, Wang Y, Jiang Y, Yan T, Cao Y (2014a) The structure and retrotransposition mechanism of LTR-retrotransposons in the asexual yeast Candida albicans. Virulence 5:655

Zhang XO, Wang HB, Zhang Y, Lu X, Chen LL, Yang L (2014b) Complementary sequence-mediated exon circularization. Cell 159:134–147

Zhang X, Wang W, Zhu W, Dong J, Cheng Y, Yin Z, Shen F (2019a) Mechanisms and functions of long non-coding RNAs at multiple regulatory levels. Int J Mol Sci 20:5573

Zhang Y, Li T, Preissl S, Amaral ML, Grinstein JD, Farah EN, Destici E, Qiu Y, Hu R, Lee AY, Chee S, Ma K, Ye Z, Zhu Q, Huang H, Fang R, Yu L, Izpisua Belmonte JC, Wu J, Evans SM, Chi NC, Ren B, Zhang Y, Li T, Preissl S, Amaral ML, Grinstein JD, Farah EN, Destici E, Qiu

Y, Hu R, Lee AY, Chee S, Ma K, Ye Z, Zhu Q, Huang H, Fang R, Yu L, Izpisua Belmonte JC, Wu J, Evans SM, Chi NC, Ren B (2019b) Transcriptionally active HERV-H retrotransposons demarcate topologically associating domains in human pluripotent stem cells. Nat Genet 51:1380

Zhang J, Hou W, Zhao Q, Xiao S, Linghu H, Zhang L, Du J, Cui H, Yang X, Ling S, Su J, Kong Q (2023a) Deep annotation of long noncoding RNAs by assembling RNA-seq and small RNA-seq data. J Biol Chem 299:105130

Zhang X, Zhang Y, Zhao J, Wu Y, Zhang N, Shen W (2023b) ARID1A mutations in cancer development: mechanism and therapy. Carcinogenesis 44:197–208

Zhou WY, Cai ZR, Liu J, Wang DS, Ju HQ, Xu RH (2020) Circular RNA: metabolism, functions and interactions with proteins. Mol Cancer 19:172

Ziats MN, Rennert OM (2013) Aberrant expression of long noncoding RNAs in autistic brain. J Mol Neurosci 49:589–593

Chapter 6
Transposable Elements: The Role of the piRNA Pathway

Jiao-Yan Ma and Xin Zhiguo Li

Abstract PIWI-interacting RNAs (piRNAs) are a class of small silencing RNAs that play a crucial role in preserving the germline genomic integrity by silencing transposable elements (TEs). As the most heterogeneous small RNAs, they comprise over one million unique sequences, enabling them to comprehensively target and repress the diverse TE families in each species and thus safeguard the genome. piRNA biogenesis is distinct from other small RNA pathways, relying on a unique set of proteins and processing from long single-stranded RNA precursors. While many core components of the piRNA pathway are evolutionarily conserved, different animal clades, including *Tetrahymena*, nematodes, arthropods, and amniotes, have evolved clade-specific strategies for piRNA production. Functionally, beyond their well-established role in TE silencing at both transcriptional and post-transcriptional levels, piRNA pathways are increasingly recognized for their broader, species-specific regulatory functions. These include controlling gene expression in germ cells and supporting diverse somatic roles, particularly in invertebrates. This chapter provides a comparative overview of piRNA biogenesis and function in *Drosophila* and mice, highlighting species-specific adaptations and detailing their molecular mechanisms in defending against TE activation. A deeper understanding of piRNA pathways not only sheds light on their fundamental role in the evolutionary arms race with TEs, a conflict that dates back to the origin of life, but also reveals their emerging relevance in human infertility.

Keywords piRNA · Transposable element · Germ cell · Genome integrity

J.-Y. Ma · X. Z. Li (✉)
Center for RNA Medicine, the Fourth Affiliated Hospital of School of Medicine, and International School of Medicine, International Institutes of Medicine, Zhejiang University, Yiwu, China
e-mail: xinlirna@zju.edu.cn

L. M. Vaschetto (ed.), *Transposable Elements, Transcriptomics, and Diseases*, Epigenetics and Human Health 14, https://doi.org/10.1007/978-3-032-04125-8_6

Abbreviations

3′ UTR	3′ untranslated region
Ago3	Argonaute 3
A-MYB	A-myeloblastosis viral oncogene homolog
Aub	Aubergine
BTBD18	BTB domain-containing 18
CB	Chromatoid body
CtBP	C-terminal binding protein
DNMT3C	DNA methyltransferase 3C
GTSF1	Gametocyte-specific factor 1
H3K4me3	H3 lysine 4 trimethylation
H3K9me3	H3 lysine 9 trimethylation
H3K23	H3 lysine 23
HP1	Heterochromatin protein 1
IMC	Intermitochondrial cement
lncRNA	Long single-stranded non-coding RNA
MAEL	Maelstrom
miRNAs	MicroRNAs
MOV10L1	Moloney leukemia virus 10-like protein 1
NMD	Nonsense-mediated decay
NuRD	Nucleosome remodeling and deacetylase
ORFs	Open reading frames
Panx	Panoramix
PiRISCs	piRNA-induced silencing complexes
piRNAs	PIWI-interacting RNAs
PLD6	Phospholipase D family member 6
PNLDC1	Poly(A)-specific ribonuclease-like domain-containing 1
Pol II	Polymerase II
RDC	Rhino-Deadlock-Cutoff
SAM	S-adenosylmethionine
siRNAs	Small interfering RNAs
Spn-E	Spindle-E
SPOCD1	SPOC domain-containing protein 1
SUMO	Small ubiquitin-like modifier
SVs	Structural variations
TDRD	Tudor domain-containing protein
TEX15	Testis-expressed 15
THOC	THO complex
TEs	Transposable elements
Trf2	TATA-box-binding protein-related factor 2
U	Uridine
Yb	Ypsilon schachtel
Zuc	Zucchini

6.1 Introduction

Transposable elements (TEs) are mobile genetic sequences capable of proliferating and relocating within the genome (Kidwell and Lisch 1997). Historically viewed as "selfish elements" due to their potential to compromise genome integrity by mutagenic insertion and ectopic recombination, TEs were long considered detrimental (Doolittle and Sapienza 1980; Orgel and Crick 1980; Charlesworth and Langley 1986). However, recent studies have highlighted their broader impacts on shaping gene expression, promoting phenotypic plasticity, and driving adaptation to environmental challenges, as well as influencing the emergence of new organs and traits (Wessler 1988; Whitelaw and Martin 2001; Sasaki et al. 2008; Kunarso et al. 2010; Chuong et al. 2013; Schrader et al. 2014; Notwell et al. 2015; Liang et al. 2023). Despite these beneficial roles, the unchecked amplification of active, young TEs—particularly in germ cells—poses a threat to reproductive fitness and species continuity. To counter this, hosts have evolved diverse mechanisms to silence TEs, especially in the germline. Among these, the PIWI-interacting RNA (piRNA) pathway has emerged as a primary defense mechanism against TEs in metazoans and some protozoans (Aravin et al. 2006; Girard et al. 2006; Grivna et al. 2006; Lau et al. 2006; Vagin et al. 2006; Houwing et al. 2007; Zhou et al. 2007; Batista et al. 2008).

6.1.1 Definition of piRNAs

piRNAs are a class of small non-coding RNAs, typically 24–35 nucleotides in length, that specifically associate with PIWI proteins—a highly conserved subfamily of Argonaute proteins essential for germ cell function across the animal kingdom (Deng and Lin 2002; Murchison et al. 2008; Chirn et al. 2015; Wynant et al. 2017). piRNAs are identified and characterized based on a combination of biochemical properties and computational criteria. Experimentally, they can be enriched by immunoprecipitation using PIWI-specific antibodies (Grivna et al. 2006; Lau et al. 2006; Vagin et al. 2006) or by exploiting their resistance to periodate oxidation due to a unique 3′ end modification (Horwich et al. 2007; Kirino and Mourelatos 2007; Saito et al. 2007). Computationally, they are defined by their characteristic size range, strong 5′ uridine preference (1U bias), and mapping to discrete piRNA-producing loci in the genome (Brennecke et al. 2007; Li et al. 2013). Because of their extreme sequence diversity and species specificity, sequence conservation is not a reliable criterion (Brennecke et al. 2007; Batista et al. 2008). These features collectively define piRNAs as a distinct class of small regulatory RNAs.

6.1.2 Comparison with Other Small RNAs

Small silencing RNAs, including microRNAs (miRNAs), small interfering RNAs (siRNAs), and piRNAs, mediate gene silencing by interacting with Argonaute proteins. Despite this commonality, they exhibit distinct biogenesis pathways,

Table 6.1 A comparison of miRNA, siRNA, and piRNA pathways

	miRNA	siRNA	piRNA
Length (nt)	21–23	20–25	24–35
Binding partner	Argonaute (AGO)	Argonaute (AGO)	PIWI (a subfamily of AGO)
3′ End modification	2′-OH	2′-OH	2′-*O*-CH_3
Origin	Hairpin-loop	Double-stranded RNA	Single-stranded RNA
Biogenesis	Dicer-dependent	Dicer-dependent	Dicer-independent
Function	Gene expression regulation	mRNA degradation, antiviral response	TE silencing
Tissue distribution	Ubiquitous	Induced	Mostly germline
Target region	Seed region (g2–8)	Entire sequence	Entire sequence
Target pairing	Partial complementarity at the seed region	Perfect complementarity	Over 15nt base pairs between the target and any part of the piRNA

functions, and regulatory mechanisms (Table 6.1). miRNAs and siRNAs are both derived from precursors with double-stranded RNA regions and processed by the Dicer protein into small duplex RNAs, typically 20–22 nucleotides in length (Bernstein et al. 2001; Hutvágner et al. 2001). In contrast, piRNAs are processed from long single-stranded precursors in a Dicer-independent manner (Vagin et al. 2006; Li et al. 2013). piRNAs are typically longer, ranging from 24 to 35 nucleotides, and are characterized by 2′-*O*-methylation at their 3′ ends—a modification absent in miRNAs and siRNAs in mammals (Horwich et al. 2007; Kirino and Mourelatos 2007; Saito et al. 2007). Primarily active in germ cells, piRNAs interact specifically with PIWI proteins (Lin and Spradling 1997; Cox et al. 1998; Aravin et al. 2006; Girard et al. 2006).

miRNAs primarily regulate gene expression by binding partially complementary sequences usually within the 3′ untranslated region (3′ UTR) of target mRNAs. This interaction most commonly involves a 6–8 nucleotide "seed region" spanning positions 2–8 from the miRNA 5′ end (Bartel 2009). This interaction leads to translational repression or mRNA degradation (Farh et al. 2005; Grimson et al. 2007; Guo et al. 2010). siRNAs, on the other hand, degrade target RNAs through precise sequence complementarity of the entire sequence (Elbashir et al. 2001). Recent studies indicate that piRNAs target TEs with stringent requirements for base pairing without defined regions functioning as seed sequences. Specifically, piRNA–TE interactions necessitate a minimal matching of 15 base pairs in any region between piRNAs and the target duplex (Gainetdinov et al. 2023; Li et al. 2025). Upon target recognition, PIWI–piRNA complexes induce silencing through post-transcriptional mRNA cleavage, an endonuclease activity harbored at the PIWI domain of PIWI protein (Gunawardane et al. 2007; Reuter et al. 2011), or recruit other cofactor to mediate transcriptional chromatin repression (Sienski et al. 2012; Le Thomas et al. 2013; Zoch et al. 2020). Such stringent target recognition ensures specific TE targeting while preventing autoimmunity by avoiding unintended interactions with endogenous genes.

6.1.3 Classification of Mammalian piRNAs

piRNAs exhibit distinct temporal expression patterns and regulatory roles during germ cell development in mammals, in contrast to *Drosophila*, where piRNAs are expressed continuously throughout germ cell development (Brennecke et al. 2007; Li et al. 2009; Marie et al. 2017). In mice, piRNAs can be classified into four subtypes: prenatal, pre-pachytene, pachytene, and hybrid piRNAs, based on their stage-specific expression during germ cell development (Aravin et al. 2007a; Li et al. 2013; Table 6.2). The pachytene stage refers to a defined subphase of meiotic prophase in which homologous chromosomes undergo full synapsis (Handel and Schimenti 2010; Li et al. 2013). Prenatal piRNAs are expressed during embryonic development in primordial germ cells. They are enriched in TE sequences and associated with the PIWI proteins MILI and MIWI2; these piRNAs play a critical role in TE silencing by directing de novo DNA methylation during the establishment of genomic imprints (Kuramochi-Miyagawa et al. 2008; De Fazio et al. 2011; Schöpp et al. 2020). Notably, prenatal piRNAs are the only known class of mammalian small RNAs with nuclear epigenetic functions. Pre-pachytene piRNAs, only associated with MILI as the expression of MIWI2 ceases right before birth, are expressed after birth and represent the least abundant piRNA subtype. Despite their low expression levels, they are essential for maintaining genomic integrity by targeting TEs (Aravin et al. 2007b; Di Giacomo et al. 2013). Pachytene piRNAs are the most

Table 6.2 Classification of mammalian piRNAs

	Prenatal piRNA	Pre-pachytene piRNA	Pachytene piRNA	Hybrid piRNA
Expression timing	Embryonic stage	After birth	Pachytene stage	After birth and induced at the pachytene stage
Origins	TE-rich sequences	TE sequences and mRNA 3′UTRs	lncRNAs from intergenic regions	mRNA 3′UTRs embedded with TE sequences
Associated PIWI proteins	MILI, MIWI2	MILI	MILI, MIWI	MILI, MIWI
Relative abundance	Moderate	Lowest	Highest	Moderate
Functions	TE silencing transcriptionally and post-transcriptionally	TE silencing post-transcriptionally	Silencing young and active TE post-transcriptionally	TE silencing post-transcriptionally and fine tune protein synthesis

abundant subtypes, representing over 90% of the total piRNAs in adult testis. They are expressed during the pachytene stage of meiosis and associate with both MILI and MIWI, with MIWI expressed at early pachytene to specifically bind pachytene piRNAs and hybrid piRNAs. These piRNAs are primarily derived from intergenic regions that are largely devoid of TE sequences compared to the rest of the genome. Initially thought to function outside of TE silencing, recent data indicate that 1% of pachytene piRNAs are essential to silence the young and active TEs (Sun et al. 2023). The function of non-TE-targeting piRNAs remains unclear (Xu and Li 2024). Hybrid piRNAs display characteristics of both pre-pachytene and pachytene piRNAs. They are expressed as early as the pre-pachytene stage but increase in abundance during the pachytene stage. Hybrid piRNAs are mainly derived from mRNA 3′ UTRs and contribute to both TE silencing and the fine-tuning of protein synthesis (Li et al. 2013; Sun et al. 2021).

6.2 piRNA Biogenesis

6.2.1 Transcription of piRNA-Producing Loci

piRNAs are transcribed from discrete genomic regions—referred to as piRNA-producing loci or, computationally, as piRNA clusters—defined by their piRNA output computationally rather than transcriptional units as conventional genes (Aravin et al. 2006; Brennecke et al. 2007; Li et al. 2013). In *Drosophila*, piRNA precursors often originate from heterochromatic regions such as pericentromeric and telomeric loci, whereas in mammals, piRNA-producing loci resemble canonical euchromatic RNA polymerase II (Pol II) transcription units (Brennecke et al. 2007; Li et al. 2013; Mohn et al. 2014). In *Drosophila melanogaster*, these loci are categorized into two main types: uni-strand clusters and dual-strand clusters. Uni-strand clusters are transcribed similarly to canonical mRNA transcription, featuring defined promoters, 5′ capping, 3′ polyadenylation, and occasional splicing (Brennecke et al. 2007; Klattenhoff et al. 2009; Mohn et al. 2014). An example is the somatic piRNA locus *flamenco*, located in the pericentromeric heterochromatin of the X chromosome. *Flamenco* undergoes conventional transcription to produce a long precursor transcript, which is processed into piRNAs. This locus is enriched with antisense TE sequences, enabling the generation of piRNAs that directly target and degrade TE mRNAs, thereby suppressing TE activity (Brennecke et al. 2007; Li et al. 2009; Malone et al. 2009).

In *Drosophila*, transcription of piRNA clusters is cell type-specific: dual-strand clusters are active in the germline during early oogenesis, whereas uni-strand clusters like *flamenco* are transcribed in somatic follicle cells. *Drosophila* dual-strand clusters are transcribed convergently from both DNA strands and lack defined promoters. They produce non-polyadenylated, unspliced RNAs through non-canonical transcription mechanisms (Brennecke et al. 2007; Mohn et al. 2014; Zhang et al.

2014). These loci are marked by histone modifications, such as H3 lysine 9 trimethylation (H3K9me3), typically associated with transcriptional repression but paradoxically required for piRNA expression (Mohn et al. 2014; Andersen et al. 2017; Fig. 6.1). A germline-specific heterochromatin protein 1 (HP1) homolog, Rhino, binds to H3K9me3 marks and recruits the Moonshiner transcription factor, forming an alternative pre-initiation complex with TATA-box-binding protein-related factor 2 (Trf2) (Zhang et al. 2014; Andersen et al. 2017; Parhad et al. 2020). Additionally, Rhino recruits Deadlock and Cutoff to bypass the need for promoter sequences following H3K9me3 binding. Subsequently, this Rhino-Deadlock-Cutoff (RDC) complex enables RNA Pol II to initiate transcription from multiple sites on both DNA strands (Mohn et al. 2014; Andersen et al. 2017). Cutoff prevents splicing and polyadenylation of nascent piRNA transcripts while collaborating with Deadlock and the co-repressor C-terminal binding protein (CtBP) to maintain transcriptional balance (Mohn et al. 2014; Parhad et al. 2020). The histone acetyltransferase Enok further enhances Rhino expression through H3 lysine 23 (H3K23) acetylation, reinforcing piRNA-producing loci transcription (Tsai and Huang 2021). Post-transcriptionally, Bootlegger recruits the nuclear export factor nuclear export factor 3 (Nxf3) to facilitate the transport of piRNA precursors from the nucleus to the "nuage" (French for "cloud"), a perinuclear structure unique to germ cells, for further processing via a chromosome region maintenance 1 (Crm1)-dependent pathway (Kneuss et al. 2019). The lysine demethylase 3 (Kdm3), a histone demethylase, further safeguards host genes from inadvertent processing into piRNA pathways (Casier et al. 2023).

In mice, piRNAs are primarily derived from long single-stranded non-coding RNA (lncRNA) precursors transcribed in intergenic regions. They constitute approximately 90% of all piRNAs in adult testes and begin to accumulate in spermatocytes during the pachytene stage of meiosis (Aravin et al. 2006; Girard et al. 2006). Similar to canonical mRNAs, piRNA loci are characterized by 5′ capping, polyadenylation signals, distinct promoters, and splice sites within their transcriptional units (Li et al. 2013). The transcription factor A-myeloblastosis viral oncogene homolog (A-MYB) initiates the transcription of pachytene piRNA precursors. It also activates the expression of mRNAs encoding proteins involved in the piRNA biogenesis pathway, forming a feedforward loop that ensures the precise and coordinated activation of the entire pathway. Loss of A-MYB results in the depletion of both piRNA precursor transcripts and mature piRNAs across the entire loci (Li et al. 2013). Additionally, A-MYB initiates transcription of *Tcfl5*, another transcription factor that reinforces its own transcription and that of *A-Myb,* establishing a positive feedback loop essential for piRNA production (Yu et al. 2023). The transcription of piRNA precursors is carried out by Pol II (Li et al. 2013). The nuclear protein BTB domain-containing 18 (BTBD18) enhances the expression of a subset of pachytene piRNA-producing loci by promoting transcription elongation (Zhou et al. 2017). The THO complex (THOC), comprising THOC1 and THOC2, facilitates the nuclear export of pachytene piRNA precursor transcripts (Yu et al. 2021). Additionally, DEAD-box helicase D1PAS1 facilitates R-loop resolution and engages with the nuclear export machinery to ensure the transport of pachytene piRNA precursors

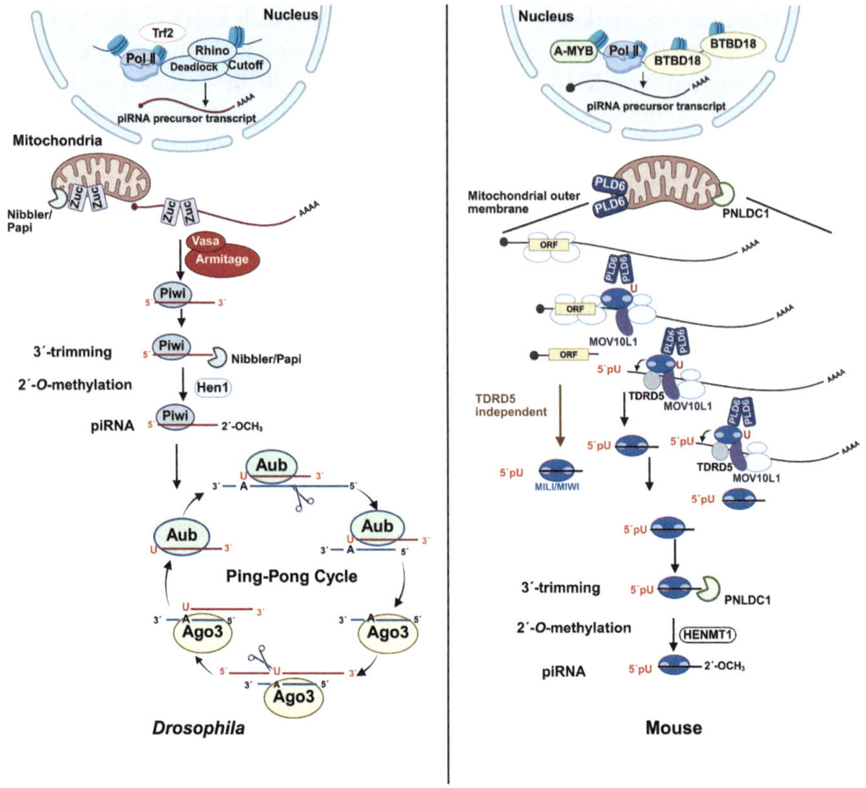

Fig. 6.1 piRNA transcription and biogenesis in *Drosophila* and mice. (Left) In *Drosophila*, dual-strand piRNA clusters are transcribed in a promoter-independent manner by RNA polymerase II. This process is initiated by the RDC complex, in which Rhino recognizes H3K9me3 marks and recruits Deadlock, Cutoff, Moonshiner, and Trf2 to form an alternative transcription initiation complex. The resulting unspliced, non-polyadenylated transcripts are exported to the nuage for processing. Vasa and Armitage assist precursor processing by promoting the remodeling and mitochondrial localization of piRNA precursors. PIWI proteins Aub and Ago3 then engage in the Ping-Pong amplification cycle, generating TE-targeting piRNAs with characteristic 1U and 10A nucleotide biases. Maturation involves 3´-end trimming by Nibbler and 2´-*O*-methylation by Hen1. (Right) In mice, pachytene piRNA precursors are transcribed by RNA polymerase II under the regulation of A-MYB, with BTBD18 selectively enhancing transcriptional elongation at a subset of loci. These capped and polyadenylated RNAs undergo splicing and are exported to the cytoplasm, where they localize to the IMC. There, PLD6 on the mitochondrial outer membrane performs phased cleavage, generating the 5´ ends of further piRNAs. In the first phase of piRNA biogenesis, ribosomes translate the open reading frame (ORF) region, and piRNAs are produced in a TDRD5-independent manner. In the second phase, ribosomes translocate to the UTR region, facilitated by MOV10L1, and guide piRNA production in a TDRD5-dependent manner. The cleaved fragments are then loaded onto MILI or MIWI, where PNLDC1 trims the 3´ ends and HENMT1 adds 2´-*O*-methylation to produce mature piRNAs. (Figure created with BioRender.com)

from the nucleus to the cytoplasm, where they are further processed into mature piRNAs (Choi et al. 2024).

In summary, the piRNA pathway exhibits striking regulatory divergence between *Drosophila* and mice: In *Drosophila*, piRNA loci reside in heterochromatic regions and rely on a chromatin-driven mechanism centered on the Rhino–Moonshiner–Trf2 axis. This system bypasses canonical promoters by recruiting Moonshiner to H3K9me3-marked loci, enabling RNA Pol II to initiate non-canonical transcription in the germline. In contrast, mouse piRNA-producing loci are euchromatic intergenic regions transcribed by RNA Pol II under the control of A-MYB/*Tcfl5* feedforward loops, which synchronize piRNA precursor synthesis and processing machinery protein expression during meiosis. These differences highlight a fundamental divergence in the regulatory strategies of the piRNA pathway across species.

6.2.2 Post-Transcriptional Processing of piRNA Precursors

The post-transcriptional maturation of piRNAs involves a series of tightly regulated steps, including endonucleolytic cleavage, PIWI loading, 3′ end trimming, and terminal methylation, which collectively determine piRNA identity and function. Among these, the initial step in directing a piRNA precursor toward piRNA production involves endonucleolytic cleavage, generating a 5′ monophosphate end as the future piRNA 5′ end. The mechanism of 5′ end formation distinguishes primary piRNAs from secondary piRNAs in the biogenesis pathway. The 5′ ends of primary piRNAs are cleaved by Zucchini (Zuc), the homolog of phospholipase D family member 6 (PLD6) in mice, an endonuclease located on the outer mitochondrial membrane (Gao and Frohman 2012; Voigt et al. 2012; Kabayama et al. 2017). In vivo, Zuc/PLD6 preferentially cleaves before uridine (U) residues, simultaneously generating the 3′ end of pre-piRNAs and the 5′ end of adjacent pre-pre-piRNAs (Izumi et al. 2020). The resulting pre-pre-piRNAs are loaded onto PIWI proteins, triggering further Zuc/PLD6-dependent cleavage events in a phased manner, producing a series of "trailing piRNAs" (Han et al. 2015; Homolka et al. 2015; Mohn et al. 2015). The 3′ ends of these piRNAs are refined by nucleases such as Nibbler and Papi in *Drosophila* and by Poly(A)-specific ribonuclease-like domain-containing 1 (PNLDC1) in mice (Han et al. 2011; Liu et al. 2011; Zhang et al. 2018), while Hen1 mediates 2′-*O*-methylation via S-adenosylmethionine (SAM), enhancing piRNA stability (Horwich et al. 2007; Saito et al. 2007; Hayashi et al. 2016).

The 5′ ends of secondary piRNAs are generated through the endonucleolytic activity of PIWI proteins guided by existing piRNAs. These piRNA-induced silencing complexes (piRISCs) cleave target transcripts, typically TEs, between the 10th and 11th nucleotides from the 5′ end of the piRNA guide. The cleaved transcripts are then loaded onto PIWI proteins and processed into secondary piRNAs (Brennecke et al. 2007; Gunawardane et al. 2007) with 3′ ends processed by the same machinery as primary piRNAs. These secondary piRNAs can target primary piRNA precursor transcripts enriched with antisense TE sequences, leading to the production of additional anti-TE piRNAs and establishing a piRNA-specific Ping-Pong amplification loop (Brennecke et al. 2007; Gunawardane et al. 2007). In *Drosophila* germ cells, the Ping-Pong amplification is robust. The PIWI protein Aubergine (Aub) binds antisense-strand piRNAs and cleaves sense-strand TE mRNAs, while Argonaute 3 (Ago3) binds sense-strand piRNAs and cleaves antisense transcripts. This reciprocal cleavage cycle continuously amplifies the piRNA pool, ensuring efficient TE silencing (Brennecke et al. 2007; Gunawardane et al. 2007; Le Thomas et al. 2013; Rozhkov et al. 2013). In mice, the generation of prenatal and pre-pachytene piRNAs heavily relies on the Ping-Pong amplification cycle. MILI-bound piRNAs initiate this loop, amplifying the piRNA pool and reinforcing post-transcriptional TE silencing (De Fazio et al. 2011). During the prenatal stage, a subset of MILI-generated secondary piRNAs is loaded onto MIWI2. Although MIWI2 lacks endonuclease activity to trigger Ping-Pong amplification, MIWI2-piRNAs are transported into the nucleus to guide DNA methylation and transcriptional silencing (Aravin et al. 2008; Kuramochi-Miyagawa et al. 2008; De Fazio et al. 2011; Schöpp et al. 2020).

In *Drosophila* germ cells, primary piRNA biogenesis is initiated by PIWI-mediated cleavage, followed by phased Zuc-dependent cleavage, thereby amplifying the piRNA pool and ensuring robust TE silencing (Watanabe et al. 2011; Voigt et al. 2012; Han et al. 2015; Homolka et al. 2015; Mohn et al. 2015; Izumi et al. 2020). In mice, most pachytene piRNAs are produced via a translation-coupled, PLD6-dependent phased pathway that operates largely independently of the Ping-Pong amplification cycle, although whether initial processing involves PIWI-guided cleavage remains unclear (Gainetdinov et al. 2018; Sun et al. 2020). Their biogenesis occurs in a translation-coupled manner in two distinct phases. Tudor domain-containing protein 5 (TDRD5) is a scaffold protein that promotes piRNA granule assembly, stabilizes precursor RNAs, and supports PIWI–piRNA complex formation (Ding et al. 2018). These two phases appear to differ in their dependence on TDRD5, reflecting distinct modes of precursor processing (Ding et al. 2018; Sun et al. 2020). In the first phase, 80S ribosomes translate open reading frames (ORFs) near the 5′ end of precursor transcripts, leading to the production of phased piRNAs in a TDRD5-independent manner (Ding et al. 2018; Sun et al. 2020, 2021). Contrary to conventional mRNAs, ribosomes bind and translocate on 3′ UTRs of piRNA precursors, a process facilitated by the RNA helicase Moloney leukemia virus 10-like protein 1 (MOV10L1) essential for piRNA production (Zheng et al. 2010; Vourekas et al. 2015). This second phase of piRNA production at 3′ UTR is guided by ribosomes in a TDRD5-dependent manner (Sun et al. 2020). Because mouse pachytene piRNAs precursors are much longer than *Drosophila*, and the processivity of phasing biogenesis is low (Gainetdinov et al. 2018), it is believed that translation coupling utilizes elongating ribosomes as helicases to unwind structures and remove RNA-binding proteins from piRNA precursors, streamlining piRNA biogenesis (Sun et al. 2020).

6.2.3 Location for piRNA Biogenesis and Function: RNA Granules

Germ cell-specific RNA granules are thought to serve as specialized subcellular platforms that concentrate piRNA processing factors and RNA-binding proteins and may transiently associate with piRNA precursors to facilitate biogenesis, TE silencing, and post-transcriptional regulation (Brennecke et al. 2007; Nishida et al. 2007; Watanabe et al. 2011; Meikar et al. 2014; Anbazhagan et al. 2020, 2022). Their organization and function differ significantly between *Drosophila* and mice.

In *Drosophila*, two distinct cytoplasmic RNA granules are prominent: Ypsilon schachtel (Yb) bodies and nuage. Yb bodies are electron-dense perinuclear structures located in somatic follicle cells, where they regulate the transcription of the *flamenco* piRNA locus and mediate somatic TE silencing. These structures contain key proteins such as Yb and Armitage (the homolog of mouse MOV10L1), essential for primary piRNA biogenesis (Szakmary et al. 2009; Handler et al. 2011). Armitage binds to PIWI and is required for localizing PIWI into Yb bodies (Qi et al. 2011). In contrast, nuage granules, situated around the nucleus of germ cells, serve as the primary site for the Ping-Pong amplification cycle, facilitating secondary piRNA biogenesis and TE silencing (Brennecke et al. 2007; Nishida et al. 2007). Armitage shuttles between nuage and mitochondria, linking the Ping-Pong amplification cycle to the phased piRNA pathway (Ge et al. 2019).

In mice, three distinct RNA granules form at specific developmental stages: (1) piP-bodies: Exclusive to embryonic germ cells, piP-bodies function in secondary piRNA biogenesis. These granules are enriched with factors including MIWI2, TDRD9, and Maelstrom (MAEL), serving as key sites for secondary piRNA production. piP-bodies progressively diminish after birth, indicating their function is restricted to early embryogenesis (Aravin et al. 2009; Shoji et al. 2009). (2) pi-bodies/intermitochondrial cement (IMC): Present during embryonic development, pi-bodies contain the MILI-TDRD1 complex and are equivalent to the IMC found in postnatal germ cells, acting as key sites for primary piRNA processing. This electron-dense granule is positioned among clustered mitochondria. It first appears in fetal prespermatogonia becomes prominent in mid-pachytene spermatocytes, and disassembles in late-pachytene stages. Key piRNA biogenic factors, such as PLD6, GASZ, and Tudor domain-containing protein KH (TDRKH), localize on the outer membrane of mitochondria and act as the founding components for the IMC to further recruit other non-membrane-embedded factors, with TDRKH recruiting MIWI and GASZ recruiting MILI to initiate piRNA processing (Ma et al. 2009; Watanabe et al. 2011; Saxe et al. 2013; Shiromoto et al. 2013; Ding et al. 2019). TDRD1 further promotes IMC formation via phase separation, utilizing its coiled-coil domains at the C-terminal (Gao et al. 2024). In the late-pachytene stage, the IMC disassembles, and its components disperse throughout the cytosol (Onohara and Yokota 2012). This disassembly likely reflects a developmental shift in piRNA pathway from biogenesis to function. (3) Chromatoid body (CB): Shortly after meiosis, some IMC components reaggregate into a single, lobulated perinuclear granule known as the CB in round spermatids (Fawcett et al. 1970; Meikar et al. 2011). This structural reorganization is essential for redirecting MIWI-bound piRNAs from the IMC to the CB, where they engage in post-meiotic mRNA regulation. Failure in this transition leads to MIWI entrapment and degradation, ultimately causing spermatogenic arrest (Wei et al. 2024). Proteomic analyses have shown that ~70% of the CB proteome comprises regulatory proteins involved in piRNA biogenesis, nonsense-mediated decay, and post-transcriptional regulation, such as DEAD-box helicase (DDX)25, MIWI, MILI, TDRD6, and poly(A)-binding protein 1 (PABP1) (Meikar et al. 2014; Lehtiniemi et al. 2018; Anbazhagan et al. 2022). These features position the CB as a potential hub for post-transcriptional gene regulation during spermiogenesis (Kotaja and Sassone-Corsi 2007; Lehtiniemi et al. 2018; Anbazhagan et al. 2022; Kaneda et al. 2024; Table 6.3). However, direct functional dissection of piRNA activities within the CB is still lacking, leaving its precise contribution to piRNA-mediated silencing an important topic for future research.

Although in *Drosophila*, primary piRNA biogenesis occurs on mitochondria, requiring Zuc (Han et al. 2015), it does not form IMC as observed in mice (Ge et al. 2019), suggesting that IMC might be specifically associated with the translation-dependent processing of ultra-long piRNA precursors. Notably, the CB is unique to mammals and has no analogous counterpart in *Drosophila*. These underscore the evolutionary divergence in piRNA granule organization and function between species.

6.3 Functions of piRNAs

6.3.1 TE Silencing

The primary function of piRNAs is to suppress TEs, thereby maintaining genomic integrity. This suppression is particularly crucial in germ cells, where global DNA demethylation during early development and chromatin remodeling during meiosis and spermiogenesis can lead to elevated TE activity, necessitating robust mechanisms to ensure genome integrity across generations (Lippman et al. 2003; Kuramochi-Miyagawa et al. 2008; Watanabe et al. 2011; Sienski et al. 2012).

6.3.1.1 Transcriptional Silencing

In the nucleus, piRISCs mediate transcriptional silencing by recognizing nascent TE transcripts and recruiting chromatin remodelers to establish epigenetic repression. Notably, this process does not require PIWI slicer activity (De Fazio et al. 2011; Darricarrère et al. 2013; Fig. 6.2). With DNA methylation playing a minimal role in *Drosophila*, piRISCs primarily silence TEs through histone modifications (Saito et al. 2009; Klenov et al. 2011; Sienski et al. 2012; Iwasaki et al. 2021). Upon

Table 6.3 Comparison of piRNA pathways in *Drosophila* germ cells and adult mammalian testes

	Drosophila	Mice
piRNA production loci	Enriched TEs	TE content lower than the genome average
Transcription factor	Moonshiner, Deadlock, Cutoff	A-MYB, TCFL5
Transcription	Convergent and unidirectional transcription	Divergent and unidirectional transcription
Chromatin	Heterochromatin (H3K9me3, HP1)	Euchromatin (H3K4me3)
Special epigenetic marks	Present (Rhino)	Absent or unknown
Splicing	No	Yes
Transcription start sites	Multiple sites, cryptic transcripts	Single site, starts from a specific promoter
Processing location	Nuage, mitochondrial outer membrane	IMC
Ribosome guide biogenesis	Absent	Present
Ping-Pong amplification	Strong	Weak
Phased production	Weak	Strong
3′ end processing	Multiple mechanisms (Papi, Nibbler, Zuc)	Single mechanism (PNLDC1)
Target	Most piRNAs target TE	Only a small fraction of piRNAs target TE
Silencing mechanism	Transcriptional silencing (chromatin modification) and post-transcriptional cleavage	Transcriptional silencing (DNA methylation) and post-transcriptional cleavage
Function	TE silencing	TE silencing and other unknown functions

recognizing nascent TE transcripts, the PIWI complex recruits Panoramix (Panx), which interacts with the Nxf2-Nxt1 complex to facilitate chromatin remodeler recruitment and reinforce transcriptional repression (Yu et al. 2015; Eastwood et al. 2021; Andreev et al. 2022). Additionally, small ubiquitin-like modifier (SUMO) modification by Su(var)2-10 and the nucleosome remodeling and deacetylase (NuRD) complex, including Mi-2, cooperatively regulates chromatin restructuring and PIWI-dependent TE silencing (Mugat et al. 2020; Ninova et al. 2020; Ninova et al. 2023). Furthermore, Maelstrom enhances chromatin remodeling and TE repression through its interaction with the SWI–SNF complex (Onishi et al. 2020).

In mouse fetal gonocytes, MIWI2-dependent transcriptional silencing predominantly relies on DNA methylation (Carmell et al. 2007; Aravin et al. 2008; Kuramochi-Miyagawa et al. 2008; Pezic et al. 2014; Kojima-Kita et al. 2016). Initially, H3 lysine 4 trimethylation (H3K4me3) and H3K9me3 histone marks on young TE loci facilitate the recruitment of Spindlin-1 (SPIN1), an epigenetic reader that preconditions TE loci for silencing (Dias Mirandela et al. 2024). SPIN1

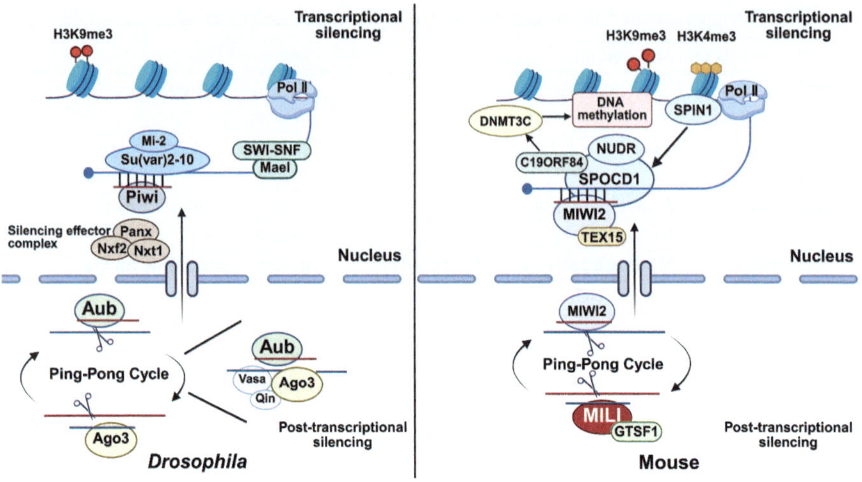

Fig. 6.2 piRNAs silence TE transcriptionally and post-transcriptionally. (Left) In *Drosophila*, the nuclear PIWI–piRNA complex binds to nascent TE transcripts and recruits multiple factors to suppress TE transcription. Panx interacts with the Nxf2–Nxt1 complex, while Mael facilitates the recruitment of chromatin remodelers by interacting with the SWI–SNF complex, thereby enhancing TE transcriptional repression. Su(var)2–10 and the nucleosome remodeling and deacetylase NuRD complex further contribute to chromatin remodeling and TE silencing. In the cytoplasm, Vasa and Qin participate in the Ping-Pong cycle, promoting TE mRNA cleavage and facilitating post-transcriptional silencing. (Right) In mice, SPIN1 interacts with SPOCD1, which subsequently associates with MIWI2 and targets nascent TE transcripts, promoting DNA methylation. Furthermore, SPOCD1 recruits C19ORF84, facilitating DNMT3C-mediated DNA methylation, thereby ensuring TE transcriptional silencing. TEX15 also contributes to DNA methylation, ensuring the stable suppression of TEs. In the cytoplasm, MILI directly cleaves TE mRNAs in a GTSF1-enhanced manner, reinforcing post-transcriptional repression. (Figure created with BioRender.com)

interacts with SPOC domain-containing protein 1 (SPOCD1) to form a complex that exists prior to MIWI2 engagement. Subsequently, SPOCD1 associates with MIWI2 and targets nascent TE transcripts, ultimately triggering piRNA-directed DNA methylation (Kojima-Kita et al. 2016; Dias Mirandela et al. 2024). Additionally, SPOCD1 not only facilitates MIWI2 recruitment but also serves as a molecular bridge linking the DNA methylation machinery with chromatin remodeling factors, such as the NuRD complex, thereby reinforcing TE silencing (Zoch et al. 2020). Furthermore, SPOCD1 recruits C19ORF84, which facilitates DNA methyltransferase 3C (DNMT3C)-mediated DNA methylation, ensuring stable TE repression (Barau et al. 2016; Dura et al. 2022; Zoch et al. 2024). Moreover, Testis-expressed 15 (TEX15) interacts with MIWI2 and MILI, further supporting piRNA-dependent DNA methylation (Schöpp et al. 2020; Yang et al. 2020).

Notably, transcriptional silencing via piRNAs has been predominantly observed in *Drosophila* and mammals, and remains undetectable or poorly characterized in birds, zebrafish, silkworms, and other invertebrates (Houwing et al. 2007; Kuramochi-Miyagawa et al. 2008; Sienski et al. 2012). Even in mice, this mechanism is restricted to the prenatal germline, particularly in fetal prospermatogonia

(Aravin et al. 2008; Kuramochi-Miyagawa et al. 2008). By contrast, post-transcriptional silencing predominates in postnatal stages and in species where transcriptional repression is lacking, indicating that the piRNA pathway employs a stage- and species-specific division of labor to ensure robust TE control.

6.3.1.2 Post-Transcriptional Silencing

Post-transcriptional silencing occurs in the cytoplasm, where TE RNAs are degraded through direct cleavage. This process is closely linked to piRNA biogenesis, as TE transcripts serve as templates for phased piRNA production and the Ping-Pong amplification cycle (Czech and Hannon 2016). In *Drosophila*, piRNAs associate with PIWI proteins such as Aub and Ago3. These complexes, along with the endonuclease Zuc, cleave target mRNAs through their endonuclease activity. Several proteins enhance piRNA-mediated TE silencing. For instance, the DEAD-box RNA helicase Vasa, localized in the nuage, facilitates the transfer of RNA fragments between Aub and Ago3, thereby improving TE silencing efficiency (Xiol et al. 2014). Spindle-E (Spn-E) and Tudor domain proteins like Qin/Kumo are indispensable for the proper function of the Ping-Pong amplification cycle. Qin is an E3 ubiquitin ligase that ensures the specificity of piRNA processing, prevents non-specific piRNA pairing, and stabilizes piRNA complexes, thereby ensuring effective degradation of TE mRNA (Zhang et al. 2011). In mice, gametocyte-specific factor 1 (GTSF1) transforms the intrinsically weak, piRNA-guided cleavage activity of PIWI proteins into efficient endonuclease function, thereby enhancing TE silencing (Arif et al. 2022; Li et al. 2024, 2025). The coordination between transcriptional and post-transcriptional silencing is essential for effective TE repression and genome integrity in fetal gonads, as the transcriptional TE silencing via nuclear MIWI2-piRNAs is generated from cytoplasmic MILI-piRNAs through Ping-Pong mechanisms (Aravin et al. 2008; De Fazio et al. 2011).

6.3.2 *piRNA Function in Regulating Gene Expression*

Beyond their established role in TE silencing, piRNAs also regulate the expression of protein-coding genes, thereby expanding their functional repertoire (Gou et al. 2014; Zhang et al. 2015). In *Drosophila* embryos, TE-derived piRNAs guide Aub proteins to target *Nanos* mRNA, leading to its destabilization (Rouget et al. 2010; Barckmann et al. 2015). In *Drosophila* testes, piRNAs silence the *Stellate* gene, which encodes a casein kinase II β-like protein located on the X chromosome. In mice, piRNA/MIWI complexes recruit CCR4-associated factor 1 (CAF1) to trigger mRNA deadenylation and degradation (Gou et al. 2014). Knockdown of MIWI or CAF1 upregulates a substantial number (~5000) of protein-coding genes, highlighting the role of this mechanism in maintaining transcriptome homeostasis in reproductive tissues (Gou et al. 2014). Additionally, knockout of abundantly expressed pachytene piRNAs leads to infertility, with some piRNAs directly guiding MIWI to cleave target mRNAs that are essential for germ cell development (Goh et al. 2015; Wu et al. 2020). Recent data also suggest that piRNAs may function to activate translation in mice (Dai et al. 2019).

6.3.3 piRNAs in Human Infertility

In terms of reproductive health, alterations in the piRNA pathway have been linked to infertility. While the association between piRNA dysfunction and male infertility is well-documented, recent studies have identified specific gene mutations that contribute to this condition. Mutations in *PIWIL1, PNLDC1, MOV10L1, TDRD1*, glycerol-3-phosphate acyltransferase 2(*GPAT2*), and *DDX4* have been shown to disrupt piRNA biogenesis, impair TE silencing, and lead to spermatogenic arrest (Gou et al. 2017; Nagirnaja et al. 2021; Ghieh et al. 2022; Li et al. 2022; Stallmeyer et al. 2024). In parallel, growing evidence also implicates piRNA pathway disruptions in female infertility (Zhang et al. 2021). For instance, studies in golden hamsters have demonstrated that loss of PIWIL1 and MOV10L1 leads to female infertility, highlighting the essential role of the piRNA pathway in female reproductive function in certain mammals (Hasuwa et al. 2021; Loubalova et al. 2021; Zhang et al. 2021). Additionally, a study identified a homozygous loss-of-function variant in TDRD12, a Tudor domain-containing protein essential for secondary piRNA production and MIWI2 loading in male germ cells (Pandey et al. 2013), in a woman diagnosed with premature ovarian insufficiency, a condition related to impaired oocyte maturation, suggesting a potential link between piRNA pathway mutations and female infertility in humans (Li et al. 2021). These findings underscore the critical role of the piRNA pathway in female fertility and suggest that disruptions in this pathway can lead to reproductive challenges.

6.3.4 Somatic piRNA Function

Recent studies have revealed diverse functions for piRNAs, especially in the somatic cells of invertebrates such as the fat body in *Drosophila*, where they play roles in TE suppression and gene regulation (Jones et al. 2016; Lewis et al. 2018). They have also been implicated in sex determination in the silkworm and *C. elegans*, as well as memory-related synaptic plasticity in *Aplysia*; piRNAs have been shown to regulate long-term memory formation. This occurs by guiding DNA methylation of memory-suppressor genes in a stimulus-dependent manner, suggesting broader regulatory roles beyond genome defense (Rajasethupathy et al. 2012; Kiuchi et al. 2014; Tang et al. 2018). Notably, somatic piRNA pathways are prevalent in invertebrates but appear to be absent in vertebrates (Lewis et al. 2018). However, several reports have identified piRNA-like molecules or the expression of PIWI proteins in vertebrate somatic tissues. Recent advances in computational pipelines, such as WIND (Geles et al. 2021), have improved the sensitivity of piRNA annotation and expanded the detectable repertoire of piRNA-like molecules in these tissues. For instance, piRNA-like molecules have been detected in various somatic tissues of mice and rhesus macaques, displaying characteristics similar to germline piRNAs (Yan et al. 2011; Ortogero et al. 2014). Despite these findings, disrupting piRNA pathways in mice does not result in phenotypes beyond reproductive defects (Deng and Lin 2002; Kuramochi-Miyagawa et al. 2004), raising questions about the functional significance of somatic piRNAs in vertebrates. This discrepancy suggests two possibilities: somatic piRNAs may exist in low abundance with underexplored functions, or mice may not be an ideal model for studying somatic piRNAs (Yan et al. 2011). Further investigation is needed to elucidate the roles of piRNAs in vertebrate somatic tissues.

6.3.5 piRNA Evolution

While the piRNA pathway is a conserved mechanism across the animal kingdom, both the piRNA pathway proteins and piRNA-producing loci exhibit signs of rapid adaptive evolution in both insect and vertebrate lineages (Obbard et al. 2009; Kolaczkowski et al. 2011; Levine et al. 2012, 2016; Simkin et al. 2013; Yi et al. 2014). This phenomenon is believed to be largely attributed to the ongoing "arms race" between host genomes and TEs, necessitating continuous adaptation of the piRNA machinery to counteract the evolving strategies of TEs (Kidwell and Lisch 1997; Aravin et al. 2007b; Daugherty and Malik 2012; Luo et al. 2020). Studies have demonstrated that piRNA pathway proteins undergo strong positive selection due to the need to rapidly adjust target specificity to cope with mutating TEs and the requirement to balance TE suppression while avoiding off-target effects that could lead to the erroneous silencing of host genes (Yi et al. 2014; Blumenstiel et al. 2016; Parhad et al. 2017; Koonin and Yutin 2020; Zhang et al. 2020). This adaptive evolution not only shapes the functionality of the PIWI complex but also drives dynamic changes in piRNA-producing loci, reflecting the pathway's critical role in maintaining genomic integrity.

Historically, it is believed that pachytene piRNAs do not function to suppress TEs and their loci are not undergoing selection (Özata et al. 2020). This view was primarily based on two observations: first, the pachytene piRNA loci are significantly depleted of TEs compared to the rest of the genome; and second, knockout of pachytene piRNAs does not lead to TE desilencing (Vourekas et al. 2015; Wu et al. 2020; Choi et al. 2021). However, this notion has been challenged by recent findings showing that pachytene piRNAs indeed contribute to silencing young and actively transposing TEs across amniotes, indicating a specialized, evolutionarily conserved silencing role (Sun et al. 2023). Recent findings further challenge this view by showing that pachytene piRNA loci are hotspots for structural variations (SVs) across amniotes (Sun et al. 2023). It has been known that SV hotspots are depleted of TEs due to their high local mutation rates (Gokcumen et al. 2013). Thus, the low TE density observed in pachytene piRNA loci is likely due to the nature of SV hotspots rather than the lack of function in TE silencing. Moreover, earlier studies that failed to detect TE derepression in pachytene piRNA-deficient models were limited in scope. Many of them relied on partial cluster knockouts or utilization of RT-PCR with consensus primers that broadly detect TE superfamilies. In contrast, recent analyses used *Mov10l1* conditional knockout (CKO) mice, in which pachytene piRNAs are nearly completely lost (Sun et al. 2023). By combining this model with transcriptome-wide RNA-seq, researchers found that although most TE families remained unchanged, a subset of young and actively transposing TEs was selectively derepressed (Sun et al. 2023). These TE families are not depleted from pachytene piRNA loci, supporting a direct silencing relationship. This function appears to be evolutionarily conserved across amniotes, suggesting that pachytene piRNAs contribute to a broader, conserved genome defense strategy (Sun et al. 2023).

The identification of piRNA loci as SV hotspots brings an additional selection force into the picture: the negative selection to avoid affecting the expression of nearby genes. Recent studies have shown that amniote piRNA loci are often located in close proximity to protein-coding genes and are significantly overlapped with pathogenic SV sites associated with human developmental diseases, such as autism (Sun et al. 2023). The detrimental effects of SV hotspots at piRNA loci lend support to the soma-germ cell conflict theory (Haig 2016; Sun et al. 2023), which posits that when a genetic trait is harmful to somatic cells but beneficial to germ cells, it may still be subject to positive selection. This framework helps explain why certain pathogenic SV hotspots persist in the human population and continue to give rise to recurrent disease-associated SVs, despite their deleterious effects on somatic development and organismal viability. Thus, the evolution of mammalian piRNA loci appears to be shaped by the interplay of three major forces: high local mutation rates, positive selection for TEs defense, and negative selection to minimize disruption of nearby protein-coding genes. This pattern also reflects the distinct selection pressures acting on germline and somatic piRNA loci.

In summary, the piRNA pathway continually adapts to the challenges posed by TEs. The rapid evolution of piRNA pathway proteins and the structural variability of piRNA loci highlight the intricate balance between genomic integrity and adaptability, ensuring the preservation of genome integrity across generations.

6.4 Conclusion and Perspectives

Despite major advances in our understanding of piRNAs, several fundamental questions remain unresolved. One of the central mysteries concerns piRNA biogenesis: how are piRNA precursors selectively recognized and processed, and how is this process spatially coordinated with ribosomes and mitochondria? The mechanisms governing selective precursor specification, processive fragmentation, and transport to these organelles—along with the assembly of precursor-PIWI protein complexes—remain poorly understood. Dissecting the interplay between ribosomes, piRNA biogenic factors, and mitochondria is a critical next step.

A second major open question concerns the function of non-TE-targeting piRNAs. While piRNAs are well established as guardians of the germline genome through TE silencing, a general assumption is that piRNAs must be regulating protein-coding genes. However, evidence to support such function only comes from a small subset of mammalian piRNAs (Gou et al. 2014; Zhang et al. 2015). The biological significance and mechanistic basis of these majority non-TE-targeting piRNAs remain largely unexplored (Xu and Li 2024). What are their targets?

In conclusion, piRNA research has uncovered a rich and dynamic system of genome regulation and RNA processing, yet much remains to be discovered. Future work will need to unravel the molecular logic of piRNA precursor selection, dissect the full spectrum of piRNA functions beyond TE silencing, and clarify their roles in development, evolution, and disease. These efforts will deepen our understanding of RNA biology and open new avenues in reproductive and biomedical research.

Acknowledgments The authors thank Luis M. Vaschetto, the Series Editor from the *Epigenetics and Human Health* book series, for help with editing the manuscript, and members of the Li Laboratory for advice and critical comments on the manuscript. This text was edited for language and formatting using generative AI tools, including ChatGPT (GPT-4, OpenAI) and DeepSeek (2025). All content remains the original work of the authors and has been fully reviewed to ensure scientific accuracy.

Contributor Information Jiao-Yan Ma, Center for RNA Medicine, the Fourth Affiliated Hospital of School of Medicine, and International School of Medicine, International Institutes of Medicine, Zhejiang University, Yiwu, China, 322000. Xin Zhiguo Li, Center for RNA Medicine, the Fourth Affiliated Hospital of School of Medicine, and International School of Medicine, International Institutes of Medicine, Zhejiang University, Yiwu, China, 322000.

Author Contributions Jiao-Yan Ma and Xin Zhiguo Li contributed to the preparation of the manuscript.

Compliance with Ethical Standards Conflict of interest: The authors declare that they have no known competing financial interests or personal relationships that could have appeared to influence the work reported in this manuscript.

Ethical approval This chapter contains no studies with human/animal participants performed by the authors.

References

Anbazhagan R, Kavarthapu R, Coon SL, Dufau ML (2020) Role of phosphorylated gonadotropin-regulated testicular RNA helicase (GRTH/DDX25) in the regulation of germ cell-specific mRNAs in chromatoid bodies during spermatogenesis. Front Cell Dev Biol 8:580019. https://doi.org/10.3389/fcell.2020.580019

Anbazhagan R, Kavarthapu R, Dufau ML (2022) Chromatoid bodies in the regulation of spermatogenesis: novel role of GRTH. Cells 11(4):613. https://doi.org/10.3390/cells11040613

Andersen PR, Tirian L, Vunjak M et al (2017) A heterochromatin-dependent transcription machinery drives piRNA expression. Nature 549(7670):54–59. https://doi.org/10.1038/nature23482

Andreev VI, Yu C, Wang J et al (2022) Panoramix SUMOylation on chromatin connects the piRNA pathway to the cellular heterochromatin machinery. Nat Struct Mol Biol 29(2):130–142. https://doi.org/10.1038/s41594-022-00721-x

Aravin AA, Gaidatzis D, Pfeffer S et al (2006) A novel class of small RNAs bind to MILI protein in mouse testes. Nature 442(7099):203–207. https://doi.org/10.1038/nature04916

Aravin AA, Sachidanandam R, Girard A et al (2007a) Developmentally regulated piRNA clusters implicate MILI in transposon control. Science 316(5825):744–747. https://doi.org/10.1126/science.1142612

Aravin AA, Hannon GJ, Brennecke J (2007b) The Piwi-piRNA pathway provides an adaptive defense in the transposon arms race. Science 318(5851):761–764. https://doi.org/10.1126/science.1146484

Aravin AA, Sachidanandam R, Bourc'his D et al (2008) A piRNA pathway primed by individual transposons is linked to de novo DNA methylation in mice. Mol Cell 31(6):785–799. https://doi.org/10.1016/j.molcel.2008.09.003

Aravin AA, van der Heijden GW, Castañeda J et al (2009) Cytoplasmic compartmentalization of the fetal piRNA pathway in mice. PLoS Genet 5(12):e1000764. https://doi.org/10.1371/journal.pgen.1000764

Arif A, Bailey S, Izumi N et al (2022) GTSF1 accelerates target RNA cleavage by PIWI-clade Argonaute proteins. Nature 608(7923):618–625. https://doi.org/10.1038/s41586-022-05009-0

Barau J, Teissandier A, Zamudio N et al (2016) The DNA methyltransferase DNMT3C protects male germ cells from transposon activity. Science 354(6314):909–912. https://doi.org/10.1126/science.aah5143

Barckmann B, Pierson S, Dufourt J et al (2015) Aubergine iCLIP reveals piRNA-dependent decay of mRNAs involved in germ cell development in the early embryo. Cell Rep 12(7):1205–1216. https://doi.org/10.1016/j.celrep.2015.07.030

Bartel DP (2009) MicroRNAs: target recognition and regulatory functions. Cell 136(2):215–233. https://doi.org/10.1016/j.cell.2009.01.002

Batista PJ, Ruby JG, Claycomb JM et al (2008) PRG-1 and 21U-RNAs interact to form the piRNA complex required for fertility in *C. elegans*. Mol Cell 31(1):67–78. https://doi.org/10.1016/j.molcel.2008.06.002

Bernstein E, Caudy AA, Hammond SM et al (2001) Role for a bidentate ribonuclease in the initiation step of RNA interference. Nature 409(6818):363–366. https://doi.org/10.1038/35053110

Blumenstiel JP, Erwin AA, Hemmer LW (2016) What drives positive selection in the *Drosophila* piRNA machinery? The genomic autoimmunity hypothesis. Yale J Biol Med 89(4):499–512

Brennecke J, Aravin AA, Stark A et al (2007) Discrete small RNA-generating loci as master regulators of transposon activity in *Drosophila*. Cell 128(6):1089–1103. https://doi.org/10.1016/j.cell.2007.01.043

Carmell MA, Girard A, Van De Kant HJ et al (2007) MIWI2 is essential for spermatogenesis and repression of transposons in the mouse male germline. Dev Cell 12(4):503–514. https://doi.org/10.1016/j.devcel.2007.03.001

Casier K, Autaa J, Gueguen N et al (2023) The histone demethylase Kdm3 prevents autoimmune piRNAs production in *Drosophila*. Sci Adv 9(14):eade3872. https://doi.org/10.1126/sciadv.ade3872

Charlesworth B, Langley CH (1986) The evolution of self-regulated transposition of transposable elements. Genetics 112(2):359–383. https://doi.org/10.1093/genetics/112.2.359

Chirn GW, Rahman R, Sytnikova YA et al (2015) Conserved piRNA expression from a distinct set of piRNA cluster loci in eutherian mammals. PLoS Genet 11(11):e1005652. https://doi.org/10.1371/journal.pgen.1005652

Choi H, Wang Z, Dean J (2021) Sperm acrosome overgrowth and infertility in mice lacking chromosome 18 pachytene piRNA. PLoS Genet 17(4):e1009485. https://doi.org/10.1371/journal.pgen.1009485

Choi H, Zhou L, Zhao Y et al (2024) RNA helicase D1PAS1 resolves R-loops and forms a complex for mouse pachytene piRNA biogenesis required for male fertility. Nucleic Acids Res 52(19):11973–11994. https://doi.org/10.1093/nar/gkae712

Chuong EB, Rumi MK, Soares MJ et al (2013) Endogenous retroviruses function as species-specific enhancer elements in the placenta. Nat Genet 45(3):325–329. https://doi.org/10.1038/ng.2553

Cox DN, Chao A, Baker J et al (1998) A novel class of evolutionarily conserved genes defined by piwi are essential for stem cell self-renewal. Genes Dev 12(23):3715–3727. https://doi.org/10.1101/gad.12.23.3715

Czech B, Hannon GJ (2016) One loop to rule them all: the ping-pong cycle and piRNA-guided silencing. Trends Biochem Sci 41(4):324–337. https://doi.org/10.1016/j.tibs.2015.12.008

Dai P, Wang X, Gou LT et al (2019) A translation-activating function of MIWI/piRNA during mouse spermiogenesis. Cell 179(7):1566–1581. https://doi.org/10.1016/j.cell.2019.11.022

Darricarrère N, Liu N, Watanabe T et al (2013) Function of Piwi, a nuclear Piwi/Argonaute protein, is independent of its slicer activity. Proc Natl Acad Sci 110(4):1297–1302. https://doi.org/10.1073/pnas.1213283110

Daugherty MD, Malik HS (2012) Rules of engagement: molecular insights from host-virus arms races. Annu Rev Genet 46(1):677–700. https://doi.org/10.1146/annurev-genet-110711-155522

De Fazio S, Bartonicek N, Di Giacomo M et al (2011) The endonuclease activity of Mili fuels piRNA amplification that silences LINE1 elements. Nature 480(7376):259–263. https://doi.org/10.1038/nature10547

Deng W, Lin HF (2002) Miwi, a murine homolog of piwi, encodes a cytoplasmic protein essential for spermatogenesis. Dev Cell 2(6):819–830. https://doi.org/10.1016/s1534-5807(02)00165-x

Di Giacomo M, Comazzetto S, Saini H et al (2013) Multiple epigenetic mechanisms and the piRNA pathway enforce LINE1 silencing during adult spermatogenesis. Mol Cell 50(4):601–608. https://doi.org/10.1016/j.molcel.2013.04.026

Dias Mirandela M, Zoch A, Leismann J et al (2024) Two-factor authentication underpins the precision of the piRNA pathway. Nature 634(8035):979–985. https://doi.org/10.1038/s41586-024-07963-3

Ding D, Liu J, Midic U et al (2018) TDRD5 binds piRNA precursors and selectively enhances pachytene piRNA processing in mice. Nat Commun 9(1):127. https://doi.org/10.1038/s41467-017-02622-w

Ding D, Liu J, Dong K et al (2019) Mitochondrial membrane-based initial separation of MIWI and MILI functions during pachytene piRNA biogenesis. Nucleic Acids Res 47(5):2594–2608. https://doi.org/10.1093/nar/gky1281

Doolittle WF, Sapienza C (1980) Selfish genes, the phenotype paradigm and genome evolution. Nature 284(5757):601–603. https://doi.org/10.1038/284601a0

Dura M, Teissandier A, Armand M et al (2022) DNMT3A-dependent DNA methylation is required for spermatogonial stem cells to commit to spermatogenesis. Nat Genet 54(4):469–480. https://doi.org/10.1038/s41588-022-01040-z

Eastwood EL, Jara KA, Bornelöv S et al (2021) Dimerisation of the PICTS complex via LC8/Cut-up drives co-transcriptional transposon silencing in *Drosophila*. elife 10:e65557. https://doi.org/10.7554/eLife.65557

Elbashir SM, Harborth J, Lendeckel W et al (2001) Duplexes of 21-nucleotide RNAs mediate RNA interference in cultured mammalian cells. Nature 411(6836):494–498. https://doi.org/10.1038/35078107

Farh KKH, Grimson A, Jan C et al (2005) The widespread impact of mammalian MicroRNAs on mRNA repression and evolution. Science 310(5755):1817–1821. https://doi.org/10.1126/science.1121158

Fawcett DW, Eddy EM, Phillips DM (1970) Observations on the fine structure and relationships of the chromatoid body in mammalian spermatogenesis. Biol Reprod 2(1):129–153. https://doi.org/10.1095/biolreprod2.1.129

Gainetdinov I, Colpan C, Arif A et al (2018) A single mechanism of biogenesis, initiated and directed by PIWI proteins, explains piRNA production in most animals. Mol Cell 71(5):775–790. https://doi.org/10.1016/j.molcel.2018.08.007

Gainetdinov I, Vega-Badillo J, Cecchini K et al (2023) Relaxed targeting rules help PIWI proteins silence transposons. Nature 619(7969):394–402. https://doi.org/10.1038/s41586-023-06257-4

Gao Q, Frohman MA (2012) Roles for the lipid-signaling enzyme MitoPLD in mitochondrial dynamics, piRNA biogenesis, and spermatogenesis. BMB Rep 45(1):7. https://doi.org/10.5483/bmbrep.2012.45.1.7

Gao J, Jing J, Shang G et al (2024) TDRD1 phase separation drives intermitochondrial cement assembly to promote piRNA biogenesis and fertility. Dev Cell 59(20):2704–2718. https://doi.org/10.1016/j.devcel.2024.06.017

Ge DT, Wang W, Tipping C et al (2019) The RNA-binding ATPase, Armitage, couples piRNA amplification in nuage to phased piRNA production on mitochondria. Mol Cell 74(5):982–995. https://doi.org/10.1016/j.molcel.2019.04.006

Geles K, Palumbo D, Sellitto A et al (2021) WIND (Workflow for pIRNAs aNd beyonD): a strategy for in-depth analysis of small RNA-seq data. F1000Res 10:1. https://doi.org/10.12688/f1000research.27868.3

Ghieh F, Barbotin AL, Swierkowski-Blanchard N et al (2022) Whole-exome sequencing in patients with maturation arrest: a potential additional diagnostic tool for prevention of recurrent negative testicular sperm extraction outcomes. Hum Reprod 37(6):1334–1350. https://doi.org/10.1093/humrep/deac057

Girard A, Sachidanandam R, Hannon GJ et al (2006) A germline-specific class of small RNAs binds mammalian Piwi proteins. Nature 442(7099):199–202. https://doi.org/10.1038/nature04917

Goh WSS, Falciatori I, Tam OH et al (2015) piRNA-directed cleavage of meiotic transcripts regulates spermatogenesis. Genes Dev 29(10):1032–1044. https://doi.org/10.1101/gad.260455.115

Gokcumen O, Tischler V, Tica J et al (2013) Primate genome architecture influences structural variation mechanisms and functional consequences. Proc Natl Acad Sci 110(39):15764–15769. https://doi.org/10.1073/pnas.1305904110

Gou LT, Dai P, Yang JH et al (2014) Pachytene piRNAs instruct massive mRNA elimination during late spermiogenesis. Cell Res 24(6):680–700. https://doi.org/10.1038/cr.2014.41

Gou LT, Kang JY, Dai P et al (2017) Ubiquitination-deficient mutations in human Piwi cause male infertility by impairing histone-to-protamine exchange during spermiogenesis. Cell 169(6):1090–1104. https://doi.org/10.1016/j.cell.2017.04.034

Grimson A, Farh KKH, Johnston WK et al (2007) MicroRNA targeting specificity in mammals: determinants beyond seed pairing. Mol Cell 27(1):91–105. https://doi.org/10.1016/j.molcel.2007.06.017

Grivna ST, Beyret E, Wang Z et al (2006) A novel class of small RNAs in mouse spermatogenic cells. Genes Dev 20(13):1709–1714. https://doi.org/10.1101/gad.1434406

Gunawardane LS, Saito K, Nishida KM et al (2007) A slicer-mediated mechanism for repeat-associated siRNA 5'end formation in *Drosophila*. Science 315(5818):1587–1590. https://doi.org/10.1126/science.1140494

Guo H, Ingolia NT, Weissman JS et al (2010) Mammalian microRNAs predominantly act to decrease target mRNA levels. Nature 466(7308):835–840. https://doi.org/10.1038/nature09267

Haig D (2016) Transposable elements: self-seekers of the germline, team-players of the soma. BioEssays 38:1158–1166. https://doi.org/10.1002/bies.201600125

Han BW, Hung JH, Weng Z et al (2011) The 3′-to-5′ exoribonuclease Nibbler shapes the 3′ ends of microRNAs bound to *Drosophila* Argonaute1. Curr Biol 21(22):1878–1887. https://doi.org/10.1016/j.cub.2011.09.034

Han BW, Wang W, Li C et al (2015) piRNA-guided transposon cleavage initiates Zucchini-dependent, phased piRNA production. Science 348(6236):817–821. https://doi.org/10.1126/science.aaa1264

Handel MA, Schimenti JC (2010) Genetics of mammalian meiosis: regulation, dynamics and impact on fertility. Nat Rev Genet 11(2):124–136. https://doi.org/10.1038/nrg2723

Handler D, Olivieri D, Novatchkova M (2011) A systematic analysis of *Drosophila* TUDOR domain-containing proteins identifies Vreteno and the Tdrd12 family as essential primary piRNA pathway factors. EMBO J 30(19):3977–3993. https://doi.org/10.1038/emboj.2011.308

Hasuwa H, Iwasaki YW, Yeung WKA et al (2021) Production of functional oocytes requires maternally expressed PIWI genes and piRNAs in golden hamsters. Nat Cell Biol 23(9):1002–1012. https://doi.org/10.1038/s41556-021-00745-3

Hayashi R, Schnabl J, Handler D et al (2016) Genetic and mechanistic diversity of piRNA 3′-end formation. Nature 539(7630):588–592. https://doi.org/10.1038/nature20162

Homolka D, Pandey RR, Goriaux C et al (2015) PIWI slicing and RNA elements in precursors instruct directional primary piRNA biogenesis. Cell Rep 12(3):418–428. https://doi.org/10.1016/j.celrep.2015.06.030

Horwich MD, Li C, Matranga C et al (2007) The *Drosophila* RNA methyltransferase, DmHen1, modifies germline piRNAs and single-stranded siRNAs in RISC. Curr Biol 17(14):1265–1272. https://doi.org/10.1016/j.cub.2007.06.030

Houwing S, Kamminga LM, Berezikov E et al (2007) A role for Piwi and piRNAs in germ cell maintenance and transposon silencing in Zebrafish. Cell 129(1):69–82. https://doi.org/10.1016/j.cell.2007.03.026

Hutvágner G, McLachlan J, Pasquinelli AE et al (2001) A cellular function for the RNA-interference enzyme Dicer in the maturation of the let-7 small temporal RNA. Science 293(5531):834–838. https://doi.org/10.1126/science.1062961

Iwasaki YW, Sriswasdi S, Kinugasa Y et al (2021) Piwi–piRNA complexes induce stepwise changes in nuclear architecture at target loci. EMBO J 40(18):e108345. https://doi.org/10.15252/embj.2021108345

Izumi N, Shoji K, Suzuki Y et al (2020) Zucchini consensus motifs determine the mechanism of pre-piRNA production. Nature 578(7794):311–316. https://doi.org/10.1038/s41586-020-1966-9

Jones BC, Wood JG, Chang C et al (2016) A somatic piRNA pathway in the *Drosophila* fat body ensures metabolic homeostasis and normal lifespan. Nat Commun 7(1):13856. https://doi.org/10.1038/ncomms13856

Kabayama Y, Toh H, Katanaya A et al (2017) Roles of MIWI, MILI and PLD6 in small RNA regulation in mouse growing oocytes. Nucleic Acids Res 45(9):5387–5398. https://doi.org/10.1093/nar/gkx027

Kaneda Y, Miyata H, Xu Z et al (2024) FBXO24 deletion causes abnormal accumulation of membraneless electron-dense granules in sperm flagella and male infertility. elife 13:RP92794. https://doi.org/10.7554/eLife.92794

Kidwell MG, Lisch D (1997) Transposable elements as sources of variation in animals and plants. Proc Natl Acad Sci 94(15):7704–7711. https://doi.org/10.1073/pnas.94.15.7704

Kirino Y, Mourelatos Z (2007) Mouse Piwi-interacting RNAs are 2′-O-methylated at their 3′ termini. Nat Struct Mol Biol 14(4):347–348. https://doi.org/10.1038/nsmb1218

Kiuchi T, Koga H, Kawamoto M et al (2014) A single female-specific piRNA is the primary determiner of sex in the silkworm. Nature 509(7502):633–636. https://doi.org/10.1038/nature13315

Klattenhoff C, Xi H, Li C et al (2009) The *Drosophila* HP1 homolog Rhino is required for transposon silencing and piRNA production by dual-strand clusters. Cell 138(6):1137–1149. https://doi.org/10.1016/j.cell.2009.07.014

Klenov MS, Sokolova OA, Yakushev EY et al (2011) Separation of stem cell maintenance and transposon silencing functions of Piwi protein. Proc Natl Acad Sci 108(46):18760–18765. https://doi.org/10.1073/pnas.1106676108

Kneuss E, Munafò M, Eastwood EL et al (2019) Specialization of the *Drosophila* nuclear export family protein Nxf3 for piRNA precursor export. Genes Dev 33(17–18):1208–1220. https://doi.org/10.1101/gad.328690.119

Kojima-Kita K, Kuramochi-Miyagawa S, Nagamori I et al (2016) MIWI2 as an effector of DNA methylation and gene silencing in embryonic male germ cells. Cell Rep 16(11):2819–2828. https://doi.org/10.1016/j.celrep.2016.08.027

Kolaczkowski B, Hupalo DN, Kern AD (2011) Recurrent adaptation in RNA interference genes across the *Drosophila* phylogeny. Mol Biol Evol 28(2):1033–1042. https://doi.org/10.1093/molbev/msq284

Koonin EV, Yutin N (2020) The crAss-like phage group: how metagenomics reshaped the human virome. Trends Microbiol 28(5):349–359. https://doi.org/10.1016/j.tim.2020.01.010

Kotaja N, Sassone-Corsi P (2007) The chromatoid body: a germ-cell-specific RNA-processing centre. Nat Rev Mol Cell Biol 8(1):85–90. https://doi.org/10.1038/nrm2081

Kunarso G, Chia NY, Jeyakani J et al (2010) Transposable elements have rewired the core regulatory network of human embryonic stem cells. Nat Genet 42(7):631–634. https://doi.org/10.1038/ng.600

Kuramochi-Miyagawa S, Kimura T, Ijiri TW et al (2004) Mili, a mammalian member of piwi family gene, is essential for spermatogenesis. Development 131(4):839–849. https://doi.org/10.1242/dev.00973

Kuramochi-Miyagawa S, Watanabe T, Gotoh K et al (2008) DNA methylation of retrotransposon genes is regulated by Piwi family members MILI and MIWI2 in murine fetal testes. Genes Dev 22(7):908–917. https://doi.org/10.1101/gad.1640708

Lau NC, Seto AG, Kim J et al (2006) Characterization of the piRNA complex from rat testes. Science 313(5785):363–367. https://doi.org/10.1126/science.1130164

Le Thomas A, Rogers AK, Webster A et al (2013) Piwi induces piRNA-guided transcriptional silencing and establishment of a repressive chromatin state. Genes Dev 27(4):390–399. https://doi.org/10.1101/gad.209841.112

Lehtiniemi T, Kotaja N (2018) Germ granule-mediated RNA regulation in male germ cells. Reproduction 155(1):R77–R91. https://doi.org/10.1530/REP-17-0356

Levine MT, McCoy C, Vermaak D et al (2012) Phylogenomic analysis reveals dynamic evolutionary history of the *Drosophila* heterochromatin protein 1 (HP1) gene family. PLoS Genet 8(6):e1002729. https://doi.org/10.1371/journal.pgen.1002729

Levine MT, Vander Wende HM, Hsieh E et al (2016) Recurrent gene duplication diversifies genome defense repertoire in Drosophila. Mol Biol Evol 33(7):1641–1653. https://doi.org/10.1093/molbev/msw053

Lewis SH, Quarles KA, Yang Y et al (2018) Pan-arthropod analysis reveals somatic piRNAs as an ancestral defence against transposable elements. Nat Ecol Evol 2(1):174–181. https://doi.org/10.1038/s41559-017-0403-4

Li C, Vagin VV, Lee S et al (2009) Collapse of germline piRNAs in the absence of Argonaute3 reveals somatic piRNAs in flies. Cell 137(3):509–521. https://doi.org/10.1016/j.cell.2009.04.027

Li XZ, Roy CK, Dong X et al (2013) An ancient transcription factor initiates the burst of piRNA production during early meiosis in mouse testes. Mol Cell 50(1):67–81. https://doi.org/10.1016/j.molcel.2013.02.016

Li Y, Liang Z, Liang Z et al (2021) Abnormal PIWI-interacting RNA profile and its association with the deformed extracellular matrix of oocytes from recurrent oocyte maturation arrest patients. Fertil Steril 115(5):1318–1326. https://doi.org/10.1016/j.fertnstert.2020.11.037

Li L, Tan YQ, Lu LY (2022) Defective piRNA processing and Azoospermia. N Engl J Med 386(17):1675–1676. https://doi.org/10.1056/NEJMc2116008

Li Z, Li Z, Zhang Y et al (2024) Mammalian PIWI–piRNA–target complexes reveal features for broad and efficient target silencing. Nat Struct Mol Biol 31(8):1222–1231. https://doi.org/10.1038/s41594-024-01287-6

Li Z, Xu Q, Zhong J et al (2025) Structural insights into RNA cleavage by PIWI Argonaute. Nature 1–10. https://doi.org/10.1038/s41586-024-08438-1

Liang L, Cao C, Ji L et al (2023) Complementary Alu sequences mediate enhancer–promoter selectivity. Nature 619(7971):868–875. https://doi.org/10.1038/s41586-023-06323-x

Lin H, Spradling AC (1997) A novel group of pumilio mutations affects the asymmetric division of germline stem cells in the *Drosophila* ovary. Development 124(12):2463–2476. https://doi.org/10.1242/dev.124.12.2463

Lippman Z, May B, Yordan C et al (2003) Distinct mechanisms determine transposon inheritance and methylation via small interfering RNA and histone modification. PLoS Biol 1(3):e67. https://doi.org/10.1371/journal.pbio.0000067

Liu N, Abe M, Sabin LR et al (2011) The exoribonuclease Nibbler controls 3′ end processing of microRNAs in *Drosophila*. Curr Biol 21(22):1888–1893. https://doi.org/10.1016/j.cub.2011.10.006

Loubalova Z, Fulka H, Horvat F et al (2021) Formation of spermatogonia and fertile oocytes in golden hamsters requires piRNAs. Nat Cell Biol 23(9):992–1001. https://doi.org/10.1038/s41556-021-00746-2

Luo S, Zhang H, Duan Y et al (2020) The evolutionary arms race between transposable elements and piRNAs in *Drosophila* melanogaster. BMC Evol Biol 20:1–18. https://doi.org/10.1186/s12862-020-1580-3

Ma L, Buchold GM, Greenbaum MP et al (2009) GASZ is essential for male meiosis and suppression of retrotransposon expression in the male germline. PLoS Genet 5(9):e1000635. https://doi.org/10.1371/journal.pgen.1000635

Malone CD, Brennecke J, Dus M et al (2009) Specialized piRNA pathways act in germline and somatic tissues of the *Drosophila* ovary. Cell 137(3):522–535. https://doi.org/10.1016/j.cell.2009.03.040

Marie PP, Ronsseray S, Boivin A (2017) From embryo to adult: piRNA-mediated silencing throughout germline development in *Drosophila*. G3 (Bethesda) 7(2):505–516. https://doi.org/10.1534/g3.116.037291

Meikar O, Da Ros M, Korhonen H, Kotaja N (2011) Chromatoid body and small RNAs in male germ cells. Reproduction 142(2):195–209. https://doi.org/10.1530/REP-11-0057

Meikar O, Vagin VV, Chalmel F (2014) An atlas of chromatoid body components. RNA 20(4):483–495. https://doi.org/10.1261/rna.043729.113

Mohn F, Sienski G, Handler D et al (2014) The rhino-deadlock-cutoff complex licenses noncanonical transcription of dual-strand piRNA clusters in *Drosophila*. Cell 157(6):1364–1379. https://doi.org/10.1016/j.cell.2014.04.031

Mohn F, Handler D, Brennecke J (2015) piRNA-guided slicing specifies transcripts for Zucchini-dependent, phased piRNA biogenesis. Science 348(6236):812–817. https://doi.org/10.1126/science.aaa1039

Mugat B, Nicot S, Varela-Chavez C et al (2020) The Mi-2 nucleosome remodeler and the Rpd3 histone deacetylase are involved in piRNA-guided heterochromatin formation. Nat Commun 11(1):2818. https://doi.org/10.1038/s41467-020-16635-5

Murchison EP, Kheradpour P, Sachidanandam R et al (2008) Conservation of small RNA pathways in platypus. Genome Res 18(6):995–1004. https://doi.org/10.1101/gr.073056.107

Nagirnaja L, Mørup N, Nielsen JE et al (2021) Variant *PNLDC1*, defective piRNA processing, and azoospermia. N Engl J Med 385(8):707–719. https://doi.org/10.1056/NEJMoa2028973

Ninova M, Chen YA, Godneeva B et al (2020) Su (var) 2-10 and the SUMO pathway link piRNA-guided target recognition to chromatin silencing. Mol Cell 77(3):556–570. https://doi.org/10.1016/j.molcel.2019.11.012

Ninova M, Holmes H, Lomenick B et al (2023) Pervasive SUMOylation of heterochromatin and piRNA pathway proteins. Cell Genom 3(7):100329. https://doi.org/10.1016/j.xgen.2023.100329

Nishida KM, Saito K, Mori T et al (2007) Gene silencing mechanisms mediated by Aubergine–piRNA complexes in *Drosophila* male gonad. RNA 13(11):1911–1922. https://doi.org/10.1261/rna.744307

Notwell JH, Chung T, Heavner W et al (2015) A family of transposable elements co-opted into developmental enhancers in the mouse neocortex. Nat Commun 6(1):6644. https://doi.org/10.1038/ncomms7644

Obbard DJ, Welch JJ, Kim KW et al (2009) Quantifying adaptive evolution in the *Drosophila* immune system. PLoS Genet 5(10):e1000698. https://doi.org/10.1371/journal.pgen.1000698

Onishi R, Sato K, Murano K et al (2020) Piwi suppresses transcription of Brahma-dependent transposons via Maelstrom in ovarian somatic cells. Sci Adv 6(50):eaaz7420. https://doi.org/10.1126/sciadv.aaz7420
Onohara Y, Yokota S (2012) Nuage components and their contents in mammalian spermatogenic cells, as revealed by immunoelectron microscopy. INTECH Open Access Publisher, Rijeka
Orgel LE, Crick FH (1980) Selfish DNA: the ultimate parasite. Nature 284(5757):604–607. https://doi.org/10.1038/284604a0
Ortogero N, Schuster AS, Oliver DK et al (2014) A novel class of somatic small RNAs similar to germ cell pachytene PIWI-interacting small RNAs. J Biol Chem 289(47):32824–32834. https://doi.org/10.1074/jbc.M114.613232
Özata DM, Yu T, Mou H et al (2020) Evolutionarily conserved pachytene piRNA loci are highly divergent among modern humans. Nat Ecol Evol 4(1):156–168. https://doi.org/10.1038/s41559-019-1065-1
Pandey RR, Tokuzawa Y, Yang Z et al (2013) Tudor domain containing 12 (TDRD12) is essential for secondary PIWI interacting RNA biogenesis in mice. Proc Natl Acad Sci 110(41):16492–16497. https://doi.org/10.1073/pnas.1316316110
Parhad SS, Tu S, Weng Z et al (2017) Adaptive evolution leads to cross-species incompatibility in the piRNA transposon silencing machinery. Dev Cell 43(1):60–70. https://doi.org/10.1016/j.devcel.2017.08.012
Parhad SS, Yu T, Zhang G et al (2020) Adaptive evolution targets a piRNA precursor transcription network. Cell Rep 30(8):2672–2685. https://doi.org/10.1016/j.celrep.2020.01.109
Pezic D, Manakov SA, Sachidanandam R et al (2014) piRNA pathway targets active LINE1 elements to establish the repressive H3K9me3 mark in germ cells. Genes Dev 28(13):1410–1428. https://doi.org/10.1101/gad.240895.114
Qi H, Watanabe T, Ku HY et al (2011) The Yb body, a major site for Piwi-associated RNA biogenesis and a gateway for Piwi expression and transport to the nucleus in somatic cells. J Biol Chem 286(5):3789–3797. https://doi.org/10.1074/jbc.M110.193888
Rajasethupathy P, Antonov I, Sheridan R et al (2012) A role for neuronal piRNAs in the epigenetic control of memory-related synaptic plasticity. Cell 149(3):693–707. https://doi.org/10.1016/j.cell.2012.02.057
Reuter M, Berninger P, Chuma S et al (2011) Miwi catalysis is required for piRNA amplification-independent LINE1 transposon silencing. Nature 480(7376):264–267. https://doi.org/10.1038/nature10672
Rouget C, Papin C, Boureux A et al (2010) Maternal mRNA deadenylation and decay by the piRNA pathway in the early *Drosophila* embryo. Nature 467(7319):1128–1132. https://doi.org/10.1038/nature09465
Rozhkov NV, Hammell M, Hannon GJ (2013) Multiple roles for Piwi in silencing *Drosophila* transposons. Genes Dev 27(4):400–412. https://doi.org/10.1101/gad.209767.112
Saito K, Sakaguchi Y, Suzuki T et al (2007) Pimet, the *Drosophila* homolog of HEN1, mediates 2′-O-methylation of Piwi-interacting RNAs at their 3′ ends. Genes Dev 21(13):1603–1608. https://doi.org/10.1101/gad.1563607
Saito K, Inagaki S, Mituyama T et al (2009) A regulatory circuit for piwi by the large Maf gene traffic jam in *Drosophila*. Nature 461(7268):1296–1299. https://doi.org/10.1038/nature08501
Sasaki T, Nishihara H, Hirakawa M et al (2008) Possible involvement of SINEs in mammalian-specific brain formation. Proc Natl Acad Sci 105(11):4220–4225. https://doi.org/10.1073/pnas.0709398105
Saxe JP, Chen M, Zhao H et al (2013) Tdrkh is essential for spermatogenesis and participates in primary piRNA biogenesis in the germline. EMBO J 32(13):1869–1885. https://doi.org/10.1038/emboj.2013.121
Schöpp T, Zoch A, Berrens RV et al (2020) TEX15 is an essential executor of MIWI2-directed transposon DNA methylation and silencing. Nat Commun 11(1):3739. https://doi.org/10.1038/s41467-020-17372-5

Schrader L, Kim JW, Ence D et al (2014) Transposable element islands facilitate adaptation to novel environments in an invasive species. Nat Commun 5(1):5495. https://doi.org/10.1038/ncomms6495

Shiromoto Y, Kuramochi-Miyagawa S, Daiba A et al (2013) GPAT2, a mitochondrial outer membrane protein, in piRNA biogenesis in germline stem cells. RNA 19(6):803–810. https://doi.org/10.1261/rna.038521.113

Shoji M, Tanaka T, Hosokawa M et al (2009) The TDRD9-MIWI2 complex is essential for piRNA-mediated retrotransposon silencing in the mouse male germline. Dev Cell 17(6):775–787. https://doi.org/10.1016/j.devcel.2009.10.012

Sienski G, Dönertas D, Brennecke J (2012) Transcriptional silencing of transposons by Piwi and maelstrom and its impact on chromatin state and gene expression. Cell 151(5):964–980. https://doi.org/10.1016/j.cell.2012.10.040

Simkin A, Wong A, Poh YP et al (2013) Recurrent and recent selective sweeps in the piRNA pathway. Evolution 67(4):1081–1090. https://doi.org/10.1111/evo.12011

Stallmeyer B, Bühlmann C, Stakaitis R et al (2024) Inherited defects of piRNA biogenesis cause transposon de-repression, impaired spermatogenesis, and human male infertility. Nat Commun 15(1):6637. https://doi.org/10.1038/s41467-024-50930-9

Sun YH, Zhu J, Xie LH et al (2020) Ribosomes guide pachytene piRNA formation on long intergenic piRNA precursors. Nat Cell Biol 22(2):200–212. https://doi.org/10.1038/s41556-019-0457-4

Sun YH, Wang RH, Du K et al (2021) Coupled protein synthesis and ribosome-guided piRNA processing on mRNAs. Nat Commun 12(1):5970. https://doi.org/10.1038/s41467-021-26233-8

Sun YH, Cui H, Song C et al (2023) Amniotes co-opt intrinsic genetic instability to protect germline genome integrity. Nat Commun 14(1):812. https://doi.org/10.1038/s41467-023-36354-x

Szakmary A, Reedy M, Qi H et al (2009) The Yb protein defines a novel organelle and regulates male germline stem cell self-renewal in *Drosophila* melanogaster. J Cell Biol 185(4):613–627. https://doi.org/10.1083/jcb.200903034

Tang W, Seth M, Tu S et al (2018) A sex chromosome piRNA promotes robust dosage compensation and sex determination in *C. elegans*. Dev Cell 44(6):762–770. https://doi.org/10.1016/j.devcel.2018.01.025

Tsai S, Huang F (2021) Acetyltransferase Enok regulates transposon silencing and piRNA cluster transcription. PLoS Genet 17(2):e1009349. https://doi.org/10.1371/journal.pgen.1009349

Vagin VV, Sigova A, Li C et al (2006) A distinct small RNA pathway silences selfish genetic elements in the germline. Science 313(5785):320–324. https://doi.org/10.1126/science.1129333

Voigt F, Reuter M, Kasaruho A et al (2012) Crystal structure of the primary piRNA biogenesis factor Zucchini reveals similarity to the bacterial PLD endonuclease Nuc. RNA 18(12):2128–2134. https://doi.org/10.1261/rna.034967.112

Vourekas A, Zheng K, Fu Q et al (2015) The RNA helicase MOV10L1 binds piRNA precursors to initiate piRNA processing. Genes Dev 29(6):617–629. https://doi.org/10.1101/gad.254631.114

Watanabe T, Chuma S, Yamamoto Y et al (2011) MITOPLD is a mitochondrial protein essential for nuage formation and piRNA biogenesis in the mouse germline. Dev Cell 20(3):364–375. https://doi.org/10.1016/j.devcel.2011.01.005

Wei H, Gao J, Lin DH et al (2024) piRNA loading triggers MIWI translocation from the intermitochondrial cement to chromatoid body during mouse spermatogenesis. Nat Commun 15(1):2343. https://doi.org/10.1038/s41467-024-46664-3

Wessler SR (1988) Phenotypic diversity mediated by the maize transposable elements Ac and Spm. Science 242(4877):399–405. https://doi.org/10.1126/science.2845581

Whitelaw E, Martin DI (2001) Retrotransposons as epigenetic mediators of phenotypic variation in mammals. Nat Genet 27(4):361–365. https://doi.org/10.1038/86850

Wu PH, Fu Y, Cecchini K et al (2020) The evolutionarily conserved piRNA-producing locus pi6 is required for male mouse fertility. Nat Genet 52(7):728–739. https://doi.org/10.1038/s41588-020-0657-7

Wynant N, Santos D, Vanden Broeck J (2017) The evolution of animal Argonautes: evidence for the absence of antiviral AGO Argonautes in vertebrates. Sci Rep 7(1):9230. https://doi.org/10.1038/s41598-017-08043-5

Xiol J, Spinelli P, Laussmann MA et al (2014) RNA clamping by Vasa assembles a piRNA amplifier complex on transposon transcripts. Cell 157(7):1698–1711. https://doi.org/10.1016/j.cell.2014.05.018

Xu MM, Li XZ (2024) Enigmatic Pachytene PIWI-Interacting RNAs. Genome Biol Evol 16(10):evae162. https://doi.org/10.1093/gbe/evae162

Yan Z, Hu HY, Jiang X et al (2011) Widespread expression of piRNA-like molecules in somatic tissues. Nucleic Acids Res 39(15):6596–6607. https://doi.org/10.1093/nar/gkr298

Yang F, Lan Y, Pandey RR et al (2020) TEX15 associates with MILI and silences transposable elements in male germ cells. Genes Dev 34(11–12):745–750. https://doi.org/10.1101/gad.335489.119

Yi M, Chen F, Luo M et al (2014) Rapid evolution of piRNA pathway in the teleost fish: implication for an adaptation to transposon diversity. Genome Biol Evol 6(6):1393–1407. https://doi.org/10.1093/gbe/evu105

Yu Y, Gu J, Jin Y et al (2015) Panoramix enforces piRNA-dependent cotranscriptional silencing. Science 350(6258):339–342. https://doi.org/10.1126/science.aab0700

Yu T, Fan K, Özata DM et al (2021) Long first exons and epigenetic marks distinguish conserved pachytene piRNA clusters from other mammalian genes. Nat Commun 12(1):73. https://doi.org/10.1038/s41467-020-20345-3

Yu T, Biasini A, Cecchini K et al (2023) A-MYB/TCFL5 regulatory architecture ensures the production of pachytene piRNAs in placental mammals. RNA 29(1):30–43. https://doi.org/10.1261/rna.079472.122

Zhang Z, Xu J, Koppetsch BS et al (2011) Heterotypic piRNA Ping-Pong requires qin, a protein with both E3 ligase and Tudor domains. Mol Cell 44(4):572–584. https://doi.org/10.1016/j.molcel.2011.10.011

Zhang Z, Wang J, Schultz N et al (2014) The HP1 homolog rhino anchors a nuclear complex that suppresses piRNA precursor splicing. Cell 157(6):1353–1363. https://doi.org/10.1016/j.cell.2014.04.030

Zhang P, Kang JY, Gou LT et al (2015) MIWI and piRNA-mediated cleavage of messenger RNAs in mouse testes. Cell Res 25(2):193–207. https://doi.org/10.1038/cr.2015.4

Zhang Y, Liu W, Li R et al (2018) Structural insights into the sequence-specific recognition of Piwi by *Drosophila* Papi. Proc Natl Acad Sci 115(13):3374–3379. https://doi.org/10.1073/pnas.1717116115

Zhang S, Pointer B, Kelleher ES (2020) Rapid evolution of piRNA-mediated silencing of an invading transposable element was driven by abundant de novo mutations. Genome Res 30(4):566–575. https://doi.org/10.1101/gr.251546.119

Zhang H, Zhang F, Chen Q et al (2021) The piRNA pathway is essential for generating functional oocytes in golden hamsters. Nat Cell Biol 23(9):1013–1022. https://doi.org/10.1038/s41556-021-00750-6

Zheng K, Xiol J, Reuter M et al (2010) Mouse MOV10L1 associates with Piwi proteins and is an essential component of the Piwi-interacting RNA (piRNA) pathway. Proc Natl Acad Sci 107(26):11841–11846. https://doi.org/10.1073/pnas.1003953107

Zhou X, Liao Z, Jia Q et al (2007) Identification and characterization of Piwi subfamily in insects. Biochem Biophys Res Commun 362(1):126–131. https://doi.org/10.1016/j.bbrc.2007.07.179

Zhou L, Canagarajah B, Zhao Y et al (2017) BTBD18 regulates a subset of piRNA-generating loci through transcription elongation in mice. Dev Cell 40(5):453–466. https://doi.org/10.1016/j.devcel.2017.02.007

Zoch A, Auchynnikava T, Berrens RV et al (2020) SPOCD1 is an essential executor of piRNA-directed de novo DNA methylation. Nature 584(7822):635–639. https://doi.org/10.1038/s41586-020-2557-5

Zoch A, Konieczny G, Auchynnikava T et al (2024) C19ORF84 connects piRNA and DNA methylation machineries to defend the mammalian germ line. Mol Cell 84(6):1021–1035. https://doi.org/10.1016/j.molcel.2024.01.014

Chapter 7
The Role of Transposable Elements in Development

John R. McCoy and Tara T. Doucet-O'Hare

Abstract Transposable elements (TEs) are DNA sequences that can move from one location in the genome to another. TEs make up a large portion of the eukaryote genome throughout evolution, TE expression has been co-opted for nonpathogenic purposes. In the realm of development, these elements have been shown to play significant roles in the growth and differentiation of stem cells. TEs and their derived proteins have also been shown to play a significant role in the formation and function of the placenta. Following cellular differentiation and successful embryo implantation, TEs are also crucial in the development of the nervous system. Some evidence suggests that long noncoding RNAs produced from TEs may contribute to epigenetic networks crucial in neuronal differentiation. TEs are mainly expressed early in development and are gradually silenced by a myriad of mechanisms, including pre-transcriptional repression by DNA and histone methylation, and post-transcriptional repression by noncoding RNAs. In this chapter, we illustrate some of the known contributions of TEs to mammalian developmental programming, with an emphasis on examples in humans.

Keywords Transposable elements · Placenta · Development · Transcription factor · Epigenetics · Evolution

Abbreviations

ALV	Avian leukosis virus
Alu	TE named for its original characterization in *Arthrobacter luteus (Alu)*

J. R. McCoy · T. T. Doucet-O'Hare (✉)
Department of Genetics and Biochemistry, and Center for Human Genetics, Clemson University, Greenwood, SC, USA
e-mail: tdoucet@clemson.edu

L. M. Vaschetto (ed.), *Transposable Elements, Transcriptomics, and Diseases*, Epigenetics and Human Health 14, https://doi.org/10.1007/978-3-032-04125-8_7

APOBEC	Apolipoprotein B mRNA editing catalytic polypeptide
Arc	Activity-regulated cytoskeleton
Bglap3	Bone gamma-carboxylglutamate protein 3
Cas9	CRISPR-associated protein 9
CNS	Central nervous system
CRISPR	Clustered regularly interspaced short palindromic repeats
CRISPRi	Clustered regularly interspaced short palindromic repeats interference
CTCF	CCCTC-binding factor
Dicer	Enzymatic gene, known as endoribonuclease Dicer or helicase with RNase motif
DNMT3A	DNA (cytosine-5) methyltransferase 3A
DNMT3B	DNA methyl transferase 3 B
dsRNA	Double-stranded RNA
DUX4	Double homeobox 4
ENV	Envelope
ERV	Endogenous retrovirus
ERVL	Endogenous retrovirus L
ESC	Embryonic stem cell
EV	Extracellular vesicle
FOA	Fetal oocyte attrition
GH	Growth hormone
GAG	Group-specific antigen
Gypsy	LTR retrotransposons, also known as Ty3
HERV	Human endogenous retrovirus
HERV-K	Human endogenous retrovirus K
HERV-K10	Human endogenous retrovirus K10
HIV	Human immunodeficiency virus
HML-2	Human endogenous MMTV-like 2
HSATII	Human satellite II repeats
IAP	Intracisternal A-particle
ICM	Inner cell mass
IPSC	Induced pluripotent stem cell
KRAB-ZFP	Krüppel-associated box domain-zinc finger protein
LINE	Long interspersed nuclear element
L1	LINE-1
lncRNA	Long noncoding RNA
LTEUX	Leucine twenty homeobox
LTR	Long terminal repeat
MaLR	Mammalian apparent LTR retrotransposons
MAVS	Mitochondrial antiviral signaling protein
MeCP	Methyl-CpG-binding protein
MER39	Medium reiterated frequency repeat 39
MER77	Medium reiterated frequency repeat 77
MER130	Medium reiterated frequency repeat 130

MERVL	Murine endogenous retrovirus L (leucine tRNA primer)
miRNA	Micro RNA
MMTV	Mouse mammary tumor virus
MT-C	An intronic retrotransposon promoter in mice
mRNA	Messenger RNA
OCT-4	Octamer-binding transcription factor 4
ORF1p	LINE-1 open reading frame 1 protein
ORF2p	LINE-1 open reading frame 2 protein
PDAC	Pancreatic ductal adenocarcinoma
PEG10	Paternally expressed gene 10
PGBD5	PiggyBac transposable element derived 5
piRNA	PIWI-interacting RNA
PIWI	P-element induced wimpy testis
PNS	Peripheral nervous system
Rec	HERV functional homolog of HIV-1 Rev and HTLV Rev
RIG-I	Retinoic acid inducible gene I
RNAi	RNA-interference
RNP	Ribonucleoproteins
RPB5	RNA polymerase II subunit 5
SCT	Syncytium
SINE	Short interspersed nuclear element
SOX2	Sex determining region Y-box 2
Sp1	Specificity protein 1 transcription factor
SVA	SINE-VNTR-Alu
T cell	Lymphocyte produced by the thymus
TE	Transposable element
THE1B	Transposon-like human element 1B
Ty3	LTR retrotransposons, also known as Gypsy
VNTR	Variable number tandem repeat
URI	Unconventional prefolding RNBP5 interactor
ZGA	Zygotic genome activation
ZSCAN4	Zinc finger and SCAN domain containing 4

7.1 Introduction

Transposable elements (TEs) are segments of DNA with the ability to move themselves throughout the genome (Fig. 7.1 and Table 7.1). These elements were first described by Barbara McClintock in 1951 (McClintock 1951). TEs were historically regarded as 'junk' DNA for decades (Doolittle and Sapienza 1980; Orgel and Crick 1980) until McClintock's discovery. Since the initial human genome assembly revealed that up to nearly 50% of the genome is comprised of TE sequences, research has revealed the crucial role of TEs in normal physiological functions, particularly in development (Lander et al. 2001; SanMiguel et al. 1996; Hoyt et al.

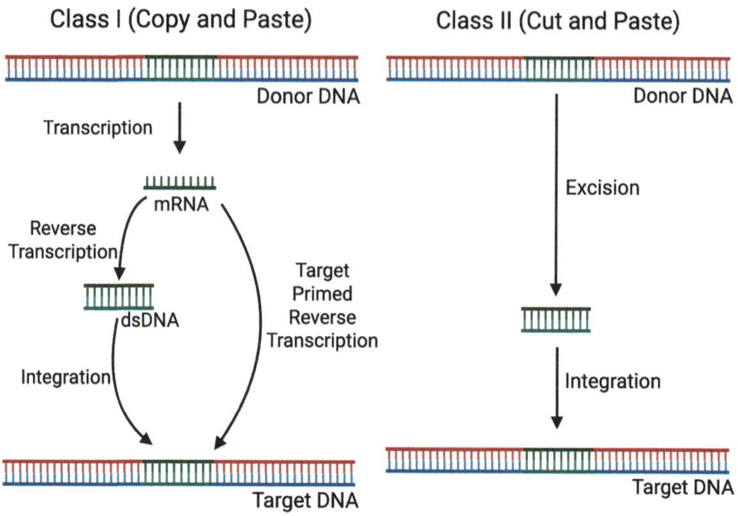

Fig. 7.1 Mechanism of transposable element (TE) integration into the genome. TEs typically fall into one of two classes. Class I TEs are characterized by an RNA-intermediate mechanism before integration into the genome through a reverse transcriptase (Copy and Paste). Class II TEs integrate into the genome through integration of an already present DNA intermediary (Cut and Paste). (Generated with BioRender)

2022; Garcia-Perez et al. 2016). TEs can mobilize in the host germ cell and subsequently be inherited by future generations like other established Mendelian genes; however, these integrations are often species-specific and can be co-opted for similar physiological purposes (Senft and Macfarlan 2021). One example of a co-opted TE gene includes syncytin, which is present in both mice and in humans, although these genes—syncytin-1 and syncytin-2 in humans and syncytin-A and syncytin-B in mice—integrated separately during evolution, they all play an essential role in placental development (Dupressoir et al. 2012). In another example, several envelope (env) genes derived from the endogenous avian leukosis viruses (ALVs) have been found to provide protection against exogenous infections from the same subgroup, thus preventing retroviral entry (Robinson et al. 1981).

TEs are classified by their mode of transposition and include DNA transposons and retrotransposons. DNA transposons use a cut-and-paste mechanism to integrate into new areas of the genome, while retrotransposons use reverse transcription to copy and paste their sequences into new locations (Fig. 7.1). Retrotransposons can further be divided by their structure, specifically whether they contain long terminal repeats (LTRs), which flank the internal sequence of elements such as endogenous retroviruses (ERVs) but are absent in elements like long and short interspersed nuclear elements (LINEs and SINEs). This chapter will cover the many types of TEs that contribute to developmental processes. It should be noted that TEs do not necessarily have to mobilize into new locations in a host genome to have both physiological and potentially pathological effects, as they often encode open reading

Table 7.1 Transposable elements relevant to human development

TE family	Class of TE	Function	References
HERVs	Type 1	Comprise a broad range of TEs that influence gene expression through LTRs, especially during early embryogenesis.	Katoh and Honda (2023), Mao et al. (2021)
LINEs	Type 1	Create novel expression patterns that influence gene expression in developmental genes.	Li et al. (2013)
SINEs	Type 1	Can act as enhancers, chromatin organizers, and generate noncoding RNA that work to modulate gene expression.	Elbarbary et al. (2016), Sun et al. (2024)
SVA	Type 1	Modulate gene expression through integration into untranslated regions and act to enhance or repress transcription.	Xing et al. (2006), Kwon et al. (2013)
Processed Pseudogenes	Type 1	Production of siRNAs can interfere with gene expression by preventing mRNA translation.	Gu et al. (2021), Roychowdhury and Abyzov (2019)
hAT	Type 2	Integration near the gene alters the transcriptional landscape of the gene.	Loreto et al. (2018)
PiggyBac	Type 2	Highly evolutionarily conserved and integrated into the genome multiple times. Evidence shows that PiggyBac plays a role in proper brain development and may aid in genetic adaptability.	Zapater et al. (2024), Newman et al. (2008)
Tc1/mariner	Type 2	Has the ability to act as a promoter in untranslated regions, and there is some evidence that TEs in this family can act as mutagens and work toward increasing genetic adaptability.	Palazzo et al. (2019), Miskey et al. (2007)

A select list of TE families that are linked to human development. Each family is classified by class, function, and references for each family

frames for viral proteins, some of which are still capable of successful translation (Doucet-O'Hare et al. 2015, 2021). Although this is not an exhaustive list of ways these elements contribute to genomic and transcriptomic regulation there are many instances in which TEs have contributed to alternative splicing, differential activation of adjacent genes via transcription factor binding sites present in their sequence, and have contributed to pathology via their double stranded RNA activating the innate immune system (Cao et al. 2020). There are likely still many undiscovered ways in which TEs contribute to cell-specific expression and or developmental pathways.

Development is a complex process that requires an undifferentiated cell to divide, migrate, and organize into diverse tissue types (Shahbazi 2020). During development, cells undergo significant changes in their gene expression and cell function, which will facilitate the formation of tissues with specific physiologic purposes in a developing organism (Shahbazi 2020). The process is enabled by chromatin remodeling proteins, changes in DNA methylation, histone modifications, RNA splicing factors, and many other mechanisms that act as precise genetic regulators in the host

(Donkin and Barres 2018). While there are many gene regulation mechanisms involved in development, TEs play a significant role. TEs can aid in the modulation of developmental genes by acting as or contributing to promoters, repressors, or chromatin modulators to enhance or prevent transcription and RNA processing (Gebrie 2023). Given the complexity of development, dysregulation of developmental regulatory genes—genes expressed during development that control the expression of at least one other gene—can disrupt development and lead to systemic issues in an individual organism's physiology, sometimes affecting the central nervous system (Zapater et al. 2024). One component of early development is the transition from cells that are totipotent to cells which are pluripotent and finally to cells that differentiate into their final form. TEs and TE-derived proteins are one mechanism that regulates the differentiation of totipotent and pluripotent stem cells across multicellular life, and this is an example of co-option, the process by which an inherited gene or trait switches function in a new host (de Souza et al. 2013; Ma et al. 2022; Rodriguez-Terrones and Torres-Padilla 2018). It has recently been shown that dysregulation of stem cells can lead to systemic developmental defects and potentially contribute to the development of embryonic tumors (Doucet-O'Hare et al. 2021; Huhtala et al. 2024; Shah et al. 2023). These mechanisms are actively being explored to better understand both developmental tumors and to find potential targeted treatment options for patients.

In addition to their role in stem cell identity, TEs can also play a role in generating immune tolerance in pregnancy. Immune tolerance is a key part of a successful pregnancy, and the immune system is integral to regulating potential interactions between the fetus and the mother. There are many mechanisms that largely involve the activation of specific T cells that allow maternal tolerance of the fetus, and while some mechanisms are more critical than others, no one mechanism is sufficient to provide fetal protection (Genebrier and Tarte 2021). One of these mechanisms is facilitated by TEs, which play a key role in the interplay between the mother and the embryo, where they are integral for implantation of the blastocyst into the uterine lining and placental development (Burton and Fowden 2015; Imakawa et al. 2022). Moreover, new evidence suggests TE expression must be carefully controlled in the placenta throughout pregnancy, in part because TEs ensure proper immune tolerance, preventing the mother's immune system from attacking the placenta or embryo (Katoh and Honda 2023).

Following successful embryo implantation, and cellular differentiation in the embryo, TEs are also crucial in the development of the nervous system (Reilly et al. 2013; Hua and Smith 2004). Recently, a gag protein originating from an ERV in zebrafish has been shown to be essential for mesoderm development, and its depletion led to defects in neural crest development (Chang et al. 2025). In *Drosophila*, the neuronal gene activity-regulated cytoskeleton (*Arc*), derived from the *gag* gene of a Ty3/Gypsy retrotransposon, facilitates intercellular RNA transfer and is secreted from neurons via extracellular vesicles (EVs) (Ashley et al. 2018). Dysfunction of the Arc protein in both *Drosophila* and mammals has been linked to neurodevelopmental disorders, highlighting its potential role in developmental processes (Pastuzyn et al. 2018; Ashley et al. 2018). In one study from Fromer et al., the authors report that small de novo variants that affect one or a few nucleotides are overrepresented in glutamatergic postsynaptic proteins like arc and that these variants may contribute to phenotypes like schizophrenia and autism (Fromer et al. 2014). Throughout this chapter, we will use examples from the literature to illustrate the integral role that TEs play in development. We will survey their structural diversity and evolutionary conservation to appreciate how intricately woven TEs are into the fabric of development. This chapter will explore some examples of both the molecular mechanisms and functional consequences of TE activity in development.

7.2 TEs and Development

TE control is essential to prevent elements from "jumping" or integrating into new locations in the host genome which can result in interrupted open reading frames for protein-coding genes, affect splice junctions, add new promoters that can lead to differential transcription, and lead to a variety of other disruptions in genetic homeostasis during cellular division and meiosis (reviewed in Garcia-Perez et al. (2016)). Mechanisms that restrict TE mobilization and expression in germ cells and pluripotent cells in early embryos, which contribute to germ cell precursors, are paramount, as any new integrations can be transmitted to the next generation (reviewed in Crichton et al. (2014)). There are numerous examples demonstrating how the aberrant expression of TEs can negatively impact normal physiological development. For example, one study reported that the overexpression of the Human Endogenous Retrovirus K (HERV-K) rec protein in mice led to abnormal germ cell development and ultimately resulted in the development of a precursor lesion for seminoma (Galli et al. 2005). In a different study, the authors disrupted Maelstrom, a murine homolog of a *Drosophila* nuage protein (Soper et al. 2008). The nuage protein is part of a membraneless organelle found within the germline; it is conserved in metazoans and hypothesized to play a role in transposon expression (Eddy 1975; Voronina et al. 2011; Lim and Kai 2007). As a result of *Maelstrom* disruption in mice, spermatocytes with nuclear LINE-1 (L1) ribonucleoproteins (RNPs) contained extreme DNA damage and displayed severe chromosome asynapsis even in the absence of double-stranded breaks (Soper et al. 2008). Another study in mice showed evidence that during the process of fetal oocyte attrition (FOA), oocytes are preferentially selected if they possess less L1 activity and L1 Open Reading Frame 1 protein (ORF1p) expression, likely due to fewer double-stranded breaks in oocytes with lower L1 activity and thus higher cell viability (Malki et al. 2014). Georgiou et al. showed in human oocytes that L1, HERV-K10, and SINE-VNTR-Alu (SVA) retrotransposons were expressed and that retrotransposition occurred in these cells (Georgiou et al. 2009). From these studies, it appears that regulation of TE expression in early germ cell development is vital for successful reproductive outcomes in several species.

Although aberrant expression of TEs has been shown to be detrimental in early developmental processes, especially in germ cell development and meiosis, there is some evidence that a measured amount of TE expression could be beneficial as well. In one study of human oocytes, the authors suggest that controlled retrotransposition events or TE expression may serve as a benefit due to the generation of antisense TE transcripts and subsequent RNA-interference (RNAi)-mediated degradation and TE methylation (Puschendorf et al. 2006; Hannon 2002). TE expression and novel integrations could also provide opportunities for new properties and functions of adjacent host genes. For example, TE promoters are studded with transcription factor binding sites for proteins like Octamer-binding transcription factor 4 (OCT-4) and Specificity protein 1 transcription factor (Sp1) (Manghera and Douville 2013). *Dicer*, a gene found in a wide range of eukaryotic species, including mice and humans, is an example of a TE-derived promoter that generates an isoform specific to a cell-type from a host gene and confers novel genetic features (Flemr et al. 2013). The mouse *Dicer1* gene encodes an RNA endonuclease that generates small regulatory RNAs; however, an intragenic insertion of an intronic retrotransposon promoter, MT-C endogenous retrovirus (ERV), in the gene generates a truncated form expressed specifically in oocytes (Flemr et al. 2013). This truncated ERV-driven form of the gene is more enzymatically active than the full-length version due to the absence of a potentially autoinhibitory helicase domain present in the N-terminal region of the full-length protein (Flemr et al. 2013). ERV-driven transcription is a feature present in somatic cells as well. In mice, transcription of bone gamma-carboxylglutamate protein 3 (*Bglap3)* depends on the activity of a Krüppel-associated box domain-Zinc Finger Protein (KRAB-ZFP), which targets repressive histone modifications to an adjacent Intracisternal A-particle (IAP) element (Ecco et al. 2016).

Exaptation of TEs as genetic regulators and alternative promoters appears in many species and often displays evidence of convergent evolution with regard to host gene control in developmental contexts. For example, in some mammals, the *prolactin* gene can be expressed from an alternative promoter in the maternal endometrial decidua during pregnancy (Jabbour and Critchley 2001). The endometrial decidua regulates immune cells, the generation of blood vessels, and the invasion of the fetal placenta into maternal tissues (Jabbour and Critchley 2001). A medium reiterated frequency repeat element 39 (MER39) ERV initiates transcription of the decidual prolactin in humans, a MER77 ERV regulates decidual prolactin expression in mice, and an elephant-specific L1 insertion promotes expression of decidual prolactin as well (Emera et al. 2012; Emera and Wagner 2012). Independent evolution of TE-derived prolactin promoters in multiple mammalian species demonstrates that TE sequences are a fertile source of transcription factor binding sites, which allow host genes to perform new cell and tissue-specific functions.

TEs can also act as regulators of mammalian chromatin organization; for example, TE sequences are highly enriched in regions that bind CCCTC-binding factor (CTCF), and in mice, B2 SINEs are highly enriched in regions with this binding motif (Bourque et al. 2008; Schmidt et al. 2012; Sundaram et al. 2014). Another notable example of developmental gene regulation influenced by a TE occurs via chromatin remodeling during the expression of growth hormone (GH) in the developing pituitary gland in mice (Lunyak et al. 2007). In this situation, the B2 SINE is essential for the gene activation of GH, and the repeat serves as a boundary that blocks potential interference from repressive chromatin modifications (Lunyak et al. 2007). Taken together, these examples suggest that TEs can impact transcription of a gene in a myriad of ways and that it is not necessary for them to reintegrate into new locations to play a role in developmental programming.

7.2.1 *Expression of Transposable Elements During Early Development*

TEs are crucial throughout all stages of early embryonic development, especially at the time of zygotic genome activation (ZGA) (Dopkins and Nixon 2024; Goke and Ng 2016; Goke et al. 2015; Grow et al. 2015). Double homeobox 4 (DUX4), a transcription factor that binds to LTRs, is vital in the process of ZGA and acts on mammalian apparent LTR retrotransposons (MaLRs), ERVs L and K (ERVL and ERVK), and pericentromeric human satellite II (HSATII) sequences (Lin et al. 2025; Young et al. 2013; Whiddon et al. 2017; Hendrickson et al. 2017; De Iaco et al. 2017). DUX4 binds retrotransposon sequences during the oocyte-to-embryo transition, during which the oocyte matures into a full embryo. DUX4-activated MaLR and ERV elements facilitate the activity of novel gene promoters, long noncoding RNAs (lncRNAs), and antisense transcripts involved in developmental regulation (Young et al. 2013). One example of transcriptional activation due to DUX4 binding to TE-derived enhancers is the Zinc Finger And SCAN Domain Containing 4 (ZSCAN4). DUX4 expression is associated with the activation of ZSCAN4, which has a promoter enriched with DUX4 binding motifs (Vuoristo et al. 2022). A putative ZSCAN4 enhancer identified recently overlaps an ERVL-MaLR repeat element and further supports the connection between TEs and early zygotic development (Vuoristo et al. 2022). ZSCAN4 expression is paramount to ZGA, thus reinforcing the importance of TE-based regulation in this process; however, this is still an area of active investigation (Chau et al. 2021).

7.2.2 *Transposable Elements in Developmental Stem Cells*

Stem cells, defined by their capacity to differentiate into all cell types of the body, are essential for early developmental processes (Ghazimoradi et al. 2022). The earliest of these cells, known as totipotent stem cells, possess the ability to give rise to all embryonic and extraembryonic lineages in response to developmental cues (Ghazimoradi et al. 2022). Totipotent cells are present in the zygote and persist through early stages of embryogenesis, where they serve as the foundation for subsequent tissue and organ formation. Stem cells also play critical roles beyond early development, contributing to physiological processes throughout the lifespan of an individual. Notably, they are involved in the formation of the placenta—a transient yet functionally distinct organ that arises during pregnancy to mediate the complex interactions at the maternal-fetal interface. Thus, the ability of totipotent cells to differentiate and develop into a myriad of different cell types is critical for proper embryo, organ, and tissue development. In recent years, it has become clear that TEs play a crucial role in the development and regulation of totipotent cells (de la Rosa et al. 2024; Oomen et al. 2025; Shah et al. 2023; Wang et al. 2020).

In 2012, one group revealed that a rare and transient cell population existed within mouse embryonic stem cell (ESCs) and induced pluripotent cell (iPSCs) cultures that expressed high levels of pluripotency proteins like Mouse Endogenous Retrovirus L (MERVL) (Macfarlan et al. 2012). A more recent study highlights that these stem cells play a role in maintaining a totipotent state by evaluating the function of MERVL-gag protein in mice (de la Rosa et al. 2024). The MERVL-gag protein actively initiates the degradation of transcription factor binding proteins OCT-4 and SOX2 by disrupting their interaction with the unconventional prefoldin RNA polymerase II subunit 5 (RPB5) interactor (URI) (de la Rosa et al. 2024). MERVL-gag provides one of the strongest examples of how TE expression plays a role in stem cell identity. MERVL-gag regulation of OCT-4 and SOX2 is significant because SOX2 and OCT-4 are well known for their role in maintaining stem cell pluripotency—this is the stage during which totipotent cells shed their ability to create extraembryonic tissues but retain the ability to differentiate into most other cell types (Rizzino 2013; Xu and Liang 2022). Interestingly, OCT-4 and SOX2 have also been implicated in the regulation of TEs, where they can form a complex known to aid in silencing genes, suggesting a regulatory feedback loop (Nikolopoulou et al. 2017; Pontis et al. 2022; Subhramanyam et al. 2020). In a recent study by Sakashita et al., the authors demonstrated that a knockdown of MERVL resulted in embryonic lethality due to defects in differentiation and genomic stability (Sakashita et al. 2023). In a different study by Yang et al., the authors found that using clustered regularly interspaced short palindromic repeats interference (CRISPRi) to perturb expression of MERVL copies resulted in preimplantation defects in mice (Yang et al. 2024). Interestingly, in human development, there are many TE sequences, including various LTR retrotransposons like LTR3B, LTR5, LTR7B, LTR12C, LTR14B, and non-LTR retrotransposons like L1, SVA, and Alu elements that are expressed at different points throughout early embryo development (Goke and Ng 2016; Goke et al. 2015; Ru Ma and Liu 2025; Gao et al. 2018; Oomen et al. 2025) (Fig. 7.2). Taken together, these data indicate this is still an active area of investigation, but TEs seem to play a significant role in stem cells at multiple stages of development in many species.

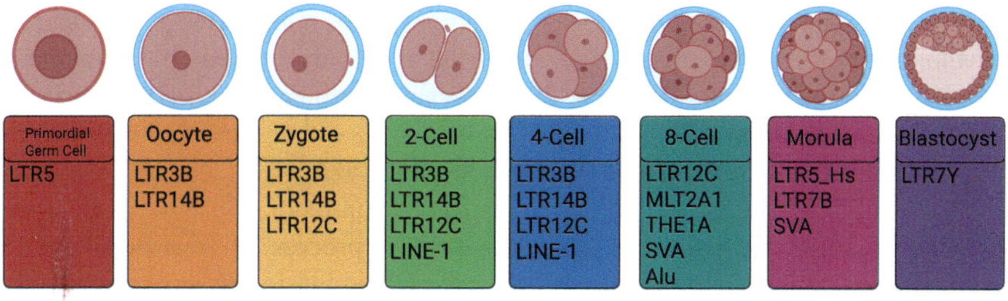

Fig. 7.2 Activity of TEs in early human development. TEs that are highly expressed in the early stages of human development are included in this figure, which shows the particular times of each element's expression. (Data in the figure was acquired from (Goke and Ng 2016; Goke et al. 2015; Ru Ma and Liu 2025; Gao et al. 2018; Oomen et al. 2025). Generated with Biorender)

7.3 Placental Development and Immune Tolerance

The placenta plays a crucial role in human development and is a unique organ that forms in the early stages of pregnancy. It acts as an intermediary between the mother and the fetus, allowing for the transfer of nutrients, gas exchange, and waste disposal (Burton and Fowden 2015). One function that some TEs serve in the placenta is gene expression regulation. The multiple integrations of ERVs in placental mammals, especially those that retain the ability to encode functional envelope proteins, have been critical in the formation of the syncytiotrophoblast through the fusion of cytotrophoblast (Imakawa et al. 2022). Although these TE-derived genes are no longer able to mobilize to new locations in the genome, their presence and expression patterns provide evidence that they have been co-opted for normal physiological purposes during developmental processes such as placentation. Interestingly, the process of placentation has evolved independently many times in vertebrates, including squamate reptiles, cartilaginous fishes, and teleost fishes, but only once in mammals, approximately 200 million years ago (Baker 2024). This aspect of development is a fantastic example of convergent evolution, where selective pressure to adapt a TE-derived envelope protein led to the formation of a multinucleated syncytial barrier between the embryo and the mother (Baker 2024).

In humans, mice, and other mammals, ERV expression is regulated via epigenetic mechanisms. For example, the RLTR13D5 ERV family has been shown to act as enhancer elements in the trophoblast (Chuong et al. 2013). The trophoblast is a thin layer of cells on the surface of the placenta that facilitates the attachment of the embryo to the uterine wall (Chuong et al. 2013). This fusion of the cytotrophoblast into the syncytiotrophoblast, a cell layer responsible for the exchange of nutrients and waste, is vital for the development of a functional placenta and thus a viable embryo (Renaud and Jeyarajah 2022). This process is facilitated by an open chromatin status and specific enhancers, which allow expression of the proteins syncytin-1 and syncytin-2, envelope proteins derived from open reading frames of immobile ERVs that promote cell fusion (Blaise et al. 2003; Wang et al. 2023; Trejbalova et al. 2011; Prudhomme et al. 2004; Yu et al. 2002). In the study by Bu et al., it was found that the depletion of syncytin-1 led to impaired angiogenesis, the development of new blood vessels to support fetal growth, and the establishment of circulation in the placenta (Bu et al. 2022). Syncytin proteins are highly conserved in all placental mammals, highlighting the importance that these proteins play in placental development in multiple species (Mi et al. 2000).

Placental development occurs over a period of approximately 8 to 12 weeks in the first trimester of pregnancy and precedes the process of fetal development (Burton and Jauniaux 2018). During this process, the blastocyst implants into the uterine endometrium via the trophoectoderm cells adjacent to the inner cell mass (ICM) (James et al. 2012). The trophoectoderm produces the syncytium (SCT), which emerges and spreads as it develops, including the rapid proliferation of cytotrophoblasts, which will form finger-like projections that penetrate the SCT (James et al. 2012). In order for these developmental processes to occur in a short span of time, phenotypic plasticity is necessary, and this plasticity is facilitated by rapidly changing epigenetic marks and processes, which include methylation, regulation via long noncoding RNAs, and changes to histone marks (Apicella et al. 2019). One TE-related instance of this phenomenon comes from the transposon-like human element 1B (THE1B), an ERV. THE1B contains an LTR, which appears to act as a promoter and enhancer on genes in close proximity (1–20 kb upstream) (Edginton-White et al. 2019). There is also evidence that alteration to THE1B methylation can alter the health of the placenta, leading to preeclampsia and intrauterine growth restriction (Norton et al. 2023). Further, THE1B is associated with gestational length through its ability to control corticotropin-releasing hormone (Dunn-Fletcher et al. 2018). However, the role of THE1B is still highly contested, and some groups have suggested it may play no role in human pregnancy. For instance, Zuo et al. showed that ZNF430, a zinc finger gene, may have evolved as a mediator that works to completely silence THE1B activity in the cell (Zuo 2023). This contradicts the findings by Dunn-Fletcher et al., suggesting we may not fully understand the role of THE1B in the context of development (Dunn-Fletcher et al. 2018; Zuo 2023). Another TE-derived gene is paternally expressed gene 10 (PEG10), which seems to play a role in the function of the trophoblast cells. While the entire mechanism has not been elucidated, studies have found that a decrease in expression of PEG10 correlates with an increased rate in the development of early-onset preeclampsia, suggesting its importance in proper placental function (Baird et al. 2023; Chen et al. 2015; Ono et al. 2006). THE1B and PEG10 are not the only TE-derived genes implicated in preeclampsia. Two recent studies from Patel et al. and Frost et al. indicated that some LTRs, ERVs, LINEs, and SINEs are upregulated during preeclampsia and may act as enhancers to activate nearby genes (Frost et al. 2023; Patel et al. 2024). Research is ongoing to further clarify the role of these TEs and their derivative proteins in preeclampsia as well as the overall development and formation of the placenta.

The placenta plays a crucial role in managing the movement of waste and nutrients between mother and fetus. However, the placenta also provides a necessary barrier between the mother's immune system and the developing fetus. The adaptive immune system is critical for survival and works by identifying foreign agents that enter the body (Chi et al. 2024). The immune system must have a method of detecting proteins, nucleic acids, and other materials, which are native to the organism versus those which have entered the body from the external environment. This allows for the identification of antigens by the identification of non-native proteins and generates signals for their destruction (Chaplin 2010). This system protects the body against bacteria, viruses, parasites, and other infectious agents, but it could become problematic in case of pregnancy (Vento-Tormo et al. 2018). Since the adaptive immune system targets proteins that are not native to the body, this puts both the placenta, which expresses both maternal and fetal genes, and the fetus at risk for detection and elimination by the immune system (Sood et al. 2006). Thus, it is necessary for the mother's immune system, the placenta, and the fetus to enter a complex state of immune tolerance to allow for the embryo to remain viable.

Immune tolerance of the developing fetus is key to successful reproduction in a species. For this immune tolerance to occur, several pathways work in tandem, likely not only to assure successful protection of the developing fetus, but also as a means of providing genetic redundancy. One of the commonly used mechanisms for fetal protection is a reduction of chemokine recruitment of leukocytes generated in the thymus (T cells) into the uterine tissue surrounding the fetus and the placenta, which has been shown in mice (Nancy et al. 2012). TE expression could be one method of achieving immune tolerance in the placenta, and several TEs have been implicated in this process. In mice, two ERV-derived proteins named syncytin-A and syncytin-B are expressed in the murine placenta, have been demonstrated to promote cell-cell fusion in the syncytiotrophoblast, and are essential to placental formation (Dupressoir et al. 2005). In humans, there are orthologous ERV-derived proteins, syncytin-1 and syncytin-2, which facilitate similar processes during placental development (Dupressoir et al. 2012). Further, syncytin-2 and a subtype of HERV-K, the human endogenous mouse mammary tumor virus (MMTV) MMTV-like 2 (HML-2), possess immunosuppressive domains in their envelope proteins and could contribute to preventing maternal rejection of the placenta (Fig. 7.3) (Roberts et al. 2021; Denner 2016).

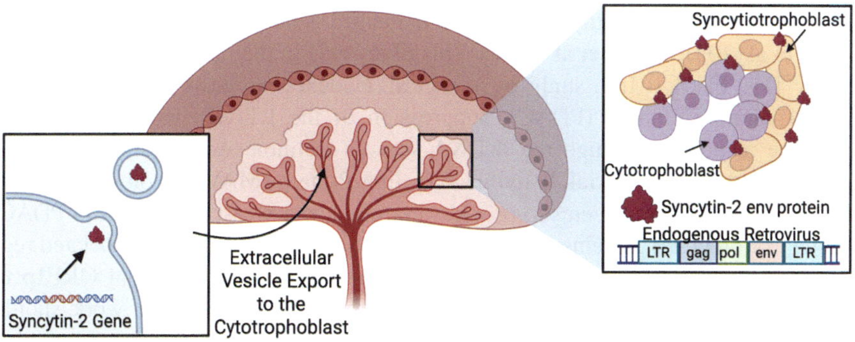

Fig. 7.3 Fusion of the cytotrophoblast to syncytiotrophoblast, mediated by syncytin-2. One of the key steps in the generation of a maternal-fetal interface in the placenta is the fusion of the cytotrophoblast to the syncytiotrophoblast. This is mediated through the activity of syncytin-2 (HERV-FRD) a TE-derived protein that uses its ancestral viral envelope to aid in cell fusion. (Generated with Biorender)

Syncytin-1 and syncytin-2 are TE-derived proteins that act to facilitate the cellular fusion necessary in the formation of the placenta. Outside of the placenta and periods of early development, TEs and the transcripts for TE-derived proteins are normally repressed through DNA methylation (Jangam et al. 2017). Syncytin-1 originated from a HERV-W locus and acts in the formation of the syncytiophoblast—the outer layer of the placenta, which is essential for successful fetal development (Renaud and Jeyarajah 2022). Syncytin-2 shares many similarities with syncytin-1. Syncytin-2 was derived from HERV-FRD, an older ERV, and also aids in the formation of the syncytiotrophoblast through the export of EVs into the cytotrophoblast (Levine et al. 2020; Blaise et al. 2003). HML-2 has open reading frames for proteins (Env and Rec) that seem to act in the same manner as both syncytins, enabling both fusogenic and immunosuppressing abilities in the placenta (Frank et al. 2022; Malassine et al. 2008; Hohn et al. 2013; Jones et al. 2012). Although there is much evidence that TE-derived proteins play a role in some aspects of placental formation, there is mounting evidence that there are multiple steps to initiating placental development including the following: (1) establishing cellular competence as cytotrophoblast cells exit the mitotic cycle and differentiate into fusion -competent cells, (2) cellular commitment characterized by intercellular adhesion and assembly of fusogenic machinery, and (3) cell fusion mediated by syncytin-1 and syncytin-2 which allow the opening of the fusion pore and complete membrane merging (Li et al. 2023). Placental development is a complex process that relies on many pathways functioning together to complete the formation of a new organ vital for successful pregnancy. Syncytin-1, syncytin-2, HML-2 env, and rec are not the only components necessary for this process to complete; however, they seem to have an essential role to play.

Another example of TE-derived modulation originates from the retrotransposon family L1 (Giorgi et al. 2018). L1 is a family of TEs that are ubiquitous in mammals, comprising vast swaths of most mammals' genomes—including in humans, where they make up 17% of the human genome (Giorgi et al. 2018; Lander et al. 2001). While L1 expression is normally repressed in most somatic tissues by epigenetic marks such as DNA methylation, in the placenta tissue, there is generally a lower level of DNA methylation across the entire genome (Schroeder et al. 2013; Yuan et al. 2021; Deniz et al. 2019). This likely aids in the increased expression of many TEs, like L1. In a study by You et al. evidence suggests that in pancreatic ductal adenocarcinoma (PDAC), suppression of the L1 ORF1p induces innate immune responses through the dsDNA sensors Retinoic Acid Inducible Gene I (RIG-I) and Mitochondrial Antiviral Signaling Protein (MAVs) (You et al. 2025). L1 ORF1p expression seems to provide an immunosuppressive function in PDAC, much like other TE proteins expressed in other tumors. The ORF1p-associated regulatory mechanisms in PDAC may be an example of the potential of ORF1p to provide an immunosuppressive function in other contexts. Further work should determine whether expression of ORF1p can provide protection from innate immune activation in the placenta.

7.4 Transposable Elements and Early Nervous System Development

The nervous system is a critical part of the body that acts as a control center. This system is split into a central nervous system (CNS) which consists of the brain and spinal cord, and the peripheral nervous system (PNS) connects disparate regions of the body to the CNS through a network of nerves. Both of these systems cooperate together to regulate all bodily functions (Hua and Smith 2004). The formation of the CNS starts early after embryo implantation, specifically around week three or four in humans (Nikolopoulou et al. 2017). As with much of the developmental process, it appears that TEs may play a crucial role in CNS development (Hua and Smith 2004).

TEs are critical in CNS development in large part due to their ability to function as regulatory elements (Reilly et al. 2013). In this role, TEs provide a DNA sequence adjacent to the gene of interest that allows for the binding of transcription factors such as activators or repressors, resulting in the modulation of gene expression (Gebrie 2023; Reilly et al. 2013). TEs can alter gene expression by acting as external alternative promoters to adjacent genes, often resulting in upregulation of the adjacent gene (Fueyo et al. 2022). TEs are also known to modify mRNAs through alternative polyadenylation or alternative splicing (Lee et al. 2008).

A key example of TE regulation during nervous system development occurs in the mouse neocortex via expression of the MER130 family (Notwell et al. 2015). MER130 transcripts, as MER130 is a noncoding DNA transposon, are highly enriched in the dorsal cerebral wall—a region that eventually gives rise to the development of the neocortex (Saito et al. 2019). MER130 is a highly conserved TE that appears to have originated in a common ancestor before the rise of tetrapods—the superclass that includes humans. Thus, MER130 integrated into the genome in a period of evolution that predates the evolution of the neocortex in mammals (Notwell et al. 2015; Bejerano et al. 2006). The conservation of MER130 in species with a neocortex could be suggestive of an evolutionary pressure to retain this gene's function, thus supporting the claim that MER130 has been co-opted by the body for the regulation of early nervous system development. MER130 repeats are strongly enriched in E14.5 (14.5-day-old murine embryos) at enhancers active in the murine neocortex as it develops and are adjacent to important cortical genes that are vital in development (Notwell et al. 2015). In an early study of *Xenopus* development, Faunes et al. found localized expression of DNA transposons from the Tc1-elements in neural tissues during development, suggesting they could play a role in nervous system development (Faunes et al. 2011). Taken together, this work suggests that DNA transposons also contribute to nervous system development in many different species.

TEs like L1 are usually transcriptionally silenced, preventing disruptions and insertional mutations in the genome. In an in vitro study in human neural progenitor cells—cells crucial for development of the early nervous system—were treated with a demethylating agent, and resultant L1 activation correlated with the upregulation of neuronal genes (Jonsson et al. 2019). Genes involved in neuronal differentiation and synaptic function were particularly affected by the activation of L1. When demethylation was reversed, expression of neuronal genes remained enhanced (Jonsson et al. 2019). In a study in mice, Mangoni et al. demonstrated that L1 expression was required for mouse brain corticogenesis as their sequences seem to operate as regulatory lncRNAs (Mangoni et al. 2023). These regulatory lncRNAs help to regulate post-mitotic neuron migration and thereby contribute to establishing varying proportions of different neuronal cell types (Mangoni et al. 2023). The results of another recent study support the idea that L1 and L1-chimeric transcripts play a role in neural differentiation programs by using CRISPRi downregulation and observing significant changes in neural differentiation programs and reduced cerebral organoid size (Adami et al. 2025). In an investigation by Garza et al., human fetal forebrain samples were analyzed with a multiomics approach and the authors surprisingly observed that evolutionarily young L1 elements were expressed at levels near housekeeping genes during forebrain development (Garza et al. 2023). The authors suggest that these L1-derived transcripts could provide a previously unappreciated source of primate and human specific transcriptional complexity during brain development which contributes to functional diversification (Garza et al. 2023). Taken together, these studies reinforce that L1 expression plays a role in the nervous system and specifically brain development in a multitude of ways. Further, these studies demonstrate that L1 transcription during brain development has also been observed in other organisms. This could suggest that TE and TE-derived transcripts and their expression in nervous system development could be a conserved mechanism to generate functional diversification.

A transposon-derived element, the PiggyBac Transposable Element Derived 5 (PGBD5), has recently been connected to normal brain development, although its expression in humans was first observed in childhood solid tumors, where it appeared to lead to oncogenic rearrangements due to its nuclease activity (Zapater et al. 2024). PGBD5 is a highly evolutionarily conserved gene derived from a vertebrate DNA transposon (Henssen et al. 2015). Due to its pervasive expression in nervous system tissues and the shared neuroectodermal developmental origin of PGBD5-expressing pediatric tumors, Zapater et al. investigated its potential role in neurodevelopment (Zapater et al. 2024). Five unrelated consanguineous families with PGDB5 mutations were exome sequenced, and the mutations segregated with family members affected with clinical phenotypes of neurodevelopmental and motor dysfunction (Zapater et al. 2024). Using PGDB5 knockout mice, they observed intellectual disability, disordered movement, and seizures (Zapater et al. 2024). In the brain cortex, loss of PGBD5 leads to abnormal gene expression and differentiation. These findings suggest that PGBD5 may be required for proper brain development in mammals (Zapater et al. 2024; Yamada et al. 2024).

Although TEs play many important roles in CNS development, elements that retain the ability to retrotranspose into new locations must still be tightly controlled. An active TE can insert into and disrupt a gene, potentially causing a frameshift mutation and leading to aberrant development (Ayarpadikannan and Kim 2014; Ilik et al. 2024). Apolipoprotein B mRNA editing catalytic polypeptide (APOBEC) proteins were first identified as restriction factors for human immunodeficiency virus (HIV) (Sheehy et al. 2002). APOBEC3B is highly expressed in human pluripotent stem cells and silences L1 during cellular differentiation (Marchetto et al. 2013). Although there are several different APOBEC3 proteins, it seems they all inhibit L1 retrotransposons; further, Alu elements are restricted by APOBEC3F and APOBEC3G (Muckenfuss et al. 2006; Salter et al. 2016; Khatua et al. 2010). Other systems responsible for guarding against active mobilization of TEs during development are chromatin-associated proteins, such as DNA (cytosine-5) methyltransferase 3A (DNMT3A), DNA methyltransferase 3B (DNMT3B), and methyl-CpG-binding proteins (MeCP) (Goodier 2016). Patients with Rett syndrome, a rare neurodevelopmental disorder caused by mutations in the gene *methyl-CpG-binding protein-2 (MECP2),* have abnormally high retrotransposition activity of L1 in their brain (Muotri et al. 2010; Skene et al. 2010). Neurodevelopment and nervous system development are complex processes with many genetic pathways, which must be regulated with precision to result in functional tissues and organisms. Although it seems that some TE expression may play a key role in neurodevelopment and nervous system development, it is still essential that TE expression be tightly regulated throughout development so that neurodevelopmental disorders and tumors do not occur.

7.5 Conclusion

As demonstrated throughout this chapter, TEs play an important role in the developmental outcomes of an organism. Though once thought of as mere "junk DNA," through the years the significance of TEs has become clearer. From their role in the placenta to stem cell differentiation and nervous system development, TEs play many key roles across development. Over evolutionary history, TEs have been co-opted into vital mechanisms for gene regulation. As seen with HML-2 and the syncytin proteins, viral elements can be co-opted for uses that become integral in development. While it seems that TEs have been found to play an outsized role in development, there is still much to be learned. We do not know the full repertoire of roles these elements play, but as sequencing technologies and spatial transcriptional mapping become more accurate, we are better able to pinpoint the expressed loci in the genome in specific cell types throughout stages of development. Understanding the regulation of TE transcription in different physiological and pathological contexts will allow us to further elucidate their role.

Due to the inability to perform many of these developmental studies on human samples, the use of model organisms has been a vital component of developmental research. Model organisms provide an accessible, easily genetically manipulated, and ethical way to perturb normal processes to understand developmental consequences. Many mechanisms of genetic programming are highly conserved between organisms, including humans, suggesting a shared evolutionary history even in the presence of divergent evolution mechanisms and physical form (Wagner 2007).

Over the years, researchers have utilized *Drosophila*, mice, chickens, *Caenorhabditis elegans*, *Xenopus*, and other organisms to study developmental processes to understand normal physiology as well as developmental pathology. In mice, both syncytin-A and syncytin-B have been identified as murine-specific fusogenic placenta-specific envelope genes that originated from retroviruses (Dupressoir et al. 2005). In *Drosophila*, the Arc1 protein, which originated from a retrotransposon, has been shown to transfer mRNA between cells and is involved in synaptic plasticity during development (Erlendsson et al. 2020; Mattaliano et al. 2007). However, in some commonly studied developmental models, such as the chicken (*gallus gallus*), it is currently unknown how ERVs contribute to their developmental process, even though it was in this species that ERVs were discovered (Weiss 2006). It is also paramount that we investigate the spatial orientation of relevant ERV proteins to understand their roles in development. As data continues to be collected, we need to share it equitably and quickly, to create a critical mass of information that can reveal further roles TEs may play in physiology and pathological conditions.

Acknowledgments John R. McCoy and Dr. Tara T Doucet-O'Hare wrote and edited this chapter together. The text of this chapter was refined with the assistance of artificial intelligence (AI) tools—OpenAI's ChatGPT and GPT-4-turbo, used in May of 2025. Every section was carefully reviewed, revised, and approved by the author(s) to ensure that the text is appropriate concerning content and scientific correctness and to maintain the integrity of the original content.

Funding This study was funded by start-up funds provided to Dr. Tara T Doucet-O'Hare.

Compliance with Ethical Standards

Conflict of Interest Dr. Tara T Doucet-O'Hare declares that she has no conflict of interest. John R. McCoy declares that he has no conflict of interest.

Ethical Approval This chapter is a review of previously published literature; as such, no animal or human studies were performed.

References

Adami A, Garza R, Gerdes P, Johansson PA, Dorazehi F, Koutounidou S, Castilla-Vallmanya L, Atacho DAM, Sharma Y, Johansson JG, Tam O, Kirkeby A, Barker RA, Gale-Hammell M, Douse CH, Jakobsson J (2025) LINE-1 retrotransposons regulate the exit of human pluripotency and early brain development. bioRxiv:202520012017633315. https://doi.org/10.1101/2025.01.17.633315

Apicella C, Ruano CSM, Mehats C, Miralles F, Vaiman D (2019) The role of epigenetics in placental development and the etiology of Preeclampsia. Int J Mol Sci 20(11). https://doi.org/10.3390/ijms20112837

Ashley J, Cordy B, Lucia D, Fradkin LG, Budnik V, Thomson T (2018) Retrovirus-like Gag Protein Arc1 Binds RNA and traffics across synaptic Boutons. Cell 172(1–2):262–274 e211. https://doi.org/10.1016/j.cell.2017.12.022

Ayarpadikannan S, Kim HS (2014) The impact of transposable elements in genome evolution and genetic instability and their implications in various diseases. Genom Inform 12(3):98–104. https://doi.org/10.5808/GI.2014.12.3.98

Baird L, Cannon P, Kandel M, Nguyen TV, Nguyen A, Wong G, Murphy C, Brownfoot FC, Kadife E, Hannan NJ, Tong S, Bartho LA, Kaitu'u-Lino TJ (2023) Paternal Expressed Gene 10 (PEG10) is decreased in early-onset preeclampsia. Reprod Biol Endocrinol 21(1):65. https://doi.org/10.1186/s12958-023-01116-3

Baker JC (2024) Placental evolution: innovating how to feed babies. Annu Rev Genet 58(1):391–408. https://doi.org/10.1146/annurev-genet-111523-102135

Bejerano G, Lowe CB, Ahituv N, King B, Siepel A, Salama SR, Rubin EM, Kent WJ, Haussler D (2006) A distal enhancer and an ultraconserved exon are derived from a novel retroposon. Nature 441(7089):87–90. https://doi.org/10.1038/nature04696

Blaise S, de Parseval N, Benit L, Heidmann T (2003) Genomewide screening for fusogenic human endogenous retrovirus envelopes identifies syncytin 2, a gene conserved on primate evolution. Proc Natl Acad Sci USA 100(22):13013–13018. https://doi.org/10.1073/pnas.2132646100

Bourque G, Leong B, Vega VB, Chen X, Lee YL, Srinivasan KG, Chew JL, Ruan Y, Wei CL, Ng HH, Liu ET (2008) Evolution of the mammalian transcription factor binding repertoire via transposable elements. Genome Res 18(11):1752–1762. https://doi.org/10.1101/gr.080663.108

Bu C, Wang Z, Ren Y, Chen D, Jiang SW (2022) Syncytin-1 nonfusogenic activities modulate inflammation and contribute to preeclampsia pathogenesis. Cell Mol Life Sci 79(6):290. https://doi.org/10.1007/s00018-022-04294-2

Burton GJ, Fowden AL (2015) The placenta: a multifaceted, transient organ. Philos Trans R Soc Lond Ser B Biol Sci 370(1663):20140066. https://doi.org/10.1098/rstb.2014.0066

Burton GJ, Jauniaux E (2018) Development of the human placenta and fetal heart: synergic or independent? Front Physiol 9:373. https://doi.org/10.3389/fphys.2018.00373

Cao X, Zhang Y, Payer LM, Lords H, Steranka JP, Burns KH, Xing J (2020) Polymorphic mobile element insertions contribute to gene expression and alternative splicing in human tissues. Genome Biol 21(1):185. https://doi.org/10.1186/s13059-020-02101-4

Chang NC, Wells JN, Wang AY, Schofield P, Huang YC, Truong VH, Simoes-Costa M, Feschotte C (2025) Gag proteins encoded by endogenous retroviruses are required for zebrafish development. Proc Natl Acad Sci USA 122(18):e2411446122. https://doi.org/10.1073/pnas.2411446122

Chaplin DD (2010) Overview of the immune response. J Allergy Clin Immunol 125(2 Suppl 2):S3–23. https://doi.org/10.1016/j.jaci.2009.12.980

Chau J, Kong X, Viet Nguyen N, Williams K, Ball M, Tawil R, Kiyono T, Mortazavi A, Yokomori K (2021) Relationship of DUX4 and target gene expression in FSHD myocytes. Hum Mutat 42(4):421–433. https://doi.org/10.1002/humu.24171

Chen H, Sun M, Liu J, Tong C, Meng T (2015) Silencing of paternally expressed gene 10 inhibits trophoblast proliferation and invasion. PLoS One 10(12):e0144845. https://doi.org/10.1371/journal.pone.0144845

Chi H, Pepper M, Thomas PG (2024) Principles and therapeutic applications of adaptive immunity. Cell 187(9):2052–2078. https://doi.org/10.1016/j.cell.2024.03.037

Chuong EB, Rumi MA, Soares MJ, Baker JC (2013) Endogenous retroviruses function as species-specific enhancer elements in the placenta. Nat Genet 45(3):325–329. https://doi.org/10.1038/ng.2553

Crichton JH, Dunican DS, Maclennan M, Meehan RR, Adams IR (2014) Defending the genome from the enemy within: mechanisms of retrotransposon suppression in the mouse germline. Cell Mol Life Sci 71(9):1581–1605. https://doi.org/10.1007/s00018-013-1468-0

De Iaco A, Planet E, Coluccio A, Verp S, Duc J, Trono D (2017) DUX-family transcription factors regulate zygotic genome activation in placental mammals. Nat Genet 49(6):941–945. https://doi.org/10.1038/ng.3858

de la Rosa S, Del Mar RM, Vargiu P, Ortega S, Djouder N (2024) Endogenous retroviruses shape pluripotency specification in mouse embryos. Sci Adv 10(4):eadk9394. https://doi.org/10.1126/sciadv.adk9394
de Souza FS, Franchini LF, Rubinstein M (2013) Exaptation of transposable elements into novel cis-regulatory elements: is the evidence always strong? Mol Biol Evol 30(6):1239–1251. https://doi.org/10.1093/molbev/mst045
Deniz O, Frost JM, Branco MR (2019) Regulation of transposable elements by DNA modifications. Nat Rev Genet 20(7):417–431. https://doi.org/10.1038/s41576-019-0106-6
Denner J (2016) Expression and function of endogenous retroviruses in the placenta. APMIS 124(1–2):31–43. https://doi.org/10.1111/apm.12474
Donkin I, Barres R (2018) Sperm epigenetics and influence of environmental factors. Mol Metab 14:1–11. https://doi.org/10.1016/j.molmet.2018.02.006
Doolittle WF, Sapienza C (1980) Selfish genes, the phenotype paradigm and genome evolution. Nature 284(5757):601–603. https://doi.org/10.1038/284601a0
Dopkins N, Nixon DF (2024) Activation of human endogenous retroviruses and its physiological consequences. Nat Rev Mol Cell Biol 25(3):212–222. https://doi.org/10.1038/s41580-023-00674-z
Doucet-O'Hare TT, DiSanza BL, DeMarino C, Atkinson AL, Rosenblum JS, Henderson LJ, Johnson KR, Kowalak J, Garcia-Montojo M, Allen SJ, Orr BA, Santi M, Wang T, Fathi S, Lee MH, Sampson K, Li W, Zhuang Z, Nath A (2021) SMARCB1 deletion in atypical teratoid rhabdoid tumors results in human endogenous retrovirus K (HML-2) expression. Sci Rep 11(1):12893. https://doi.org/10.1038/s41598-021-92223-x
Doucet-O'Hare TT, Rodic N, Sharma R, Darbari I, Abril G, Choi JA, Young Ahn J, Cheng Y, Anders RA, Burns KH, Meltzer SJ, Kazazian HH Jr (2015) LINE-1 expression and retrotransposition in Barrett's esophagus and esophageal carcinoma. Proc Natl Acad Sci USA 112(35):E4894–E4900. https://doi.org/10.1073/pnas.1502474112
Dunn-Fletcher CE, Muglia LM, Pavlicev M, Wolf G, Sun MA, Hu YC, Huffman E, Tumukuntala S, Thiele K, Mukherjee A, Zoubovsky S, Zhang X, Swaggart KA, Lamm KYB, Jones H, Macfarlan TS, Muglia LJ (2018) Anthropoid primate-specific retroviral element THE1B controls expression of CRH in placenta and alters gestation length. PLoS Biol 16(9):e2006337. https://doi.org/10.1371/journal.pbio.2006337
Dupressoir A, Lavialle C, Heidmann T (2012) From ancestral infectious retroviruses to bona fide cellular genes: role of the captured syncytins in placentation. Placenta 33(9):663–671. https://doi.org/10.1016/j.placenta.2012.05.005
Dupressoir A, Marceau G, Vernochet C, Benit L, Kanellopoulos C, Sapin V, Heidmann T (2005) Syncytin-A and syncytin-B, two fusogenic placenta-specific murine envelope genes of retroviral origin conserved in Muridae. Proc Natl Acad Sci USA 102(3):725–730. https://doi.org/10.1073/pnas.0406509102
Ecco G, Cassano M, Kauzlaric A, Duc J, Coluccio A, Offner S, Imbeault M, Rowe HM, Turelli P, Trono D (2016) Transposable elements and their KRAB-ZFP controllers regulate gene expression in adult tissues. Dev Cell 36(6):611–623. https://doi.org/10.1016/j.devcel.2016.02.024
Eddy EM (1975) Germ plasm and the differentiation of the germ cell line. Int Rev Cytol 43:229–280. https://doi.org/10.1016/s0074-7696(08)60070-4
Edginton-White B, Cauchy P, Assi SA, Hartmann S, Riggs AG, Mathas S, Cockerill PN, Bonifer C (2019) Global long terminal repeat activation participates in establishing the unique gene expression programme of classical Hodgkin lymphoma. Leukemia 33(6):1463–1474. https://doi.org/10.1038/s41375-018-0311-x
Emera D, Casola C, Lynch VJ, Wildman DE, Agnew D, Wagner GP (2012) Convergent evolution of endometrial prolactin expression in primates, mice, and elephants through the independent recruitment of transposable elements. Mol Biol Evol 29(1):239–247. https://doi.org/10.1093/molbev/msr189

Emera D, Wagner GP (2012) Transformation of a transposon into a derived prolactin promoter with function during human pregnancy. Proc Natl Acad Sci USA 109(28):11246–11251. https://doi.org/10.1073/pnas.1118566109
Erlendsson S, Morado DR, Cullen HB, Feschotte C, Shepherd JD, Briggs JAG (2020) Structures of virus-like capsids formed by the Drosophila neuronal Arc proteins. Nat Neurosci 23(2):172–175. https://doi.org/10.1038/s41593-019-0569-y
Faunes F, Sanchez N, Moreno M, Olivares GH, Lee-Liu D, Almonacid L, Slater AW, Norambuena T, Taft RJ, Mattick JS, Melo F, Larrain J (2011) Expression of transposable elements in neural tissues during Xenopus development. PLoS One 6(7):e22569. https://doi.org/10.1371/journal.pone.0022569
Flemr M, Malik R, Franke V, Nejepinska J, Sedlacek R, Vlahovicek K, Svoboda P (2013) A retrotransposon-driven dicer isoform directs endogenous small interfering RNA production in mouse oocytes. Cell 155(4):807–816. https://doi.org/10.1016/j.cell.2013.10.001
Frank JA, Singh M, Cullen HB, Kirou RA, Benkaddour-Boumzaouad M, Cortes JL, Garcia Perez J, Coyne CB, Feschotte C (2022) Evolution and antiviral activity of a human protein of retroviral origin. Science 378(6618):422–428. https://doi.org/10.1126/science.abq7871
Fromer M, Pocklington AJ, Kavanagh DH, Williams HJ, Dwyer S, Gormley P, Georgieva L, Rees E, Palta P, Ruderfer DM, Carrera N, Humphreys I, Johnson JS, Roussos P, Barker DD, Banks E, Milanova V, Grant SG, Hannon E, Rose SA, Chambert K, Mahajan M, Scolnick EM, Moran JL, Kirov G, Palotie A, McCarroll SA, Holmans P, Sklar P, Owen MJ, Purcell SM, O'Donovan MC (2014) De novo mutations in schizophrenia implicate synaptic networks. Nature 506(7487):179–184. https://doi.org/10.1038/nature12929
Frost JM, Amante SM, Okae H, Jones EM, Ashley B, Lewis RM, Cleal JK, Caley MP, Arima T, Maffucci T, Branco MR (2023) Regulation of human trophoblast gene expression by endogenous retroviruses. Nat Struct Mol Biol 30(4):527–538. https://doi.org/10.1038/s41594-023-00960-6
Fueyo R, Judd J, Feschotte C, Wysocka J (2022) Roles of transposable elements in the regulation of mammalian transcription. Nat Rev Mol Cell Biol 23(7):481–497. https://doi.org/10.1038/s41580-022-00457-y
Galli UM, Sauter M, Lecher B, Maurer S, Herbst H, Roemer K, Mueller-Lantzsch N (2005) Human endogenous retrovirus rec interferes with germ cell development in mice and may cause carcinoma in situ, the predecessor lesion of germ cell tumors. Oncogene 24(19):3223–3228. https://doi.org/10.1038/sj.onc.1208543
Gao L, Wu K, Liu Z, Yao X, Yuan S, Tao W, Yi L, Yu G, Hou Z, Fan D, Tian Y, Liu J, Chen ZJ, Liu J (2018) Chromatin accessibility landscape in human early embryos and its association with evolution. Cell 173(1):248–259 e215. https://doi.org/10.1016/j.cell.2018.02.028
Garcia-Perez JL, Widmann TJ, Adams IR (2016) The impact of transposable elements on mammalian development. Development 143(22):4101–4114. https://doi.org/10.1242/dev.132639
Garza R, Atacho DAM, Adami A, Gerdes P, Vinod M, Hsieh P, Karlsson O, Horvath V, Johansson PA, Pandiloski N, Matas-Fuentes J, Quaegebeur A, Kouli A, Sharma Y, Jonsson ME, Monni E, Englund E, Eichler EE, Gale Hammell M, Barker RA, Kokaia Z, Douse CH, Jakobsson J (2023) LINE-1 retrotransposons drive human neuronal transcriptome complexity and functional diversification. Sci Adv 9(44):eadh9543. https://doi.org/10.1126/sciadv.adh9543
Gebrie A (2023) Transposable elements as essential elements in the control of gene expression. Mob DNA 14(1):9. https://doi.org/10.1186/s13100-023-00297-3
Genebrier S, Tarte K (2021) The flawless immune tolerance of pregnancy. Joint Bone Spine 88(5):105205. https://doi.org/10.1016/j.jbspin.2021.105205
Georgiou I, Noutsopoulos D, Dimitriadou E, Markopoulos G, Apergi A, Lazaros L, Vaxevanoglou T, Pantos K, Syrrou M, Tzavaras T (2009) Retrotransposon RNA expression and evidence for retrotransposition events in human oocytes. Hum Mol Genet 18(7):1221–1228. https://doi.org/10.1093/hmg/ddp022
Ghazimoradi MH, Khalafizadeh A, Babashah S (2022) A critical review on induced totipotent stem cells: types and methods. Stem Cell Res 63:102857. https://doi.org/10.1016/j.scr.2022.102857

Giorgi G, Virgili M, Monti B, Del Re B (2018) Long INterspersed nuclear Elements (LINEs) in brain and non-brain tissues of the rat. Cell Tissue Res 374(1):17–24. https://doi.org/10.1007/s00441-018-2843-9
Goke J, Lu X, Chan YS, Ng HH, Ly LH, Sachs F, Szczerbinska I (2015) Dynamic transcription of distinct classes of endogenous retroviral elements marks specific populations of early human embryonic cells. Cell Stem Cell 16(2):135–141. https://doi.org/10.1016/j.stem.2015.01.005
Goke J, Ng HH (2016) CTRL+INSERT: retrotransposons and their contribution to regulation and innovation of the transcriptome. EMBO Rep 17(8):1131–1144. https://doi.org/10.15252/embr.201642743
Goodier JL (2016) Restricting retrotransposons: a review. Mob. DNA 7:16. https://doi.org/10.1186/s13100-016-0070-z
Grow EJ, Flynn RA, Chavez SL, Bayless NL, Wossidlo M, Wesche DJ, Martin L, Ware CB, Blish CA, Chang HY, Pera RA, Wysocka J (2015) Intrinsic retroviral reactivation in human preimplantation embryos and pluripotent cells. Nature 522(7555):221–225. https://doi.org/10.1038/nature14308
Hannon GJ (2002) RNA interference. Nature 418(6894):244–251. https://doi.org/10.1038/418244a
Hendrickson PG, Dorais JA, Grow EJ, Whiddon JL, Lim JW, Wike CL, Weaver BD, Pflueger C, Emery BR, Wilcox AL, Nix DA, Peterson CM, Tapscott SJ, Carrell DT, Cairns BR (2017) Conserved roles of mouse DUX and human DUX4 in activating cleavage-stage genes and MERVL/HERVL retrotransposons. Nat Genet 49(6):925–934. https://doi.org/10.1038/ng.3844
Henssen AG, Henaff E, Jiang E, Eisenberg AR, Carson JR, Villasante CM, Ray M, Still E, Burns M, Gandara J, Feschotte C, Mason CE, Kentsis A (2015) Genomic DNA transposition induced by human PGBD5. elife 4. https://doi.org/10.7554/eLife.10565
Hohn O, Hanke K, Bannert N (2013) HERV-K(HML-2), the best preserved family of HERVs: endogenization, expression, and implications in health and disease. Front Oncol 3:246. https://doi.org/10.3389/fonc.2013.00246
Hoyt SJ, Storer JM, Hartley GA, Grady PGS, Gershman A, de Lima LG, Limouse C, Halabian R, Wojenski L, Rodriguez M, Altemose N, Rhie A, Core LJ, Gerton JL, Makalowski W, Olson D, Rosen J, Smit AFA, Straight AF, Vollger MR, Wheeler TJ, Schatz MC, Eichler EE, Phillippy AM, Timp W, Miga KH, O'Neill RJ (2022) From telomere to telomere: the transcriptional and epigenetic state of human repeat elements. Science 376(6588):eabk3112. https://doi.org/10.1126/science.abk3112
Hua JY, Smith SJ (2004) Neural activity and the dynamics of central nervous system development. Nat Neurosci 7(4):327–332. https://doi.org/10.1038/nn1218
Huhtala L, Karabiyik G, Rautajoki KJ (2024) Development and epigenetic regulation of Atypical teratoid/rhabdoid tumors in the context of cell-of-origin and halted cell differentiation. Neurooncol Adv 6(1):vdae162. https://doi.org/10.1093/noajnl/vdae162
Ilik IA, Glazar P, Tse K, Brandl B, Meierhofer D, Muller FJ, Smith ZD, Aktas T (2024) Autonomous transposons tune their sequences to ensure somatic suppression. Nature 626(8001):1116–1124. https://doi.org/10.1038/s41586-024-07081-0
Imakawa K, Kusama K, Kaneko-Ishino T, Nakagawa S, Kitao K, Miyazawa T, Ishino F (2022) Endogenous Retroviruses and placental evolution, development, and diversity. Cells 11(15). https://doi.org/10.3390/cells11152458
Jabbour HN, Critchley HO (2001) Potential roles of decidual prolactin in early pregnancy. Reproduction 121(2):197–205. https://doi.org/10.1530/rep.0.1210197
James JL, Carter AM, Chamley LW (2012) Human placentation from nidation to 5 weeks of gestation. Part I: what do we know about formative placental development following implantation? Placenta 33(5):327–334. https://doi.org/10.1016/j.placenta.2012.01.020
Jangam D, Feschotte C, Betran E (2017) Transposable element domestication As an adaptation to evolutionary conflicts. Trends Genet 33(11):817–831. https://doi.org/10.1016/j.tig.2017.07.011
Jones RB, John VM, Hunter DV, Martin E, Mujib S, Mihajlovic V, Burgers PC, Luider TM, Gyenes G, Sheppard NC, Sengupta D, Tandon R, Yue FY, Benko E, Kovacs C, Nixon DF, Ostrowski MA (2012) Human endogenous retrovirus K(HML-2) Gag- and Env-specific T-cell

responses are infrequently detected in HIV-1-infected subjects using standard peptide matrix-based screening. Clin Vaccine Immunol 19(2):288–292. https://doi.org/10.1128/CVI.05583-11

Jonsson ME, Ludvik Brattas P, Gustafsson C, Petri R, Yudovich D, Pircs K, Verschuere S, Madsen S, Hansson J, Larsson J, Mansson R, Meissner A, Jakobsson J (2019) Activation of neuronal genes via LINE-1 elements upon global DNA demethylation in human neural progenitors. Nat Commun 10(1):3182. https://doi.org/10.1038/s41467-019-11150-8

Katoh H, Honda T (2023) Roles of human endogenous Retroviruses and endogenous virus-like elements in cancer development and innate immunity. Biomolecules 13(12). https://doi.org/10.3390/biom13121706

Khatua AK, Taylor HE, Hildreth JE, Popik W (2010) Inhibition of LINE-1 and Alu retrotransposition by exosomes encapsidating APOBEC3G and APOBEC3F. Virology 400(1):68–75. https://doi.org/10.1016/j.virol.2010.01.021

Lander ES, Linton LM, Birren B, Nusbaum C, Zody MC, Baldwin J, Devon K, Dewar K, Doyle M, Fitz Hugh W, Funke R, Gage D, Harris K, Heaford A, Howland J, Kann L, Lehoczky J, Le Vine R, Mc Ewan P, Mc Kernan K, Meldrim J, Mesirov JP, Miranda C, Morris W, Naylor J, Raymond C, Rosetti M, Santos R, Sheridan A, Sougnez C, Stange-Thomann Y, Stojanovic N, Subramanian A, Wyman D, Rogers J, Sulston J, Ainscough R, Beck S, Bentley D, Burton J, Clee C, Carter N, Coulson A, Deadman R, Deloukas P, Dunham A, Dunham I, Durbin R, French L, Grafham D, Gregory S, Hubbard T, Humphray S, Hunt A, Jones M, Lloyd C, Mc Murray A, Matthews L, Mercer S, Milne S, Mullikin JC, Mungall A, Plumb R, Ross M, Shownkeen R, Sims S, Waterston RH, Wilson RK, Hillier LW, McPherson JD, Marra MA, Mardis ER, Fulton LA, Chinwalla AT, Pepin KH, Gish WR, Chissoe SL, Wendl MC, Delehaunty KD, Miner TL, Delehaunty A, Kramer JB, Cook LL, Fulton RS, Johnson DL, Minx PJ, Clifton SW, Hawkins T, Branscomb E, Predki P, Richardson P, Wenning S, Slezak T, Doggett N, Cheng JF, Olsen A, Lucas S, Elkin C, Uberbacher E, Frazier M, Gibbs RA, Muzny DM, Scherer SE, Bouck JB, Sodergren EJ, Worley KC, Rives CM, Gorrell JH, Metzker ML, Naylor SL, Kucherlapati RS, Nelson DL, Weinstock GM, Sakaki Y, Fujiyama A, Hattori M, Yada T, Toyoda A, Itoh T, Kawagoe C, Watanabe H, Totoki Y, Taylor T, Weissenbach J, Heilig R, Saurin W, Artiguenave F, Brottier P, Bruls T, Pelletier E, Robert C, Wincker P, Smith DR, Doucette-Stamm L, Rubenfield M, Weinstock K, Lee HM, Dubois J, Rosenthal A, Platzer M, Nyakatura G, Taudien S, Rump A, Yang H, Yu J, Wang J, Huang G, Gu J, Hood L, Rowen L, Madan A, Qin S, Davis RW, Federspiel NA, Abola AP, Proctor MJ, Myers RM, Schmutz J, Dickson M, Grimwood J, Cox DR, Olson MV, Kaul R, Raymond C, Shimizu N, Kawasaki K, Minoshima S, Evans GA, Athanasiou M, Schultz R, Roe BA, Chen F, Pan H, Ramser J, Lehrach H, Reinhardt R, McCombie WR, de la Bastide M, Dedhia N, Blocker H, Hornischer K, Nordsiek G, Agarwala R, Aravind L, Bailey JA, Bateman A, Batzoglou S, Birney E, Bork P, Brown DG, Burge CB, Cerutti L, Chen HC, Church D, Clamp M, Copley RR, Doerks T, Eddy SR, Eichler EE, Furey TS, Galagan J, Gilbert JG, Harmon C, Hayashizaki Y, Haussler D, Hermjakob H, Hokamp K, Jang W, Johnson LS, Jones TA, Kasif S, Kaspryzk A, Kennedy S, Kent WJ, Kitts P, Koonin EV, Korf I, Kulp D, Lancet D, Lowe TM, Mc Lysaght A, Mikkelsen T, Moran JV, Mulder N, Pollara VJ, Ponting CP, Schuler G, Schultz J, Slater G, Smit AF, Stupka E, Szustakowki J, Thierry-Mieg D, Thierry-Mieg J, Wagner L, Wallis J, Wheeler R, Williams A, Wolf YI, Wolfe KH, Yang SP, Yeh RF, Collins F, Guyer MS, Peterson J, Felsenfeld A, Wetterstrand KA, Patrinos A, Morgan MJ, de Jong P, Catanese JJ, Osoegawa K, Shizuya H, Choi S, Chen YJ, Szustakowki J, International Human Genome Sequencing C (2001) Initial sequencing and analysis of the human genome. Nature 409(6822):860–921. https://doi.org/10.1038/35057062

Lee JY, Ji Z, Tian B (2008) Phylogenetic analysis of mRNA polyadenylation sites reveals a role of transposable elements in evolution of the 3′-end of genes. Nucleic Acids Res 36(17):5581–5590. https://doi.org/10.1093/nar/gkn540

Levine L, Habertheuer A, Ram C, Korutla L, Schwartz N, Hu RW, Reddy S, Freas A, Zielinski PD, Harmon J, Molugu SK, Parry S, Vallabhajosyula P (2020) Syncytiotrophoblast extracellular microvesicle profiles in maternal circulation for noninvasive diagnosis of preeclampsia. Sci Rep 10(1):6398. https://doi.org/10.1038/s41598-020-62193-7

Li X, Li ZH, Wang YX, Liu TH (2023) A comprehensive review of human trophoblast fusion models: recent developments and challenges. Cell Death Discov 9(1):372. https://doi.org/10.1038/s41420-023-01670-0

Lim AK, Kai T (2007) Unique germ-line organelle, nuage, functions to repress selfish genetic elements in Drosophila melanogaster. Proc Natl Acad Sci USA 104(16):6714–6719. https://doi.org/10.1073/pnas.0701920104

Lin M, Du Z, Guo D, Zhang Y, Dan J (2025) Dux cluster duplication ensures full activation of totipotent genes. Proc Natl Acad Sci USA 122(10):e2421594122. https://doi.org/10.1073/pnas.2421594122

Lunyak VV, Prefontaine GG, Nunez E, Cramer T, Ju BG, Ohgi KA, Hutt K, Roy R, Garcia-Diaz A, Zhu X, Yung Y, Montoliu L, Glass CK, Rosenfeld MG (2007) Developmentally regulated activation of a SINE B2 repeat as a domain boundary in organogenesis. Science 317(5835):248–251. https://doi.org/10.1126/science.1140871

Ma G, Babarinde IA, Zhou X, Hutchins AP (2022) Transposable elements in Pluripotent stem cells and human disease. Front Genet 13:902541. https://doi.org/10.3389/fgene.2022.902541

Macfarlan TS, Gifford WD, Driscoll S, Lettieri K, Rowe HM, Bonanomi D, Firth A, Singer O, Trono D, Pfaff SL (2012) Embryonic stem cell potency fluctuates with endogenous retrovirus activity. Nature 487(7405):57–63. https://doi.org/10.1038/nature11244

Malassine A, Frendo JL, Blaise S, Handschuh K, Gerbaud P, Tsatsaris V, Heidmann T, Evain-Brion D (2008) Human endogenous retrovirus-FRD envelope protein (syncytin 2) expression in normal and trisomy 21-affected placenta. Retrovirology 5:6. https://doi.org/10.1186/1742-4690-5-6

Malki S, van der Heijden GW, O'Donnell KA, Martin SL, Bortvin A (2014) A role for retrotransposon LINE-1 in fetal oocyte attrition in mice. Dev Cell 29(5):521–533. https://doi.org/10.1016/j.devcel.2014.04.027

Manghera M, Douville RN (2013) Endogenous retrovirus-K promoter: a landing strip for inflammatory transcription factors? Retrovirology 10:16. https://doi.org/10.1186/1742-4690-10-16

Mangoni D, Simi A, Lau P, Armaos A, Ansaloni F, Codino A, Damiani D, Floreani L, Di Carlo V, Vozzi D, Persichetti F, Santoro C, Pandolfini L, Tartaglia GG, Sanges R, Gustincich S (2023) LINE-1 regulates cortical development by acting as long non-coding RNAs. Nat Commun 14(1):4974. https://doi.org/10.1038/s41467-023-40743-7

Marchetto MCN, Narvaiza I, Denli AM, Benner C, Lazzarini TA, Nathanson JL, Paquola ACM, Desai KN, Herai RH, Weitzman MD, Yeo GW, Muotri AR, Gage FH (2013) Differential L1 regulation in pluripotent stem cells of humans and apes. Nature 503(7477):525–529. https://doi.org/10.1038/nature12686

Mattaliano MD, Montana ES, Parisky KM, Littleton JT, Griffith LC (2007) The Drosophila ARC homolog regulates behavioral responses to starvation. Mol Cell Neurosci 36(2):211–221. https://doi.org/10.1016/j.mcn.2007.06.008

McClintock B (1951) Chromosome organization and genic expression. Cold Spring Harb Symp Quant Biol 16:13–47. https://doi.org/10.1101/sqb.1951.016.01.004

Mi S, Lee X, Li X, Veldman GM, Finnerty H, Racie L, LaVallie E, Tang XY, Edouard P, Howes S, Keith JC Jr, McCoy JM (2000) Syncytin is a captive retroviral envelope protein involved in human placental morphogenesis. Nature 403(6771):785–789. https://doi.org/10.1038/35001608

Muckenfuss H, Hamdorf M, Held U, Perkovic M, Lower J, Cichutek K, Flory E, Schumann GG, Munk C (2006) APOBEC3 proteins inhibit human LINE-1 retrotransposition. J Biol Chem 281(31):22161–22172. https://doi.org/10.1074/jbc.M601716200

Muotri AR, Marchetto MC, Coufal NG, Oefner R, Yeo G, Nakashima K, Gage FH (2010) L1 retrotransposition in neurons is modulated by MeCP2. Nature 468(7322):443–446. https://doi.org/10.1038/nature09544

Nancy P, Tagliani E, Tay CS, Asp P, Levy DE, Erlebacher A (2012) Chemokine gene silencing in decidual stromal cells limits T cell access to the maternal-fetal interface. Science 336(6086):1317–1321. https://doi.org/10.1126/science.1220030

Nikolopoulou E, Galea GL, Rolo A, Greene ND, Copp AJ (2017) Neural tube closure: cellular, molecular and biomechanical mechanisms. Development 144(4):552–566. https://doi.org/10.1242/dev.145904

Norton C, Clarke D, Holmstrom J, Stirland I, Reynolds PR, Jenkins TG, Arroyo JA (2023) Altered epigenetic profiles in the placenta of Preeclamptic and intrauterine growth restriction patients. Cells 12(8). https://doi.org/10.3390/cells12081130

Notwell JH, Chung T, Heavner W, Bejerano G (2015) A family of transposable elements co-opted into developmental enhancers in the mouse neocortex. Nat Commun 6:6644. https://doi.org/10.1038/ncomms7644

Ono R, Nakamura K, Inoue K, Naruse M, Usami T, Wakisaka-Saito N, Hino T, Suzuki-Migishima R, Ogonuki N, Miki H, Kohda T, Ogura A, Yokoyama M, Kaneko-Ishino T, Ishino F (2006) Deletion of Peg10, an imprinted gene acquired from a retrotransposon, causes early embryonic lethality. Nat Genet 38(1):101–106. https://doi.org/10.1038/ng1699

Oomen ME, Rodriguez-Terrones D, Kurome M, Zakhartchenko V, Mottes L, Simmet K, Noll C, Nakatani T, Mourra-Diaz CM, Aksoy I, Savatier P, Goke J, Wolf E, Kaessmann H, Torres-Padilla ME (2025) An atlas of transcription initiation reveals regulatory principles of gene and transposable element expression in early mammalian development. Cell 188(4):1156–1174. e1120. https://doi.org/10.1016/j.cell.2024.12.013

Orgel LE, Crick FH (1980) Selfish DNA: the ultimate parasite. Nature 284(5757):604–607. https://doi.org/10.1038/284604a0

Pastuzyn ED, Day CE, Kearns RB, Kyrke-Smith M, Taibi AV, McCormick J, Yoder N, Belnap DM, Erlendsson S, Morado DR, Briggs JAG, Feschotte C, Shepherd JD (2018) The neuronal gene arc encodes a repurposed Retrotransposon Gag Protein that mediates intercellular RNA transfer. Cell 172(1-2):275–288 e218. https://doi.org/10.1016/j.cell.2017.12.024

Patel M, Ali AS, Dabrowska A, Boulet F, Sinha AK, Pradeepa MM (2024) Early prediction of Preeclampsia based on transposable elements signature in cell-free RNA. bioRxiv:20242011 2008622691. https://doi.org/10.1101/2024.11.08.622691

Pontis J, Pulver C, Playfoot CJ, Planet E, Grun D, Offner S, Duc J, Manfrin A, Lutolf MP, Trono D (2022) Primate-specific transposable elements shape transcriptional networks during human development. Nat Commun 13(1):7178. https://doi.org/10.1038/s41467-022-34800-w

Prudhomme S, Oriol G, Mallet F (2004) A retroviral promoter and a cellular enhancer define a bipartite element which controls env ERVWE1 placental expression. J Virol 78(22):12157–12168. https://doi.org/10.1128/JVI.78.22.12157-12168.2004

Puschendorf M, Stein P, Oakeley EJ, Schultz RM, Peters AH, Svoboda P (2006) Abundant transcripts from retrotransposons are unstable in fully grown mouse oocytes. Biochem Biophys Res Commun 347(1):36–43. https://doi.org/10.1016/j.bbrc.2006.06.106

Reilly MT, Faulkner GJ, Dubnau J, Ponomarev I, Gage FH (2013) The role of transposable elements in health and diseases of the central nervous system. J Neurosci 33(45):17577–17586. https://doi.org/10.1523/JNEUROSCI.3369-13.2013

Renaud SJ, Jeyarajah MJ (2022) How trophoblasts fuse: an in-depth look into placental syncytiotrophoblast formation. Cell Mol Life Sci 79(8):433. https://doi.org/10.1007/s00018-022-04475-z

Rizzino A (2013) Concise review: the Sox2-Oct4 connection: critical players in a much larger interdependent network integrated at multiple levels. Stem Cells 31(6):1033–1039. https://doi.org/10.1002/stem.1352

Roberts RM, Ezashi T, Schulz LC, Sugimoto J, Schust DJ, Khan T, Zhou J (2021) Syncytins expressed in human placental trophoblast. Placenta 113:8–14. https://doi.org/10.1016/j.placenta.2021.01.006

Robinson HL, Astrin SM, Senior AM, Salazar FH (1981) Host Susceptibility to endogenous viruses: defective, glycoprotein-expressing proviruses interfere with infections. J Virol 40(3):745–751. https://doi.org/10.1128/JVI.40.3.745-751.1981

Rodriguez-Terrones D, Torres-Padilla ME (2018) Nimble and ready to mingle: Transposon outbursts of early development. Trends Genet 34(10):806–820. https://doi.org/10.1016/j.tig.2018.06.006

Ru Ma NX, Liu N (2025) Expression of LINE-1 elements is required for preimplantation development and totipotency. Genes Dis. https://doi.org/10.1016/j.gendis.2025.101555

Saito K, Okamoto M, Watanabe Y, Noguchi N, Nagasaka A, Nishina Y, Shinoda T, Sakakibara A, Miyata T (2019) Dorsal-to-ventral cortical expansion is physically primed by ventral

streaming of early embryonic Preplate neurons. Cell Rep 29(6):1555–1567. e1555. https://doi.org/10.1016/j.celrep.2019.09.075
Sakashita A, Kitano T, Ishizu H, Guo Y, Masuda H, Ariura M, Murano K, Siomi H (2023) Transcription of MERVL retrotransposons is required for preimplantation embryo development. Nat Genet 55(3):484–495. https://doi.org/10.1038/s41588-023-01324-y
Salter JD, Bennett RP, Smith HC (2016) The APOBEC protein family: united by structure, divergent in function. Trends Biochem Sci 41(7):578–594. https://doi.org/10.1016/j.tibs.2016.05.001
SanMiguel P, Tikhonov A, Jin YK, Motchoulskaia N, Zakharov D, Melake-Berhan A, Springer PS, Edwards KJ, Lee M, Avramova Z, Bennetzen JL (1996) Nested retrotransposons in the intergenic regions of the maize genome. Science 274(5288):765–768. https://doi.org/10.1126/science.274.5288.765
Schmidt D, Schwalie PC, Wilson MD, Ballester B, Goncalves A, Kutter C, Brown GD, Marshall A, Flicek P, Odom DT (2012) Waves of retrotransposon expansion remodel genome organization and CTCF binding in multiple mammalian lineages. Cell 148(1–2):335–348. https://doi.org/10.1016/j.cell.2011.11.058
Schroeder DI, Blair JD, Lott P, Yu HO, Hong D, Crary F, Ashwood P, Walker C, Korf I, Robinson WP, LaSalle JM (2013) The human placenta methylome. Proc Natl Acad Sci USA 110(15):6037–6042. https://doi.org/10.1073/pnas.1215145110
Senft AD, Macfarlan TS (2021) Transposable elements shape the evolution of mammalian development. Nat Rev Genet 22(11):691–711. https://doi.org/10.1038/s41576-021-00385-1
Shah AH, Rivas SR, Doucet-O'Hare TT, Govindarajan V, DeMarino C, Wang T, Ampie L, Zhang Y, Banasavadi-Siddegowda YK, Walbridge S, Maric D, Garcia-Montojo M, Suter RK, Lee MH, Zaghloul KA, Steiner J, Elkahloun AG, Chandar J, Seetharam D, Desgraves J, Li W, Johnson K, Ivan ME, Komotar RJ, Gilbert MR, Heiss JD, Nath A (2023) Human endogenous retrovirus K contributes to a stem cell niche in glioblastoma. J Clin Invest 133(13). https://doi.org/10.1172/JCI167929
Shahbazi MN (2020) Mechanisms of human embryo development: from cell fate to tissue shape and back. Development 147(14). https://doi.org/10.1242/dev.190629
Sheehy AM, Gaddis NC, Choi JD, Malim MH (2002) Isolation of a human gene that inhibits HIV-1 infection and is suppressed by the viral Vif protein. Nature 418(6898):646–650. https://doi.org/10.1038/nature00939
Skene PJ, Illingworth RS, Webb S, Kerr AR, James KD, Turner DJ, Andrews R, Bird AP (2010) Neuronal MeCP2 is expressed at near histone-octamer levels and globally alters the chromatin state. Mol Cell 37(4):457–468. https://doi.org/10.1016/j.molcel.2010.01.030
Sood R, Zehnder JL, Druzin ML, Brown PO (2006) Gene expression patterns in human placenta. Proc Natl Acad Sci USA 103(14):5478–5483. https://doi.org/10.1073/pnas.0508035103
Soper SF, van der Heijden GW, Hardiman TC, Goodheart M, Martin SL, de Boer P, Bortvin A (2008) Mouse maelstrom, a component of nuage, is essential for spermatogenesis and transposon repression in meiosis. Dev Cell 15(2):285–297. https://doi.org/10.1016/j.devcel.2008.05.015
Subhramanyam CS, Cao Q, Wang C, Heng ZSL, Zhou Z, Hu Q (2020) Role of PIWI-like 4 in modulating neuronal differentiation from human embryonal carcinoma cells. RNA Biol 17(11):1613–1624. https://doi.org/10.1080/15476286.2020.1757896
Sundaram V, Cheng Y, Ma Z, Li D, Xing X, Edge P, Snyder MP, Wang T (2014) Widespread contribution of transposable elements to the innovation of gene regulatory networks. Genome Res 24(12):1963–1976. https://doi.org/10.1101/gr.168872.113
Trejbalova K, Blazkova J, Matouskova M, Kucerova D, Pecnova L, Vernerova Z, Heracek J, Hirsch I, Hejnar J (2011) Epigenetic regulation of transcription and splicing of syncytins, fusogenic glycoproteins of retroviral origin. Nucleic Acids Res 39(20):8728–8739. https://doi.org/10.1093/nar/gkr562
Vento-Tormo R, Efremova M, Botting RA, Turco MY, Vento-Tormo M, Meyer KB, Park JE, Stephenson E, Polanski K, Goncalves A, Gardner L, Holmqvist S, Henriksson J, Zou A, Sharkey AM, Millar B, Innes B, Wood L, Wilbrey-Clark A, Payne RP, Ivarsson MA, Lisgo S, Filby A, Rowitch DH, Bulmer JN, Wright GJ, Stubbington MJT, Haniffa M, Moffett A, Teichmann SA (2018) Single-cell reconstruction of the early maternal-fetal interface in humans. Nature 563(7731):347–353. https://doi.org/10.1038/s41586-018-0698-6

Voronina E, Seydoux G, Sassone-Corsi P, Nagamori I (2011) RNA granules in germ cells. Cold Spring Harb Perspect Biol 3(12). https://doi.org/10.1101/cshperspect.a002774

Vuoristo S, Bhagat S, Hyden-Granskog C, Yoshihara M, Gawriyski L, Jouhilahti EM, Ranga V, Tamirat M, Huhtala M, Kirjanov I, Nykanen S, Krjutskov K, Damdimopoulos A, Weltner J, Hashimoto K, Recher G, Ezer S, Paluoja P, Paloviita P, Takegami Y, Kanemaru A, Lundin K, Airenne TT, Otonkoski T, Tapanainen JS, Kawaji H, Murakawa Y, Burglin TR, Varjosalo M, Johnson MS, Tuuri T, Katayama S, Kere J (2022) DUX4 is a multifunctional factor priming human embryonic genome activation. iScience 25(4):104137. https://doi.org/10.1016/j.isci.2022.104137

Wagner GP (2007) The developmental genetics of homology. Nat Rev Genet 8(6):473–479. https://doi.org/10.1038/nrg2099

Wang Q, Shi Y, Bian Q, Zhang N, Wang M, Wang J, Li X, Lai L, Zhao Z, Yu H (2023) Molecular mechanisms of syncytin-1 in tumors and placental development related diseases. Discov Oncol 14(1):104. https://doi.org/10.1007/s12672-023-00702-6

Wang T, Medynets M, Johnson KR, Doucet-O'Hare TT, DiSanza B, Li W, Xu Y, Bagnell A, Tyagi R, Sampson K, Malik N, Steiner J, Hadegan A, Kowalak J, O'Malley J, Maric D, Nath A (2020) Regulation of stem cell function and neuronal differentiation by HERV-K via mTOR pathway. Proc Natl Acad Sci USA 117(30):17842–17853. https://doi.org/10.1073/pnas.2002427117

Weiss RA (2006) The discovery of endogenous retroviruses. Retrovirology 3:67. https://doi.org/10.1186/1742-4690-3-67

Whiddon JL, Langford AT, Wong CJ, Zhong JW, Tapscott SJ (2017) Conservation and innovation in the DUX4-family gene network. Nat Genet 49(6):935–940. https://doi.org/10.1038/ng.3846

Xu H, Liang H (2022) The regulation of totipotency transcription: perspective from in vitro and in vivo totipotency. Front Cell Dev Biol 10:1024093. https://doi.org/10.3389/fcell.2022.1024093

Yamada M, Keller RR, Gutierrez RL, Cameron D, Suzuki H, Sanghrajka R, Vaynshteyn J, Gerwin J, Maura F, Hooper W, Shah M, Robine N, Demarest P, Bayin NS, Zapater LJ, Reed C, Hebert S, Masilionis I, Chaligne R, Socci ND, Taylor MD, Kleinman CL, Joyner AL, Raju GP, Kentsis A (2024) Childhood cancer mutagenesis caused by transposase-derived PGBD5. Sci Adv 10(12):eadn4649. https://doi.org/10.1126/sciadv.adn4649

Yang J, Cook L, Chen Z (2024) Systematic evaluation of retroviral LTRs as cis-regulatory elements in mouse embryos. Cell Rep 43(3):113775. https://doi.org/10.1016/j.celrep.2024.113775

You E, Patel BK, Rojas AS, Sun S, Danaher P, Ho NI, Phillips IE, Raabe MJ, Song Y, Xu KH, Kocher JR, Richieri PM, Shin P, Taylor MS, Nieman LT, Greenbaum BD, Ting DT (2025) LINE-1 ORF1p mimics viral innate immune evasion mechanisms in pancreatic ductal adenocarcinoma. Cancer Discov. https://doi.org/10.1158/2159-8290.CD-24-1317

Young JM, Whiddon JL, Yao Z, Kasinathan B, Snider L, Geng LN, Balog J, Tawil R, van der Maarel SM, Tapscott SJ (2013) DUX4 binding to retroelements creates promoters that are active in FSHD muscle and testis. PLoS Genet 9(11):e1003947. https://doi.org/10.1371/journal.pgen.1003947

Yu C, Shen K, Lin M, Chen P, Lin C, Chang GD, Chen H (2002) GCMa regulates the syncytin-mediated trophoblastic fusion. J Biol Chem 277(51):50062–50068. https://doi.org/10.1074/jbc.M209316200

Yuan V, Hui D, Yin Y, Penaherrera MS, Beristain AG, Robinson WP (2021) Cell-specific characterization of the placental methylome. BMC Genomics 22(1):6. https://doi.org/10.1186/s12864-020-07186-6

Zapater LJ, Lewis SA, Gutierrez RL, Yamada M, Rodriguez-Fos E, Planas-Felix M, Cameron D, Demarest P, Nabila A, Mueller H, Zhao J, Bergin P, Reed C, Chwat-Edelstein T, Pagnozzi A, Nava C, Bourel-Ponchel E, Cornejo P, Dursun A, Ozgul RK, Akar HT, Maroofian R, Houlden H, Cheema HA, Anjum MN, Zifarelli G, Essid M, Ben Hafsa M, Benrhouma H, Montoya CIG, Proekt A, Zhao X, Socci ND, Hayes M, Bigot Y, Rabadan R, Torrents D, Kleinmann CL, Kruer MC, Toth M, Kentsis A (2024) A transposase-derived gene required for human brain development. bioRxiv. https://doi.org/10.1101/2023.04.28.538770

Zuo Z (2023) THE1B may have no role in human pregnancy due to ZNF430-mediated silencing. Mob DNA 14(1):6. https://doi.org/10.1186/s13100-023-00294-6

Chapter 8
The Impact of Transposable Elements on Immune Pathways

Mónica Salinas-Pena, Pau Homs-Aguadé, and Albert Jordan

Abstract This chapter explores the intricate interplay between transposable elements (TEs) and the immune system, emphasizing their dual role as regulators and activators within immune pathways. TEs have been co-opted to play critical roles in immune cell function and the regulation of immune-related genes, serving as cis-regulatory elements that shape transcriptional networks. Beyond these regulatory functions, the activation of endogenous TEs can profoundly impact immune responses, particularly through their ability to initiate immune signaling cascades. This process, termed viral mimicry, occurs when endogenous TE-derived nucleic acids are expressed and detected in the cytosol, triggering antiviral interferon responses. Such TE-driven immune modulation represents a promising approach for enhancing anti-cancer immunity. Furthermore, TE-derived transcripts demonstrate immunogenic potential and act as a reservoir for neoantigen generation. Overall, this chapter provides a comprehensive overview of the multifaceted roles of TEs in shaping immune responses, with a focus on their translational potential in cancer therapy.

Keywords Transposable elements · Viral mimicry · Cytosolic sensing · dsRNA sensing · TE-derived-RNAs · Innate immunity · Neoantigens · Interferon response · TE exonization · Syncytins

M. Salinas-Pena
Molecular Biology Institute of Barcelona (IBMB-CSIC), Barcelona, Spain

Centre for Genomic Regulation (CRG), The Barcelona Institute of Science and Technology (BIST), Barcelona, Spain

P. Homs-Aguadé · A. Jordan (✉)
Molecular Biology Institute of Barcelona (IBMB-CSIC), Barcelona, Spain
e-mail: Albert.jordan@ibmb.csic.es

L. M. Vaschetto (ed.), *Transposable Elements, Transcriptomics, and Diseases*, Epigenetics and Human Health 14, https://doi.org/10.1007/978-3-032-04125-8_8

Abbreviations

5-aza-CdR	5-aza-2-deoxycytidine
5hmC	5-hydroxymethylcytosine
5mC	5-methylcytosine
Aza	5-azacytidine
AML	Acute myeloid leukemia
APCs	Antigen-presenting cells
bp	Base pairs
ccRCC	Clear cell renal cell carcinoma
cGAS	Cyclic GMP-AMP synthase
DDX3X	DEAD-box RNA helicase 3X
DHX9	DExD/H-box helicase 9
DNMTi	DNA-methyltransferase inhibitors
dsRNA	Double-stranded RNA
ERVs	Endogenous retroviruses
env	Envelope
HDACi	Histone deacetylase inhibitors
KDM	Histone (lysine) demethylase
HLA	Human leukocyte antigen
ICIs	Immune checkpoint inhibitors
IFN	Interferon
IRFs	Interferon regulatory factors
ISGs	Interferon-stimulated genes
ISRE	Interferon-stimulated response elements
IR-Alu	Inverted Alu
JAK-STAT	Janus kinase-signal transducer and activator of transcription
JETs	Junctions between exons and TEs
KZNFPs	KRAB zinc-finger proteins
LINEs	Long interspersed nuclear elements
MHC	Major histocompatibility complex
MIRs	Mammalian-wide interspersed repeats
MDA5	Melanoma differentiation-associated protein 5
MAVS	Mitochondrial antiviral signaling protein
m6A	N6-methyladenosine
ORFs	Open reading frames
PAMPs	Pathogen-associated molecular patterns
PRRs	Pattern-recognition receptors
PRC2	Polycomb repressive complex 2
PRMTs	Protein arginine methyltransferases
RIG-I	Retinoic acid-inducible gene I
RT-qPCR	Reverse-transcriptase, quantitative PCR
SAM	S-adenosylmethionine
SINEs	Short interspersed nuclear elements

TET	Ten–eleven translocation
TCGA	The Cancer Genome Atlas
TLRs	Toll-like receptors
TFBSs	Transcription factor binding sites
TEs	Transposable elements
TAAs	Tumor-associated antigens
TSAs	Tumor-specific antigens
TS-TEPs	Tumor-specific TE-derived proteins

8.1 Introduction

Throughout evolution, waves of retrotransposon insertions have invaded mammalian genomes, to the extent that nearly half of the human genome is comprised of TEs. Retrotransposons, or Class I TEs, transpose through an RNA intermediate in a "copy and paste" mechanism. They include long interspersed nuclear elements (LINEs) and short interspersed nuclear elements (SINEs), the two dominant repeat classes in the human genome. Other retrotransposons include SINE-VNTR-Alu elements (SVAs) and long-terminal-repeat (LTR) elements, also known as endogenous retroviruses (ERVs). Among LINEs, the LINE-1 (L1) family is the most common. A full-length L1 contains a 5′ untranslated region (5′UTR) with an RNA polymerase II promoter, two open reading frames (ORF1p and ORF2p), and a 3′UTR with a poly-A tail, allowing it to autonomously propagate. However, most L1 repeats in humans are inactive, with activity restricted to specific, recently evolved copies like the L1PA subfamily or L1HS (Freeman et al. 2022). SINEs, a prevalent class of non-autonomous TEs transcribed by RNA polymerase III, include Alu elements, which have successfully expanded throughout the primate lineage. Along with SVAs, Alu elements rely on L1 machinery for mobilization (Deininger 2011). Multiple Alu subfamilies have evolved, with AluY (e.g., AluYa5, AluYb8) being the most recent and active in humans (Giordano et al. 2007). Similarly, SVAs, restricted to hominids, display a composite structure and exist in six families, named SVA_A to SVA_F, reflecting evolutionary age (Wang et al. 2005). In contrast to these non-LTR retrotransposons, full-length ERVs contain LTRs and encode three core genes (*gag*, *pol*, and *env*) flanked by LTR sequences. ERVs make up $\approx 8\%$ of the human genome and are the remnants of exogenous retroviruses that integrated into the germline millions of years ago.

Most of human TEs have lost their ability to transpose due to a long-standing evolutionary "arms race" between these parasitic sequences and the host genome and its silencing mechanism (i.e., DNA methylation) (Feschotte and Gilbert 2012; Jacobs et al. 2014). However, far from being mere "junk DNA," many regulatory TE sequences have been co-opted by the host genome, integrating into its regulatory networks (Feschotte 2008; Sundaram et al. 2014; Chuong et al. 2017; Trizzino et al. 2017). In fact, a substantial portion of the sequences with regulatory potential in primate genomes originates from TEs (Jacques et al. 2013; Trizzino et al. 2017). In this context, immune pathways are no exception, as TEs have also been "domesticated" to rewire the regulatory frameworks that control immune responses.

On the other hand, the loss of the acquired mechanisms that mediate TE silencing has a direct impact on the cellular context. Expression of endogenous TE-derived RNAs may be intracellularly identified as an exogenous infection, triggering an innate immune cascade, underscoring TE origin as foreign viral nucleic acids. Recent research has explored how this process, named "viral mimicry," could be exploited in clinics. In this regard, particular attention has been directed toward cancer, when promising results show that this immune modulation could synergize with existing treatments to increase tumor immunosensitivity (discussed in Sect. 8.4).

Innate immunity is the coordinated defense system that is commonly activated upon the recognition of non-self-particles, allowing for a quick and non-specific response against pathogen infections. Cells of the innate immune system, such as dendritic cells, macrophages, or neutrophils, as well as epithelial cells, express specialized sensors capable of detecting pathogens that induce a broad repertoire of immune signaling pathways. These sensors, known as pattern-recognition receptors (PRRs), recognize conserved molecular structures referred to as pathogen-associated molecular patterns (PAMPs). PAMPs include a diverse range of molecules such as DNA and RNA, lipopolysaccharides, peptidoglycan, and structural proteins, among others (Katze et al. 2008; Schneider et al. 2014; Li and Wu 2021).

Toll-like receptors (TLRs) are a large family of membrane-bound PRRs and are expressed either on the cell surface or within endosomes. Cell surface TLRs recognize non-nucleic acid molecular patterns, whereas endosomal TLRs primarily detect nucleic acids (Kawai et al. 2024). For example, TLR3 is an endosomal TLR that senses viruses and virus-like dsRNA (Takeuchi and Akira 2010; Chen and Hur 2022). Other PRRs are cytosolic rather than membrane-bound. For instance, cyclic GMP-AMP synthase (cGAS) is a cytosolic receptor that recognizes double-stranded DNA and activates the cGAS-STING pathway (Ablasser and Chen 2019). Similarly, retinoic acid-inducible gene I (RIG-I)-like receptors (RLRs), including RIG-I (encoded by the *DDX58* gene) and melanoma differentiation-associated protein 5 (MDA5) (encoded by the *IFIH1* gene), function as cytosolic RNA sensors. Whereas RIG-I senses short dsRNA (around 500 bp) and 5′-triphosphate RNA from RNA and DNA viruses, MDA5 senses long dsRNA (around 1000 bp) from RNA viruses, with its activity progressively increasing with dsRNA length (Takeuchi and Akira 2010; Dias Junior et al. 2019; Rehwinkel and Gack 2020; Chen and Hur 2022). While TLRs and RLRs detect pathogenic DNA and RNA to initiate immune responses, they can also recognize mislocalized cellular DNA or aberrant transcripts resulting from TEs. This detection triggers an immune response through a phenomenon known as "viral mimicry," which will be discussed later in this chapter.

Upon activation of RIG-I or MDA5, these proteins interact with the adapter protein mitochondrial antiviral signaling protein (MAVS), located on the outer membrane of mitochondria. This interaction triggers the oligomerization of multiple MAVS proteins, which subsequently leads to the phosphorylation of TANK-binding kinase 1 (TBK1) and the transcription factors IRF3 and IRF7. The phosphorylated transcription factors are then translocated into the nucleus, resulting in the expression of type-I interferons (IFNs) (Wang et al. 2025). Type-I IFNs bind to type-I interferon receptors (IFNARs) in a paracrine or endocrine manner, activating the Janus kinase-signal transducer and activator of transcription (JAK-STAT) pathway. This pathway induces the expression of interferon-stimulated genes (ISGs), whose products, known as restriction factors, inhibit pathogenic transcription, translation, and replication. Collectively, these processes establish an antimicrobial cellular state (Ivashkiv and Donlin 2014; Schneider et al. 2014). The type-I IFN response not only serves as an immediate line of defense against pathogens but also sets the stage for antigen presentation, a crucial mechanism through which innate immune cells communicate with the adaptive immune system to ensure a coordinated response. By shaping the local immune environment, IFNs enhance antigen processing and presentation, as well as the production of immune mediators such as cytokines and chemokines. Antigen presentation is mediated by professional antigen-presenting cells (APCs), such as dendritic cells, macrophages, and B cells, which process antigens and present them on their surface using major histocompatibility complex (MHC) molecules, also known as human leukocyte antigen (HLA) in humans (Jakubzick et al. 2017; Rastogi et al. 2022). Antigens are produced by the degradation of proteins and peptides by cytosolic and nuclear proteasomes. These antigens can originate from pathogens or endogenous sources, such as transposable elements. Intracellular antigens are displayed on MHC class I molecules to activate cytotoxic CD8+ T cells, while extracellular antigens are processed and presented on MHC class II molecules to stimulate helper CD4+ T cells. Type-I IFN enhances the expression of MHC molecules and the antigen-processing machinery to amplify immune recognition (Neefjes et al. 2011). Additionally, type-I IFNs influence adaptive immunity by enhancing antibody production in B cells and amplifying T-cell effector functions. The expression of aberrant transcripts from transposable elements can generate TE-derived peptides, which are subsequently presented as antigens and may trigger an immune response.

This chapter reviews the diverse roles of TEs as regulators of the immune gene expression and modulators of the immune response. Certain TEs have been co-opted to influence immune gene regulatory networks or generate novel transcripts through exonization of TE sequences. TEs may play a dual role in modulating innate immune responses: they can act as suppressors but more notably as activators, triggering innate signaling and IFN pathways through viral mimicry. Furthermore, TEs serve as a source of neoantigens, enhancing antigen presentation and bridging innate and adaptive immunity. Overall, TE activation might converge on the activation of diverse immune pathways, offering opportunities to harness their activity for therapeutic strategies, particularly in cancer immunotherapy (Fig. 8.1).

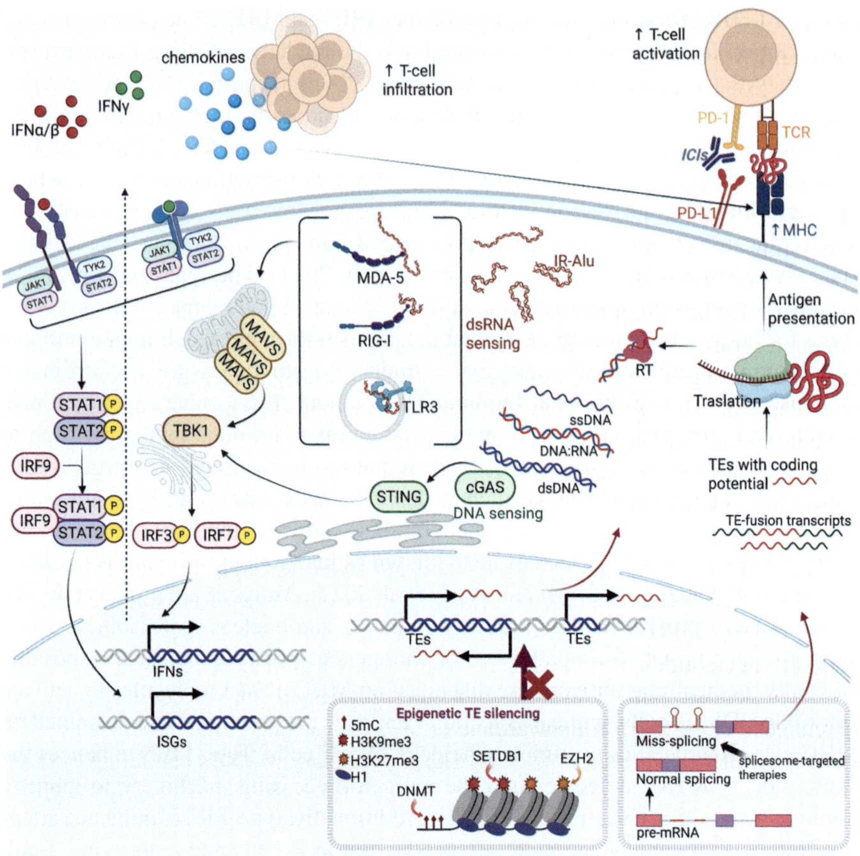

Fig. 8.1 TE-mediated immune activation. Epigenetic silencing mechanisms, including DNA methylation (5mC), histone modifications (H3K9me3, H3K27me3), and the binding of histone H1, among others, maintain TE repression under normal conditions. Loss of TE silencing through various mechanisms can lead to TE expression. TE-derived noncoding RNAs can be sensed by cytosolic PRRs, such as MDA5, RIG-I, and TLR3. This culminates in the production of type-I and type II interferons (IFNα/β and IFNγ), as well as ISGs and chemokines, driving antitumor immune responses. The resulting genome instability and R-loop accumulation can activate the cGAS/STING sensing pathway, further amplifying immune signaling. Additionally, L1-mediated reverse transcription can enhance the production of cytosolic DNA species, further stimulating cGAS/STING signaling. TEs with coding potential contribute to the generation of novel proteins and TE-fusion transcripts, which enhance antigen presentation via MHC and promote T-cell activation. Moreover, spliceosome-targeted therapies and splicing dysregulation can increase immune activation through distinct pathways. On the one hand, TE-containing introns may serve as a source of immunogenic RNA. On the other hand, TE exonization can lead to the generation of novel transcripts, which could potentially be processed and presented as neoantigens. Altogether, TE expression modulates innate immune responses by activating cell-intrinsic nucleic acid sensors and adaptive immune responses by increasing antigen presentation. Importantly, these pathways collectively boost T-cell infiltration, potentially promoting anti-cancer immunity and enhancing responsiveness to ICIs. (Figure created with BioRender)

8.2 TEs as Regulators of Immune Cell Function and Gene Expression

TEs can serve as binding platforms for multiple transcription factors, highlighting their potential to act as regulatory regions such as promoters or enhancers (Bourque et al. 2008; Sundaram et al. 2014; Thompson et al. 2016; Hermant and Torres-Padilla 2021). The cis-regulatory potential of TEs has been explored across various cellular contexts, impacting a wide range of transcriptional outcomes. TEs can play a role in cell-type-specific gene expression regulation of immune cells, contributing to the diversity and functionality of immune responses. Indeed, higher occurrences of TEs are found in the genomic loci of immune genes, and a higher enrichment of TE-derived enhancers has been observed in immune cells compared to other cell-type-specific enhancers (Van De Lagemaat et al. 2003; Ye et al. 2020), suggesting a preferential cooption of TEs to modulate immunity. Aligned with this, single-cell profiling of tissue-resident immune cells (e.g., T regulatory cells) has shown that tissue adaptation-specific enhancers are enriched within specific TE families (Simon et al. 2024). Using a similar approach, a recent preprint demonstrated that TEs are overrepresented within cis-regulatory elements of human peripheral blood mononuclear cells, displaying high cell-type specificity (Du et al. 2024). Overall, these studies highlight how immune cells exploit their TE repertoire to shape distinct regulatory landscapes.

This reported evidence of a closer association of TEs with immune pathways may reflect the unique evolutionary pressures faced by immune systems. The use of TEs as cis-regulatory sequences could provide an advantageous mechanism for host genomes to enhance diversity and develop adaptive strategies for immune network innovation. Notably, immune networks evolve more rapidly than other gene networks, primarily due to pathogen-driven pressures, benefiting from adaptive responses to evolutionary challenges, which often differ in a lineage-specific manner (Barreiro and Quintana-Murci 2010; Avraham et al. 2015; Hagai et al. 2018). Indeed, although the molecular signaling machinery underlying innate immune responses is highly conserved among mammalian species, significant interspecies differences emerge at the level of their transcriptional outputs (Barreiro et al. 2010; Schroder et al. 2012; Shay et al. 2013; Hagai et al. 2018). In this context, lineage-specific TEs can introduce unique regulatory elements, enabling tailored immune responses to specific evolutionary contexts. Furthermore, immune genes often require coordinated regulation. The use of related TE families, with multiple copies dispersed throughout the genome, could serve as an evolutionary resource to regulate genes within related signaling pathways by providing shared regulatory sequences, thereby ensuring their coordinated transcriptional activity.

In this section, we will explore the growing body of evidence supporting the role of TEs as regulatory sequences impacting immune gene expression and shaping immune responses.

8.2.1 TEs as Regulatory Elements Modulating Transcription of Immune Genes: LTRs

Among TEs, ERVs are the most well-characterized for their cis-regulatory functions (Fueyo et al. 2022). Originally, the LTR sequences in ancient ERVs served as promoters and enhancers, enabling viral replication by hijacking the host's cellular machinery. Over evolutionary time, regulatory sequences originally used by retroviruses for infection were co-opted by host genomes, integrating into transcriptional networks and contributing to gene regulation, a process termed exaptation (Rebollo et al. 2012; Xie et al. 2013). Similarly, during cancer evolution, TEs can be reactivated as cryptic regulatory elements and repurposed to drive oncogenesis, a phenomenon referred to as onco-exaptation (Babaian and Mager 2016; Babaian et al. 2016; Jang et al. 2019).

As a first indication of their putative regulatory potential, ERVs can serve as transcription binding site platforms more frequently than other TE classes (Jacques et al. 2013; Sundaram et al. 2014; Ito et al. 2017). Notably, ERVs containing transcription factor binding sites (TFBSs) are often located near genes involved in human innate immune responses, compared to other non-ERV TFBSs (Ito et al. 2017). In this line, a study of mouse CD8+ T cells revealed that core accessible regions of enhancers were found enriched with ERVs hosting motifs for transcription factors essential for T-cell function (Ye et al. 2020).

HERVK 5′LTR consensus sequence contains two interferon-stimulated response elements (ISRE) capable of binding IRF-1, IRF-3, or IRF-7, along with several predicted NF-Kb or STAT binding sites (Manghera and Douville 2013). This is suggestive of its regulation upon an inflammatory or innate immune response. Notably, functional evidence of HERVK regulation upon induction of IFN signaling was later demonstrated for MER41 repeats. MER41, a primate-specific ERV family, contains inducible STAT1 binding sites (Schmid and Bucher 2010). Analysis of IFNg-stimulated human cells revealed a significant IRF1 and STAT1 binding to a diversity of ERVs, predominantly located near ISGs (Chuong et al. 2016). Multiple MER41 elements act as cis-regulatory elements for several immune genes. CRISPR-mediated depletion of a specific MER41 copy near the AIM2 gene in HeLa cells impaired inflammasome activation, pointing to this ERV as an essential IFN-induced enhancer. This observed cooption of the MER41 family may be the reflection of an evolutionary strategy to coordinately regulate a transcriptional response of multiple immune-related genes. In mice (which lack MER41), a mouse-specific ERV was also found to be STAT1-bound and exhibit enhancer-like features upon IFN stimulation. These analogous analyses from different species illustrate how different ERVs, including lineage-specific ones, similarly rewire IFN responses in mammalian host genomes. CRISPR editing functional studies also identified another ERV-derived distal enhancer, in this case from an LTR71 element, essential for HLA-G expression (Ferreira et al. 2016). Importantly, HLA-G is a non-classical major histocompatibility complex (MHC) molecule almost exclusively expressed in the placenta. It is crucial for maintaining maternal immune tolerance toward the fetus, a topic that will be further discussed, as TEs play an important role in regulating immune adaptations to pregnancy. More generally, ERVs have been found to act as enhancers in the placenta in a species-specific manner, thus contributing to the evolutionary diversification of placental development (Chuong et al. 2013).

In acute myeloid leukemia (AML), specific ERV-derived enhancers have been found to have regulatory potential, influencing disease outcome (Deniz et al. 2020). Six LTR families (LTR2B, LTR2C, LTR5B, LTR5_Hs, LTR12C, and LTR13A) were identified as AML-specific accessible chromatin regions using different AML cell lines and patient datasets. These regions harbor an enhancer-like chromatin signature, are bound by AML-related transcription factors, and modulate the expression of nearby genes. Depletion of these AML-ERV-enhancers impacts cell proliferation by inducing apoptosis. Of note, most of these ERVs are also active in other immune progenitor cells, suggesting that, rather than being cancer-specific, they are linked to an undifferentiated state, which may confer a growth advantage to AML cells (Fueyo et al. 2022).

Apart from acting as enhancer regions, ERVs can also be co-opted as alternative promoters. In humans, an alternative isoform of the ACE2 gene originates from intronic LTR16 and MIRb elements (which belong to the MIR family, a non-Alu SINE family), which act as a cryptic IFN-inducible promoter (Ng et al. 2020; Onabajo et al. 2020; Blume et al. 2021). The ACE2 gene plays a critical role in SARS-CoV-2 infection, as has been reported as the primary receptor in host cells for viral entry (Yan et al. 2020). The TE-derived isoform is predicted to translate into a truncated protein that lacks domains required for SARS-CoV-2 binding, highlighting putative regulatory roles during viral infections that need to be further addressed. In Hodgkin lymphoma (HL), IRF5 upregulation occurs through onco-exaptation of LOR1a, a solo-LTR from the ERV1 family located upstream of its canonical TSS (Babaian et al. 2016). While LOR1a remains silenced in normal cells, its hypomethylation in HL allows it to function as a promoter, driving the formation of LOR1a-IRF5 chimeric transcripts and leading to increased IRF5 expression.

8.2.2 *TEs as Regulatory Elements Modulating Transcription of Immune Genes: LINEs and SINEs*

Although less extensively studied, other non-LTR TEs also impact gene programs and are a source of TFB hubs. LINEs are the only transposable elements in the human genome capable of autonomous retrotransposition. However, most LINE copies are inactive and unable to transpose. While the 5′ untranslated regions of LINEs can function as bidirectional promoters, they are often truncated in most copies. As a result, compared to ERVs and their LTR sequences, LINEs are less commonly recognized as major sources of regulatory activity. Despite these, several examples of L1 elements acting as enhancers of immune genes exist. In immune primary human cells, an L1M2a element located within an intronic region of the Interferon Alpha/Beta Receptor 1 (IFNAR1) gene regulates its expression under basal conditions and its upregulation upon IFN treatment (Buttler et al. 2023). This L1M2a copy was marked both by repressive H3K9me3 and enhancer-associated histone marks. This characteristic epigenetic status was also found in a subset of LINE elements that tend to be located near immune genes. This suggests that these sequences may act as immune cell-type-specific "bivalent" regulators, allowing for rapid inducible innate immune responses.

Regarding SINEs, it has been reported that Alu elements can act as enhancers, interacting with gene promoters through long-range interactions (Su et al. 2014). Enhancer-like features were more prominent for evolutionary-old Alu elements but also show tissue-specific variation (which, per se, is also considered an enhancer feature, as enhancers are generally tissue-specific). Alu repeats show an enrichment of the H3K4me1 mark in immune cell populations, while Alus marked by both H3K4me1 and H3K27ac (combinatorial histone mark status characteristics for active enhancers) are often nearby immunity and inflammation-related genes (Yan et al. 2020). While Alu repeats are primate-specific and successfully expanded throughout primate genomes (in humans they represent the 11% of our genome), other SINE elements populate non-primate species. In mouse CD8+ T cells, 25% of the enhancer flanking regions are enriched for SINE repeats and show enrichment for the H3K4me1 enhancer-associated histone mark (Ye et al. 2020). A subset of murine SINE repeats belonging to the B2 family (B2_Mm2 subfamily) contains strong binding sites for both STAT1 and CTCF, is marked by H3K27ac and behave as IFNG-inducible enhancers (Horton et al. 2023). This behavior of SINEs as the predominant source of TE-derived inducible enhancers in the murine genome matches the observation in ruminant genomes, where SINE subfamily Bov-A2 is predicted to act as IFN-inducible enhancers in bovine cells (Kelly et al. 2022). Altogether, these reports support how TEs from different families can regulate immunity in different species, contributing to delineating lineage-specific immunity and being responsible for interspecies divergence on immune responses (Junhee Seok et al. 2013; Shay et al. 2013; Shaw et al. 2017). MIRs, an ancient SINE family present in vertebrates (Korotkov et al. 2000), have been involved in mammalian maternal immunotolerance during pregnancy, functioning as an enhancer of an essential gene (transcription factor Foxp3) required for the generation of regulatory T cells outside the thymus (Samstein et al. 2012).

Insulators are DNA sequences that regulate chromatin organization by acting as boundaries (i.e., preventing heterochromatin spreading) or restricting enhancer-promoter interactions (i.e., enhancer-blocking elements) (Phillips-Cremins and Corces 2013). TEs can also serve as insulator sequences (Lunyak et al. 2007; Román et al. 2011; Ichiyanagi et al. 2021; Gebrie 2023). In human CD4+ T cells, MIRs are predicted to act as insulator regions (Wang et al. 2015). Chromatin barrier activity of these MIR-derived insulators seems to be cell-type specific and delineates chromatin domains enriched for genes involved in T-cell functional pathways. While enhancer-blocking activity of MIR-derived insulators appears to be universal between cell lines, their chromatin barrier activity seems to be cell-type specific and delineates chromatin domains enriched for genes involved in T-cell functional pathways (Wang et al. 2015).

8.2.3 TE Exonization in Immune-Related Genes

TEs can influence mRNA splicing through various mechanisms. They may be incorporated as exons (a process known as exonization), introduce premature polyadenylation signals, or create or disrupt splice sites, among others. These alterations can result in the production of alternative transcript variants, potentially modifying gene function or regulation (Yu et al. 2013; Clayton et al. 2020; Arribas et al. 2024). TE-driven exonization dynamics can shape and diversify immunological responses by modulating CD4+ T lymphocyte functionality (Marasca et al. 2022; Burattin et al. 2025) or generating novel protein isoforms that ultimately regulate IFN pathways (Arribas et al. 2024; Pasquesi et al. 2024).

More recently, L1 elements have been reported to regulate the functional state of human CD4+ T lymphocytes through splicing modulation of L1-containing transcripts (Marasca et al. 2022; Burattin et al. 2025). The expression of these L1-containing transcripts controls T-cell functionality and maintains quiescence of naïve T cells (Marasca et al. 2022). Mechanistically, intronic L1 are spliced as exons of noncanonical transcript variants of genes related to T-cell activation. Of note, these L1 elements belong to certain L1M subfamilies (L1ME4a, L1MC4, and L1ME4b), which are old L1s present in Eutheria (placental mammals) species. The expression of these alternative splicing variants harboring L1-containing exons was regulated by IRF4. Moreover, these L1-containing transcripts are bound to chromatin, influencing the expression of canonical transcripts by preventing H3K36me3 deposition and thus maintaining gene silencing in quiescent CD4+ T cells. Upon T-cell activation, inclusion of L1 in exons is suppressed (by binding of splicing-related factors PTBP1 and GTF2F1), so L1-containing transcripts are downregulated, promoting the expression of the corresponding genes. Interestingly, in exhausted T cells, these L1-containing transcripts are re-expressed. A follow-up study revealed that L1 influences not just adult T-cell function but also neonatal CD4+ T cells (Burattin et al. 2025). Furthermore, L1 expression was observed to increase gradually with age, emphasizing L1's role as a key modulator of human T-cell functionality throughout the lifespan.

The hominid-specific exonization of an intronic LINE L2A retroelement in the CD274 gene produces the CD274-L2A transcript, representing another example of LINEs splicing impacting T cell functionality (Ng et al. 2019). The CD274 gene encodes PD-L1, a key membrane protein that inhibits T-cell activation by binding to the PD-1 receptor on T cells (Ai et al. 2020). This splicing event generates a truncated isoform lacking the transmembrane domain, resulting in a soluble PD-L1 form (sPD-L1). Unlike membrane-bound PD-L1, sPD-L1 functions as a receptor antagonist, blocking the suppressive signaling of membrane PD-L1 (Ng et al. 2019).

In the above-mentioned example of an alternative ACE2 isoform originating from LTR16 and MIRb elements (Ng et al. 2020), these TEs not only act as cryptic promoters but were also used as alternative exons, and the resulting truncated protein product resulted in a nonfunctional receptor. In line with this, alternative splicing of TE-derived exons, which results in the generation of a decoy receptor, has been described in a recent work by Pasquesi et al. (2024). Long-read transcriptome data revealed events of TE exonization that potentially originate alternatively expressed mRNAs of human immune genes. Exonization of a primate-specific Alu element generates a truncated Interferon Alpha and Beta Receptor subunit 2 (IFNAR2) receptor, which lacks the intracellular signaling domain. Expression of this truncated protein hinders phosphorylation of STAT proteins, inhibiting type-I IFN cascade and even compromising antiviral cell responses to SARS-CoV-2 and dengue virus.

In general, TEs are underrepresented at exon boundaries, probably resulting from an evolutionary selection of TE insertions with deleterious effects on mRNA splicing regulation (Lev-Maor et al. 2008; Zhang et al. 2011). However, analysis of some polymorphic Alu elements (belonging to the youngest AluY family, which are not fixed in the human genome, such as AluYa5 or AluYb8) located near exon boundaries reveals their implication in directly affecting exon usage, promoting both exon skipping and inclusion (Payer et al. 2019). One example of this mechanistic effect on immune genes was reported for CD58. CD58 gene encodes a cell surface protein (LFA-3) involved in facilitating T-cell activation and adhesion (Zhang et al. 2021b). A polymorphic AluY insertion located within the 3′ intronic region near exon 3 of CD58, increases skipping of that exon. Importantly, the presence of the polymorphic Alu and the associated alternative transcript variant has been associated with an increased multiple sclerosis (MS) risk (De Jager et al. 2009; Payer et al. 2019).

8.3 TEs as Immunosupresors: Syncytins

TEs are not only key regulators of gene expression in immune pathways, but its own expression can also directly modulate immune responses. This dual role stems from their ability to be recognized by the immune system, generally leading to the activation of the host immunity, as we will cover in the next section. Nevertheless, TEs also play important roles under specific contexts in promoting immune suppression, as occurs with syncytins during mammalian pregnancy.

Certain ERVs have been co-opted by mammals to promote maternal immune tolerance, preventing the maternal immune system from attacking the fetus. During pregnancy, maintaining immune balance is critical: fetal trophoblasts expressing paternal antigens must invade maternal tissues without triggering a destructive immune response, while maternal immunity must still protect against infections (Mor et al. 2017).

Although most ERVs are rendered nonfunctional due to accumulated mutations, a subset retains intact open reading frames (ORFs), particularly within their envelope (*env*) genes (de Parseval et al. 2003; Villesen et al. 2004). In the retroviral lifecycle, Env proteins mediate receptor recognition and facilitate viral entry by enabling membrane fusion. In the human genome, a limited number of *env* genes with potential functionality have been identified (de Parseval et al. 2003; Blaise et al. 2005). Among them, syncytins are almost exclusively expressed in the placenta, with syncytin-1 and syncytin-2 (encoded by ERVW-1 and ERVFRD-1 genes, respectively) being the most studied (Sha et al. 2000; de Parseval et al. 2003; Malassiné et al. 2007; Roberts et al. 2021). They are critical for placental development and immune regulation during pregnancy.

These retroviral envelope proteins promote cell fusion to form the syncytiotrophoblast (hence their name syncytins), a multinucleated placental layer that acts as the maternal-fetal interface (Lokossou et al. 2014; Priščáková et al. 2023). This structure facilitates nutrient and gas exchange while serving as a protective barrier for the fetus. That is, the ancestral viral function of *env* genes, originally enabling cell-cell fusion for viral entry, has been repurposed in mammals for placental development.

Beyond their structural role—primarily attributed to syncytin-1—syncytins, particularly syncytin-2, promote maternal immune tolerance through their immunosuppressive domain, conserved from their retroviral origin. This domain suppresses maternal immune responses to prevent rejection of the fetus, which carries paternal antigens. However, the precise mechanisms by which syncytins promote immune tolerance remain incompletely understood (Mangeney et al. 2007; Lokossou et al. 2014, 2020). Other ERVs also contribute to maternal tolerance, such as the already-mentioned LTR71 element, which acts as a placenta-specific enhancer of HLA-G expression (Ferreira et al. 2016).

During the second trimester of pregnancy, an immune shift must take place, from proinflammatory (Th1) responses toward anti-inflammatory (Th2) cytokines (Mor et al. 2017). Briefly, Th1 responses involve macrophage activation and cytotoxic T cells, driven by cytokines such as IFN-γ, IL-2, and TNF-α, to combat intracellular pathogens. In contrast, Th2 responses promote antibody production and anti-inflammatory processes through cytokines such as IL-4, IL-5, IL-10, and IL-13. Syncytins are implicated in this Th1-to-Th2 shift, reducing inflammation and safeguarding pregnancy (Hummel et al. 2015; Lokossou et al. 2020).

The immunosuppressive domains of Syncytin-1 or Syncytin-2 induce molecular changes in T cells, including activation of the MAP kinase pathway and inhibition of TNF-α production, both of which contribute to a diminished Th1 immune response (Tolosa et al. 2012; Lokossou et al. 2020). Additionally, both syncytin-1 and syncytin-2 are present on the surface of extracellular vesicles, including exosomes (Tolosa et al. 2012; Lokossou et al. 2020). Deletion of synsytin-2 immunosuppressive domain from exosome surfaces significantly reduces their ability to inhibit IFN-γ or TNF-α. In summary, syncytins, through their immunosuppressive domain and their presence in placental exosomes, modulate the fetal microenvironment by attenuating Th1 responses (Lokossou et al. 2020).

Both fusogenic and non-fusogenic functions of placental syncytins have been linked to disease pathogenesis, including pregnancy-related diseases, as well as tumorigenesis or autoimmune diseases such as multiple sclerosis, as reviewed in (Gröger and Cynis 2018; Bu et al. 2022; Priščáková et al. 2023; Wang et al. 2023).

8.4 TE Transcripts Activating Innate Immune Responses: Viral Mimicry

Compared to their role in immunosuppression, the activation of innate immune responses by self-TEs has been extensively documented across various cellular contexts. TE-derived noncoding RNAs can be recognized by the host immune system and sensed as non-self by pattern-recognition receptors (PRRs), thereby activating innate immunity pathways. This process mimics the cellular response triggered by the detection of foreign viral nucleic acids, for which the phenomenon was termed "viral mimicry." This concept reflects the evolutionary origin of many TEs, which, although typically silenced in the genome, retain sufficient similarity to viral sequences to elicit antiviral responses when their repression is compromised.

Dissecting the mechanistic steps through which aberrant TE expression triggers innate immunity involves several key milestones. First, the epigenetic silencing mechanism is being disrupted, which is often linked to a particular disease scenario or tumoral context. Second, TEs' classes or families are being deregulated. And third, the cytosolic machinery, which recognizes these aberrantly expressed elements as exogenous sequences. While the specific factors involved may vary across studies, they converge on a common outcome: the activation of innate immune responses. For this reason, in recent years, viral mimicry has garnered significant attention for its implications in cancer immunity and therapy. Table 8.1 summarizes the various scenarios that have been reported where induced TE expression activated the innate immune response (described in more detail below).

Immune evasion is a hallmark of cancer (Hanahan 2022; Galassi et al. 2024). It refers to the ability of tumor cells to escape immune surveillance by suppressing or avoiding immune responses. Key mechanisms include the tumor expression of immune checkpoint proteins such as PD-L1, which binds to its ligand PD-1 on T cells to inhibit their activity, and the recruitment of regulatory T cells expressing CTLA-4 to suppress effector and natural killer cell functions. Immunotherapy, particularly immune checkpoint inhibitors (ICIs) targeting PD-1 and CTLA-4, has revolutionized cancer treatment by restoring the immune system's capacity to recognize and attack tumor cells (Topalian et al. 2015; Sharma et al. 2023). These therapies have shown durable responses in patients, particularly in solid tumors with high neoantigen burden and immune infiltration. However, resistance to ICIs remains a significant challenge, with many patients either failing to respond initially or developing resistance over time (Morad et al. 2021).

Table 8.1 TEs driving viral mimicry

		TEs expression driving innate immune responses													
		TEs activation					Cytosolic sensing					Immune response			
				TE family			RNA				DNA	in vitro	in vivo		Reference
		Molecular trigger	Cellular context	ERVs	SINE/Alu	LINE/L1	MDA-5	RIG-I	PKR	TLR-3	cGAS/STING	IFN response	anti-tumor immunity	synergy ICIs	
DNA methylation		DNMTi	Ovarian and colon CCLs. Melanoma MM	■			■			■		■	■	■	Chiappinelli et al, 2015
		DNMTi	Colon CCLs	■			■	■				■			Roulois et al, 2015
		DNMTi	Hematological cancer patient data	■	■	■	x					■			Ohtani et al, 2020
		Low DNMT1	Blind mole rat		■	■					■	■	■		Zhao et al, 2021
		CDK4/6i (indirect DNMT1i)	MM breast, colon tumours	■			■	■				■	■	■	Goel et al, 2017
		PRMT7i (indirect DNMT1i)	Melanoma MM	■			■	■				■	■	■	Srour et al, 2022
		VitaminC + DNMTi	CCLs	■	■	■	■	■				■			Liu et al, 2016
RNA editing		ADAR1 depletion + DNMTi	Patient-derived colorectal cancer cells		■		■					■	■		Mehdipour et al, 2020
		ADAR1 depletion + DNMTi	Ovarian CCLs	■	■	■	x					■	■		Gomez et al, 2022
		ADAR1 depletion	Melanoma MM		■		■		■			■	■	■	Ishizuka et al, 2019
Histone-modifying enzymes & Histones	Histone methyltransferases	SETDB1 depletion	Acute myeloid leukemia CCLs	■		■	■	■				■			Cuellar et al, 2017
		SETDB1 KO	Melanoma and lung cancer MM	■			x					■	■	■	Griffin et al, 2021
		SETDB1 depletion	CCLs. Melanoma and lung cancer MM	■			■	■				■	■		Pan et al, 2022
		HUSH complex depletion	Human cell lines, including CCLs	■		■	■	■				■			Tunbak et al, 2020
		KAP1 depletion	Adult human cells	■			■	■				■			Tie et al, 2018
		TNRC18 depletion	Human cell lines, including CCLs	■			x					■			Zhao et al, 2022
		G9Ai + DNMTi	Ovarian CCLs with high G9A expression	■								■			Liu et al, 2018a
		H3.3K27M + DNMTi	Pediatric High-grade glioma	■	■	■	■	■	■			■	■		Krug et al, 2019
		EZH2 depletion	Prostate cancer MM and organoids	■							■	■	■	■	Morel et al, 2021
		EZH2i	Atypical teratoid rhabdoid CCLs		■	■	■	■			■	■			Feng et al, 2024
		EZH2i	Taxane-resistant triple negative breast CCLs	■			■	■				■	■		Deblois et al, 2020
		PRC2i + DNMT1i	Malignant peripheral nerve sheath tumor	■	■	■			■			■	■		Patel et al, 2022
	Histone demethylas	KDM5B depletion	Melanoma MM	■		■	■	■			■	■	■		Zhang et al, 2021
		PHF8 depletion	Mouse CCLs. MM	■	■	■	■	■			■	■	■	■	Liu et al, 2023
		LSD1 depletion	CCLs. Melanoma MM	■			■			■		■	■	■	Sheng et al, 2018
		p53-mediated LSD1i + DNMTi	CCLs. Melanoma MM	■			■			■		■	■	■	Zhou et al, 2021
	H1	H1.2+H1.4 depletion	Breast CCLs	■			■	■			■	■			Izquierdo et al, 2017
RNA processing and Splicing		SF3B1i	Myc-driven triple negative breast CCLs, MM		■ *	■ *	■	■				■	■		Bowling et al, 2021
		PRMTi	Triple negative breast CCLs		■ *				■	■		■	■		Wu et al, 2022
		SETD2 loss + DNMTi	Clear cell renal cell carcinoma CCLs, MM			■ *		■				■	■	■	Li et al, 2023
		hnRNPM depletion	CCLs		■ *	■ *	■	■		■		■			Zheng et al, 2024
		DDX3X depletion	Breast CCLs	■			■					■	■		Choi et al, 2021
		DHX9 depletion	Small cell lung cancer CCLs, MM	■	■	■	■				■	■	■	■	Murayama et al, 2024

Summary table showcasing various scenarios where TE-driven viral mimicry has been reported. The table highlights the main TE families expressed, the primary DNA or RNA sensors involved, and the downstream consequences of TE sensing in the context of the innate immune response. In this regard, it includes in vitro and in vivo evidence, as well as instances of synergy with immunotherapy. *i* pharmacological inhibition; *CCLs* Cancer Cell Lines; *MM* Mouse Models; *x* not assessed. * indicates that TEs are located within the retained introns

In this context, overcoming tumor immunosuppression is critical. Epigenetic therapies that induce the production of IFN and proinflammatory cytokines through the activation of TEs could synergize with existing immunotherapies, boosting immune responses and countering tumor immune evasion. The phenomenon of viral mimicry, extensively studied in tumor contexts, is supported by compelling in vivo evidence demonstrating that activating immune signatures can effectively enhance antitumor immunity. This section will explore the mechanistic molecular insights into how self-TEs drive cell-intrinsic innate immune activation while discussing their relevance to tumor immunity and the therapeutic opportunities this phenomenon presents.

8.4.1 Induction of Viral Mimicry by Targeting DNA Methylation and RNA Editing

The term "viral mimicry" was coined back in 2015 when two subsequent studies identified it as a mechanism underlying the anti-tumoral effect of DNA-methyltransferase inhibitors (DNMTi) in clinics. Treatment of cancer cell lines with DNA-demethylating agents 5-aza-2-deoxycytidine (5-aza-CdR) (Roulois et al. 2015) or 5-azacytidine (Aza) (Chiappinelli et al. 2015) results in ERV-derived-dsRNAs expression. These dsRNAs activate the MDA5/MAVS RNA-sensing pathway or TLR3 receptor, leading to downstream activation of ISGs and IFN signaling. Among the upregulated ERVs, there were several *env* genes (e.g., the previously mentioned Syncytin-1), although the upregulation occurred only at the transcript level, and demethylation did not increase their protein levels. Importantly, genetic ablation of the RNA sensors impaired the long-term anti-proliferative effects of DNMTi treatments, pointing to TE-derived dsRNAs as the main cause for the antiviral response (Chiappinelli et al. 2015; Roulois et al. 2015).

CDK4/6 inhibitors (e.g., abemaciclib) have been used as antitumor drugs (Morrison et al. 2024). Although their main mechanism of action was thought to be through cell cycle inhibition and G1 arrest, they also promote antitumor immunity through activation of TEs (Goel et al. 2017). Mechanistically, the induced type III IFN response is mediated by suppression of DNMT1 activity, inducing DNA demethylation. The 5′ LTR of ERV3-1 is demethylated and expressed upon abemaciclib treatment in both in vitro (cancer cell lines) and in vivo (PDX tumors) models of breast cancer.

Analysis of expression profiles of hematologic cancer patients before, during, and after Aza treatment revealed that the activation of a subset of evolutionarily young TEs was linked to the patient response to DNMTi (Ohtani et al. 2020). Both the innate immune signature and the expression of these TEs were specifically upregulated in the responder group. These results are in line with the observation of a switching silencing mechanism of LTRs depending on their evolutionary age, being young TEs mainly suppressed by DNA methylation, while intermediate-age

LTRs are more associated with H3K9me3 repression (Ohtani et al. 2018). One of the Aza responder-specific upregulated TE was LTR12C (Ohtani et al. 2020), which has been found to act as a cryptic promoter of non-annotated transcripts with immunogenic potential following combined DNMTi and histone deacetylase inhibitors (HDACi) treatment (Brocks et al. 2017). This highlights how a given TE and its TE-derived-RNAs could enhance the immune signature through different mechanisms. In a similar way, combinatorial treatments targeting different epigenetic TE silencing mechanisms have proven a synergistic effect on eliciting an antiviral response, which opens new therapeutic opportunities.

Although 5-aza-CdR treatment led to young LTRs expression, their upregulation was increased when 5-aza-CdR was combined with histone methyltransferase inhibitors (Ohtani et al. 2020). Accordingly, dual inhibition of DNA methylation (5-aza-CdR treatment) and H3K9me2 (through inhibition of histone methyltransferase G9A) results in synergistic upregulation of ERVs, resulting in an increased antiviral response in ovarian cancer cell lines with high levels of G9A expression (Liu et al. 2018a). In high-grade gliomas, histone H3K27M is a driver mutation, leading to a global loss of H3K27me3 and concomitant gain of H3K27ac deposition within TEs (Krug et al. 2019). Alteration of this histone modification status leads to a general TE activation (from LTR, LINE, SINE, and DNA classes) and predisposes H3K27M tumors to viral mimicry responses. Importantly, treatment with DNMTi or HDACi significantly further amplifies this TE activation and their dsRNA sensing via MAVS and PKR pathways (Krug et al. 2019).

A synergistic anti-proliferative response of 5-aza-CdR treatment has also been seen with Vitamin C (Liu et al. 2016). Vitamin C is a cofactor of ten–eleven translocation (TET) enzymes, which are responsible for the conversion of 5-methylcytosine (5mC) into oxidized derivatives such as 5-hydroxymethylcytosine (5hmC). Importantly, the synergistic effect was due to the enhanced viral mimicry caused by the upregulation of multiple TEs, probably due to the loss of DNA methylation at their genomic loci (direct DNA methylation loss due to DNMTi and also passive, as 5hmC was also increased). Those ERVs transcribed from bidirectional promoters (e.g., LTR5, LTR26, PRIMA41, and LTR12C) were identified as the main putative cause of cytosolic dsRNA accumulation.

Most reports focus on bidirectional transcription of ERVs as the major TE source of immunogenic dsRNAs. However, among the evolutionary young TEs associated with a positive response to Aza treatment, apart from ERVs, SINEs, and LINEs upregulation was also detected (Ohtani et al. 2020). Strand-specific RNA-Seq of 5-aza-CdR and Vitamin C-treated-cells detected increased transcription not only of ERV1 and ERVL-MaLR families (LTR class), but also SINE and LINE classes (Liu et al. 2016). Despite this, inverted Alu (IR-Alu) rather than bidirectional ERV transcripts were identified as the main origin of immunogenic dsRNA induced by DNMTi (Mehdipour et al. 2020). IR-Alu are pairs of nearby Alu elements arranged in opposite orientations within the genome and whose transcription generates dsRNAs via hybridization of both sense and antisense transcripts (Lee et al. 2024). Alternatively, IR-Alu dsRNAs can also be generated via formation of intramolecular pairing and stem-loop formation, resulting in a hairpin structure (Mehdipour et al. 2020). These dsRNAs structures can be substrates for A-to-I RNA editing, a process whereby adenosine (A) is enzymatically deaminated to inosine (I) by double-stranded RNA-specific adenosine deaminases (ADARs) (Eisenberg and Levanon 2018). Although these IR-Alu can be recognized by MDA5, ADAR1-mediated RNA editing of IR-Alu prevents recognition by RNA sensors (Ahmad et al. 2018; Mehdipour et al. 2020). Thus, upon 5-aza-CdR treatment, although transcription of IR-Alu is enhanced, ADAR1 limits its capacity to trigger an immune response. In fact, a combination of ADAR1 depletion with DNMTi results in a synergistic antitumor effect in vivo, overcoming resistance to 5-aza-CdR treatment (Mehdipour et al. 2020). In line with this, in a mouse model of ovarian cancer, a combination of DNMTi and ADAR1 depletion increases type-I IFN signaling and proinflammatory tumor microenvironment, reducing tumor burden and prolonging survival (Gomez et al. 2022). In addition, ADAR1 inhibition also sensitizes tumors to ICIs and to irradiation (Ishizuka et al. 2019). These studies illustrate how the combination of ADAR1 with pre-existing treatments, through boosting viral mimicry responses, represents a promising clinical approach.

Altogether, the limited DNA methylation at TE loci, achieved through chemical or genetic inhibition of DNMTs, induces TE-triggered immune responses across various biological contexts, which can be further increased by synergy with other interventions, including ADAR1-mediated limited RNA editing. These responses often result in anti-cancer effects, supported by correlations between patient survival and the viral mimicry process, as well as functional in vivo evidence. In this context, an elegant study on blind mole rats, rodents naturally resistant to cancer, suggests that retrotransposons may act as tumor suppressors in these rodents, driving their innate resistance to tumorigenesis (Zhao et al. 2021). This species exhibits low DNMT1 expression, leading to TE derepression specifically in hyperplastic and premalignant cells. The activated TEs, including SINEs (B1 and B2 families) and LINE L1s, accumulate as RNA-DNA hybrids in the cytosol, activating the cGAS-STING pathway and triggering IFN responses to prevent cellular progression. Notably, no dsRNA accumulation or involvement of TLR3, MAVS, MDA5, or RIG-I was detected in this IFN response. This natural cooption of TEs as anti-cancer mediators in blind mole rats provides further evidence of their therapeutic potential.

8.4.2 Induction of Viral Mimicry by Targeting Histone-Modifying Enzymes and Histone Variants

Histone post-translational modifications and histone variants constitute two key layers of chromatin regulation. Deposition of repressive histone modifications, such as H3K9me3 and H3K27me3, is a common TE silencing mechanism (Fukuda and Shinkai 2020). Consequently, targeting histone-modifying enzymes has been shown to induce viral mimicry across multiple cellular contexts.

In AML cell lines, SETDB1 deficiency is a critical regulator of cell survival in an RNA-sensing machinery-dependent manner. SETDB1 is an H3K9 methyltransferase, so upon SETDB1 loss, TEs show a reduction of the H3K9me3 histone mark that triggers their expression, which results in a type-I IFN response (Cuellar et al. 2017). Expressed TEs include multiple repeats from LTR (ERV1, ERVK, and ERVL families), LINEs, and satellite (especially SST1 satellite) classes. Although some ERVs were transcribed unidirectionally, for many others, as well as for LINE L1s and satellites, overlapping transcripts were found in both senses, suggesting they are the main source of dsRNA cytosolic accumulation. Silencing of SETDB1 sensitizes tumors to ICIs (Griffin et al. 2021) and radiotherapy (Pan et al. 2022). High efficacy of radiotherapy treatment in lung cancer and melanoma patients correlates with low SETDB1 expression, which at the same time correlates with higher expression of ERV-1 and ERVK family (Pan et al. 2022). Indeed, SETDB1 inhibition radiosensitizes tumors in vivo, by enhancing the type-I IFN response via activation of MDA5/MAVS-mediated sensing of ERVs. However, only a subset of ERVs were checked by qPCR (Pan et al. 2022). All this evidence points to SETDB1, often overexpressed in human tumors (Griffin et al. 2021), as a promising druggable therapeutic target, in combination with existing therapies.

SETDB1 can be recruited to chromatin by several proteins. As a consequence, depletion of those proteins can also trigger a viral mimicry response due to the loss of H3K9me3 status within TEs, as has been described for KAP1 (Tie et al. 2018), the HUSH complex (Tunbak et al. 2020) or histone demethylases (Zhang et al. 2021a; Liu et al. 2023).

KAP1 (also known as TRIM28), through recruitment by KRAB zinc-finger proteins (KZNFPs) to repetitive sequences, at the same time recruits repressive factors such as SETDB1 or HP1alpha. This repressive axis is crucial for retrotransposon silencing, both in development and adult tissues (Rowe et al. 2010; Turelli et al. 2014; Tie et al. 2018). Depletion of KAP1 in adult human cells elicits an innate immune response depending on MAVS signaling, potentially originating from the H3K9me3 loss within TEs. Concretely, upregulation of HERVK14C and SVA_D was found upon KAP1 loss (Tie et al. 2018).

SETDB1 can be recruited by the HUSH complex, a repressor that has been particularly implicated in epigenetic control of LINEs, preventing retrotransposition of L1s (especially evolutionary young L1s, including L1Hs and other young L1PA subfamilies such as L1PA2) (Liu et al. 2018b; Robbez-Masson et al. 2018; Gu et al. 2021). HUSH depletion in human cultured cells leads to the activation of an IFN and inflammatory gene signature in a MAVS-dependent manner. Of note, MDA5 contributed more than RIG-I to that response, while cGAS/STING DNA-sensing pathway was not activated. Regarding TEs, at a family level, multiple young (primate-specific) LTRs were upregulated. However, extended analysis focusing on single TE copies (that is, using only uniquely mapped RNA-Seq reads) revealed that numerous copies of LTR, SINEs, but especially evolutionary-recent LINE L1s (which, as mentioned, are bound by HUSH) were upregulated. Indeed, L1Hs and L1PA2 bidirectional transcription was identified as the potential source of accumulated dsRNA. In fact, the use of an shRNA targeting these young L1s dramatically reduced the ISGs induction (Tunbak et al. 2020). Interestingly, recent work identified a second HUSH complex, named HUSH2 (Danac et al. 2024). Instead of binding to L1 sequences, HUSH2 is enriched within gene promoters, particularly of ISGs. Both complexes share some subunits, then coupling TE silencing and ISGs induction by a new competitive mechanism (Danac et al. 2024).

While HUSH is important for H3K9me-mediated repression of LINEs, TNRC18 is an H3K9me3-reader within evolutionary young ERV1 elements (e.g., LTR12C, LTR12D, LTR12E, etc.). In multiple human cell lines, TNRC18 depletion leads to the upregulation of ERV1 repeats, ISGs, and immunity-related genes (Zhao et al. 2023).

KDM5B-mediated SETDB1 recruitment to TEs mediates epigenetic silencing of those and suppresses antitumor immunity (Zhang et al. 2021a). KDM5B is a H3K4 demethylase, but derepression of TEs upon KDM5B is independent of its demethylase activity and occurs with no changes in H3K4me3 status. Conversely, it is mediated by loss of SETDB1 recruitment to those TE loci, resulting in an elevated expression of bidirectional LTR and LINE transcripts in a mouse melanoma cell line. Thus, KDM5B depletion activates the type-I IFN response through both cytosolic RNA-sensing (MDA5/MAVS) and DNA-sensing (cGAS/STING) pathways. Treatment with reverse transcriptase (RT) inhibitors abolished the IFN signaling, suggesting that reverse transcription of expressed TEs may be responsible for cytosolic DNA accumulation that triggers cGAS/STING cascade. Importantly, the study shows how the depletion of one of the most upregulated TEs, MMVL30 (ERV1 murine family), significantly reduced ISGs induction (Zhang et al. 2021a). Also in a demethylase-independent manner, PHF8 (KDM7B) promotes a viral mimicry response and antitumor immunity in colon cancer (Liu et al. 2023). Although the main targets of PHF8-mediated demethylation are H3K9me1/2 and H4K20me1, the viral mimicry response upon PHF8 KD was mechanistically initiated by H3K9me3 loss at TE loci, originating from promoted SETDB1 destabilization. Importantly, PHF8 ablation results in tumor cell immunogenicity and improves anti-PD1 response to reduce tumor growth in vivo. PHF8 KD mouse cells induce multiple LTR, LINEs, and SINEs, which presumably form cytosolic dsRNA or are retrotranscribed, as both RNA- and DNA-sensing pathways seem to play a role (Liu et al. 2023).

Other histone demethylases have been implicated in triggering viral mimicry in a demethylase-dependent manner. Inhibition of LSD1 (KDM1A), an H3K4 demethylase, promotes type-I IFN signaling, enhances antitumor immunity, and increases responsiveness to anti-PD-1 blockade in vivo in a melanoma mouse model by boosting tumor immunogenicity and T-cell infiltration (Sheng et al. 2018). In human cells, LSD1 ablation results in H3K4me2 accumulation that triggers the bidirectional transcription of ERVs, leading to the production of dsRNA. This dsRNA is sensed by TLR3 and MDA5, but not by RIG-I or the cGAS/STING pathway, suggesting that long dsRNAs, rather than ssRNAs or short dsRNAs, drive ISG induction. Similarly, antitumor immunity via viral mimicry was found in tumors upon pharmacologic activation of p53, as it induces the inhibition of LSD1 (and DNMT1) within LTRs (Zhou et al. 2021). Indeed, a combination of a small-molecule MDM2 inhibitor (which mediates p53 activation) in combination with immunotherapy has shown good results in a phase II study (Shaheen et al. 2023).

Histone demethylases (KDMs) exemplify how the immune response can be activated through diverse mechanisms that ultimately converge on common pathways. For instance, KDM5B depletion has been shown to suppress STING expression by directly binding to its promoter and limiting H3K4me3 levels (Wu et al. 2018). PHF8 binding to a subset of IFNγ-responsive promoters restricts their expression, thereby modulating the immune response (Asensio-Juan et al. 2017). Furthermore, LSD1 depletion reduces the expression of RISC components, which are crucial for RNA interference through dsRNA cleavage (Sheng et al. 2018). Consequently, through both increased ERV transcription and diminished RISC activity, LSD1 depletion prevents the suppression of dsRNA accumulation, amplifying the immune response (Sheng et al. 2018).

Referring back to histone methyltransferases, EZH2 is the catalytic subunit of Polycomb repressive complex 2 (PRC2), responsible for the deposition of the H3K27me3 mark, a key repressive histone modification involved in TE silencing. Overexpression of EZH2 can be found in different types of tumors and is associated with an immunosuppressive tumor microenvironment and resistance to ICIs (Kang et al. 2020). In prostate cancer murine in vivo models, EZH2 inhibition sensitizes tumors to anti-PD-1 therapy (Morel et al. 2021). Upregulation of a subset of ERVs located in an antisense orientation within the 3′UTR of ISGs was found in patients with low-expressed EZH2 tumors (Cañadas et al. 2018; Morel et al. 2021). In atypical teratoid rhabdoid tumors, a highly aggressive childhood cancer with no effective treatment, EZH2 inhibition triggers viral mimicry (Feng et al. 2024). IR-Alu elements were identified as the main source of dsRNA accumulation. These IR-Alu were located within introns and 3′UTR sequences, and upon H3K27me3 loss in the promoter of their host genes, transcription of these genes and their containing IR-Alu occurred. Notably, some of these genes were ISGs, thus establishing a positive feedback loop by both their own expression and the generation of more Alu-derived-dsRNA. However, the immune response was not totally dependent on MAVS-mediated RNA-sensing pathway but also cGAS-STING DNA-sensing pathway played a role, as only co-depletion of both MAVS and cGAS completely abrogated ISG induction. A subset of full-length L1 (L1PA subfamily) was also upregulated and, potentially, could be reverse-transcribed, activating STING sensing. Indeed, activation of ISGs upon EZH2 inhibition was significantly abolished with the inactivation of L1 elements (using sgRNAs targeting the 5′UTR or ORF1 of L1 and in silico predicted to target multiple L1s, including L1H, L1PA2, and L1PA3), and treatment with RT inhibitors reduced the levels of cytosolic ssDNA (Feng et al. 2024). PRC2-inactivating mutations in high-grade malignant peripheral nerve sheath tumors increase the antitumor response of DNMTi (Patel et al. 2022). In chemotherapy-resistant triple-negative breast cancer, the metabolism of S-adenosylmethionine (SAM), the primary donor of methyl groups, is disrupted. This metabolic alteration leads to a loss of DNA methylation at repressive loci. However, resistant cells compensate by reorganizing their H3K27me3 landscape, depositing this repressive mark at the affected loci. These loci are enriched with L1 and LTR elements, and their activation can be triggered by pharmacological inhibition of EZH2 in cell lines or PDX-derived organoids, ultimately triggering a viral mimicry response. (Deblois et al. 2020). Overall, the enhanced viral mimicry elicited upon EZH2 inhibition has been observed across multiple cancer types, although the TE families responsible for the final immune response are context dependent. Thus, targeting EZH2 is a promising approach to enhance the efficacy of existing treatments. Indeed, the use of an EZH2 inhibitor (tazemetostat) is being explored in multiple clinical trials, especially in lymphoma patients (e.g., NCT03456726, NCT01897571, NCT02601937, NCT03213665).

Histone variants represent an additional layer of epigenetic regulation and have been implicated in triggering viral mimicry. The linker histone H1 has 11 variants or subtypes in mammals, 7 of which are present in somatic cells: H1.0, H1.1 to H1.5, and H1X. H1 variants display different functionality and binding profiles, including within TEs (Millán-Ariño et al. 2016; Salinas-Pena et al. 2024). Combined depletion of H1.2 and H1.4 in breast cancer cells induces a type-I IFN response mediated by the upregulation of repetitive elements (Izquierdo-Bouldstridge et al. 2017). The repeats activated upon H1 loss include satellites and LTRs, triggering an antiviral immune response primarily through RIG-I/MDA5, with a putative contribution from STING. This response is further amplified in a feedback loop via IFN-β production. Interestingly, while downstream induction of ISGs and an IFN signature has not been observed with other H1 depletion conditions, H1X and H1.4 are notably enriched at young TEs from various families, including SVA and younger SINE-Alu, LINE L1, and LTRs (Salinas-Pena et al. 2024). Depletion of these variants leads to a modest upregulation of the TEs where they are enriched, an effect that becomes more pronounced when combined with 5-aza-CdR treatment (Salinas-Pena et al. 2024). This observation opens doors to explore the possibility of combining H1 ablation (targeting single or multiple variants) with other targets or anti-cancer drugs to synergistically enhance the induction of viral mimicry.

8.4.3 Induction of Viral Mimicry by Targeting RNA Processing

Disruption of RNA processing programs, including RNA splicing or deposition of RNA post-transcriptional modifications, can also lead to an antiviral immune response. Splicing is a tightly regulated process in which hundreds of accessory proteins take part. Aberrant splicing is a hallmark of many cancers, and therapies targeting the spliceosome have demonstrated clinical efficacy. Interestingly, in Myc-activated triple-negative breast cancer, splicesome-targeted therapies lead to the activation of dsRNA-sensing pathways and trigger an immune response. Spliceosome-targeted therapies in this context result in the retention of introns, including introns containing LINE and Alu elements, which serve as endogenous sources of dsRNA, triggering viral mimicry and tumor cell death (Bowling et al. 2021). Similarly, kidney cancers with low expression of the H3K36me3 methyltransferase SETD2 exhibit a robust innate immune response when treated with

DNMTi. This response is driven by the transcription of intronic TEs, primarily LINE-derived sequences. In SETD2-deficient renal cancer, DNMT inhibitors induce TE-containing transcripts and sensitize tumors to the combination of 5-aza-CdR and ICIs in vivo (Li et al. 2023).

Type-I protein arginine methyltransferases (PRMTs) modify multiple splicing-related factors. Chemical inhibition of PRMTs (targeting mainly PRMT1) suppresses tumor growth in triple-negative breast cancer due to an increased IFN response (Wu et al. 2022). The underlying molecular cause for dsRNA accumulation (sensed by RIG-I and TLR-3) is the deregulation of RNA splicing, particularly inducing intron retention. Notably, the main immunogenetic source of dsRNA is where the forming IR-Alu located within the retained introns. Depletion of another PRMT family member, PRMT7, has also been described to promote viral mimicry, promoting antitumor immunity in melanoma (Srour et al. 2022). However, in this case, the PRMT7-mediated IFN response originated by loss of DNA methylation within ERV loci, by indirect inhibition of DNMTs (Srour et al. 2022), in contrast to the altered-splicing profile observed upon PRMT1 pharmacologic inhibition (Wu et al. 2022). hnRNPM is an RNA-binding protein that binds to evolutionarily young LINEs located within deep introns and suppresses the usage of cryptic splice sites (Zheng et al. 2024). Loss of hnRNPM results in splicing of cryptic exons that can generate long dsRNAs through base-pairing of intronic IR-Alu interspersed among LINEs, triggering a RIG-I/MDA5 and TLR3-mediated type-I IFN response (Zheng et al. 2024).

Depletion of RNA helicases, a critical player for RNA processing, can also trigger an innate immune response (Choi et al. 2021; Murayama et al. 2024). Depletion of DEAD-box RNA helicase 3X (DDX3X) in breast cancer cells triggered a tumor-intrinsic type-I IFN response due to the aberrant cytoplasmic accumulation of ERV-derived dsRNAs sensed by MDA5, which also led to an enhanced antitumor activity, antigen presentation, and tumor infiltration of cytotoxic T cells and dendritic cells in mouse mammary tumors (Choi et al. 2021). Interestingly, DDX3X interacts with ADAR1 and promotes ADAR1-mediated dsRNA editing, and simultaneous depletion of both proteins results in a synergistic accumulation of dsRNA. In spite of the strong relationship between ADAR1 editing and IR-Alu repeats (Mehdipour et al. 2020; Lee et al. 2024), only induction of some ERVs was checked by RT-qPCR in this context. On the other hand, DExD/H-box helicase 9 (DHX9) has been shown to be a critical regulator of dsRNA accumulation in small cell lung cancer, and its ablation promotes sensitivity to anti-PD-1 therapy in vivo (Murayama et al. 2024). Mechanistically, the helicase activity of DHX9 allows for dsRNA unwinding, limiting dsRNA accumulation in basal conditions. Upon DX9 depletion, dsRNAs derived from LINE, SINE, and LTR families accumulate in the cytosol and are bound mainly by MDA5 rather than RIG-I. DHX9 ablation also increases cGAS/STING pathway due to the accumulation of R-loops and DNA/RNA hybrids, which also enhances the IFN response.

RNA modifications are also critical for RNA processing. The most abundant RNA modification is N6-methyladenosine (m6A), co-transcriptionally deposited by METTL3. METTL3 has also been implicated in regulating innate immune response

(Durbin et al. 2016; Rubio et al. 2018; Winkler et al. 2019; Gao et al. 2020; Chelmicki et al. 2021; Guirguis et al. 2023) and potentiates response to anti-PD-1 therapy in preclinical mouse models (Guirguis et al. 2023). Mechanistically, the study proposes that, in addition to promoting double-stranded RNA (dsRNA) formation, loss of m6A modification on MHC-I-related genes leads to an increased antitumor immunity, favoring antigen presentation. Although this study did not investigate whether TEs were the source of the increased dsRNA, m6a has been shown to regulate ERVs (Chelmicki et al. 2021), and viral RNA can be modified my m6A, which hinders their recognition by RIG-I (Lu et al. 2020). Illustrating the potential of targeting METTL3, STC-15 is the first METTL3 inhibitor in Phase 1 clinical studies for patients with solid tumors (NCT05584111).

In summary, viral mimicry has been seen to occur in response to many epigenetic alterations that affect TE expression, giving a range of possibilities for intervention against tumor progression in conjunction with existing immunotherapy.

8.5 TEs Driving Antigen Presentation: Impact of TE Activation on Tumor Immunogenicity

A growing body of evidence has reported a strong correlation between the expression levels of TEs and the activation of the immune system in cancer, including increased immune cell infiltration into tumors. An extensive analysis of transposable element expression in The Cancer Genome Atlas (TCGA) data revealed that among various cancer types, stomach, bladder, liver, and head and neck cancers exhibited the most notable elevated TE expression when compared to corresponding healthy tissues (Kong et al. 2019). In this study, the majority of recurrently overexpressed TEs belonged to the evolutionarily youngest groups in the human genome, including L1HS, L1PA2, HERVK, and HERVH. The expression of these TEs was associated with an IFN response, CD8+ T-cell activity, immune checkpoint activity, and immune infiltration. Another study found two HERVK and one HERVE element to be expressed in multiple cancer types but not in normal tissues (Rooney et al. 2015). Other types of cancer, such as AML, have been found to exhibit their own subset of specifically expressed ERVs, particularly the LTR2B family (Deniz et al. 2020), and different subtypes of breast cancer have been reported to express HERVK (Johanning et al. 2017). These findings suggest that the expression of TEs in tumors is influenced by the specific type of tissue involved and that the initiation of the innate immune response will depend on the source of expressed TEs.

Cancer cells may express tumor-associated antigens (TAAs), which are typically present at low levels in normal tissues, and tumor-specific antigens (TSAs), also known as neoantigens, which often arise from mutations in cancer cell DNA that lead to the synthesis of mutated proteins uniquely expressed in tumor tissue. However, tumor specificity can also result from the unique expression of non-mutated genomic regions that are normally silenced, such as TEs or cancer testis antigens. Moreover, although TEs are generally categorized as noncoding RNA, some contain open reading frames, such as ORF1/ORF2 within L1 elements or the *gag*, *pol*, and *env* genes found in intact ERVs. These protein-coding regions of TEs have the potential to be translated, presented on MHC molecules, and serve as targets for antitumor immune responses (Reid Cahn et al. 2022; Xie et al. 2023).

A proteogenomic analysis of MHC-I-associated peptides from human B cells reported that approximately 10% of these peptides originate from noncoding genomic sequences and out-of-frame translations (Laumont et al. 2016). In a later study, the same group identified 40 neoantigens in several human primary tumors using a novel proteogenomic approach. About 90% of these neoantigens originated from allegedly noncoding regions and would have been missed by standard exome-based approaches. Most of these TSAs derived from non-mutated but aberrantly expressed transcripts, such as ERVs (Laumont et al. 2018). A comprehensive pan-cancer analysis of 33 tumor types revealed a landscape of tumor-specific TE-gene chimeric transcripts. These transcripts, often driven by reactivated TE-derived promoters, can serve as alternative promoters for canonical genes, leading to the aberrant activation and overexpression of oncogenes. This study identified 2297 tumor-specific TE-gene chimeric transcripts enriched for LTR and SVA elements. Importantly, many of these chimeric transcripts encoded unique peptide sequences, producing truncated or chimeric proteins presented by MHC molecules. Such tumor-specific TE-derived proteins (TS-TEPs) were validated using mass spectrometry and shown to be expressed on the extracellular surface, making them accessible for antibody and cellular therapies (Shah et al. 2023). A separate study employing de novo transcript assembly profiled cancer-specific transcripts overlapping with LTR elements across various cancers. In a melanoma model, these cancer-specific LTR-overlapping transcripts were found to predict melanoma subtypes and possess protein-coding potential. These transcripts were enriched with sequences uniquely mapped to cancer-specific LTR-overlapping transcripts. MHC-eluted peptides derived from these transcripts showed antigenic potential, highlighting their importance in the cancer immunopeptidome (Attig et al. 2019). A combination of single-cell transcriptomics and proteogenomics analysis identified 370 TE-derived peptides encoded by differentially expressed TEs in glioblastoma. Single-cell transcriptomics revealed that these TEs originated from younger L1, LTR, and SVA subfamilies. Some TEs shared MHC-I-presented peptide-coding sequences, while others, particularly those from older subfamilies, were more tumor-specific and may serve as potential targets for cancer immunotherapy. The analysis also identified nonredundant, peptide-coding TEs that were highly enriched in glioblastoma tumors and recurrent in cancer patients. Additionally, some of these TE-derived peptides were found to be immunogenic in vitro, indicating their potential to induce T-cell responses (Bonté et al. 2022).

Further insights into the immunogenic potential of TE-derived transcripts came from epigenetic treatment studies. In human non-small cell lung carcinoma cells, DNMT and HDAC inhibition with 5-aza-CdR and SB939, respectively, induced transcripts predominantly arising from LTRs, particularly LTR12C. These treatments led to the identification of 272 proteins from predicted ORFs, many of which were presented on MHC-I molecules. The resulting neoantigens, often derived from short ORFs, elicited immune responses characterized by CD8+ T cells producing IFN-γ and TNF. This phenomenon was consistent across lung cancer, AML, glioblastoma, and colon cancer models, with LTR12 elements showing the strongest induction across cell lines (Goyal et al. 2023). In a mouse model, depletion of SETDB1 led to the emergence of new chromatin-accessible sites, most of which had TEs. Among the upregulated TEs, some were ERVs with coding potential. In this case, the induction of ERVs did not trigger a viral mimicry response but instead resulted in the generation of putative MHC-I peptides and the expansion of CD8+ T cells expressing cytotoxic genes (Griffin et al. 2021).

Multiple studies have reported the expression of neoantigens originating from proteins or peptides encoded by HERVs, which can bind to MHC molecules and activate a CD8+ T-cell response (Smith et al. 2018; Bonaventura et al. 2022). In renal cell carcinoma, the MHC-I immunopeptidome included a small subset of peptides derived from HERVs, one of which was immunogenic and capable of triggering a CD8+ T-cell response (Kobayashi et al. 2023). Ovarian cancer cells have been shown to express HERV transcripts and the HERV-K Env protein. The Env protein induced antigen-specific T-cell proliferation, IFN-γ production, and cytotoxic T lymphocyte activity. These patients also had high reverse-transcriptase activity and HERV-K immunoreactive antibodies in their plasma, suggesting potential biomarkers for detection and diagnosis (Rycaj et al. 2015). In breast cancer, RNA-Seq data revealed that the expression of HERV-K Env protein is associated with the promotion of tumorigenesis and is linked to key signaling pathways, including p53, Myc, and Ras. Functional studies demonstrated that knockdown of the HERVK Env protein inhibited breast cancer cell proliferation, migration, and invasion, reducing tumor formation and preventing metastasis. Interestingly, cell cycle arrest at the S phase was observed in some breast cancer cells, suggesting a potential reversion to a non-tumorigenic phenotype. Conversely, overexpression of the HERVK *env* gene enhanced the migration, invasion, and transformation of breast cancer cells, further supporting its role in promoting malignancy (Zhou et al. 2016). HERVK Env protein expression has also been detected in various breast cancer tissues and cell lines, distinguishing them from normal breast tissues. Some patients have also been found to produce anti-HERVK serum antibodies, indicating that the immune system recognizes these viral-like proteins as foreign antigens, which could be leveraged for immunotherapeutic approaches (Wang-Johanning et al. 2008). In melanoma, malignant phenotypes in vitro were accompanied by the activation of HERVK expression and a massive production of viral-like particles (Serafino et al. 2009). Other subfamilies of HERVs, such as HERVE and HERVH, have also been observed to be expressed in malignant cells, generating neoantigens capable of triggering a T-cell response (Takahashi et al. 2008; Mullins and Linnebacher 2012). For example, HERVE expression was identifiable at the earliest stage of tumor development of clear cell renal cell carcinoma and was selectively expressed in clear cell renal cell carcinoma (ccRCC) but not in normal kidney tissues or other tumor types, including pancreatic, prostate, breast, ovarian, colon, lung cancers, melanoma, neuroblastoma, and hematopoietic malignancies. The HERVE-detected transcripts represented splice variants of the endogenous retrovirus containing a common region from the 5′ LTR and a non-shared region from the *pro*, *pol*, and *env* genes. However, only envelope peptides were found to be present on the surface of ccRCC cells. Some of those peptides, acting as neoantigens, had high affinity for MHC-I molecules, making them capable of stimulating specific CD8+ T cells (Cherkasova et al. 2016). Although the majority of described neoantigens are derived from ERVs, other TE classes can also contribute. Other TE-derived peptides presented on MHC-I molecules have been reported to originate from intronic or intergenic regions of LINEs, SINEs, and SVAs (Kong et al. 2019).

TEs significantly contribute to transcriptome diversity in cancer through their involvement in alternative splicing. Over 4800 alternative splicing events across 723 cancer-associated genes have been attributed to TEs, highlighting their critical role in cancer biology (Clayton et al. 2020). Among these, Alu and L1 elements are the primary contributors to TE-generated splice sites in cancer genes, both serving as sources of splice acceptor and donor sites (Belancio et al. 2006; Lev-Maor et al. 2008). A comprehensive analysis of 8705 cancer patients identified 930 tumor-specific alternative splicing events, some of which were predicted to produce peptides capable of acting as neoantigens (Kahles et al. 2018). Two consecutive studies reported TE involvement in noncanonical splicing events with coding exons, which were named junctions between exons and TEs (JETs) (Burbage et al. 2023; Merlotti et al. 2023). JETs have been found to be tumor-specific and often absent from healthy tissues. They preferentially involve intronic TEs with canonical splicing sites in exons and internal, cryptic splicing sites. In murine tumor cell lines, LTRs, particularly the ERVL-MaLR family, were found to be frequently enriched as both donor and acceptor splice sites in JETs, although they were more prominently enriched as acceptor sites. These JETs can encode peptides presented on MHC-I molecules, serving as neoantigens that elicit immune responses. In mice bearing MC38 tumor cells, JET-specific peptides stimulated T cells that infiltrated tumors. Immunization with long peptides containing predicted JET-encoded MHC-I-binding sequences generated T-cell responses, delayed tumor growth, and reduced tumor invasion (Burbage et al. 2023). In non-small cell lung cancer, 41 TE subfamilies were enriched among donor sites and 54 among acceptor sites, with nine subfamilies (HERVS71-int, L1PA4, L1PB3, MER1A, MER4-int, MER65A, MER70-int, Tigger 1, and Tigger 2) enriched in both categories. Peptides derived from JETs can stimulate immune responses in healthy donors. After in vitro stimulation, CD8+ T-cell precursors specifically recognized JET-derived neoantigens presented on MHC-I molecules. These JET-specific CD8+ T cells were capable of targeting tumor cells expressing the corresponding neoantigens (Merlotti et al. 2023).

In general, the activation of the immune system due to TE expression and the resulting IFN response via viral mimicry has also been concurrently associated with an increase in the machinery and pathways involved in antigen presentation and the stimulation of CD8+ T cells. This suggests that immune activation can occur through multiple pathways simultaneously, acting synergistically to enhance the overall immune response (Goel et al. 2017; Kong et al. 2019; Zhang et al. 2021a; Zhou et al. 2021; Guirguis et al. 2023).

8.6 Conclusion

TEs have played a fundamental role in shaping the evolution of mammalian genomes, contributing not only to genome architecture but also to gene regulation and immune system modulation. Their capacity to act as regulatory elements, to influence immune signaling pathways, and to contribute to antigen presentation underscores their significance beyond being merely "selfish DNA." This perspective echoes Barbara McClintock's pioneer work (McClintock 1956), which, though largely overlooked for decades due to the prevailing "junk DNA" paradigm, emphasized the functional evolutionary importance of TEs.

The interplay between TEs and the immune system is complex and multifaceted, encompassing both their domesticated functions in immune gene regulation and their potential to trigger immune responses. The presence of lineage-specific TEs emphasizes their role in driving species-specific immune adaptations. In this context, the contribution of more recently evolved TE families to immune responses may be underestimated, as they may be absent in commonly used animal model species. For instance, SVA repeats are unique to hominids, and Alu elements are primate-specific, meaning that studies conducted in mice cannot capture the potential involvement of these TEs in immune regulation or activation.

Ultimately, TEs represent key evolutionary resources that drive diversity within immune networks and fine-tune the expression of immune-related genes. Their role as cis-regulatory elements enhances the capacity of host genomes to respond to changing environmental and pathogenic challenges, reinforcing their importance in immune system evolution across mammalian species. In addition, recent work highlights the role of TEs in expanding the diversity of the immune proteome, with over a hundred isoforms generated through TE exonization reported to be translated into immune-related proteins (Arribas et al. 2024; Pasquesi et al. 2024). The emergence of these uncharacterized proteins may have profound implications for immune-related diseases, positioning TE exonization as a crucial yet underappreciated mechanism in the evolution of immune regulation. Also in relation to their translation, TEs contribute to antigen presentation by generating neoantigens that can elicit adaptive immune responses. TE-derived peptides and fusion transcripts have been detected in various cancers and shown to be presented on MHC molecules, offering promising targets for immunotherapy or biomarker discovery.

Beyond their regulatory functions, TEs can also be potent inducers of immune activation. Importantly, TE activation can act at the interface of both innate and adaptive responses (Fig. 8.1). On the one hand, TE-derived RNAs can be sensed in the cytosol, leading to innate immune activation via the IFN response, a phenomenon termed viral mimicry. On the other hand, TEs can enhance antigen presentation and T-cell infiltration. Notably, disruption of epigenetic mechanisms that normally silence TEs has emerged as a powerful strategy for inducing viral mimicry and enhancing antitumor immunity (Table 8.1). Further mechanistic studies are needed to precisely determine which specific TE copies or families drive viral mimicry and how this scenario changes across different biological contexts. Methodologically, detecting cytosolic TE-derived RNA has traditionally relied on J2 (anti-dsRNA) immunofluorescence, while TE expression is typically assessed via RT-qPCR or RNA-Seq. The use of RT-qPCR implies the evaluation of a very limited subset of TE families and hinders the exploration of a "full TE picture," whereas RNA-Seq faces challenges in uniquely mapping short reads. In this regard, the development of long-read sequencing techniques and the use of recent complete genome assemblies (Nurk et al. 2022) will be instrumental in overcoming these technical limitations.

While this chapter focuses on how TE activation can enhance antitumor immunity, the clinical exploitation of TE-driven immune responses extends beyond cancer. TEs have been implicated in aging, autoimmune, infectious, and neurodegenerative diseases. For instance, L1 repeats promote age-associated inflammation and IFN signaling in senescence (De Cecco et al. 2019), while TE expression mediates IFN responses in systemic lupus erythematosus (Wang et al. 2024) and Alzheimer's disease (Scopa et al. 2023).

Looking ahead, the challenge lies in harnessing the beneficial aspects of TE activation while mitigating its potential deleterious effects, such as chronic inflammation and autoimmunity. Understanding the context-dependent regulation of TEs will be crucial for developing targeted therapies that exploit their immunomodulatory potential. As research in this field advances, integrating TE biology into cancer immunotherapy and other immune-related diseases holds great promise, potentially paving the way for novel therapeutic strategies that exploit TEs as both biomarkers and therapeutic targets.

Funding This work was funded by the Spanish Ministry of Science, Innovation and Universities and FEDER, EU (grant number PID2023-146239OB-I00/AEI/10.13039/501100011033). We also acknowledge the Generalitat de Catalunya Suport Grups de Recerca AGAUR (grant number 2017-SGR-597).

Compliance with Ethical Standards

Conflicts of Interest The authors declare that they have no conflicts of interest.

Ethical Approval This chapter is a review of previously published accounts; as such, no animal or human studies were performed.

Corrected with the assistance of artificial intelligence (AI) tools developed by OpenAI, ChatGPT (version GPT-4), 2025. Every section was carefully reviewed, revised, and approved by the author(s) to ensure that the text is appropriate concerning content and scientific correctness and to maintain the integrity of the original content.

References

Ablasser A, Chen ZJ (2019) CGAS in action: expanding roles in immunity and inflammation. Science (80-) 363. https://doi.org/10.1126/science.aat8657

Ahmad S, Mu X, Yang F et al (2018) Breaching self-tolerance to Alu Duplex RNA underlies MDA5-mediated inflammation. Cell 172. https://doi.org/10.1016/j.cell.2017.12.016

Ai L, Xu A, Xu J (2020) Roles of PD-1/PD-L1 pathway: signaling, cancer, and beyond. Adv Exp Med Biol 1248:33–59. https://doi.org/10.1007/978-981-15-3266-5_3

Arribas YA, Baudon B, Rotival M et al (2024) Transposable element exonization generates a reservoir of evolving and functional protein isoforms. Cell 187:7603–7620.e22. https://doi.org/10.1016/j.cell.2024.11.011

Asensio-Juan E, Fueyo R, Pappa S et al (2017) The histone demethylase PHF8 is a molecular safeguard of the IFNκ response. Nucleic Acids Res 45. https://doi.org/10.1093/nar/gkw1346

Attig J, Young GR, Hosie L et al (2019) LTR retroelement expansion of the human cancer transcriptome and immunopeptidome revealed by de novo transcript assembly. Genome Res 29. https://doi.org/10.1101/gr.248922.119

Avraham R, Haseley N, Brown D et al (2015) Pathogen cell-to-cell variability drives heterogeneity in host immune responses. Cell 162. https://doi.org/10.1016/j.cell.2015.08.027

Babaian A, Mager DL (2016) Endogenous retroviral promoter exaptation in human cancer. Mob DNA 7:1–21. https://doi.org/10.1186/s13100-016-0080-x

Babaian A, Romanish MT, Gagnier L et al (2016) Onco-exaptation of an endogenous retroviral LTR drives IRF5 expression in Hodgkin lymphoma. Oncogene 35. https://doi.org/10.1038/onc.2015.308

Barreiro LB, Marioni JC, Blekhman R et al (2010) Functional comparison of innate immune signaling pathways in primates. PLoS Genet 6. https://doi.org/10.1371/journal.pgen.1001249

Barreiro LB, Quintana-Murci L (2010) From evolutionary genetics to human immunology: how selection shapes host defence genes. Nat Rev Genet 11. https://doi.org/10.1038/nrg2698

Belancio VP, Hedges DJ, Deininger P (2006) LINE-1 RNA splicing and influences on mammalian gene expression. Nucleic Acids Res 34. https://doi.org/10.1093/nar/gkl027

Blaise S, de Parseval N, Heidmann T (2005) Functional characterization of two newly identified human endogenous retrovirus coding envelope genes. Retrovirology 2. https://doi.org/10.1186/1742-4690-2-19

Blume C, Jackson CL, Spalluto CM et al (2021) A novel ACE2 isoform is expressed in human respiratory epithelia and is upregulated in response to interferons and RNA respiratory virus infection. Nat Genet 53. https://doi.org/10.1038/s41588-020-00759-x

Bonaventura P, Alcazer V, Mutez V et al (2022) Identification of shared tumor epitopes from endogenous retroviruses inducing high-avidity cytotoxic T cells for cancer immunotherapy. Sci Adv 8. https://doi.org/10.1126/sciadv.abj3671

Bonté PE, Arribas YA, Merlotti A et al (2022) Single-cell RNA-seq-based proteogenomics identifies glioblastoma-specific transposable elements encoding HLA-I-presented peptides. Cell Rep 39. https://doi.org/10.1016/j.celrep.2022.110916

Bourque G, Leong B, Vega VB et al (2008) Evolution of the mammalian transcription factor binding repertoire via transposable elements. Genome Res 18. https://doi.org/10.1101/gr.080663.108

Bowling EA, Wang JH, Gong F et al (2021) Spliceosome-targeted therapies trigger an antiviral immune response in triple-negative breast cancer. Cell 184:384–403. https://doi.org/10.1016/j.cell.2020.12.031

Brocks D, Schmidt CR, Daskalakis M et al (2017) DNMT and HDAC inhibitors induce cryptic transcription start sites encoded in long terminal repeats. Nat Genet 49:1052–1060. https://doi.org/10.1038/ng.3889

Bu C, Wang Z, Ren Y et al (2022) Syncytin-1 nonfusogenic activities modulate inflammation and contribute to preeclampsia pathogenesis. Cell Mol Life Sci 79. https://doi.org/10.1007/s00018-022-04294-2

Burattin FV, Vadalà R, Panepuccia M et al (2025) LINE1 modulate human T cell function by regulating protein synthesis during the life span. Sci Adv 10:eado2134. https://doi.org/10.1126/sciadv.ado2134

Burbage M, Rocañín-Arjó A, Baudon B et al (2023) Epigenetically controlled tumor antigens derived from splice junctions between exons and transposable elements. Sci Immunol 8. https://doi.org/10.1126/sciimmunol.abm6360

Buttler CA, Ramirez D, Dowell RD, Chuong EB (2023) An intronic LINE-1 regulates IFNAR1 expression in human immune cells. Mob DNA 14. https://doi.org/10.1186/s13100-023-00308-3
Cañadas I, Thummalapalli R, Kim JW et al (2018) Tumor innate immunity primed by specific interferon-stimulated endogenous retroviruses. Nat Med 24. https://doi.org/10.1038/s41591-018-0116-5
Chelmicki T, Roger E, Teissandier A et al (2021) m6A RNA methylation regulates the fate of endogenous retroviruses. Nature 591. https://doi.org/10.1038/s41586-020-03135-1
Chen YG, Hur S (2022) Cellular origins of dsRNA, their recognition and consequences. Nat Rev Mol Cell Biol 23. https://doi.org/10.1038/s41580-021-00430-1
Cherkasova E, Scrivani C, Doh S et al (2016) Detection of an immunogenic HERV-E envelope with selective expression in clear cell kidney cancer. Cancer Res 76. https://doi.org/10.1158/0008-5472.CAN-15-3139
Chiappinelli KB, Strissel PL, Desrichard A et al (2015) Inhibiting DNA methylation causes an interferon response in cancer via dsRNA including endogenous Retroviruses. Cell 162:974–986. https://doi.org/10.1016/j.cell.2015.07.011
Choi H, Kwon J, Soon Cho M et al (2021) Targeting ddx3x triggers antitumor immunity via a dsrna-mediated tumor-intrinsic type i interferon response. Cancer Res 81. https://doi.org/10.1158/0008-5472.CAN-20-3790
Chuong EB, Elde NC, Feschotte C (2017) Regulatory activities of transposable elements: from conflicts to benefits. Nat Rev Genet 18. https://doi.org/10.1038/nrg.2016.139
Chuong EB, Elde NC, Feschotte C (2016) Regulatory evolution of innate immunity through co-option of endogenous retroviruses. Science (80-) 351. https://doi.org/10.1126/science.aad5497
Chuong EB, Rumi MAK, Soares MJ, Baker JC (2013) Endogenous retroviruses function as species-specific enhancer elements in the placenta. Nat Genet 45. https://doi.org/10.1038/ng.2553
Clayton EA, Rishishwar L, Huang TC et al (2020) An atlas of transposable element-derived alternative splicing in cancer. Philos Trans R Soc B Biol Sci 375. https://doi.org/10.1098/rstb.2019.0342
Cuellar TL, Herzner A-M, Zhang X et al (2017) Silencing of retrotransposons by SETDB1 inhibits the interferon response in acute myeloid leukemia. J Cell Biol 216:3535–3549. https://doi.org/10.1083/jcb.201612160
Danac JMC, Matthews RE, Gungi A et al (2024) Competition between two HUSH complexes orchestrates the immune response to retroelement invasion. Mol Cell 84:2870–2881.e5. https://doi.org/10.1016/j.molcel.2024.06.020
De Cecco M, Ito T, Petrashen AP et al (2019) L1 drives IFN in senescent cells and promotes age-associated inflammation. Nature 566. https://doi.org/10.1038/s41586-018-0784-9
De Jager PL, Baecher-Allan C, Maier LM et al (2009) The role of the CD58 locus in multiple sclerosis. Proc Natl Acad Sci USA 106. https://doi.org/10.1073/pnas.0813310106
de Parseval N, Lazar V, Casella J-F et al (2003) Survey of human genes of Retroviral origin: identification and transcriptome of the genes with coding capacity for complete envelope proteins. J Virol 77. https://doi.org/10.1128/jvi.77.19.10414-10422.2003
Deblois G, Tonekaboni SAM, Grillo G et al (2020) Epigenetic switch–induced viral mimicry evasion in chemotherapy-resistant breast cancer. Cancer Discov 10:1312–1329. https://doi.org/10.1158/2159-8290.CD-19-1493
Deininger P (2011) Alu elements: know the SINEs. Genome Biol 12:236. https://doi.org/10.1186/gb-2011-12-12-236
Deniz Ö, Ahmed M, Todd CD et al (2020) Endogenous retroviruses are a source of enhancers with oncogenic potential in acute myeloid leukaemia. Nat Commun 11. https://doi.org/10.1038/s41467-020-17206-4
Dias Junior AG, Sampaio NG, Rehwinkel J (2019) A balancing act: MDA5 in antiviral immunity and autoinflammation. Trends Microbiol 27. https://doi.org/10.1016/j.tim.2018.08.007
Du C, Fan H, Jiang J et al (2024) Unraveling transposable element-mediated regulatory landscape in diverse human immune cells. bioRxiv 20241024620010. https://doi.org/10.1101/2024.10.24.620010

Durbin AF, Wang C, Marcotrigiano J, Gehrke L (2016) RNAs containing modified nucleotides fail to trigger RIG-I conformational changes for innate immune signaling. MBio 7. https://doi.org/10.1128/mBio.00833-16
Eisenberg E, Levanon EY (2018) A-to-I RNA editing - immune protector and transcriptome diversifier. Nat Rev Genet:19
Feng S, Marhon SA, Sokolowski DJ et al (2024) Inhibiting EZH2 targets atypical teratoid rhabdoid tumor by triggering viral mimicry via both RNA and DNA sensing pathways. Nat Commun 15:9321. https://doi.org/10.1038/s41467-024-53515-8
Ferreira LMR, Meissner TB, Mikkelsen TS et al (2016) A distant trophoblast-specific enhancer controls HLA-G expression at the maternal-fetal interface. Proc Natl Acad Sci USA 113. https://doi.org/10.1073/pnas.1602886113
Feschotte C (2008) Transposable elements and the evolution of regulatory networks. Nat Rev Genet 9. https://doi.org/10.1038/nrg2337
Feschotte C, Gilbert C (2012) Endogenous viruses: insights into viral evolution and impact on host biology. Nat Rev Genet 13. https://doi.org/10.1038/nrg3199
Freeman B, White T, Kaul T et al (2022) Analysis of epigenetic features characteristic of L1 loci expressed in human cells. Nucleic Acids Res 50:1888–1907. https://doi.org/10.1093/nar/gkac013
Fueyo R, Judd J, Feschotte C, Wysocka J (2022) Roles of transposable elements in the regulation of mammalian transcription. Nat Rev Mol Cell Biol 23:481–497. https://doi.org/10.1038/s41580-022-00457-y
Fukuda K, Shinkai Y (2020) SETDB1-mediated silencing of retroelements. Viruses 12. https://doi.org/10.3390/v12060596
Galassi C, Chan TA, Vitale I, Galluzzi L (2024) The hallmarks of cancer immune evasion. Cancer Cell 42:1825–1863. https://doi.org/10.1016/j.ccell.2024.09.010
Gao Y, Vasic R, Song Y et al (2020) m6A modification prevents formation of endogenous double-stranded RNAs and deleterious innate immune responses during hematopoietic development. Immunity 52. https://doi.org/10.1016/j.immuni.2020.05.003
Gebrie A (2023) Transposable elements as essential elements in the control of gene expression. Mob DNA 14. https://doi.org/10.1186/s13100-023-00297-3
Giordano J, Ge Y, Gelfand Y et al (2007) Evolutionary history of mammalian transposons determined by genome-wide defragmentation. PLoS Comput Biol 3:e137. https://doi.org/10.1371/journal.pcbi.0030137
Goel S, Decristo MJ, Watt AC et al (2017) CDK4/6 inhibition triggers anti-tumour immunity. Nature 548. https://doi.org/10.1038/nature23465
Gomez S, Cox OL, Walker RR et al (2022) Inhibiting DNA methylation and RNA editing upregulates immunogenic RNA to transform the tumor microenvironment and prolong survival in ovarian cancer. J Immunother Cancer 10. https://doi.org/10.1136/jitc-2022-004974
Goyal A, Bauer J, Hey J et al (2023) DNMT and HDAC inhibition induces immunogenic neoantigens from human endogenous retroviral element-derived transcripts. Nat Commun 14. https://doi.org/10.1038/s41467-023-42417-w
Griffin GK, Wu J, Iracheta-Vellve A et al (2021) Epigenetic silencing by SETDB1 suppresses tumour intrinsic immunogenicity. Nature 595. https://doi.org/10.1038/s41586-021-03520-4
Gröger V, Cynis H (2018) Human endogenous retroviruses and their putative role in the development of autoimmune disorders such as multiple sclerosis. Front Microbiol 9. https://doi.org/10.3389/fmicb.2018.00265
Gu Z, Liu Y, Zhang Y et al (2021) Silencing of LINE-1 retrotransposons is a selective dependency of myeloid leukemia. Nat Genet 53. https://doi.org/10.1038/s41588-021-00829-8
Guirguis AA, Ofir-Rosenfeld Y, Knezevic K et al (2023) Inhibition of METTL3 results in a cell-intrinsic interferon response that enhances antitumor immunity. Cancer Discov 13. https://doi.org/10.1158/2159-8290.CD-23-0007
Hagai T, Chen X, Miragaia RJ et al (2018) Gene expression variability across cells and species shapes innate immunity. Nature 563. https://doi.org/10.1038/s41586-018-0657-2

Hanahan D (2022) Hallmarks of cancer: new dimensions. Cancer Discov 12. https://doi.org/10.1158/2159-8290.CD-21-1059
Hermant C, Torres-Padilla ME (2021) TFs for TEs: the transcription factor repertoire of mammalian transposable elements. Genes Dev 35. https://doi.org/10.1101/GAD.344473.120
Horton I, Kelly CJ, Dziulko A et al (2023) Mouse B2 SINE elements function as IFN-inducible enhancers. elife 12. https://doi.org/10.7554/eLife.82617
Hummel J, Kämmerer U, Müller N et al (2015) Human endogenous retrovirus envelope proteins target dendritic cells to suppress T-cell activation. Eur J Immunol 45. https://doi.org/10.1002/eji.201445366
Ichiyanagi T, Katoh H, Mori Y et al (2021) B2 SINE copies serve as a transposable boundary of DNA methylation and histone modifications in the mouse. Mol Biol Evol 38. https://doi.org/10.1093/molbev/msab033
Ishizuka JJ, Manguso RT, Cheruiyot CK et al (2019) Loss of ADAR1 in tumours overcomes resistance to immune checkpoint blockade. Nature 565:43–48. https://doi.org/10.1038/s41586-018-0768-9
Ito J, Sugimoto R, Nakaoka H et al (2017) Systematic identification and characterization of regulatory elements derived from human endogenous retroviruses. PLoS Genet 13. https://doi.org/10.1371/journal.pgen.1006883
Ivashkiv LB, Donlin LT (2014) Regulation of type i interferon responses. Nat Rev Immunol 14. https://doi.org/10.1038/nri3581
Izquierdo-Bouldstridge A, Bustillos A, Bonet-Costa C et al (2017) Histone H1 depletion triggers an interferon response in cancer cells via activation of heterochromatic repeats. Nucleic Acids Res 45:11622–11642. https://doi.org/10.1093/nar/gkx746
Jacobs FMJ, Greenberg D, Nguyen N et al (2014) An evolutionary arms race between KRAB zinc-finger genes ZNF91/93 and SVA/L1 retrotransposons. Nature 516:242–245. https://doi.org/10.1038/nature13760
Jacques PÉ, Jeyakani J, Bourque G (2013) The majority of primate-specific regulatory sequences are derived from transposable elements. PLoS Genet 9:e1003504. https://doi.org/10.1371/journal.pgen.1003504
Jakubzick CV, Randolph GJ, Henson PM (2017) Monocyte differentiation and antigen-presenting functions. Nat Rev Immunol 17. https://doi.org/10.1038/nri.2017.28
Jang HS, Shah NM, Du AY et al (2019) Transposable elements drive widespread expression of oncogenes in human cancers. Nat Genet 51. https://doi.org/10.1038/s41588-019-0373-3
Johanning GL, Malouf GG, Zheng X et al (2017) Expression of human endogenous retrovirus-K is strongly associated with the basal-like breast cancer phenotype. Sci Rep 7. https://doi.org/10.1038/srep41960
Junhee Seok H, Warren S, Alex GC et al (2013) Genomic responses in mouse models poorly mimic human inflammatory diseases. Proc Natl Acad Sci USA 110. https://doi.org/10.1073/pnas.1222878110
Kahles A, Van LK, Toussaint NC et al (2018) Comprehensive analysis of alternative splicing across tumors from 8,705 patients. Cancer Cell 34. https://doi.org/10.1016/j.ccell.2018.07.001
Kang N, Eccleston M, Clermont PL et al (2020) EZH2 inhibition: a promising strategy to prevent cancer immune editing. Epigenomics 12. https://doi.org/10.2217/epi-2020-0186
Katze MG, Fornek JL, Palermo RE et al (2008) Innate immune modulation by RNA viruses: emerging insights from functional genomics. Nat Rev Immunol 8. https://doi.org/10.1038/nri2377
Kawai T, Ikegawa M, Ori D, Akira S (2024) Decoding Toll-like receptors: recent insights and perspectives in innate immunity. Immunity 57:649–673. https://doi.org/10.1016/J.IMMUNI.2024.03.004
Kelly CJ, Chitko-McKown CG, Chuong EB (2022) Ruminant-specific retrotransposons shape regulatory evolution of bovine immunity. Genome Res 32. https://doi.org/10.1101/gr.276241.121
Kobayashi S, Tokita S, Moniwa K et al (2023) Proteogenomic identification of an immunogenic antigen derived from human endogenous retrovirus in renal cell carcinoma. JCI Insight 8. https://doi.org/10.1172/jci.insight.167712

Kong Y, Rose CM, Cass AA et al (2019) Transposable element expression in tumors is associated with immune infiltration and increased antigenicity. Nat Commun 10:5228. https://doi.org/10.1038/s41467-019-13035-2
Korotkov EV, Korotkova MA, Rudenko VM (2000) MIR: family of repeats common to vertebrate genomes. Mol Biol 34. https://doi.org/10.1007/BF02759556
Krug B, De Jay N, Harutyunyan AS et al (2019) Pervasive H3K27 Acetylation leads to ERV expression and a therapeutic vulnerability in H3K27M Gliomas. Cancer Cell 35. https://doi.org/10.1016/j.ccell.2019.04.004
Laumont CM, Daouda T, Laverdure JP et al (2016) Global proteogenomic analysis of human MHC class I-associated peptides derived from non-canonical reading frames. Nat Commun 7. https://doi.org/10.1038/ncomms10238
Laumont CM, Vincent K, Hesnard L et al (2018) Noncoding regions are the main source of targetable tumor-specific antigens. Sci Transl Med 10. https://doi.org/10.1126/scitranslmed.aau5516
Lee K, Ku J, Ku D, Kim Y (2024) Inverted Alu repeats: friends or foes in the human transcriptome. Exp Mol Med 56:1250–1262. https://doi.org/10.1038/s12276-024-01177-3
Lev-Maor G, Ram O, Kim E et al (2008) Intronic Alus influence alternative splicing. PLoS Genet 4. https://doi.org/10.1371/journal.pgen.1000204
Li D, Wu M (2021) Pattern recognition receptors in health and diseases. Signal Transduct Target Ther 6. https://doi.org/10.1038/s41392-021-00687-0
Li HT, Jang HJ, Rohena-Rivera K et al (2023) RNA mis-splicing drives viral mimicry response after DNMTi therapy in SETD2-mutant kidney cancer. Cell Rep 42. https://doi.org/10.1016/j.celrep.2023.112016
Liu M, Ohtani H, Zhou W et al (2016) Vitamin C increases viral mimicry induced by 5-aza-2′-deoxycytidine. Proc Natl Acad Sci USA 113:10238–10244. https://doi.org/10.1073/pnas.1612262113
Liu M, Thomas SL, DeWitt AK et al (2018a) Dual inhibition of DNA and histone methyltransferases increases viral mimicry in ovarian cancer cells. Cancer Res 78:5754–5766. https://doi.org/10.1158/0008-5472.CAN-17-3953
Liu N, Lee CH, Swigut T et al (2018b) Selective silencing of euchromatic L1s revealed by genome-wide screens for L1 regulators. Nature 553. https://doi.org/10.1038/nature25179
Liu Y, Hu L, Wu Z et al (2023) Loss of PHF8 induces a viral mimicry response by activating endogenous retrotransposons. Nat Commun 14. https://doi.org/10.1038/s41467-023-39943-y
Lokossou AG, Toudic C, Barbeau B (2014) Implication of human endogenous retrovirus envelope proteins in placental functions. Viruses 6. https://doi.org/10.3390/v6114609
Lokossou AG, Toudic C, Nguyen PT et al (2020) Endogenous retrovirus-encoded Syncytin-2 contributes to exosome-mediated immunosuppression of T cells. Biol Reprod 102. https://doi.org/10.1093/biolre/ioz124
Lu M, Zhang Z, Xue M et al (2020) N 6-methyladenosine modification enables viral RNA to escape recognition by RNA sensor RIG-I. Nat Microbiol 5. https://doi.org/10.1038/s41564-019-0653-9
Lunyak VV, Prefontaine GG, Núñez E et al (2007) Developmentally regulated activation of a SINE B2 repeat as a domain boundary in organogenesis. Science (80-) 317. https://doi.org/10.1126/science.1140871
Malassiné A, Blaise S, Handschuh K et al (2007) Expression of the Fusogenic HERV-FRD Env glycoprotein (Syncytin 2) in human placenta is restricted to Villous Cytotrophoblastic cells. Placenta 28. https://doi.org/10.1016/j.placenta.2006.03.001
Mangeney M, Renard M, Schlecht-Louf G et al (2007) Placental syncytins: genetic disjunction between the fusogenic and immunosuppressive activity of retroviral envelope proteins. Proc Natl Acad Sci USA 104. https://doi.org/10.1073/pnas.0707873105
Manghera M, Douville RN (2013) Endogenous retrovirus-K promoter: a landing strip for inflammatory transcription factors? Retrovirology 10. https://doi.org/10.1186/1742-4690-10-16

Marasca F, Sinha S, Vadalà R et al (2022) LINE1 are spliced in non-canonical transcript variants to regulate T cell quiescence and exhaustion. Nat Genet 54. https://doi.org/10.1038/s41588-021-00989-7

McClintock B (1956) Intranuclear systems controlling gene action and mutation. Brookhaven Symp Biol, pp 58–74

Mehdipour P, Marhon SA, Ettayebi I et al (2020) Epigenetic therapy induces transcription of inverted SINEs and ADAR1 dependency. Nature 588. https://doi.org/10.1038/s41586-020-2844-1

Merlotti A, Sadacca B, Arribas YA et al (2023) Noncanonical splicing junctions between exons and transposable elements represent a source of immunogenic recurrent neo-antigens in patients with lung cancer. Sci Immunol 8. https://doi.org/10.1126/sciimmunol.abm6359

Millán-Ariño L, Izquierdo-Bouldstridge A, Jordan A (2016) Specificities and genomic distribution of somatic mammalian histone H1 subtypes. Biochim Biophys Acta - Gene Regul Mech 1859:510–519. https://doi.org/10.1016/j.bbagrm.2015.10.013

Mor G, Aldo P, Alvero AB (2017) The unique immunological and microbial aspects of pregnancy. Nat Rev Immunol 17. https://doi.org/10.1038/nri.2017.64

Morad G, Helmink BA, Sharma P, Wargo JA (2021) Hallmarks of response, resistance, and toxicity to immune checkpoint blockade. Cell 184. https://doi.org/10.1016/j.cell.2021.09.020

Morel KL, Sheahan AV, Burkhart DL et al (2021) EZH2 inhibition activates a dsRNA–STING–interferon stress axis that potentiates response to PD-1 checkpoint blockade in prostate cancer. Nat Cancer 2:444–456. https://doi.org/10.1038/s43018-021-00185-w

Morrison L, Loibl S, Turner NC (2024) The CDK4/6 inhibitor revolution — a game-changing era for breast cancer treatment. Nat Rev Clin Oncol 21. https://doi.org/10.1038/s41571-023-00840-4

Mullins CS, Linnebacher M (2012) Endogenous retrovirus sequences as a novel class of tumor-specific antigens: an example of HERV-H env encoding strong CTL epitopes. Cancer Immunol Immunother 61. https://doi.org/10.1007/s00262-011-1183-3

Murayama T, Nakayama J, Jiang X et al (2024) Targeting DHX9 triggers tumor-intrinsic interferon response and replication stress in small cell lung cancer. Cancer Discov 14. https://doi.org/10.1158/2159-8290.CD-23-0486

Neefjes J, Jongsma MLM, Paul P, Bakke O (2011) Towards a systems understanding of MHC class i and MHC class II antigen presentation. Nat Rev Immunol 11. https://doi.org/10.1038/nri3084

Ng KW, Attig J, Bolland W et al (2020) Tissue-specific and interferon-inducible expression of nonfunctional ACE2 through endogenous retroelement co-option. Nat Genet 52. https://doi.org/10.1038/s41588-020-00732-8

Ng KW, Attig J, Young GR et al (2019) Soluble PD-L1 generated by endogenous retroelement exaptation is a receptor antagonist. elife 8. https://doi.org/10.7554/eLife.50256

Nurk S, Koren S, Rhie A et al (2022) The complete sequence of a human genome. Science (80-) 376:44–53. https://doi.org/10.1126/science.abj6987

Ohtani H, Liu M, Zhou W et al (2018) Switching roles for DNA and histone methylation depend on evolutionary ages of human endogenous retroviruses. Genome Res 28:1147–1157. https://doi.org/10.1101/gr.234229.118

Ohtani H, Ørskov AD, Helbo AS et al (2020) Activation of a subset of evolutionarily young transposable elements and innate immunity are linked to clinical responses to 5-azacytidine. Cancer Res 80:2441–2450. https://doi.org/10.1158/0008-5472.CAN-19-1696

Onabajo OO, Banday AR, Stanifer ML et al (2020) Interferons and viruses induce a novel truncated ACE2 isoform and not the full-length SARS-CoV-2 receptor. Nat Genet 52. https://doi.org/10.1038/s41588-020-00731-9

Pan D, Bao X, Hu M et al (2022) SETDB1 restrains endogenous Retrovirus expression and antitumor immunity during radiotherapy. Cancer Res 82. https://doi.org/10.1158/0008-5472.CAN-21-3523

Pasquesi GIM, Allen H, Ivancevic A et al (2024) Regulation of human interferon signaling by transposon exonization. Cell 187:7621–7636.e19. https://doi.org/10.1016/j.cell.2024.11.016

Patel AJ, Warda S, Maag JLV et al (2022) PRC2-inactivating mutations in cancer enhance cytotoxic response to DNMT1-targeted therapy via enhanced viral mimicry. Cancer Discov 12. https://doi.org/10.1158/2159-8290.CD-21-1671

Payer LM, Steranka JP, Ardeljan D et al (2019) Alu insertion variants alter mRNA splicing. Nucleic Acids Res 47. https://doi.org/10.1093/nar/gky1086

Phillips-Cremins JE, Corces VG (2013) Chromatin insulators: linking genome organization to cellular function. Mol Cell 50. https://doi.org/10.1016/j.molcel.2013.04.018

Priščáková P, Svoboda M, Feketová Z et al (2023) Syncytin-1, syncytin-2 and suppressyn in human health and disease. J Mol Med 101. https://doi.org/10.1007/s00109-023-02385-6

Rastogi I, Jeon D, Moseman JE et al (2022) Role of B cells as antigen presenting cells. Front Immunol 13. https://doi.org/10.3389/fimmu.2022.954936

Rebollo R, Romanish MT, Mager DL (2012) Transposable elements: an abundant and natural source of regulatory sequences for host genes. Annu Rev Genet 46. https://doi.org/10.1146/annurev-genet-110711-155621

Rehwinkel J, Gack MU (2020) RIG-I-like receptors: their regulation and roles in RNA sensing. Nat Rev Immunol 20. https://doi.org/10.1038/s41577-020-0288-3

Reid Cahn A, Bhardwaj N, Vabret N (2022) Dark genome, bright ideas: recent approaches to harness transposable elements in immunotherapies. Cancer Cell 40. https://doi.org/10.1016/j.ccell.2022.07.003

Robbez-Masson L, Tie CHC, Conde L et al (2018) The hush complex cooperates with trim28 to repress young retrotransposons and new genes. Genome Res 28. https://doi.org/10.1101/gr.228171.117

Roberts RM, Ezashi T, Schulz LC et al (2021) Syncytins expressed in human placental trophoblast: human trophoblast syncytins. Placenta 113. https://doi.org/10.1016/j.placenta.2021.01.006

Román AC, González-Rico FJ, Moltó E et al (2011) Dioxin receptor and SLUG transcription factors regulate the insulator activity of B1 SINE retrotransposons via an RNA polymerase switch. Genome Res 21. https://doi.org/10.1101/gr.111203.110

Rooney MS, Shukla SA, Wu CJ et al (2015) Molecular and genetic properties of tumors associated with local immune cytolytic activity. Cell 160. https://doi.org/10.1016/j.cell.2014.12.033

Roulois D, Loo Yau H, Singhania R et al (2015) DNA-Demethylating agents target colorectal cancer cells by inducing viral mimicry by endogenous transcripts. Cell 162:961–973. https://doi.org/10.1016/j.cell.2015.07.056

Rowe HM, Jakobsson J, Mesnard D et al (2010) KAP1 controls endogenous retroviruses in embryonic stem cells. Nature 463:237–240. https://doi.org/10.1038/nature08674

Rubio RM, Depledge DP, Bianco C et al (2018) RNA m(6) A modification enzymes shape innate responses to DNA by regulating interferon β. Genes Dev 32:1472–1484. https://doi.org/10.1101/gad.319475.118

Rycaj K, Plummer JB, Yin B et al (2015) Cytotoxicity of human endogenous retrovirus K-specific T cells toward autologous ovarian cancer cells. Clin Cancer Res 21. https://doi.org/10.1158/1078-0432.CCR-14-0388

Salinas-Pena M, Serna-Pujol N, Jordan A (2024) Genomic profiling of six human somatic histone H1 variants denotes that H1X accumulates at recently incorporated transposable elements. Nucleic Acids Res 52. https://doi.org/10.1093/nar/gkae014

Samstein RM, Josefowicz SZ, Arvey A et al (2012) Extrathymic generation of regulatory T cells in placental mammals mitigates maternal-fetal conflict. Cell 150. https://doi.org/10.1016/j.cell.2012.05.031

Schmid CD, Bucher P (2010) MER41 repeat sequences contain inducible STAT1 binding sites. PLoS One 5. https://doi.org/10.1371/journal.pone.0011425

Schneider WM, Chevillotte MD, Rice CM (2014) Interferon-stimulated genes: a complex web of host defenses. Annu Rev Immunol 32. https://doi.org/10.1146/annurev-immunol-032713-120231

Schroder K, Irvine KM, Taylor MS et al (2012) Conservation and divergence in Toll-like receptor 4-regulated gene expression in primary human versus mouse macrophages. Proc Natl Acad Sci USA 109. https://doi.org/10.1073/pnas.1110156109

Scopa C, Barnada SM, Cicardi ME et al (2023) JUN upregulation drives aberrant transposable element mobilization, associated innate immune response, and impaired neurogenesis in Alzheimer's disease. Nat Commun 14. https://doi.org/10.1038/s41467-023-43728-8

Serafino A, Balestrieri E, Pierimarchi P et al (2009) The activation of human endogenous retrovirus K (HERV-K) is implicated in melanoma cell malignant transformation. Exp Cell Res 315. https://doi.org/10.1016/j.yexcr.2008.12.023

Sha M, Lee X, Li X, Ping et al (2000) Syncytin is a captive retroviral envelope protein involved in human placental morphogenesis. Nature 403. https://doi.org/10.1038/35001608

Shah NM, Jang HJ, Liang Y et al (2023) Pan-cancer analysis identifies tumor-specific antigens derived from transposable elements. Nat Genet 55. https://doi.org/10.1038/s41588-023-01349-3

Shaheen MF, Segar JM, Chmielowski B et al (2023) A phase 2 study of alrizomadlin (APG-115) in combination with pembrolizumab in patients with unresectable or metastatic cutaneous melanoma that has failed immuno-oncologic (IO) drugs. J Clin Oncol 41. https://doi.org/10.1200/jco.2023.41.16_suppl.9559

Sharma P, Goswami S, Raychaudhuri D et al (2023) Immune checkpoint therapy—current perspectives and future directions. Cell 186. https://doi.org/10.1016/j.cell.2023.03.006

Shaw AE, Hughes J, Gu Q et al (2017) Fundamental properties of the mammalian innate immune system revealed by multispecies comparison of type I interferon responses. PLoS Biol 15. https://doi.org/10.1371/journal.pbio.2004086

Shay T, Jojic V, Zuk O et al (2013) Conservation and divergence in the transcriptional programs of the human and mouse immune systems. Proc Natl Acad Sci USA 110. https://doi.org/10.1073/pnas.1222738110

Sheng W, LaFleur MW, Nguyen TH et al (2018) LSD1 ablation stimulates anti-tumor immunity and enables checkpoint blockade. Cell 174. https://doi.org/10.1016/j.cell.2018.05.052

Simon M, Stüve P, Schmidleithner L et al (2024) Single-cell chromatin accessibility and transposable element landscapes reveal shared features of tissue-residing immune cells. Immunity 57:1975–1993.e10. https://doi.org/10.1016/j.immuni.2024.06.015

Smith CC, Beckermann KE, Bortone DS et al (2018) Endogenous retroviral signatures predict immunotherapy response in clear cell renal cell carcinoma. J Clin Invest 128. https://doi.org/10.1172/JCI121476

Srour N, Villarreal OD, Hardikar S et al (2022) PRMT7 ablation stimulates anti-tumor immunity and sensitizes melanoma to immune checkpoint blockade. Cell Rep 38:110582. https://doi.org/10.1016/j.celrep.2022.110582

Su M, Han D, Boyd-Kirkup J et al (2014) Evolution of Alu elements toward enhancers. Cell Rep 7:376–385. https://doi.org/10.1016/j.celrep.2014.03.011

Sundaram V, Cheng Y, Ma Z et al (2014) Widespread contribution of transposable elements to the innovation of gene regulatory networks. Genome Res 24. https://doi.org/10.1101/gr.168872.113

Takahashi Y, Harashima N, Kajigaya S et al (2008) Regression of human kidney cancer following allogeneic stem cell transplantation is associated with recognition of an HERV-E antigen by T cells. J Clin Invest 118. https://doi.org/10.1172/JCI34409

Takeuchi O, Akira S (2010) Pattern recognition receptors and inflammation. Cell 140. https://doi.org/10.1016/j.cell.2010.01.022

Thompson PJ, Macfarlan TS, Lorincz MC (2016) Long terminal repeats: from parasitic elements to building blocks of the transcriptional regulatory repertoire. Mol Cell 62. https://doi.org/10.1016/j.molcel.2016.03.029

Tie CH, Fernandes L, Conde L et al (2018) KAP 1 regulates endogenous retroviruses in adult human cells and contributes to innate immune control. EMBO Rep 19:e45000. https://doi.org/10.15252/embr.201745000

Tolosa JM, Schjenken JE, Clifton VL et al (2012) The endogenous retroviral envelope protein syncytin-1 inhibits LPS/PHA-stimulated cytokine responses in human blood and is sorted into placental exosomes. Placenta 33. https://doi.org/10.1016/j.placenta.2012.08.004

Topalian SL, Drake CG, Pardoll DM (2015) Immune checkpoint blockade: a common denominator approach to cancer therapy. Cancer Cell 27. https://doi.org/10.1016/j.ccell.2015.03.001

Trizzino M, Park YS, Holsbach-Beltrame M et al (2017) Transposable elements are the primary source of novelty in primate gene regulation. Genome Res 27:1623–1633. https://doi.org/10.1101/gr.218149.116

Tunbak H, Enriquez-Gasca R, Tie CHC et al (2020) The HUSH complex is a gatekeeper of type I interferon through epigenetic regulation of LINE-1s. Nat Commun 11. https://doi.org/10.1038/s41467-020-19170-5

Turelli P, Castro-Diaz N, Marzetta F et al (2014) Interplay of TRIM28 and DNA methylation in controlling human endogenous retroelements. Genome Res 24:1260–1270. https://doi.org/10.1101/gr.172833.114

Van De Lagemaat LN, Landry JR, Mager DL, Medstrand P (2003) Transposable elements in mammals promote regulatory variation and diversification of genes with specialized functions. Trends Genet 19. https://doi.org/10.1016/j.tig.2003.08.004

Villesen P, Aagaard L, Wiuf C, Pedersen FS (2004) Identification of endogenous retroviral reading frames in the human genome. Retrovirology 1. https://doi.org/10.1186/1742-4690-1-32

Wang-Johanning F, Radvanyi L, Rycaj K et al (2008) Human endogenous retrovirus K triggers an antigen-specific immune response in breast cancer patients. Cancer Res 68. https://doi.org/10.1158/0008-5472.CAN-07-6838

Wang FQ, Dang X, Su H et al (2024) Association of hyperactivated transposon expression with exacerbated immune activation in systemic lupus erythematosus. Mob DNA 15:23. https://doi.org/10.1186/s13100-024-00335-8

Wang H, Xing J, Grover D et al (2005) SVA elements: a hominid-specific retroposon family. J Mol Biol 354:994–1007. https://doi.org/10.1016/j.jmb.2005.09.085

Wang J, Vicente-García C, Seruggia D et al (2015) MIR retrotransposon sequences provide insulators to the human genome. Proc Natl Acad Sci USA 112. https://doi.org/10.1073/pnas.1507253112

Wang Q, Shi Y, Bian Q et al (2023) Molecular mechanisms of syncytin-1 in tumors and placental development related diseases. Discov Oncol 14. https://doi.org/10.1007/s12672-023-00702-6

Wang X, Wang Q, Zheng C, Wang L (2025) MAVS: the next STING in cancers and other diseases. Crit Rev Oncol Hematol 207:104610. https://doi.org/10.1016/J.CRITREVONC.2024.104610

Winkler R, Gillis E, Lasman L et al (2019) Correction to: m6A modification controls the innate immune response to infection by targeting type I interferons. Nat Immunol 20(2):173–182. https://doi.org/10.1038/s41590-018-0275-z

Wu L, Cao J, Cai WL et al (2018) KDM5 histone demethylases repress immune response via suppression of STING. PLoS Biol 16. https://doi.org/10.1371/journal.pbio.2006134

Wu Q, Nie DY, Ba-alawi W et al (2022) PRMT inhibition induces a viral mimicry response in triple-negative breast cancer. Nat Chem Biol 18. https://doi.org/10.1038/s41589-022-01024-4

Xie M, Hong C, Zhang B et al (2013) DNA hypomethylation within specific transposable element families associates with tissue-specific enhancer landscape. Nat Genet 45:836–841. https://doi.org/10.1038/ng.2649

Xie N, Shen G, Gao W et al (2023) Neoantigens: promising targets for cancer therapy. Signal Transduct Target Ther 8. https://doi.org/10.1038/s41392-022-01270-x

Yan R, Zhang Y, Li Y et al (2020) Structural basis for the recognition of SARS-CoV-2 by full-length human ACE2. Science (80-):367. https://doi.org/10.1126/science.abb2762

Ye M, Goudot C, Hoyler T et al (2020) Specific subfamilies of transposable elements contribute to different domains of T lymphocyte enhancers. Proc Natl Acad Sci USA 117. https://doi.org/10.1073/pnas.1912008117

Yu HL, Zhao ZK, Zhu F (2013) The role of human endogenous retroviral long terminal repeat sequences in human cancer (review). Int J Mol Med 32. https://doi.org/10.3892/ijmm.2013.1460

Zhang SM, Cai WL, Liu X et al (2021a) KDM5B promotes immune evasion by recruiting SETDB1 to silence retroelements. Nature 598:682–687. https://doi.org/10.1038/s41586-021-03994-2
Zhang Y, Liu Q, Yang S, Liao Q (2021b) CD58 Immunobiology at a glance. Front Immunol 12. https://doi.org/10.3389/fimmu.2021.705260
Zhang Y, Romanish MT, Mager DL (2011) Distributions of transposable elements reveal hazardous zones in mammalian introns. PLoS Comput Biol 7. https://doi.org/10.1371/journal.pcbi.1002046
Zhao S, Lu J, Pan B et al (2023) TNRC18 engages H3K9me3 to mediate silencing of endogenous retrotransposons. Nature 623. https://doi.org/10.1038/s41586-023-06688-z
Zhao Y, Oreskovic E, Zhang Q et al (2021) Transposon-triggered innate immune response confers cancer resistance to the blind mole rat. Nat Immunol 22. https://doi.org/10.1038/s41590-021-01027-8
Zheng R, Dunlap M, Bobkov GOM et al (2024) hnRNPM protects against the dsRNA-mediated interferon response by repressing LINE-associated cryptic splicing. Mol Cell 84:2087–2103.e8. https://doi.org/10.1016/j.molcel.2024.05.004
Zhou F, Li M, Wei Y et al (2016) Activation of HERV-K Env protein is essential for tumorigenesis and metastasis of breast cancer cells. Oncotarget 7. https://doi.org/10.18632/oncotarget.11455
Zhou X, Singh M, Santos GS et al (2021) Pharmacologic activation of p53 triggers viral mimicry response thereby abolishing tumor immune evasion and promoting antitumor immunity. Cancer Discov 11:3090–3105. https://doi.org/10.1158/2159-8290.CD-20-1741

Chapter 9
Transposable Elements and Hereditary Diseases

Sulev Kõks

Abstract Hereditary diseases, or genetic diseases, are generally considered to be caused only by single-nucleotide mutations. Typically, the assumption is that a single mutation is solely responsible for the clinical syndrome. This perspective is common, as geneticists search for one specific nucleotide mutation or variant that can account for the clinical phenotype. However, this approach has led to an oversimplified understanding of the genetic pathology or causes of these diseases, resulting in a diagnostic gap for many hereditary conditions. Consequently, we encounter patients with apparent hereditary issues yet lack a meaningful genetic explanation. This has led to reduced diagnostic yield and is one of the main challenges for genetic counsellors. Moreover, GWAS studies have shown limited success in fully elucidating the inheritance of most of the complex traits studied, typically reaching the explained heritability of not more than 30%. These limitations have given rise to the term "missing heritability", which signifies our inability to clarify genetic inheritance using mainstream methods. This chapter presents a potential solution to the missing heritability problem. It demonstrates that much of the heritability remains unaccounted for due to its complexity, arising from transposable or repetitive elements in the human genome. The genetic basis for heritable conditions is intricate and necessitates a sophisticated approach to reveal the parts of missing heritability.

Keywords Repetitive elements · Transposable elements · Genomic variations · Hereditary disease · Genetic pathology

---

S. Kõks (✉)
Medical Genomics Laboratory, Murdoch University, Personalised Medicine Centre, Perron Institute, Murdoch, WA, Australia

Genomic Therapeutics Pty Ltd, Prion OÜ, Perth, WA, Australia
e-mail: Sulev.Koks@murdoch.edu.au

L. M. Vaschetto (ed.), *Transposable Elements, Transcriptomics, and Diseases*, Epigenetics and Human Health 14, https://doi.org/10.1007/978-3-032-04125-8_9

Abbreviations

ACE	Angiotensin I converting enzyme
Alu	Arthrobacter luteus
APC	APC regulator of the WNT signalling pathway
ASO	Antisense oligonucleotide
BamHI	Bacillus amyloliquefaciens HI
Cas9	CRISPR-associated protein 9
CMT1A	Charcot-Marie-Tooth disease type 1A
CRISPR	Clustered regularly interspaced short palindromic repeats
DMD	Duchenne muscular dystrophy
FMR1	Fragile X messenger ribonucleoprotein 1
HBB	Haemoglobin subunit beta
HBD	Haemoglobin subunit delta
HBG1	Haemoglobin subunit gamma 1
HBG2	Haemoglobin subunit gamma 2
HD	Huntington's disease
HERV	Human endogenous retrovirus
HNPP	Hereditary neuropathy with liability to pressure palsies
HpaI	Haemophilus parainfluenzae gene HpaI
HPRT	Hypoxanthine phosphoribosyl transferase
KAP1	TRIM28 tripartite motif containing 28
KRAB-ZFP	KRAB domain-containing zinc finger protein
LINE1	Long interspersed nuclear element-1
LTR	Long terminal repeat
MAST2	Microtubule-associated serine/threonine kinase 2
MFSD8	Major facilitator superfamily domain containing 8
NANOG	Nanog homeobox
NF1	Neurofibromin 1
OCT4	POU5F1, POU class 5 homeobox 1
PD	Parkinson's disease
Prp	Prion protein
RFLP	Restriction-fragment-length polymorphism
RIP	Retrotransposon insertion polymorphisms
RT-PCR	Reverse transcription-polymerase chain reaction
SETDB1	SET domain bifurcated histone lysine methyltransferase 1
SINE	Short interspersed nuclear element
SNP	Single-nucleotide polymorphism
SOX2	SRY-box transcription factor 2
SVA	SINE-VNTR-Alu elements
TAF1	TATA-binding protein-associated factor 1 gene
TC1	Transposon *C. elegans* number 1
TE	Transposable element
TIRs	Terminal inverted repeats

VNTR	Variable number tandem repeat
XDP	X-linked dystonia-parkinsonism
XLDCM	X-linked dilated cardiomyopathy

9.1 Introduction

Any diagnostic endeavour aims to obtain a correct disease diagnosis and verify possible causes. This is especially true for heritable diseases, where the connection between family history and disease is ascertained and documented. But even for these diseases, the road to definitive diagnosis is full of challenges. Quite often, we end with inconclusive evidence for the connection between the mutated gene and disease. There is a substantial unmet medical need and diagnostic gap for heritable diseases. This gap is often referred to as a "missing heritability". Within the following lines in this chapter, I cover some aspects of improving the diagnostic yield and closing this gap.

Transposable elements (TEs) form a large proportion of many eukaryotic genomes, and we are beginning to understand their function (Kõks and Kõks 2018; Kõks et al. 2022b; Lander et al. 2001). TEs are a large and diverse family of elements that form part of the repetitive genome or genomic dark matter (Table 9.1). However, the majority of genetic studies have not addressed them in detail. These repetitive and significant elements are impossible to identify from SNP-based genotyping data, which is the main factor limiting research in this field thus far. With the abundance of sequencing technologies and data, the tools have been developed to correctly and reliably call TEs from these data. One term best suited for repetitive DNA analysis is structural variation analysis of the genome because it can be detected as tandem repeat polymorphisms, minisatellites, or variable number tandem repeats (VNTR). TEs are a special case of repetitive elements, and these are elements that can move around in the genome. TEs can be classified as SINE-VNTR-Alu (SVA), long interspersed nuclear element-1 (LINE-1) elements, Alu-repeats and human endogenous retrovirus (HERV). The role of repetitive elements cannot be overestimated in the structural variants because these elements (including TEs) generate structural variations. Therefore, there is a natural connection between these different elements in the genome, which reflects the dynamic complexity of the genome.

The significance of TEs in human genetics was recognised long ago when the first studies using restriction-fragment-length polymorphism (RFLP) described large fragment polymorphisms of gene mapping between individuals. Early studies on beta-globin (HBB) mapping identified a 5 kb insertion after *HpaI* restriction as a cause for sickle cell anaemia (Kan and Dozy 1978). This insertion caused the detection of 7.0 kb and 13.0 kb variants instead of a 7.6 kb fragment of the beta-globin structural gene, only in the samples of African origin. The long 13.0 kb variant was proposed to be a valuable marker for the sickle cell gene in prenatal diagnosis and was later confirmed to be a LINE1 element. This study also proposed that this

Table 9.1 Overview of the repetitive elements and their classification. Transposable elements are a subgroup of repetitive elements

<table>
<tr><td colspan="13">Repetitive elements</td></tr>
<tr><td colspan="5">Interspersed repeats</td><td rowspan="8">Segmental duplications</td><td rowspan="8">Simple repeats</td><td colspan="6">Tandem repeats</td></tr>
<tr><td rowspan="7">Pseudogenes</td><td colspan="4">Transposable elements</td><td colspan="5">Satellite DNA</td><td rowspan="7">rDNA</td></tr>
<tr><td rowspan="6">Class II DNA transposons</td><td colspan="3">Class I retrotransposons</td><td rowspan="6">Macrosatellites</td><td rowspan="6">Minisatellites</td><td rowspan="6">Microsatellites</td><td rowspan="6">Centromeric</td><td rowspan="6">Telomeric</td></tr>
<tr><td colspan="2">Non-LTR</td><td>LTR</td></tr>
<tr><td>SINE</td><td>LINE</td><td>ERV1</td></tr>
<tr><td>Alu</td><td>L1</td><td>ERV2</td></tr>
<tr><td>SVA</td><td>L2</td><td>ERV3</td></tr>
<tr><td>7SLRNA</td><td>R1</td><td>ERV4</td></tr>
<tr><td>snoRNA</td><td>RTE</td><td>MaLR</td></tr>
</table>

polymorphism is a new genetic marker class and useful for linkage analysis and population studies.

RFLP mapping has been used successfully to map the human gamma-globin gene, with recognition of the repetitive elements as a cause for the hereditary persistence of foetal haemoglobin (Tuan et al. 1979). Similarly, mapping of the HBG2, HBG1, HBD and HBB genes identified variable intervening sequences in this gene cluster of the human population in healthy individuals (Jeffreys 1979). The publications on these length polymorphism reports were eventually proposed as the beginning of "new genetics" (Lewin 1981). And the logic is that these markers can be used for systemic linkage screening for larger population samples. Some of the TEs caused the specific differences in RFLP bands, which were observable on the Southern blots (Kan and Dozy 1978; Dlott et al. 1990). This new genetics concept enables us to build linkage maps that help to discover the "DNA marker loci" responsible for the trait (Botstein et al. 1980). This approach is based on the genomic DNA probe set that describes variability based on genomic repetitive elements and uses the other set that is based on messenger RNA and the cDNA probe set (Botstein et al. 1980). The best example of this new genetics is the forensic DNA analysis that relies on analysing the highly specific fingerprint of the selected short tandem repeats (Gill et al. 1985; Jeffreys et al. 1985a, b, c). This example illustrates the power of using repetitive DNA sequences in genomic studies.

9.2 Transposable Elements and Heritable Diseases

9.2.1 Transposable Elements as Causes of Diseases

Literature is full of evidence that TEs cause genetic diseases, and frequently, the unknown cause for the disease is discovered as an inserted TE. There are multiple mechanisms for how TEs can cause diseases. They can induce alternative splicing or exon skipping, or they can be exonised (pseudoexonisation) and cause nonsense-mediated decay (Kõks et al. 2022b; Pfaff et al. 2022). Inserted TEs also regulate the expression of surrounding genes (*cis*) or even at a distance (*trans*) (Payer and Burns 2019; Koks et al. 2021). The advantage of using TEs in genetic studies is that when a TE is identified as a mutation behind the condition, its functional impact is relatively easy to predict compared to SNPs, where often the hits are in non-coding regions, or the identified variants are nonsynonymous without precise mechanisms for amino acid change. When TEs are identified, the mechanism can be validated using RNA analysis because of their impact on transcriptional alterations (Fig. 9.1).

Early genetic studies identified structural variants in genes related to cancer. The first non-cancer genes for structural variant analysis were globin genes, followed by the hypoxanthine phosphoribosyl transferase (HPRT) (Nussbaum et al. 1983; Kan and Dozy 1978; Jeffreys 1979; Tuan et al. 1979). The first genetic condition that was identified as being caused by the 6 kb insertion in the beta-globin gene was

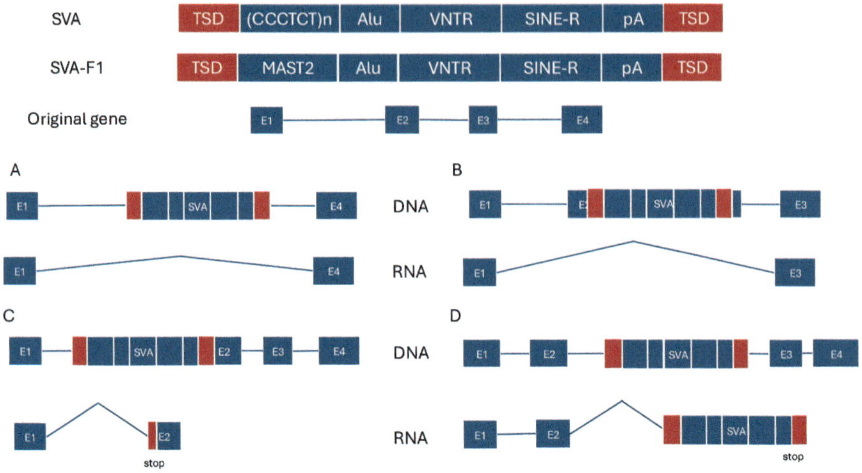

Fig. 9.1 Description of the pathogenicity of transposable elements. Four different mechanisms are described for SVAs. Two types of SVAs are illustrated, one with the hexamer and the other with the MAST2 gene. SVAS can induce deletion (**a**), be inserted into an exon and cause exon skipping (**b**), be inserted into an exon and create a stop codon inside the exon (**c**), or SVAS can be located in the intron and pseudoexonised during transcription, causing truncation of the RNA (**d**)

sickle cell disease (Kan and Dozy 1978). At the human HPRT locus, a three-allele restriction-fragment-length polymorphism for the restriction endonuclease *BamHI* using cloned cDNA sequences was discovered (Nussbaum et al. 1983). These three alleles were evident as distinct pairs on Southern blots of fragments that hybridised to HPRT cDNA: 22 kb/25 kb pair, a 12 kb/25 kb pair and a 22 kb/18 kb pair. The HPRT gene is related to the Lesch-Nyhan syndrome, and while TEs don't directly cause the syndrome, Alu-Alu recombination has caused duplication of the segment with exons 7 and 8 of the gene (Marcus et al. 1993).

Thereafter, speculations emerged that fragile X syndrome can be caused by the insertion of TE in two independent papers (Hoegerman and Rary 1986; Friedman and Howard-Peebles 1986). This hypothesis was later proved correct; fragile X is a CGG triplet expansion disease with a very large span of the repeat length. The range in healthy subjects is 5–40, and the pathological range is more than 200 repeats in the FMR1 gene, but the repeats can be up to 2.4 kb in length (Steyaert et al. 1996; Eichler et al. 1996; Gurling et al. 1997). Interestingly, the role of TEs in fragile X syndrome is also functional. In the Drosophila model for fragile X, the silencing of the FMR1 gene caused activation of TEs and significant overexpression of transposons (De Donno et al. 2023). Therefore, for fragile X, the disease mutation is a structural variant that leads to the activation of TEs, indicating the pathophysiological impact of TEs.

Haemophilia A was probably the first disease where the insertion of LINE1 in the gene coding area was documented (Kazazian et al. 1988). The authors reported LINE1 insertions into exon 14 of the factor VIII gene in two patients with haemophilia A. These insertions were of different sizes, 3.8 and 2.3 kb, and contained 3′

ends of the LINE1 sequence. Both of these insertions created target site duplications of at least 12 and 13 nucleotides of the factor VIII gene (Kazazian et al. 1988). This report is the first study showing the "jumping genes" in action (Holzman 1991). Later, the same group isolated a truncated de novo LINE1 insertion from one of the patients, JH-27 (Dombroski et al. 1991). The original, full-length LINE1 was identified at chromosome 22, and the authors showed that the de novo mutation mechanism indeed resulted in retrotransposition of LINE1 from chromosome 22 to the factor VIII gene (Dombroski et al. 1991; Mathias et al. 1991). The same group also confirmed that the original full-length LINE1 has reverse transcriptase activity (Mathias et al. 1991). These three papers were fundamental studies proposing and confirming the lasting activity of retrotransposons and their role in de novo mutagenesis.

Afterwards, several smaller insertions were discovered, causing the Tay-Sachs disease, alpha 1-antitrypsin deficiency, osteogenesis imperfecta, hereditary elliptocytosis and juvenile Sandhoff disease (Myerowitz and Costigan 1988; Curiel et al. 1989; Genovese et al. 1989; Roux et al. 1989; Dlott et al. 1990). These mutations have larger-than-usual insertions, and while they are not directly insertions from TEs, the link between structural variants and TEs intrigues their mechanism.

One of the most well-studied variants in human genetics is the insertion/deletion polymorphism in the angiotensin I converting enzyme gene (ACE) (Rigat et al. 1990; Tiret et al. 1992). This polymorphism is an Alu repeat, which is very commonly inserted in the human population, leading to reduced ACE activity. It is defined as the presence (insertion) or absence (deletion) of a 250-bp DNA fragment in intron 16 of the ACE gene. This polymorphism was used as a marker genotype in a study involving 80 healthy subjects, whose serum ACE levels were concomitantly measured. Absence (or deletion, D) was found with a frequency of 0.6, and the presence (or insertion, I) was found with a frequency of 0.4. In serum, ACE levels were markedly different between subjects in each of the three ACE genotypes (DD, ID and II) classes. Serum immunoreactive ACE concentrations were 299.3 +/− 49 μg/l for homozygotes with the I/I ($n = 14$), 392.6 +/− 66.8 μg/l for heterozygotes with the D/I ($n = 37$) and 494.1 +/− 88.3 μg/l for homozygotes with the D/D ($n = 29$). The I/D polymorphism accounts for 47% of the total variance of serum ACE, showing that the Alu repeat insertion is the major determinant for the serum ACE concentration (Rigat et al. 1990). This polymorphism has been extensively studied for very different diseases and physiological conditions, most notably for athletic endurance (Mägi et al. 2016). This polymorphism is, however, linked to different health conditions like idiopathic dilated cardiomyopathy. The frequency of the D allele was 0.757 in patients with dilated cardiomyopathy compared to 0.6 in controls, and this difference was significant (x2 = 3.96; $P < 0.05$) (Harn et al. 1995). ACE is responsible for the production of angiotensin II and the breakdown of kinins, leading to increased blood pressure (BP), induction of vascular smooth muscle cell proliferation and the stimulation of myocardial-cell hypertrophy (Harn et al. 1995). While ACE is also related to the risk of hypertension, its polymorphism is not related to the salt sensitivity of blood pressure (Kojima et al. 1994). The ACE gene is also a well-characterised locus associated with the pathogenesis and progression of chronic kidney disease. The I/I genotype is protective against the development and progression of nephropathy and is associated with a slower progression of nondiabetic proteinuric kidney disease (Ruggenenti et al. 2008). The DD genotype is associated with a better response to therapy with angiotensin II receptor antagonists in nephropathy and to ACE inhibitors in male patients with nondiabetic nephropathies. Therefore, evaluating the ACE I/D polymorphism is a reliable pharmacogenomic tool to identify patients at risk and those who may benefit the most from renoprotective therapy with ACE inhibitors or angiotensin II receptor antagonists (Ruggenenti et al. 2008). Interestingly, while the association between the ACE activity and the ACE Alu polymorphism was confirmed in another study, this study did not find an association with the severity of an autosomal dominant polycystic kidney disease (Schiavello et al. 2001).

Alu repeats have been found to cause many neurological diseases. Mutations in the NF1 gene in the region 17q11.2 cause von Recklinghausen neurofibromatosis. One type of mutation that has been described is autosomal translocation with breakpoints at 17q11.2 (Wallace et al. 1990). In some patients, de novo Alu insertion has been described as a cause of the disease, and in many cases, different sizes of insertions have been described that affect the clinical severity and gene expression (Wallace et al. 1991; Martin-Gallardo et al. 1992; Colman et al. 1997; Evans et al. 1998).

Another example is for Alzheimer's disease, where 101 individuals with atypical dementias were screened, identifying five individuals with insertion repeats in the prion protein (Prp) gene (Owen et al. 1991). One of these five identified insertions was larger (166 bp) than that described previously, suggesting that the individuals with these mutations are unlikely to be all lineally related and that insertions in the Prp gene may not be uncommon in prion diseases. This is an octapeptide repeat region in the Prp gene, and more than four octapeptides are considered to be pathogenic (Owen et al. 1991). Similarly, a case of familial Creutzfeldt-Jakob disease associated with a 144 base-pair insertion in the open reading frame of the Prp gene has been described (Nicholl et al. 1995). The brother of the proband had died from "Huntington's disease", but re-analysis of the neuropathology confirmed spongiform encephalopathy with a positive anti-prion protein.

Huntington's disease (HD) is a late-onset, autosomal dominant, fatal neurodegenerative disorder that presents in mid-adult life with personality disturbances and involuntary movements. Using a direct complementary DNA selection strategy, seven transcriptional units were identified within the minimal region believed to contain the HD gene (Goldberg et al. 1993). Screening with cDNA clones identified an Alu insertion in genomic DNA from two individuals with HD, which showed complete cosegregation with the disease in these families but was not found in 1000 control chromosomes. This study identified a 12-kilobase transcript that mapped near the Alu insertion site, and this newly discovered transcript was described as a strong candidate for causing HD. (Goldberg et al. 1993) Another study that searched for the gene causing HD detected an Alu retrotransposition event in two families in the HD region 4p16.3 (Hutchinson et al. 1993). Sequence analysis demonstrated that this Alu element is similar to five other Alu elements in the GenBank database that are characterised by a seven-base-pair internal duplication, forming a distinct Sb2 subfamily. This study provided support for multiple source genes for Alu retrotransposition in the human genome (Hutchinson et al. 1993).

The next disease with the vast data for the involvement of TEs is progressive muscular dystrophy of the Duchenne type (DMD), which is caused by different variable mutations in the dystrophin gene. One such study identified the deleted junction fragment from a DMD patient, revealing that the proximal breakpoint of the deletion in intron 43 fell within the sequence of a transposon-like element. This segment, belonging to the THE-1 family of human TEs, is usually present in a complete form in intron 43 of the dystrophin gene. This identified deletion was maternally transmitted and resulted in the elimination of two-thirds of the THE-1 element. DNA analysis from additional DMD patients revealed a second deletion with the proximal breakpoint mapping within the same THE-1 element (Pizzuti et al. 1992). In another report, the evidence of retrotransposition of LINE1 was provided and found inserted into a patient's dystrophin gene, causing DMD (Narita et al. 1993). In their analysis, a 600 bp insertion was found in exon 44 from the genomic DNA of two Japanese brothers with DMD. Interestingly, the mother of the boys had both normal DMD and DMD with the insertion, but the maternal grandparents had only the normal DMD gene. The LINE1 insertion must therefore have occurred de novo in the mother. The insertion was present zero to two bases upstream from the 3′ end of exon 44 and caused the deletion of two to four bases from the insertion site of the exon sequence (Narita et al. 1993). This LINE1 insertion caused exon 44 skipping during splicing, as was identified using RT-PCR and sequencing of the amplification product. This novel dystrophin gene mutation has important implications regarding retrotransposition of an active LINE1 element.

In a study of 454 families with patients affected with DMD, four families were found with two or more affected patients who were related through paternal lines (Zatz et al. 1991). Two affected cousins showed different DNA deletions in one of these families, suggesting two independent and possibly de novo mutations. In the case of the other two, DNA deletions were also detected in the dystrophin gene. In the fourth family, with three affected patients, no DNA deletions were found, but muscle biopsies showed a negative dystrophin immunohistochemical staining. Although one of these families might have occurred by chance, the probability of finding the other three families with DMD patients constitutes a rare event (Zatz et al. 1991, 1993). These reports led to the conclusion that other mechanisms, such as TEs in other sites of the genome, could be responsible in some families for a greater predisposition for pathogenic deletions, duplications or mutations. The analysis of 967 DMD families revealed an atypical (paternal) inheritance pattern in 3% of familial cases, again supporting TEs' involvement (Zatz et al. 1998). These studies show the importance of testing all affected patients within each family to prevent possible mistakes in carrier detection, genetic counselling and prenatal diagnosis (Zatz et al. 1998).

The study, which sequenced and characterised the 5′ end of intron 11, provides more evidence of TE involvement in DMD (Ferlini and Muntoni 1998). A 136 bp sequence rearranged in intron 11 was found in a patient affected by X-linked dilated cardiomyopathy due to a dystrophinopathy. This was a multiple copy sequence with features of a repetitive element, and its comparative analysis showed a very high homology with Alu, LINE1 and THE elements. The pattern of homology suggests that it may represent a novel Alu-like, transcriptionally active sequence with a possible retrotransposable capacity (Ferlini and Muntoni 1998). The authors hypothesised that the 5′ region of the dystrophin intron 11 may have a role in transposing this novel Alu-like sequence, as it contains common target areas for the insertion of mobile elements (Ferlini and Muntoni 1998).

A different disease, but with the same locus and gene, is X-linked dilated cardiomyopathy (XLDCM). This is a clinical phenotype of dystrophinopathy, which is characterised by preferential myocardial involvement without any signs of skeletal myopathy (Yoshida et al. 1998). A unique de novo LINE1 insertion in the exon one dystrophin gene in three unrelated XLDCM patients was identified (Yoshida et al. 1998). The LINE1 was inversely integrated in the 5′-untranslated region in exon 1, and it affected the transcription and the stability of the muscular form of dystrophin transcripts but not the brain form. The authors proposed that the identified LINE1 insertion in the dystrophin gene is causative for some of the Japanese patients with XLDCM (Yoshida et al. 1998).

The family has been reported to have hereditary elliptocytosis and pyropoikilocytosis associated with a truncated protein (Hassoun et al. 1994). This study found the biochemical characteristics of abnormal protein and showed that a mobile element disrupted the alpha-spectrin gene, resulting in exon skipping. This element was 632 bp long, duplicated the insertion site and terminated by a long poly-A tail downstream of multiple consensus polyadenylation signals. One to three copies per individual were found using Southern blot analysis, and this element was used as a probe (Hassoun et al. 1994).

The Charcot-Marie-Tooth disease type 1A (CMT1A) duplication and hereditary neuropathy with liability to pressure palsies (HNPP) deletion are reciprocal products of an unequal crossing-over event between misaligned flanking CMT1A-REP repeats (Reiter et al. 1996). Examining this apparently homologous recombination event detected novel junction fragments from the recombinant CMT1A-REPs in both CMT1A and HNPP patients, with a 1.7-kb recombination hotspot. The sequence analysis revealed a mariner transposon-like element near the hotspot, which was suggested to be a causative event and mediate strand exchange by a transposase (Reiter et al. 1996).

In addition to the monogenic diseases, many multifactorial diseases have been identified to be caused or modified by TEs. Parkinson's disease (PD) is a good example, with many identified single-nucleotide variants leading to the disease (Bandres-Ciga et al. 2019; Nalls et al. 2019). More recent studies using next-generation sequencing and whole-genome sequencing have revealed the complex and diverse impact of TEs in the progression of PD (Pfaff et al. 2021; Kõks et al. 2022a; Fröhlich et al. 2023). Some of the TEs have a protective influence, and they can slow down the neurodegeneration in Parkinson's patients (Kõks et al. 2022a). At the same time, some TES were related to a much faster progression of the disease (Fröhlich et al. 2023).

X-linked dystonia-parkinsonism (XDP) has become an exemplary disease to show the pathophysiology of the TEs. XDP is a movement disorder endemic to the Philippines (Makino et al. 2007). The disease is caused by a disease-specific SVA (composite element combining SINE, VNTR and Alu repeat) retrotransposon insertion in an intron of the TATA-binding protein-associated factor 1 gene (TAF1) (Makino et al. 2007). The TAF1 gene encodes the most significant component of the TFIID complex. This study identified significantly decreased expression levels of TAF1 and the dopamine receptor D2 gene (DRD2) in the caudate nucleus caused by the SVA. In addition to the reduced expression, an abnormal methylation pattern was also described in the DNA of the patient's caudate. Later studies identified intron retention caused by the SVA in the TAF1 gene which was finally considered to be a pathogenetic mechanism of XDP (Aneichyk et al. 2018; Bragg et al. 2017, 2019; Capponi et al. 2021). Knowing the molecular mechanism causing diseases would be essential for the understanding of disease progression and for drug development. Current technologies offer opportunities to correct the splicing errors, and I address some examples later in this chapter.

9.2.2 *Transcriptional Studies*

After identifying the DNA variants in the genome, the next step in the investigation is to find the functional consequences of the variant. The first and most reliable step is to analyse alterations on the RNA level, the expression level and splicing events or alternative splicing. As I described with the role of SVA on the TAF1 locus to generate XDP, similar mechanisms are very common for other diseases and need to be identified by using RNA-seq technology (Aneichyk et al. 2018). This is a genuinely functional genomic approach where the effect of DNA variants is investigated in the context of the transcriptome or gene expression. It does not need to be a whole transcriptome as the interpretation and analysis can be challenging. Therefore, a useful alternative is the approach focused on functionally connected genetic networks like co-regulated pathways for specific diseases, biomarker discovery, or even in experimental models (Kingo et al. 2008; Lill et al. 2013; Koks et al. 2004). These focused approaches can be performed using qualitative RT-PCR assays and assay combinations. It is essential to remember that inheritance patterns can be atypical even in very well-known Mendelian diseases. The focused gene expression studies are relevant for conditions with clear inheritance patterns and more complex conditions, like psoriasis (Loite et al. 2021; Keermann et al. 2015). Transcript diversification or alternative splicing is driven by TEs, indicating their crucial role in the disease mechanisms (Ayarpadikannan et al. 2015). Another aspect related to the RNA analysis is the combination of RNA analysis with the DNA analysis. The combination of DNA analysis with the RNA expression or whole transcriptome will drastically improve the diagnostic yield and improve the therapeutic options for patients (Deyell et al. 2024). This has also been proven for cancer patients, where the survival improved when molecular analysis combined DNA and RNA data

(Rodon et al. 2019). Targeted gene reporting was identified as therapeutically actionable in 32% of participants compared to 62% of participants with whole-genome sequencing of cancer patients (Deyell et al. 2024). This yield improved even up to 96% after integrating the whole transcriptome data (Deyell et al. 2024). In addition to the cancer examples, these complex risk scores are useful for mental disorders and skin pathologies (Koido et al. 2012; Kingo et al. 2007).

Transcriptome analysis is an invaluable source for biomarker discovery, especially if it is blood-based and can easily be repeated during the course of the disease. Blood analysis has to consider many confounding factors, such as sex, the time of day, habits like smoking, or even seasonal fluctuations, to achieve the high sensitivity of the biomarkers (Koks et al. 2000; Rull et al. 2013; Rothwell et al. 2019; Koks et al. 2015, 2016). One fundamental factor that needs to be taken into account is the population stratification that has been studied very well with Parkinson's disease and population-specific genetic architecture, with characterising population-specific risk scores and haplotype structures, has been described (Bandres-Ciga et al. 2019; Lill et al. 2015).

TEs have been shown to participate in nuclear organisation and to function as insulators and boundary elements that partition the genome into transcriptionally active and inactive domains. In mouse embryonic stem cells, satellite repeat and LINE-1 RNA transcripts were found to scaffold heterochromatin compartmentalisation, contributing to global A/B compartment integrity and transcriptional insulation (Lu et al. 2021; Novo et al. 2022). Evolutionarily, younger elements are significantly enriched in the repressed regions, and older elements are in active areas (Trizzino et al. 2018; Sundaram et al. 2014). LTR retrotransposons, particularly ERVs, profoundly influence regulatory landscape evolution and chromatin architecture. TEs frequently function as alternative promoters and *cis*-regulatory enhancers harbouring dense clusters of transcription factor (TF) binding motifs and can regulate genes over distances exceeding 100 kb in a lineage- or cell-type-specific manner. For example, ERV-derived LTRs are highly enriched in immune-related regulatory regions and modulate interferon-stimulated genes in a stimulus-dependent manner (Chuong et al. 2016). LTR elements also contribute to transcriptional innovation during embryogenesis and reprogramming, as demonstrated by their recruitment of key pluripotency TFs such as OCT4, SOX2 and NANOG (Kunarso et al. 2010). Notably, TEs have been implicated in establishing or reinforcing topologically associating domain (TAD) boundaries, a central feature of 3D genome organisation. High-resolution Hi-C datasets have revealed that LTR elements, including ERV1 and ERVK families, are significantly enriched at TAD borders across various mouse and human tissues (Choudhary et al. 2022; Diehl et al. 2020). These boundaries often coincide with CTCF binding and cohesin complex recruitment, suggesting that LTRs may act as insulator-like elements. The co-option (exaptation) of these mobile elements during evolution has played a major role in regulatory innovation (Sundaram et al. 2014). In human embryonic stem cells, HERV-H elements, a primate-specific ERV subfamily, have been shown to function as core regulatory hubs essential for maintaining pluripotency and chromatin structure (Pontis et al. 2019). These TE-derived elements integrate with canonical transcriptional circuits and facilitate long-range enhancer–promoter interactions through chromatin looping. These findings underscore the dual role of retrotransposons as both architectural elements and regulatory modules with activating or repressing effects. In an extensive eQTL analysis, SVA elements have been shown to have widespread activating effects both in *cis* and *trans*, while LINE1 elements have repressing effects on many genes in *cis* and *trans* positions (Koks et al. 2021). The regulatory activity of TEs is tightly modulated by epigenetic silencing mechanisms, such as KRAB-ZFP/KAP1/SETDB1-mediated repression, and can be reactivated during development, disease, or epigenetic remodelling (Pontis et al. 2019). Their capacity to reorganise chromatin domains and mediate long-distance gene regulation suggests a previously underappreciated role in normal physiology and pathological states such as cancer or neurodevelopmental disorders. The roles of TEs in genome architecture are discussed in Chap. 3.

TEs offer undiscovered diagnostic potential by detecting their insertions or deletions. Retrotransposon insertion polymorphisms (RIPs), particularly those involving LINE1, Alu, SVAs and HERVs, can disrupt gene function or regulation (Rebollo et al. 2012). The stable and heritable nature of RIPs, along with their influence on gene expression, supports their utility in diagnosing monogenic and complex genetic disorders. RIP patterns present a promising class of diagnostic biomarkers for hereditary diseases due to their polymorphic, stable and sometimes pathogenic characteristics. As detection and interpretation technology has advanced recently, RIPs have become a valuable tool in clinical genomics. RIPs can alter gene function through various mechanisms. As seen in L1 insertions into tumour suppressor genes like APC in familial adenomatous polyposis (Miki et al. 1992), they can disrupt coding sequences. RIPs can affect splicing patterns, illustrated by an Alu insertion disrupting the normal processing of NF1 transcripts (Wallace et al. 1991). Furthermore, RIPs can also have epigenetic effects, such as heterochromatin spreading or enhancer hijacking, leading to dysregulation of nearby gene expression (Rebollo et al. 2012). RIPs are usually detected by DNA sequencing, but RNA sequencing is also a very useful tool to understand the effects of the detected RIP.

9.3 Treatment Opportunities

After the diagnostic odyssey's successful conclusion, the next significant question is what to do with this information. A genomic pathology summary with identified mutations is a starting point for a precision therapeutic approach. Gene therapy has become a reality, and gene replacement therapy is already available for many of the diseases described (Northington et al. 2025; Malay et al. 2024; Kwiatkowski et al. 2024). These are all therapies that use viral gene transfer (adenoviruses or lentiviruses) or even replicating associated viral transfer that allows a one-time therapeutic injection to be delivered, like in the case of Zolgensma (Fisher et al. 1991). RNA-based therapies have opened an entirely new avenue, providing tools that can either completely fix the problem or at least improve the patients' quality of life and slow the disease's progression.

Antisense oligonucleotides (ASOs) help to fix the transcriptional problems, where the endogenous gene is active but is underperforming or malfunctioning. One of the first and most remarkable successes was with Duchenne muscular dystrophy (Lu et al. 2003; Wilton et al. 1999). In the mdx mouse, which carries a nonsense mutation in exon 23 of the dystrophin gene and is an animal model of Duchenne muscular dystrophy, the naturally occurring exon-skipping mechanism was identified that led to the dystrophin-positive fibres in the muscles (Wilton et al. 1999). The authors designed and tested ASOs to re-direct normal processing of the dystrophin pre-mRNA and induce specific, mutated exon skipping. They used 2′-O-methyl-ASOs, targeting the 3′ and 5′ splice sites of introns 22 and 23 and found that the 5′ ASO appeared to efficiently displace factors usually involved in the removal of intron 23, and exon 23 was also removed during the splicing of the dystrophin pre-mRNA. As a result, all dystrophin gene transcripts showed skipping of exon 23 after 24 h. Deletion of exon 23 did not disrupt the reading frame and should allow the synthesis of a shorter but presumably functional Becker-like dystrophin (Wilton et al. 1999; Mann et al. 2001; Lu et al. 2003). This discovery led to the development of several therapeutics (eteplirsen for exon 51 skipping) for Duchenne patients. These therapeutics act at the dystrophin pre-mRNA splicing level, and they reduce the severity of a Duchenne mutation to the milder Becker phenotype. These ASOs have been a tremendous success, have an excellent safety profile and are a great example of how targeting the RNA can be a viable drug development option (Aung-Htut et al. 2019; Lu et al. 2003).

In addition to the ASO therapeutic opportunity, TEs themselves generate chimeric transcripts with the genes they have activated. This is well known and described in the case of cancers (Shah et al. 2023). These chimeric transcripts can be used as highly specific disease markers related to the pathogenesis of the condition. Similar joint transcripts have been found in other complex diseases, like ALS (Pfaff et al. 2023). This shows great potential as disease-specific, pathogenetically relevant transcripts that can be used as drug targets.

CRISPR/Cas technology offers additional solutions to modify the effects of TEs or even completely cure the disease by excising the TE from the genome. This technology can provide highly efficient and seamless correction of structural variants caused by TEs, as seen in the case of thalassemia (Xie et al. 2014). It has been successfully demonstrated for the XDP, where SVA excision with CRISPR/Cas completely rescued the normal transcriptional signature (Aneichyk et al. 2018; Petrozziello et al. 2020).

The development of antisense-based therapeutics is truly booming, and as we identify new mutations, new RNAs suitable for targeting are also discovered. One remarkable study comes from Boston, Massachusetts, where genome sequencing of a Batten's disease patient identified an SVA deep in the intron 6 of the MFSD8 gene. This SVA induced mis-splicing and pseudoexonisation of parts of intron six, which in turn lead to premature translational termination and disease (Kim et al. 2019). The authors designed a set of ASOs, tested them by using the patient's fibroblasts and identified an ASO that could increase the concentration of normal mRNA more than threefold (Kim et al. 2019). This ASO was named milasen, and FDA approval was obtained to conduct an investigational clinical trial by using N-of-1 design.

The drug was effective, significantly reducing the symptoms of the disease without any significant side effects. This study offered a possible workflow and path for the rapid development of personalised therapeutics and provided the proof of concept of genomic medicine.

9.4 Conclusions

TEs play a substantial role in hereditary diseases and are the direct cause of many diseases. Moreover, structural variants, the genomic variants that are so hard to detect, are directly related to the activity of TEs and are the result of TE activities in the genome. Variations in the TE carry a massive amount of genomic variability in the human population and are responsible for the missing heritability problem because they were so difficult to analyse in the past. The emergence of new genomic technologies, like ultra-long sequencing, the combination of DNA and RNA analysis and success with the ASO therapeutics has generated an unprecedented momentum in experimental and clinical medicine, where the genomic information can be analysed very fast with actionable conclusions. Genomics and transcriptomics have certainly become mainstream analytical tools for medical research and drug development.

Acknowledgements The funding and support from the Johanna Sewell Memorial Fund and Perron Institute are appreciated.

Compliance with Ethical Standards

Conflict of Interest Sulev Kõks is a founder and a director of Genomic Therapeutics Pty Ltd. and Prion OÜ, private companies dedicated to genomic medicine.

Ethical Approval This chapter contains no studies with human/animal participants performed by the author.

References

Aneichyk T, Hendriks WT, Yadav R, Shin D, Gao D, Vaine CA, Collins RL, Domingo A, Currall B, Stortchevoi A, Multhaupt-Buell T, Penney EB, Cruz L, Dhakal J, Brand H, Hanscom C, Antolik C, Dy M, Ragavendran A, Underwood J, Cantsilieris S, Munson KM, Eichler EE, Acuña P, Go C, Jamora RDG, Rosales RL, Church DM, Williams SR, Garcia S, Klein C, Müller U, Wilhelmsen KC, Timmers HTM, Sapir Y, Wainger BJ, Henderson D, Ito N, Weisenfeld N, Jaffe D, Sharma N, Breakefield XO, Ozelius LJ, Bragg DC, Talkowski ME (2018) Dissecting the causal mechanism of X-linked dystonia-parkinsonism by integrating genome and transcriptome assembly. Cell 172(5):897–909.e821. https://doi.org/10.1016/j.cell.2018.02.011

Aung-Htut MT, McIntosh CS, Ham KA, Pitout IL, Flynn LL, Greer K, Fletcher S, Wilton SD (2019) Systematic approach to developing splice modulating antisense oligonucleotides. Int J Mol Sci 20(20). https://doi.org/10.3390/ijms20205030

Ayarpadikannan S, Lee HE, Han K, Kim HS (2015) Transposable element-driven transcript diversification and its relevance to genetic disorders. Gene 558(2):187–194. https://doi.org/10.1016/j.gene.2015.01.039

Bandres-Ciga S, Ahmed S, Sabir MS, Blauwendraat C, Adarmes-Gomez AD, Bernal-Bernal I, Bonilla-Toribio M, Buiza-Rueda D, Carrillo F, Carrion-Claro M, Gomez-Garre P, Jesus S, Labrador-Espinosa MA, Macias D, Mendez-Del-Barrio C, Perinan-Tocino T, Tejera-Parrado C, Vargas-Gonzalez L, Diez-Fairen M, Alvarez I, Tartari JP, Buongiorno M, Aguilar M, Gorostidi A, Bergareche JA, Mondragon E, Vinagre-Aragon A, Croitoru I, Ruiz-Martinez J, Dols-Icardo O, Kulisevsky J, Marin-Lahoz J, Pagonabarraga J, Pascual-Sedano B, Ezquerra M, Camara A, Compta Y, Fernandez M, Fernandez-Santiago R, Munoz E, Tolosa E, Valldeoriola F, Gonzalez-Aramburu I, Sanchez Rodriguez A, Sierra M, Menendez-Gonzalez M, Blazquez M, Garcia C, Suarez-San Martin E, Garcia-Ruiz P, Martinez-Castrillo JC, Vela-Desojo L, Ruz C, Barrero FJ, Escamilla-Sevilla F, Minguez-Castellanos A, Cerdan D, Tabernero C, Gomez Heredia MJ, Perez Errazquin F, Romero-Acebal M, Feliz C, Lopez-Sendon JL, Mata M, Martinez Torres I, Kim JJ, Dalgard CL, American Genome C, Brooks J, Saez-Atienzar S, Gibbs JR, Jorda R, Botia JA, Bonet-Ponce L, Morrison KE, Clarke C, Tan M, Morris H, Edsall C, Hernandez D, Simon-Sanchez J, Nalls MA, Scholz SW, Jimenez-Escrig A, Duarte J, Vives F, Duran R, Hoenicka J, Alvarez V, Infante J, Marti MJ, Clarimon J, Lopez de Munain A, Pastor P, Mir P, Singleton A, International Parkinson Disease Genomics C (2019) The genetic architecture of Parkinson disease in Spain: characterizing population-specific risk, differential haplotype structures, and providing etiologic insight. Mov Disord 34(12):1851–1863. https://doi.org/10.1002/mds.27864

Botstein D, White RL, Skolnick M, Davis RW (1980) Construction of a genetic linkage map in man using restriction fragment length polymorphisms. Am J Hum Genet 32(3):314–331

Bragg DC, Mangkalaphiban K, Vaine CA, Kulkarni NJ, Shin D, Yadav R, Dhakal J, Ton ML, Cheng A, Russo CT, Ang M, Acuña P, Go C, Franceour TN, Multhaupt-Buell T, Ito N, Müller U, Hendriks WT, Breakefield XO, Sharma N, Ozelius LJ (2017) Disease onset in X-linked dystonia-parkinsonism correlates with expansion of a hexameric repeat within an SVA retrotransposon in TAF1. Proc Natl Acad Sci USA 114(51):E11020–e11028. https://doi.org/10.1073/pnas.1712526114

Bragg DC, Sharma N, Ozelius LJ (2019) X-linked dystonia-parkinsonism: recent advances. Curr Opin Neurol 32(4):604–609. https://doi.org/10.1097/wco.0000000000000708

Capponi S, Stöffler N, Penney EB, Grütz K, Nizamuddin S, Vermunt MW, Castelijns B, Fernandez-Cerado C, Legarda GP, Velasco-Andrada MS, Muñoz EL, Ang MA, Diesta CCE, Creyghton MP, Klein C, Bragg DC, De Rijk P, Timmers HTM (2021) Dissection of TAF1 neuronal splicing and implications for neurodegeneration in X-linked dystonia-parkinsonism. Brain Commun 3(4):fcab253. https://doi.org/10.1093/braincomms/fcab253

Choudhary MN, Quaid K, Xing X, Schmidt H, Wang T (2022) Widespread contribution of transposable elements to the rewiring of mammalian 3D genomes and gene regulation. bioRxiv:2022.2002.2001.475239. https://doi.org/10.1101/2022.02.01.475239

Chuong EB, Elde NC, Feschotte C (2016) Regulatory evolution of innate immunity through co-option of endogenous retroviruses. Science 351(6277):1083–1087. https://doi.org/10.1126/science.aad5497

Colman SD, Abernathy CR, Ho VT, Wallace MR (1997) Four frameshift mutations in neurofibromatosis type 1 caused by small insertions. J Med Genet 34(7):579–581. https://doi.org/10.1136/jmg.34.7.579

Curiel D, Brantly M, Curiel E, Stier L, Crystal RG (1989) Alpha 1-antitrypsin deficiency caused by the alpha 1-antitrypsin Nullmattawa gene. An insertion mutation rendering the alpha 1-antitrypsin gene incapable of producing alpha 1-antitrypsin. J Clin Invest 83(4):1144–1152. https://doi.org/10.1172/jci113994

De Donno MD, Puricella A, D'Attis S, Specchia V, Bozzetti MP (2023) Expression of transposable elements in the brain of the Drosophila melanogaster model for fragile X syndrome. Genes (Basel) 14(5). https://doi.org/10.3390/genes14051060

Deyell RJ, Shen Y, Titmuss E, Dixon K, Williamson LM, Pleasance E, Nelson JMT, Abbasi S, Krzywinski M, Armstrong L, Bonakdar M, Ch'ng C, Chuah E, Dunham C, Fok A, Jones M, Lee AF, Ma Y, Moore RA, Mungall AJ, Mungall KL, Rogers PC, Schrader KA, Virani A, Wee K, Young SS, Zhao Y, Jones SJM, Laskin J, Marra MA, Rassekh SR (2024) Whole genome and transcriptome integrated analyses guide clinical care of pediatric poor prognosis cancers. Nat Commun 15(1):4165. https://doi.org/10.1038/s41467-024-48363-5

Diehl AG, Ouyang N, Boyle AP (2020) Transposable elements contribute to cell and species-specific chromatin looping and gene regulation in mammalian genomes. Nat Commun 11(1):1796. https://doi.org/10.1038/s41467-020-15520-5

Dlott B, d'Azzo A, Quon DV, Neufeld EF (1990) Two mutations produce intron insertion in mRNA and elongated beta-subunit of human beta-hexosaminidase. J Biol Chem 265(29):17921–17927

Dombroski BA, Mathias SL, Nanthakumar E, Scott AF, Kazazian HH Jr (1991) Isolation of an active human transposable element. Science 254(5039):1805–1808. https://doi.org/10.1126/science.1662412

Eichler EE, Macpherson JN, Murray A, Jacobs PA, Chakravarti A, Nelson DL (1996) Haplotype and interspersion analysis of the FMR1 CGG repeat identifies two different mutational pathways for the origin of the fragile X syndrome. Hum Mol Genet 5(3):319–330. https://doi.org/10.1093/hmg/5.3.319

Evans DG, Trueman L, Wallace A, Collins S, Strachan T (1998) Genotype/phenotype correlations in type 2 neurofibromatosis (NF2): evidence for more severe disease associated with truncating mutations. J Med Genet 35(6):450–455. https://doi.org/10.1136/jmg.35.6.450

Ferlini A, Muntoni F (1998) The 5′ region of intron 11 of the dystrophin gene contains target sequences for mobile elements and three overlapping ORFs. Biochem Biophys Res Commun 242(2):401–406. https://doi.org/10.1006/bbrc.1997.7976

Fisher CW, Lau KS, Fisher CR, Wynn RM, Cox RP, Chuang DT (1991) A 17-bp insertion and a Phe215–Cys missense mutation in the dihydrolipoyl transacylase (E2) mRNA from a thiamine-responsive maple syrup urine disease patient WG-34. Biochem Biophys Res Commun 174(2):804–809. https://doi.org/10.1016/0006-291x(91)91489-y

Friedman JM, Howard-Peebles PN (1986) Inheritance of fragile X syndrome: an hypothesis. Am J Med Genet 23(1–2):701–713. https://doi.org/10.1002/ajmg.1320230161

Fröhlich A, Pfaff AL, Bubb VJ, Quinn JP, Koks S (2023) Reference LINE-1 insertion polymorphisms correlate with Parkinson's disease progression and differential transcript expression in the PPMI cohort. Sci Rep 13(1):13857. https://doi.org/10.1038/s41598-023-41052-1

Genovese C, Brufsky A, Shapiro J, Rowe D (1989) Detection of mutations in human type I collagen mRNA in osteogenesis imperfecta by indirect RNase protection. J Biol Chem 264(16):9632–9637

Gill P, Jeffreys AJ, Werrett DJ (1985) Forensic application of DNA 'fingerprints'. Nature 318(6046):577–579. https://doi.org/10.1038/318577a0

Goldberg YP, Rommens JM, Andrew SE, Hutchinson GB, Lin B, Theilmann J, Graham R, Glaves ML, Starr E, McDonald H et al (1993) Identification of an Alu retrotransposition event in close proximity to a strong candidate gene for Huntington's disease. Nature 362(6418):370–373. https://doi.org/10.1038/362370a0

Gurling HM, Bolton PF, Vincent J, Melmer G, Rutter M (1997) Molecular and cytogenetic investigations of the fragile X region including the Frax A and Frax E CGG trinucleotide repeat sequences in families multiplex for autism and related phenotypes. Hum Hered 47(5):254–262. https://doi.org/10.1159/000154421

Harn HJ, Chang CY, Ho LI, Liu CA, Jeng JR, Lin FG, Jent W (1995) Evidence that polymorphism of the angiotensin I converting enzyme gene may be related to idiopathic dilated cardiomyopathy in the Chinese population. Biochem Mol Biol Int 35(6):1175–1181

Hassoun H, Coetzer TL, Vassiliadis JN, Sahr KE, Maalouf GJ, Saad ST, Catanzariti L, Palek J (1994) A novel mobile element inserted in the alpha spectrin gene: spectrin dayton. A truncated alpha spectrin associated with hereditary elliptocytosis. J Clin Invest 94(2):643–648. https://doi.org/10.1172/jci117380

Hoegerman SF, Rary JM (1986) Speculation on the role of transposable elements in human genetic disease with particular attention to achondroplasia and the fragile X syndrome. Am J Med Genet 23(1–2):685–699. https://doi.org/10.1002/ajmg.1320230160

Holzman D (1991) A "jumping gene" caught in the act. Science 254(5039):1728–1729. https://doi.org/10.1126/science.1662411

Hutchinson GB, Andrew SE, McDonald H, Goldberg YP, Graham R, Rommens JM, Hayden MR (1993) An Alu element retroposition in two families with Huntington disease defines a new active Alu subfamily. Nucleic Acids Res 21(15):3379–3383. https://doi.org/10.1093/nar/21.15.3379

Jeffreys AJ (1979) DNA sequence variants in the G gamma-, A gamma-, delta- and beta-globin genes of man. Cell 18(1):1–10. https://doi.org/10.1016/0092-8674(79)90348-9

Jeffreys AJ, Brookfield JF, Semeonoff R (1985a) Positive identification of an immigration test-case using human DNA fingerprints. Nature 317(6040):818–819. https://doi.org/10.1038/317818a0

Jeffreys AJ, Wilson V, Thein SL (1985b) Hypervariable 'minisatellite' regions in human DNA. Nature 314(6006):67–73. https://doi.org/10.1038/314067a0

Jeffreys AJ, Wilson V, Thein SL (1985c) Individual-specific 'fingerprints' of human DNA. Nature 316(6023):76–79. https://doi.org/10.1038/316076a0

Kan YW, Dozy AM (1978) Polymorphism of DNA sequence adjacent to human beta-globin structural gene: relationship to sickle mutation. Proc Natl Acad Sci USA 75(11):5631–5635. https://doi.org/10.1073/pnas.75.11.5631

Kazazian HH Jr, Wong C, Youssoufian H, Scott AF, Phillips DG, Antonarakis SE (1988) Haemophilia A resulting from de novo insertion of L1 sequences represents a novel mechanism for mutation in man. Nature 332(6160):164–166. https://doi.org/10.1038/332164a0

Keermann M, Kõks S, Reimann E, Prans E, Abram K, Kingo K (2015) Transcriptional landscape of psoriasis identifies the involvement of IL36 and IL36RN. BMC Genomics 16(1):322. https://doi.org/10.1186/s12864-015-1508-2

Kim J, Hu C, Moufawad El Achkar C, Black LE, Douville J, Larson A, Pendergast MK, Goldkind SF, Lee EA, Kuniholm A, Soucy A, Vaze J, Belur NR, Fredriksen K, Stojkovska I, Tsytsykova A, Armant M, DiDonato RL, Choi J, Cornelissen L, Pereira LM, Augustine EF, Genetti CA, Dies K, Barton B, Williams L, Goodlett BD, Riley BL, Pasternak A, Berry ER, Pflock KA, Chu S, Reed C, Tyndall K, Agrawal PB, Beggs AH, Grant PE, Urion DK, Snyder RO, Waisbren SE, Poduri A, Park PJ, Patterson A, Biffi A, Mazzulli JR, Bodamer O, Berde CB, Yu TW (2019) Patient-customized oligonucleotide therapy for a rare genetic disease. N Engl J Med 381(17):1644–1652. https://doi.org/10.1056/NEJMoa1813279

Kingo K, Mossner R, Koks S, Ratsep R, Kruger U, Vasar E, Reich K, Silm H (2007) Association analysis of IL19, IL20 and IL24 genes in palmoplantar pustulosis. Br J Dermatol 156(4):646–652. https://doi.org/10.1111/j.1365-2133.2006.07731.x

Kingo K, Aunin E, Karelson M, Ratsep R, Silm H, Vasar E, Koks S (2008) Expressional changes in the intracellular melanogenesis pathways and their possible role in the pathogenesis of vitiligo. J Dermatol Sci 52(1):39–46. https://doi.org/10.1016/j.jdermsci.2008.03.013

Koido K, Traks T, Balõtšev R, Eller T, Must A, Koks S, Maron E, Tõru I, Shlik J, Vasar V, Vasar E (2012) Associations between LSAMP gene polymorphisms and major depressive disorder and panic disorder. Transl Psychiatry 2(8):e152. https://doi.org/10.1038/tp.2012.74

Kojima S, Inenaga T, Matsuoka H, Kuramochi M, Omae T, Nara Y, Yamori Y (1994) The association between salt sensitivity of blood pressure and some polymorphic factors. J Hypertens 12(7):797–801

Kõks S, Kõks G (2018) Chapter 22: The role of human endogenous retroviruses (HERVs) in the pathologies of the nervous system. In: Gerlai RT (ed) Molecular-genetic and statistical techniques for behavioral and neural research. Academic, San Diego, pp 519–533. https://doi.org/10.1016/B978-0-12-804078-2.00022-2

Koks S, Mannisto PT, Bourin M, Shlik J, Vasar V, Vasar E (2000) Cholecystokinin-induced anxiety in rats: relevance of pre-experimental stress and seasonal variations. J Psychiatry Neurosci 25(1):33–42

Koks S, Luuk H, Nelovkov A, Areda T, Vasar E (2004) A screen for genes induced in the amygdaloid area during cat odor exposure. Genes Brain Behav 3(2):80–89. https://doi.org/10.1046/j.1601-183x.2003.00047.x
Koks G, Uudelepp ML, Limbach M, Peterson P, Reimann E, Koks S (2015) Smoking-induced expression of the GPR15 gene indicates its potential role in chronic inflammatory pathologies. Am J Pathol 185(11):2898–2906. https://doi.org/10.1016/j.ajpath.2015.07.006
Kõks S, Keermann M, Reimann E, Prans E, Abram K, Silm H, Kõks G, Kingo K (2016) Psoriasis-specific RNA isoforms identified by RNA-Seq analysis of 173,446 transcripts. Front Med (Lausanne) 3:46. https://doi.org/10.3389/fmed.2016.00046
Koks S, Pfaff AL, Bubb VJ, Quinn JP (2021) Expression quantitative trait loci (eQTLs) associated with retrotransposons demonstrate their modulatory effect on the transcriptome. Int J Mol Sci 22(12). https://doi.org/10.3390/ijms22126319
Kõks S, Pfaff AL, Singleton LM, Bubb VJ, Quinn JP (2022a) Non-reference genome transposable elements (TEs) have a significant impact on the progression of the Parkinson's disease. Exp Biol Med (Maywood) 247(18):1680–1690. https://doi.org/10.1177/15353702221117147
Kõks S, Singleton LM, Quinn JP, Bubb VJ, Pfaff AL (2022b) Analysis of the retrotransposon SINE-VNTR-Alu (SVA) polymorphisms in the genetics and pathophysiology of complex diseases. In: Proukakis C (ed) Genomic structural variants in nervous system disorders. Springer US, New York, pp 63–77. https://doi.org/10.1007/978-1-0716-2357-2_4
Kunarso G, Chia NY, Jeyakani J, Hwang C, Lu X, Chan YS, Ng HH, Bourque G (2010) Transposable elements have rewired the core regulatory network of human embryonic stem cells. Nat Genet 42(7):631–634. https://doi.org/10.1038/ng.600
Kwiatkowski JL, Walters MC, Hongeng S, Yannaki E, Kulozik AE, Kunz JB, Sauer MG, Thrasher AJ, Thuret I, Lal A, Tao G, Ali S, Thakar HL, Elliot H, Lodaya A, Lee J, Colvin RA, Locatelli F, Thompson AA (2024) Betibeglogene autotemcel gene therapy in patients with transfusion-dependent, severe genotype β-thalassaemia (HGB-212): a non-randomised, multicentre, single-arm, open-label, single-dose, phase 3 trial. Lancet 404(10468):2175–2186. https://doi.org/10.1016/s0140-6736(24)01884-1
Lander ES, Linton LM, Birren B, Nusbaum C et al (2001) Initial sequencing and analysis of the human genome. Nature 409(6822):860–921. https://doi.org/10.1038/35057062
Lewin R (1981) Jumping genes help trace inherited diseases. Science 211(4483):690–692. https://doi.org/10.1126/science.6256858
Lill M, Koks S, Soomets U, Schalkwyk LC, Fernandes C, Lutsar I, Taba P (2013) Peripheral blood RNA gene expression profiling in patients with bacterial meningitis. Front Neurosci 7:33. https://doi.org/10.3389/fnins.2013.00033
Lill CM, Hansen J, Olsen JH, Binder H, Ritz B, Bertram L (2015) Impact of Parkinson's disease risk loci on age at onset. Mov Disord 30(6):847–850. https://doi.org/10.1002/mds.26237
Loite U, Raam L, Reimann E, Reemann P, Prans E, Traks T, Vasar E, Silm H, Kingo K, Kõks S (2021) The expression pattern of genes related to Melanogenesis and endogenous opioids in psoriasis. Int J Mol Sci 22(23). https://doi.org/10.3390/ijms222313056
Lu QL, Mann CJ, Lou F, Bou-Gharios G, Morris GE, Xue SA, Fletcher S, Partridge TA, Wilton SD (2003) Functional amounts of dystrophin produced by skipping the mutated exon in the mdx dystrophic mouse. Nat Med 9(8):1009–1014. https://doi.org/10.1038/nm897
Lu JY, Chang L, Li T, Wang T, Yin Y, Zhan G, Han X, Zhang K, Tao Y, Percharde M, Wang L, Peng Q, Yan P, Zhang H, Bi X, Shao W, Hong Y, Wu Z, Ma R, Wang P, Li W, Zhang J, Chang Z, Hou Y, Zhu B, Ramalho-Santos M, Li P, Xie W, Na J, Sun Y, Shen X (2021) Homotypic clustering of L1 and B1/Alu repeats compartmentalizes the 3D genome. Cell Res 31(6):613–630. https://doi.org/10.1038/s41422-020-00466-6
Mägi A, Unt E, Prans E, Raus L, Eha J, Veraksitš A, Kingo K, Kõks S (2016) The association analysis between ACE and ACTN3 genes polymorphisms and endurance capacity in Young cross-country skiers: longitudinal study. J Sports Sci Med 15(2):287–294
Makino S, Kaji R, Ando S, Tomizawa M, Yasuno K, Goto S, Matsumoto S, Tabuena MD, Maranon E, Dantes M, Lee LV, Ogasawara K, Tooyama I, Akatsu H, Nishimura M, Tamiya G (2007)

Reduced neuron-specific expression of the TAF1 gene is associated with X-linked dystonia-parkinsonism. Am J Hum Genet 80(3):393–406. https://doi.org/10.1086/512129
Malay J, Salama RAA, Alam Qureshi GS, Ammar A, Janardhan G, Safdar M, Elshamy HAH (2024) Gene therapy: a revolutionary step in treating thalassemia. Hematol Rep 16(4):656–668. https://doi.org/10.3390/hematolrep16040064
Mann CJ, Honeyman K, Cheng AJ, Ly T, Lloyd F, Fletcher S, Morgan JE, Partridge TA, Wilton SD (2001) Antisense-induced exon skipping and synthesis of dystrophin in the mdx mouse. Proc Natl Acad Sci USA 98(1):42–47. https://doi.org/10.1073/pnas.98.1.42
Marcus S, Hellgren D, Lambert B, Fällström SP, Wahlström J (1993) Duplication in the hypoxanthine phosphoribosyl-transferase gene caused by Alu-Alu recombination in a patient with Lesch Nyhan syndrome. Hum Genet 90(5):477–482. https://doi.org/10.1007/bf00217444
Martin-Gallardo A, Marchuk DA, Gocayne J, Kerlavage AR, McCombie WR, Venter JC, Collins FS, Wallace MR (1992) Sequencing and analysis of genomic fragments from the NF1 locus. DNA Seq 3(4):237–243. https://doi.org/10.3109/10425179209034023
Mathias SL, Scott AF, Kazazian HH Jr, Boeke JD, Gabriel A (1991) Reverse transcriptase encoded by a human transposable element. Science 254(5039):1808–1810. https://doi.org/10.1126/science.1722352
Miki Y, Nishisho I, Horii A, Miyoshi Y, Utsunomiya J, Kinzler KW, Vogelstein B, Nakamura Y (1992) Disruption of the APC gene by a retrotransposal insertion of L1 sequence in a colon cancer. Cancer Res 52(3):643–645
Myerowitz R, Costigan FC (1988) The major defect in Ashkenazi Jews with Tay-Sachs disease is an insertion in the gene for the alpha-chain of beta-hexosaminidase. J Biol Chem 263(35):18587–18589
Nalls MA, Blauwendraat C, Vallerga CL, Heilbron K, Bandres-Ciga S, Chang D, Tan M, Kia DA, Noyce AJ, Xue A, Bras J, Young E, von Coelln R, Simón-Sánchez J, Schulte C, Sharma M, Krohn L, Pihlstrøm L, Siitonen A, Iwaki H, Leonard H, Faghri F, Gibbs JR, Hernandez DG, Scholz SW, Botia JA, Martinez M, Corvol JC, Lesage S, Jankovic J, Shulman LM, Sutherland M, Tienari P, Majamaa K, Toft M, Andreassen OA, Bangale T, Brice A, Yang J, Gan-Or Z, Gasser T, Heutink P, Shulman JM, Wood NW, Hinds DA, Hardy JA, Morris HR, Gratten J, Visscher PM, Graham RR, Singleton AB (2019) Identification of novel risk loci, causal insights, and heritable risk for Parkinson's disease: a meta-analysis of genome-wide association studies. Lancet Neurol 18(12):1091–1102. https://doi.org/10.1016/s1474-4422(19)30320-5
Narita N, Nishio H, Kitoh Y, Ishikawa Y, Ishikawa Y, Minami R, Nakamura H, Matsuo M (1993) Insertion of a 5′ truncated L1 element into the 3′ end of exon 44 of the dystrophin gene resulted in skipping of the exon during splicing in a case of Duchenne muscular dystrophy. J Clin Invest 91(5):1862–1867. https://doi.org/10.1172/jci116402
Nicholl D, Windl O, de Silva R, Sawcer S, Dempster M, Ironside JW, Estibeiro JP, Yuill GM, Lathe R, Will RG (1995) Inherited Creutzfeldt-Jakob disease in a British family associated with a novel 144 base pair insertion of the prion protein gene. J Neurol Neurosurg Psychiatry 58(1):65–69. https://doi.org/10.1136/jnnp.58.1.65
Northington MW, Rice SE, Holmes AL, Watts Alexander CS (2025) Gene-ius at work: hemophilia B treatment enters a new era. Am J Health Syst Pharm. https://doi.org/10.1093/ajhp/zxaf005
Novo CL, Wong EV, Hockings C, Poudel C, Sheekey E, Wiese M, Okkenhaug H, Boulton SJ, Basu S, Walker S, Kaminski Schierle GS, Narlikar GJ, Rugg-Gunn PJ (2022) Satellite repeat transcripts modulate heterochromatin condensates and safeguard chromosome stability in mouse embryonic stem cells. Nat Commun 13(1):3525. https://doi.org/10.1038/s41467-022-31198-3
Nussbaum RL, Crowder WE, Nyhan WL, Caskey CT (1983) A three-allele restriction-fragment-length polymorphism at the hypoxanthine phosphoribosyltransferase locus in man. Proc Natl Acad Sci USA 80(13):4035–4039. https://doi.org/10.1073/pnas.80.13.4035
Owen F, Poulter M, Collinge J, Leach M, Shah T, Lofthouse R, Chen YF, Crow TJ, Harding AE, Hardy J et al (1991) Insertions in the prion protein gene in atypical dementias. Exp Neurol 112(2):240–242. https://doi.org/10.1016/0014-4886(91)90075-n

Payer LM, Burns KH (2019) Transposable elements in human genetic disease. Nat Rev Genet 20(12):760–772. https://doi.org/10.1038/s41576-019-0165-8

Petrozziello T, Dios AM, Mueller KA, Vaine CA, Hendriks WT, Glajch KE, Mills AN, Mangkalaphiban K, Penney EB, Ito N, Fernandez-Cerado C, Legarda GPA, Velasco-Andrada MS, Acuña PJ, Ang MA, Muñoz EL, Diesta CCE, Macalintal-Canlas R, Acuña G, Sharma N, Ozelius LJ, Bragg DC, Sadri-Vakili G (2020) SVA insertion in X-linked dystonia parkinsonism alters histone H3 acetylation associated with TAF1 gene. PLoS One 15(12):e0243655. https://doi.org/10.1371/journal.pone.0243655

Pfaff AL, Bubb VJ, Quinn JP, Koks S (2021) Reference SVA insertion polymorphisms are associated with Parkinson's disease progression and differential gene expression. NPJ Parkinsons Dis 7(1):44. https://doi.org/10.1038/s41531-021-00189-4

Pfaff AL, Singleton LM, Kõks S (2022) Mechanisms of disease-associated SINE-VNTR-Alus. Exp Biol Med (Maywood) 247(9):756–764. https://doi.org/10.1177/15353702221082612

Pfaff AL, Bubb VJ, Quinn JP, Koks S (2023) A genome-wide screen for the Exonisation of reference SINE-VNTR-Alus and their expression in CNS tissues of individuals with amyotrophic lateral sclerosis. Int J Mol Sci 24(14). https://doi.org/10.3390/ijms241411548

Pizzuti A, Pieretti M, Fenwick RG, Gibbs RA, Caskey CT (1992) A transposon-like element in the deletion-prone region of the dystrophin gene. Genomics 13(3):594–600. https://doi.org/10.1016/0888-7543(92)90129-g

Pontis J, Planet E, Offner S, Turelli P, Duc J, Coudray A, Theunissen TW, Jaenisch R, Trono D (2019) Hominoid-specific transposable elements and KZFPs facilitate human embryonic genome activation and control transcription in naive human ESCs. Cell Stem Cell 24(5):724–735.e725. https://doi.org/10.1016/j.stem.2019.03.012

Rebollo R, Romanish MT, Mager DL (2012) Transposable elements: an abundant and natural source of regulatory sequences for host genes. Annu Rev Genet 46:21–42. https://doi.org/10.1146/annurev-genet-110711-155621

Reiter LT, Murakami T, Koeuth T, Pentao L, Muzny DM, Gibbs RA, Lupski JR (1996) A recombination hotspot responsible for two inherited peripheral neuropathies is located near a mariner transposon-like element. Nat Genet 12(3):288–297. https://doi.org/10.1038/ng0396-288

Rigat B, Hubert C, Alhenc-Gelas F, Cambien F, Corvol P, Soubrier F (1990) An insertion/deletion polymorphism in the angiotensin I-converting enzyme gene accounting for half the variance of serum enzyme levels. J Clin Invest 86(4):1343–1346. https://doi.org/10.1172/jci114844

Rodon J, Soria JC, Berger R, Miller WH, Rubin E, Kugel A, Tsimberidou A, Saintigny P, Ackerstein A, Braña I, Loriot Y, Afshar M, Miller V, Wunder F, Bresson C, Martini JF, Raynaud J, Mendelsohn J, Batist G, Onn A, Tabernero J, Schilsky RL, Lazar V, Lee JJ, Kurzrock R (2019) Genomic and transcriptomic profiling expands precision cancer medicine: the WINTHER trial. Nat Med 25(5):751–758. https://doi.org/10.1038/s41591-019-0424-4

Rothwell DG, Ayub M, Cook N, Thistlethwaite F, Carter L, Dean E, Smith N, Villa S, Dransfield J, Clipson A, White D, Nessa K, Ferdous S, Howell M, Gupta A, Kilerci B, Mohan S, Frese K, Gulati S, Miller C, Jordan A, Eaton H, Hickson N, O'Brien C, Graham D, Kelly C, Aruketty S, Metcalf R, Chiramel J, Tinsley N, Vickers AJ, Kurup R, Frost H, Stevenson J, Southam S, Landers D, Wallace A, Marais R, Hughes AM, Brady G, Dive C, Krebs MG (2019) Utility of ctDNA to support patient selection for early phase clinical trials: the TARGET study. Nat Med 25(5):738–743. https://doi.org/10.1038/s41591-019-0380-z

Roux AF, Morlé F, Guetarni D, Colonna P, Sahr K, Forget BG, Delaunay J, Godet J (1989) Molecular basis of Sp alpha I/65 hereditary elliptocytosis in North Africa: insertion of a TTG triplet between codons 147 and 149 in the alpha-spectrin gene from five unrelated families. Blood 73(8):2196–2201

Ruggenenti P, Bettinaglio P, Pinares F, Remuzzi G (2008) Angiotensin converting enzyme insertion/deletion polymorphism and renoprotection in diabetic and nondiabetic nephropathies. Clin J Am Soc Nephrol 3(5):1511–1525. https://doi.org/10.2215/cjn.04140907

Rull K, Tomberg K, Koks S, Mannik J, Mols M, Sirotkina M, Varv S, Laan M (2013) Increased placental expression and maternal serum levels of apoptosis-inducing TRAIL in recurrent miscarriage. Placenta 34(2):141–148. https://doi.org/10.1016/j.placenta.2012.11.032

Schiavello T, Burke V, Bogdanova N, Jasik P, Melsom S, Boudville N, Robertson K, Angelicheva D, Dworniczak B, Lemmens M, Horst J, Todorov V, Dimitrakov D, Sulowicz W, Krasniak A, Stompor T, Beilin L, Hallmayer J, Kalaydjieva L, Thomas M (2001) Angiotensin-converting enzyme activity and the ACE Alu polymorphism in autosomal dominant polycystic kidney disease. Nephrol Dial Transplant 16(12):2323–2327. https://doi.org/10.1093/ndt/16.12.2323

Shah NM, Jang HJ, Liang Y, Maeng JH, Tzeng SC, Wu A, Basri NL, Qu X, Fan C, Li A, Katz B, Li D, Xing X, Evans BS, Wang T (2023) Pan-cancer analysis identifies tumor-specific antigens derived from transposable elements. Nat Genet 55(4):631–639. https://doi.org/10.1038/s41588-023-01349-3

Steyaert J, Borghgraef M, Legius E, Fryns JP (1996) Molecular-intelligence correlations in young fragile X males with a mild CGG repeat expansion in the FMR1 gene. Am J Med Genet 64(2):274–277. https://doi.org/10.1002/(sici)1096-8628(19960809)64:2<274::Aid-ajmg8>3.0.Co;2-u

Sundaram V, Cheng Y, Ma Z, Li D, Xing X, Edge P, Snyder MP, Wang T (2014) Widespread contribution of transposable elements to the innovation of gene regulatory networks. Genome Res 24(12):1963–1976. https://doi.org/10.1101/gr.168872.113

Tiret L, Rigat B, Visvikis S, Breda C, Corvol P, Cambien F, Soubrier F (1992) Evidence, from combined segregation and linkage analysis, that a variant of the angiotensin I-converting enzyme (ACE) gene controls plasma ACE levels. Am J Hum Genet 51(1):197–205

Trizzino M, Kapusta A, Brown CD (2018) Transposable elements generate regulatory novelty in a tissue-specific fashion. BMC Genomics 19(1):468. https://doi.org/10.1186/s12864-018-4850-3

Tuan D, Biro PA, deRiel JK, Lazarus H, Forget BG (1979) Restriction endonuclease mapping of the human gamma globin gene loci. Nucleic Acids Res 6(7):2519–2544. https://doi.org/10.1093/nar/6.7.2519

Wallace MR, Andersen LB, Fountain JW, Odeh HM, Viskochil D, Marchuk DA, O'Connell P, White R, Collins FS (1990) A chromosome jump crosses a translocation breakpoint in the von Recklinghausen neurofibromatosis region. Genes Chromosomes Cancer 2(4):271–277. https://doi.org/10.1002/gcc.2870020404

Wallace MR, Andersen LB, Saulino AM, Gregory PE, Glover TW, Collins FS (1991) A de novo Alu insertion results in neurofibromatosis type 1. Nature 353(6347):864–866. https://doi.org/10.1038/353864a0

Wilton SD, Lloyd F, Carville K, Fletcher S, Honeyman K, Agrawal S, Kole R (1999) Specific removal of the nonsense mutation from the mdx dystrophin mRNA using antisense oligonucleotides. Neuromuscul Disord 9(5):330–338. https://doi.org/10.1016/s0960-8966(99)00010-3

Xie F, Ye L, Chang JC, Beyer AI, Wang J, Muench MO, Kan YW (2014) Seamless gene correction of β-thalassemia mutations in patient-specific iPSCs using CRISPR/Cas9 and piggyBac. Genome Res 24(9):1526–1533. https://doi.org/10.1101/gr.173427.114

Yoshida K, Nakamura A, Yazaki M, Ikeda S, Takeda S (1998) Insertional mutation by transposable element, L1, in the DMD gene results in X-linked dilated cardiomyopathy. Hum Mol Genet 7(7):1129–1132. https://doi.org/10.1093/hmg/7.7.1129

Zatz M, Passos-Bueno MR, Rapaport D, Vainzof M (1991) Familial occurrence of Duchenne dystrophy through paternal lines in four families. Am J Med Genet 38(1):80–84. https://doi.org/10.1002/ajmg.1320380118

Zatz M, Passos-Bueno MR, Vainzof M (1993) Transposon-like element in the dystrophin gene. Am J Med Genet 46(5):601–602. https://doi.org/10.1002/ajmg.1320460528

Zatz M, Sumita D, Campiotto S, Canovas M, Cerqueira A, Vainzof M, Passos-Bueno MR (1998) Paternal inheritance or different mutations in maternally related patients occur in about 3% of Duchenne familial cases. Am J Med Genet 78(4):361–365. https://doi.org/10.1002/(sici)1096-8628(19980724)78:4<361::aid-ajmg11>3.0.co;2-g

Chapter 10
Transposable Elements in Cancer

Sulev Kõks

Abstract Carcinogenesis is based on acquiring the mutations that will trigger and maintain the malignant cellular growth. According to the widely accepted concept, these mutations are single-nucleotide variants, and environmental factors mainly generate them. In this chapter, we describe an alternative understanding based on the human genome's internal capacity to generate structural variants within the "dark matter" of the human genome. The "dark matter" of the genome contains transposable elements (TEs) that can transpose and generate mutations in the genomes. These de novo variants affect the functioning of the genome and are the leading cause of the aberrant transcription that is the basis of many, if not most, of the complex diseases. In this chapter, I focus on the impact of TEs on cancer development.

Keywords Repetitive elements · Transposable elements · Genomic variations · Cancer · Carcinogenesis

Abbreviations

ABL1	Abelson murine leukaemia viral oncogene homolog 1
AIDs	Activation-induced deaminases
Alu	Arthrobacter luteus
APOBEC3	Apolipoprotein B mRNA editing enzyme catalytic polypeptide 3
BCR	Breakpoint Cluster Region
CAR-T	Chimeric antigen receptor T-cell

S. Kõks (✉)
Medical Genomics Laboratory, Murdoch University, Personalised Medicine Centre, Perron Institute, Murdoch, WA, Australia

Genomic Therapeutics Pty Ltd, Prion OÜ, Perth, WA, Australia
e-mail: Sulev.Koks@murdoch.edu.au

L. M. Vaschetto (ed.), *Transposable Elements, Transcriptomics, and Diseases*, Epigenetics and Human Health 14, https://doi.org/10.1007/978-3-032-04125-8_10

Cas9	CRISPR-associated protein 9
CRISPR	Clustered regularly interspaced short palindromic repeats
CTVT	Canine transmissible venereal tumour
DDX58	DExD/H-box helicase 58
DFT1	Devil facial tumour 1
DFT2	Devil facial tumour 2
DFTD	Devil facial tumour disease
DNMT	DNA methyltransferase
EBV	Ebstein-Barr virus
ERV	Endogenous retrovirus
FAM	Fossil Alu monomer
FLAM	Free left Alu monomer
FRAM	Free right Alu monomer
HBV	Hepatitis B virus
HCV	Hepatitis C virus
HDAC	Histone deacetylase
HepG2T14.1	Hepatocellular carcinoma cell line from 15-year-old male (HepG2) variant transfected with hepatitis B virus (HBV)
HERV	Human endogenous retrovirus
HHV-8	Human herpesvirus-8
HPV	Human papillomavirus
HSATII	Human Satellite II
HTLV-1	Human T-cell lymphotropic virus-1
IFIH1	Interferon induced with helicase C domain 1
IFN	Interferon
ISG	Interferon-stimulated gene
JAK1	Janus kinase 1
KAP1	KRAB-associated protein-1
KRAB	Krüppel-associated box
KRAB-ZFP	Krüppel-associated box domain-containing zinc finger protein
LIN28B	Lin-28 homolog B protein
LINE-1	Long interspersed nuclear element-1
LTR	Long terminal repeat
Mariner	Drosophila mauritiana (mariner)
MCPV	Merkel cell polyomavirus
MDA5	Melanoma differentiation-associated protein 5
MEI	Mobile element insertions
MHC	Major histocompatibility complex
ORF1	Open reading frame 1
ORF2	Open reading frame 2
PRC2	Polycomb repressive complex 2
PRRS	Pattern recognition receptors
RIG1	RNA sensor RIG-I
RSV	Rous Sarcoma Virus
SETDB1	SET domain bifurcated histone lysine methyltransferase 1

SINE	Short interspersed nuclear element
SRP	Signal recognition particle
STAT1	Signal transducer and activator of transcription 1
SV	Structural variant
SVA	SINE-VNTR-Alu elements
TC1	Transposon C. elegans number 1
TE	Transposable element
TIRs	Terminal inverted repeats
VNTR	Variable number tandem repeat
ZFP	Zinc finger protein
7SL	RNA component of signal recognition particle 7SL1

10.1 Introduction

Carcinogenesis is a complex and multi-step process based on the accumulation of multiple complementary genetic mutations or other molecular events in the cells that lead to cancer. Malignant neoplasms are characterised by several hallmarks, which are the phenotypic attributes of cancers (Hanahan and Weinberg 2000). Our current understanding of carcinogenesis is based on a multiple-hit model, meaning it takes more than one single mutation for cancer cells to generate a clinically significant cancer (Armitage and Doll 1954; Hanahan and Weinberg 2011, 2000). Mutations that contribute to the development of malignant phenotypes are known as driver mutations, and the first driver mutation is the initiating mutation. However, the time to accumulate all the necessary mutations can be long; therefore, cancers arise after a long prodromal period. As cancer development can be very long, gaining information about cancer development can be challenging.

The development of the "multiple-hit" theory was not straightforward, as not all the data supported this conclusion at the beginning. The initial hypothesis made by Armitage and Doll suggested that at least six events are necessary for typical cancers that are not affected by hormonal or environmental factors (Armitage and Doll 1954). In their later paper, they found that, possibly, getting "two hits" is a minimal requirement for cancer initiation and development. These two theories, "two-hit" and "multiple-hit", were taken as opposing each other for some time (Armitage and Doll 1954, 1957, 2004). However, a detailed mathematical and histological analysis indicated that the "multiple-hit" theory explains clinical, epidemiological and histological data much better and that these two theories were found not to be mutually exclusive (Ashley 1969). It was proposed that at least six hits are required for most cancers, while malignancies may require only two or three events, mutations, to develop a cancer. Another analysis of the time of the diagnosis of retinoblastoma indicated that two mutations are necessary for the tumour to develop and led to the proposal of a two-hit theory (Knudson 1971). The development of these different theories indicates that there are other ways for the tumours to develop, depending on the tissue of origin. Still, it always involves two or more genetic mutations. As a result, the multiple-hit theory has been widely accepted, and by now, many experimental and clinical studies have validated the multiple-hit theory (Hanahan and Weinberg 2011). The central mechanism for cancer development is the accumulation of random somatic mutations during the lifetime. One primary source for these random somatic mutations is transposable elements (TEs) that cause the transposition of genomic regions, and their effect is clearly observable in the familiar cancer syndromes. The idea here is that these multiple hits do not need to be environmentally caused but can be acquired by the genomic instability and generated endogenously. When the first hit has been inherited, the cancer requires fewer mutations (and a shorter time) to develop clinically active cancer. Examples can be faster development of retinoblastoma in patients with a deletion in chromosome 13 or Wilms' tumour in patients with a deletion in chromosome 11 (Huff 1996; Hsu 1983; Strong et al. 1981; Riccardi et al. 1978). This two-hit hypothesis has been experimentally proven using the induced renal carcinoma rat model, where carcinoma development was only induced after acquiring the second mutation (Kobayashi et al. 1997). Accumulation of the genetic mutations in the HepG2T14.1 cell line after the hepatitis B virus (HBV) transfection and the increase in mutation rate as a sign of genomic instability have also been shown (Livezey and Simon 1997).

The speed of the development of these mutations can also vary. Sometimes, the mutations accumulate quickly, have a powerful effect and malignancy develops much faster, like in the case of specific chromosomal translocations (Rowley 1979; Klein and Lenoir 1982; Hecht et al. 1985). The Philadelphia chromosome is a classic example here, with a translocation between chromosomes 9 and 22, resulting in a fusion between the *ABL1* and the *BCR* genes that is a very powerful oncogene (Hecht et al. 1985). The transposition of the *ABL1* gene to chromosome 22 has also been found in patients without the Philadelphia chromosome (Stopera et al. 1990; Morris et al. 1990; Rassool et al. 1990). Similarly, Philadelphia chromosome-negative patients have been described as having *BCR* rearrangements (Weinstein et al. 1988). These are only some examples of significant aberrations that require a certain extent of genomic or chromosome instability for carcinogenesis (Lim 1979; Hsu 1983; Chorazy 1985). This way, accumulating the number of required mutations (two hits or six hits) takes a much shorter time; therefore, the term "genomic instability syndromes" has been proposed. Indeed, ataxia telangiectasia, Bloom's syndrome, Lynch syndrome or Fanconi anaemia are typical examples of genomic instability syndromes (Hsu 1983).

10.2 From Viral Transformation to Transposons

The instability in the genome required for carcinogenesis can be triggered by viruses that integrate into the genome and generate structural variations, which can cause malignant cell transformation (Temin 1970). The best-known viruses that induce malignancies are the Rous sarcoma virus, the murine sarcoma virus and the Moloney murine leukaemia virus (Temin 1970). The viral transformation has also been shown in osteosarcoma, and hepatitis B virus has been detected in hepatocellular carcinoma (Pritchard et al. 1975; Mitreiter et al. 1994; Brechot et al. 1981; Sherman and Shafritz 1984; Chen et al. 1986). This viral transformation is necessary for oncogenesis, as it generates structural variants fundamental to cancer formation (Cairns 1981).

Moreover, significant structural variants or rearrangements involving multiple genes caused by genetic transpositions are considered the origin of human cancers (Cairns 1981). Therefore, it is interesting to propose that there is a connection between viral oncogenic transformation and human genomic rearrangements. Indeed, the sequence analysis of the transforming region from the Moloney murine sarcoma virus showed its similarity to the transposons with two LTRs at each end (Reddy et al. 1980; Reddy et al. 1981). Moreover, sequencing of the Molony leukaemia virus revealed that the Molony provirus is a transposon (Shinnick et al. 1981; Sutcliffe et al. 1980, 1981). Similarly, LTR elements have been found in other retroviruses (Ju and Skalka 1980; Skalka et al. 1981; Hishinuma et al. 1981). Therefore, there is a clear connection between viral-induced malignant transformants and transposons, showing the necessity of transposons for these transformative abilities (Deniz et al. 2020; Ahmed and Jha 2023).

Viral integration has a transformative capacity in different cell models (Steffen 1984; Vijaya et al. 1987; Hauptmann and Schmitt 2006). This transformation is related to the virus's insertion, the generation of genomic rearrangements and the activation of proto-oncogenes (Cleveland et al. 1986; Ihle et al. 1989; Bartholomew et al. 1989). It is interesting to discuss here the possibility that retroviruses originate from the cellular mobile elements (Temin 1980). Several lines of evidence support this theory. First, the infection with retroviruses and its progression are dependent on specific host genes, indicating a close relationship between the retrovirus and the host. Second, for many retroviruses, a substantial sequence homology between viruses and uninfected cells exists. And finally, all higher organisms have endogenous retroviruses in their genomes. Even though these endogenous viruses are non-pathogenic and inactive, they are actively expressed and can produce proteins needed for the invading virus. This all resulted in the provirus hypothesis that reverse transcriptase activity exists in normal animal cells and plays a role in normal cellular differentiation, and the RNA tumour viruses have evolved based on these endogenous activities (Telesnitsky and Goff 1997; Su et al. 2019). RNA viruses have evolved to an independent genetic status from a typical cellular system of DNA to RNA to DNA information transfer, and the relationships between viruses and cells reflect this origin. The parallel evolution from cellular components, derived from common evolutionary precursors, is reflected in the relationships between viruses (Temin 1974, 1980, 1986). Cancer arises when a tumour-inducing RNA virus induces variations from these normal functions needed for cell development, differentiation or growth. The provirus hypothesis is an excellent summary that connects viral oncogenesis with the endogenous genomic elements with malignant transformation capacity (Chuong et al. 2017; Pasquesi et al. 2024; Nguyen et al. 2024).

10.2.1 Transposable Elements in the Human Genome

The human genome is mainly made of repetitive sequences. Because the function of this part of the genome is unknown, it is usually referred to as the "dark matter" of the genome. Genomic dark matter is highly complex and contains variable repetitive sequences. It occupies at least two-thirds of the human genome and actively participates in its functions (de Koning et al. 2011). The genomic dark matter mainly consists of TEs.

TEs are crucial, or the only, source of enhanced cell mutagenesis. They are types of repetitive elements or repeat sequences in the human genome, and they are highly abundant (Kõks and Kõks 2018; Kõks et al. 2022). Initial human genome sequencing revealed that only 1.5% of the human genome consists of coding sequences, and therefore, 98.5% of the genome functions as non-coding (Lander et al. 2001). Repetitive elements comprise most of the human genome, many of which are evolutionarily young, hominid and human-specific. All repetitive elements can be divided into interspersed repeats, low-complexity repeats, segmental duplications and tandem repeats (Storer et al. 2021). TEs belong to the category of interspersed repeats. They comprise more than half of the human genome and are the most abundant repetitive elements (Lander et al. 2001). TEs can be further classified as Class I (DNA transposons) or Class II (retrotransposons) TEs (Table 10.1). Retrotransposons include LTR elements (e.g. endogenous retroviruses) and non-LTR elements (e.g. LINE-1, SVA and Alu repeats) (Lander et al. 2001). LINE-1, SVA, Alu repeats and HERVs have been shown to have the ability to transpose and generate mobile element insertions (MEI) (Hehir-Kwa et al. 2016; Nelson et al. 2003, 2004). These MEIs are a frequent form of structural variation (SV) and comprise 23% of SVs larger than 20 bp (Hehir-Kwa et al. 2016).

The functional capacities of different TEs are determined by their structures (Fig. 10.1). LINE-1 elements are autonomous, meaning that they can propagate and relocate independently, as they have ORF1 and ORF2 to encode all the necessary proteins. The human genome contains hundreds of thousands of copies of LINE-1 elements, and they make up around 17% of the human genome (Lander et al. 2001). SINE elements make up almost a similar fraction of the human genome, but there are millions of copies, as they are shorter (Burns 2017). These include Alu repeats, which constitute a significant source of structural variation in the human genome. MIRs (mammalian-wide interspersed repeats) are also very common. Alu repeats, like SVAs (composite SINE-VNTR-Alu elements), require LINE-1 for the mobilisation, and these are called non-autonomous elements. Alu repeats are composed of two 7SL RNA monomers, a free left Alu monomer (FLAM) and a free right Alu monomer (FRAM). SVA is a composite element combining the CCCTCT hexamer, Alu repeat, VNTR and SINE-R. SVAs are evolutionary, the youngest elements; they are hominid-specific, while SVA-E and SVA-F are human-specific. VNTR and CCCTCT are variable in SVA, and this variation has been shown to modify the activity of SVAs (Savage et al. 2013; Fröhlich et al. 2022; Kulski et al. 2023). And finally, we have DNA transposons, characterised by the terminal inverted repeats

Table 10.1 Overview of the transposable elements and their classification

Interspersed repeats						
Transposable elements						
Class I retrotransposons			Class II DNA transposons			
LINE	LINE-dependent retrotransposon	Retrotransposon	DNA polymerase	Helicase	Transposase	Tyrosine recombinase
Non-LTR		LTR	Casposon	Helitron1Nb	TC1	Eulor5A
L1	Alu	ERV	Maverick	Helitron2	Mariner	Eulor5B
L2	SVA	MaLR	Mavirus	Helitron3Na	Charlie	Eulor6A

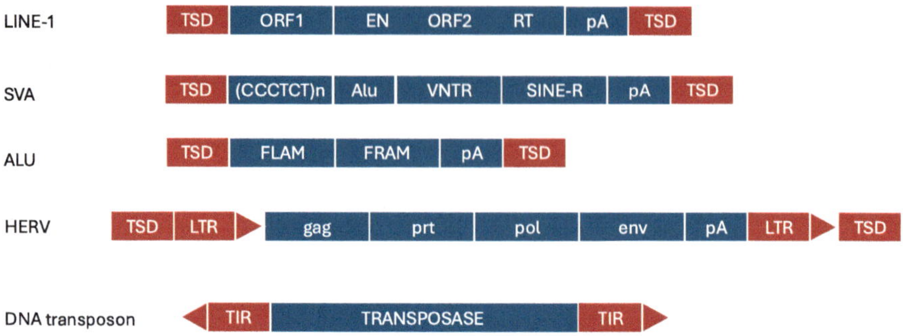

Fig. 10.1 Structural comparison of different transposable elements

(TIRs). While DNA transposons are considered to be no longer active in the human genome, they have been reported to be transcribed in cancers and inflammatory conditions (Krishnan and Kõks 2023; Valdebenito-Maturana 2024). Moreover, 40 active TC1/Mariner subclass integration-competent DNA transposons have been identified to induce transposition in human cells (Zhang et al. 2024). These elements can potentially be used in genomic engineering for CAR-T cells or solid tumours for therapeutic purposes. Interestingly, Mariner2_AG outperformed other currently used vectors in clinical practice.

Tandem repeats are another widely abundant group of repetitive elements in the genome. They are divided by length into medium-to-long satellite repeats and shorter simple repeats. Satellite repeats are, in turn, classified based on their chromosomal location. While this chapter does not have space to discuss satellite elements in depth, it is essential to mention that their evolution, development and change are products of TEs (Belyayev et al. 2020; Sharma et al. 2013; Meštrović et al. 2015). This shows a clear functional connection between TEs and tandem repeats, describing another mechanism for how TEs are involved in genomic instability.

10.2.2 *Transposable Elements, Genomic Instability and Carcinogenesis*

TEs can induce genomic instability, which is considered one of the most intriguing findings regarding the stability of the genome (Lim 1979; Sager 1979). The discovery of genomic instability in Drosophila indicated the pervasiveness and spread of this phenomenon—spontaneous X-linked lethal mutations were detected in 34 of 993 sampled sperms from 50 males (Lim 1979). These unstable chromosomes originated as clusters and are based on site-specific chromosome breaks, and they are caused by TEs (Lim 1979). It was soon proposed that TEs participate in the genomic instability and generation of mutations in cancer cells (Sager 1979).

Genomic instability has been proposed to be very common in the human population. It is expressed to different degrees, with some patients having clearly pronounced syndromes while others may have hidden ones (Hsu 1983). This background instability in the population is random. Still, it makes some subjects more vulnerable to cancer, as it is an initial phase in the multi-step process of carcinogenesis.

Intriguingly, chromosomal rearrangements have been described for most cancers, and their link to variable TEs has been documented for a long time (Chorazy 1985; Ramel 1989; Ramel et al. 1996). The best examples are from hepatocellular carcinoma, where HBV has been indicated to cause instability and changes in transcription (Sherman and Shafritz 1984; Chen et al. 1986; Twu and Schloemer 1987). Similarly, HCV has been shown to cause hepatocellular carcinoma with direct LINE-1 involvement by increasing the expression of LINE-1 and intracellular sequestration of L1ORF1p (Schöbel et al. 2021). Genomic instability is the mechanism by which oncogenic viruses initiate carcinogenesis (Lichtenberg et al. 1987; van Haaften et al. 2004). Hepatitis B virus (HBV), hepatitis C virus (HCV), Epstein-Barr virus (EBV), human papillomavirus (HPV), Rous sarcoma virus (RSV), human herpesvirus-8 (HHV-8), human T-cell lymphotropic virus-1 (HTLV-1) and Merkel cell polyomavirus (MCPV) are the most well-known viruses to cause cancer (Bouza et al. 2021; Ahmed and Jha 2023). It has been estimated that at least one in six cancer cases in the world is caused by viral infection, and at least 15% of cancers have viral integrations (Bouza et al. 2021; Zapatka et al. 2020). Therefore, the role of viral infections in genomic instability is well-documented.

Like viruses, TE mobilisation affects the host genome's stability and causes genetic disorders and cancer. Host genomes have adopted several mechanisms to fight the proliferation of viruses and TEs to limit the potential adverse effects of retrotransposition. Activation-induced deaminases (AIDs) are the enzymes that cause C to U deamination, which is central to both the adaptive and innate immune systems. In humans, the members of the APOBEC3 (apolipoprotein B mRNA editing enzyme catalytic polypeptide 3) protein family provide innate immunity against different viruses (Burns et al. 2013). They also significantly inhibit the mobilisation of the non-LTR retrotransposons LINE-1 and Alu and participate in the intracellular defence against retrotransposition by unknown mechanisms. In the human genome, around nine APOBEC family members (APOBEC1, APOBEC2, APOBEC3 and APOBEC4) have been shown to possess DNA deaminase activity (Burns et al. 2013). These genes may have been expanded to prevent genomic instability caused by endogenous retroelements (Schumann 2007). Interestingly, as these enzymes primarily target viruses and activate retrotransposons, they also target oncogenes, and the role of APOBEC3 in cancer mutagenesis has been identified (Burns et al. 2013). APOBEC3 genes form an evolutionarily young cluster of genes that have evolved as retrocopies with the help of LINE-1 (Yang et al. 2020). These retrocopies do not suppress the activity of LINE-1 but restrict the infectivity of exogenous viruses (Yang et al. 2020). Therefore, the interaction between TEs, viruses and endogenous mechanisms in genomic instability is complex.

TEs have different ways of influencing carcinogenesis. One is related to the direct regulation of gene expression, as endogenous retroviruses can act as promoters for the network of oncogenes (Babaian and Mager 2016). SVAs, LINE-1 and Alus have a widespread and general capacity to direct the expression of different genes, sometimes even in complex clusters (Kulski et al. 2024a, b, 2025). TEs also have the capacity to alter the splicing or become inserted into the transcripts as pseudo-exons. The pseudoexonisation of SVAs is a widespread phenomenon and was found in all analysed samples and for multiple genes in many cases (Pfaff et al. 2023). Moreover, these SVAs are not only in the introns and are transcribed as pseudoexons. A genome-wide analysis identified that SVAs inside the exons are very common and are transcribed (Pfaff et al. 2023).

Many recent studies have identified the impact of TEs in carcinogenesis. One study described somatic retrotranspositions in 50% of cancer cases (Tubio et al. 2014). Another found the altered expression of repetitive elements in all the studied samples (Annapragada et al. 2024). The analysis of 2837 tissue and plasma samples from 1975 patients with lung, breast, colorectal, ovarian, liver, gastric, head and neck, bladder, cervical, thyroid, or prostate cancer identified widespread activation of repetitive elements (Annapragada et al. 2024). The authors identified tumour-specific changes in 1280 repeat element types from the LINE, SINE, LTR, HERVs and human satellite elements. The identified variations included alterations to known repeats and 820 elements not previously known to be changed in human cancer. The comparison between the tumour and matched normal tissue revealed that 807 repeat elements were increased or decreased in tumour samples (Annapragada et al. 2024). The most commonly altered elements were satellites, LINEs and SINEs, although changes were also frequently observed in elements within LTRs, TEs and RNA elements. Interestingly, DNA transposons (Tc1/Mariner and Charlie) were also amongst the highly altered elements. A more detailed analysis of the location of changed repetitive elements revealed their enrichment in regions of driver genes, indicating the causative role of TEs in cancer. Their representation was affected by the structural changes and epigenetic states. The enrichment of TEs was different for different elements and in different cancers and varied substantially, but in all cancer samples, altered repetitive elements were found, and more than 40% of the 300 LTR ERV1 elements were altered in all 12 types of tumours studied (Annapragada et al. 2024). Moreover, the sample scores based on the expression of repetitive elements were able not only to classify the type of cancer correctly but also the stage of the disease (Annapragada et al. 2024). Analysis of spatial data from gastric cancer also confirms TE expression, with 204 TEs being spatially enriched in the tumour regions and the tumour microenvironment, strongly suggesting a role of TEs in tumorigenesis. In addition, a network of TE-mediated gene regulation indicated that 2000 genes could be modulated by TEs, with ca. 500 of them already implicated in cancer (Valdebenito-Maturana 2024). A pan-cancer study based on 2954 cancer genomes discovered retrotransposition in 35% of samples that resulted in variable molecular effects like structural variations, duplications, deletions, amplification of oncogenes or activation of gene fusions (Rodriguez-Martin et al. 2020). TEs can also be transcriptionally reactivated in tumours to create new TE-chimeric transcripts, which can produce immunogenic antigens. A comprehensive screen for these TE exaptation events in 33 TCGA tumour types, 30 GTEx adult tissues and 675 cancer cell lines identified 1068 TE-exapted candidates with the potential to generate shared tumour-specific TE-chimeric antigens (TS-TEAs) (Shah et al. 2023). TS-TEAs were presented on the surface of cancer cells, and tumour-specific membrane proteins were transcribed from TE promoters that constitute aberrant epitopes on the extracellular surface of cancer cells.

Another study, using whole transcriptome analysis on 7769 tumours and 625 normal samples from 15 cancer types, identified 129 cryptic promoter-activation events induced by TEs (Jang et al. 2019). This activation involved 106 oncogenes across 3864 tumours. Most intriguingly, AluJb-LIN28B was identified as a candidate for the inducer of oncogene expression. Indeed, after CRISPR-Cas9 eliminated this Alu repeat, oncogene expression was also eliminated, proving the functional link between the Alu TE and oncogene expression (Jang et al. 2019).

10.2.3 Mechanisms of Silencing of Transposable Elements

DNA methyltransferases (DNMTS), histone deacetylases (HDACS) and other epigenetic modifiers play critical roles in silencing TEs and maintaining genomic stability. DNMTS (e.g. DNMT1, DNMT3A and DNMT3B) catalyse the addition of methyl groups to cytosines at Cpg sites, establishing a repressive chromatin environment that prevents TE transcription. HDACS removes acetyl groups from histone tails, promoting chromatin condensation and the transcriptional repression of TEs. Additional modifiers, such as SETDB1 (a histone methyltransferase for H3k9me3) and components of the polycomb repressive complex 2 (PRC2), responsible for H3k27me3, further reinforce TE silencing. Dysregulation of these enzymes in cancer, including mutations, deletions or altered expression levels, leads to global DNA hypomethylation, loss of repressive histone marks and chromatin relaxation. This can reactivate TEs, promoting genomic instability, insertional mutagenesis, inflammation and tumour progression (Jones and Baylin 2002; Deniz et al. 2019; Timp and Feinberg 2013; Schulz et al. 2006).

The KRAB (Krüppel-associated box) domain-containing zinc finger proteins (KRAB-ZFPS) form one of the largest families of transcriptional repressors in mammals and are pivotal in TE silencing. KRAB-ZFPS recognise specific TE sequences and recruit the co-repressor KAP1 (also known as TRIM28), which orchestrates the assembly of a repressive chromatin environment by recruiting SETDB1 (for H3k9me3 deposition), DNMTS (for DNA methylation) and HDACS (for histone deacetylation). This multilayered epigenetic repression effectively silences TEs and preserves genome stability. In cancers, mutations, downregulation or functional impairment of KRAB-ZFPS, KAP1 or their associated partners can lead to TE reactivation, contributing to genomic instability, inflammation and tumorigenesis (Rowe and Trono 2011; Bruno et al. 2019; Martins et al. 2024; Turelli et al. 2020).

10.2.4 Reactivation of Transposable Elements and Immune Interactions

TE reactivation in cancer leads to the production of double-stranded RNAs (dsRNAs), primarily derived from endogenous retroelements such as LINE-1 and ERVs. These dsRNAs mimic viral infections and are detected by innate immune pathways, primarily through pattern recognition receptors (PRRS) like MDA5 (IFIH1) and RIG-I (DDX58). The activation of these receptors triggers type I interferon (IFN) responses, resulting in the upregulation of interferon-stimulated genes (ISGS), inflammatory cytokine production and potentially immune-mediated elimination of tumour cells (Chiappinelli et al. 2015; Roulois et al. 2015). However, tumours develop mechanisms to evade this antiviral immune response. Common evasion strategies include the upregulation of ADAR 1, an RNA-editing enzyme that modifies dsRNA to prevent its recognition by PRRS and loss-of-function mutations in key components of the IFN signalling pathway (e.g. JAK 1 and STAT 1). Some tumours also selectively reduce MHC class I expression to escape immune recognition despite innate activation. Therefore, while TE reactivation initially provokes an antiviral state that could inhibit tumour growth, cancer cells frequently acquire adaptations that dampen or bypass innate immune surveillance, facilitating continued proliferation (Ishizuka et al. 2019; Leonova et al. 2013).

Recent studies have demonstrated that TE reactivation in tumours can generate neoantigens, offering a novel avenue for cancer immunotherapy. When transcribed and translated due to loss of epigenetic silencing, TE-derived sequences can produce aberrant peptides that are processed and presented on MHC class I molecules, acting as tumour-specific neoantigens recognisable by cytotoxic T cells. For example, ERV-derived peptides were presented on the cell surface in ovarian cancer and elicited immune responses (Hu et al. 2024; Mombach et al. 2022, 2024). Similarly, Panda et al. (2018) identified TE-derived peptides presented by MHC molecules in several cancers, suggesting that TE reactivation contributes substantially to the antigenic landscape of tumours. Furthermore, therapeutic approaches like epigenetic drugs (e.g. DNA methyltransferase inhibitors) can enhance TE expression, thereby boosting the production of TE-derived neoantigens and improving tumour immunogenicity. These findings suggest that combining epigenetic therapy with immune checkpoint blockade could synergistically enhance anti-tumour immunity by unmasking TE-driven neoantigens (Chiappinelli et al. 2015; Mardis 2019; Panda et al. 2018).

10.2.5 Transmissible Tumours and Transposable Elements

Involvement of the TEs in carcinogenesis has also been shown in the case of transmissible tumours. Tasmanian devil facial tumour disease, DFTD, is the most well-known transmissible tumour; two different lineages are described as DFT1 (emerged in 1986) and DFT2 (emerged in 2011). DFT2 has been shown to have faster mutation rates than DFT1 across all variant classes, including TE insertions, and a hypermutated DFT1 lineage with defective DNA mismatch repair has been identified. In the case of DFT1, no clear evidence of LINE-1 activity has been found (Stammnitz et al. 2023). However, LINE-1 mobilisation events were clearly observed throughout the DFT2 phylogenetic tree and accumulated linearly with time (Stammnitz et al. 2023).

A canine transmissible venereal tumour (CTVT) is linked to the TEs, and a long interspersed nuclear element (LINE-1) insertion near c-myc has been found in all tumours examined so far. It can be used as a diagnostic marker to confirm a CTVT tumour (Katzir et al. 1985, 1987). Most likely, that particular LINE-1 insertion induced this tumour.

Multiple tumours (sarcomas) with independent origins occur in golden Syrian hamsters. When these tumour cells are present in the bloodstream, mosquito bites can transmit the tumours between animals (Ostrander et al. 2016). A potentially infectious transmissible agent related to the insertion of adenovirus into the vicinity of an intracisternal A particle has been described in these tumours (Lichtenberg et al. 1987).

Finally, bivalve molluscs, most prominently the soft-shell clam *Mya arenaria*, are afflicted with an epidemic transmissible circulatory system disease closely resembling leukaemia, and it can be transmitted by the retrovirus-like particles isolated from affected clams (Walker et al. 2009; Oprandy and Chang 1983). The disease is characterised by a dramatic expansion of blast-like cells in the hemolymph with a high mitotic index. The tumour is characterised by a dramatic expansion of blast-like cells in the hemolymph with a high mitotic index. The transcripts of a novel retroelement, Steamer, were identified in RNA sequencing of hemolymph (Arriagada et al. 2014). This element encodes a single large protein similar to mammalian retroviral Gag-Pol proteins and is bordered by LTRs. Steamer was highly upregulated in diseased haemocytes, and its expression was correlated with disease status (Arriagada et al. 2014). Diseased haemocytes also had evidence of extensive reverse transcription and retrotransposition. Therefore, steamer activation is an example of an induction of genetic instability that can initiate and advance leukaemia in *M. arenaria* (Arriagada et al. 2014).

10.2.6 Seasonality of Cancers and Transposable Elements

Cancers have a well-recognised seasonality pattern (Lambe et al. 2003; Samia and Nenow 2022; Hysaj et al. 2022). These differences are across all kinds of cancers—breast, lung, stomach and skin. Usually, fewer cancers occur during summer, which could indicate the effect of seasonal infections (Lambe et al. 2003; Moan et al. 2010; Tulchiner et al. 2021). In parallel, there is a "cancer-cold" hypothesis stating that cancer incidences are higher in cold countries (Voskarides 2019). Seasonality of cancers adds more complexity to the susceptibility of cancers. The variations in the seasons have been shown to affect many diseases, and seasonality has also been shown to affect cancer (Bieuville et al. 2024). The biological mechanisms affecting the seasonal difference in cancers are unclear. The seasonal fluctuation in the level of vitamin D has been suggested (Hysaj et al. 2022; Moan et al. 2010). The seasonality is in line with the evidence that the stress response in animals is seasonally affected (Koks et al. 2000). Moreover, seasonality affects hormones, especially growth hormones, which vary between seasons and are involved in many different cancers (Koks et al. 2000). Osteosarcoma is a cancer affected by growth hormones, IGF-1 and the leptin pathway (Märtson et al. 2013; Huang et al. 2015; Rothzerg et al. 2020). Targeting the IGF1R has been proposed as a potential therapeutic option (Huang et al. 2015). The activation of the growth hormone in osteosarcoma explains its highest incidence during puberty, the fast growth period. Osteosarcoma has a very high degree of genomic instability without clear reasons, but it is possibly related to the activation of LINE-1 or satellite elements (Wolf et al. 2015; Smida et al. 2017; Ho et al. 2017b). In addition, osteosarcoma has a particular fingerprint of the genes and TEs based on transcriptome analysis (Ho et al. 2017a, b; Poudel and Koks 2024). The whole osteosarcoma transcriptome analysis (Ho et al. 2017a) identified significant upregulation of satellite elements (HSATII) and DNA transposons (Ho et al. 2017a). TC1/Mariner and many HERVs were downregulated in osteosarcoma, suggesting specific and coordinated response changes related to the malignancy (Ho et al. 2017a). Therefore, seasonal differences in the incidence of cancers can be caused by seasonal hormonal variations. Immune response and connection with the TEs explain the seasonal variation in cancer incidence. A large-scale genome-wide study using 454 PBMC samples identified 5137 protein-encoding genes that showed identical seasonal differences, with inverted patterns observed between Europe and Oceania (Dopico et al. 2015). This study found that the winter season is characterised by the upregulation of pro-inflammatory and metabolic pathways with increased expression of 2826 genes. During summer, a different set of 2311 genes is upregulated (Dopico et al. 2015). Most remarkably, in the same study, the authors found similar seasonality in the subcutaneous adipose tissue with 4027 genes seasonally differentially expressed. As with the PBMC expression, in winter, the adipose tissue had the most affected inflammatory and metabolic pathways..

However, another mechanism to explain seasonality is related to the dynamics of methylation patterns of the TEs related to infections and immune activation (Alkailani and Gibbings 2023; Gasparotto et al. 2023; Jiang et al. 2023). For instance, the methylation of the tumour suppressor VTRNA2–1 is affected by the season of conception and, therefore, affects the microenvironment for the tumours (Silver et al. 2015). Smoking, one of the most significant carcinogenic environmental factors, also affects methylation at the whole-genome level (Koks et al. 2015). Moreover, the LINE-1 elements were significantly hypomethylated in the autumn and winter compared to the spring and summer periods (Ricceri et al. 2014). While the mechanisms are unknown, the seasonality has also been described in model organisms like laboratory rats or Tasmanian devils (Koks et al. 2000; Jones et al. 2008; Hamilton et al. 2020). Tasmanian devils have somatic malignant cells that they transmit to each other (DFTD), and the transmission rate has seasonal variation, which has been attributed to the variation in stress level and immune response (Peck et al. 2015; Raven et al. 2022). Tasmanian devils have significant seasonal fluctuations in their activity and social behaviour that would certainly contribute to the seasonality in the transmission of the DFTD.

10.3 Heritable Cancer Syndromes

But what about the role of TEs in heritable cancer syndromes where the single gene mutations have been well-established and implemented in clinical use? One of the most well-known heritable cancer syndromes is Lynch syndrome (LS), which is also known as hereditary non-polyposis colorectal cancer (HNPCC). It is a common hereditary cancer syndrome caused by heterozygous germline inactivation in any of the DNA mismatch repair (MMR) genes MLH1, MSH2, MSH6 and PMS2 (Nolano et al. 2022). LS patients are susceptible to colorectal, endometrial and other cancers, which are recognised by microsatellite instability (MSI-high), a hallmark of mismatch repair defects. In addition, deletions at the end of the EPCAM gene have also been shown to cause LS by generating a fused read-through transcript with the MSH2 (Ligtenberg et al. 2009). LS is the most common hereditary cause of colorectal and endometrial cancers, but pathogenic variant carriers are also at increased risk of developing ovarian, prostate and ureteral cancers (Biller et al. 2019). Prevalence has been estimated to be as high as 1/279 individuals (Win et al. 2017). However, the genetic risk of LS is not entirely defined by these known MMRs, and the existence of significant unidentified genes that can explain half of the missing heritability in LS has been proposed (Win et al. 2017). This indicates a significant share of missing heritability, even in the so-common heritable cancer syndrome and illustrates the limitations of our current knowledge.

For instance, a novel SINE-VNTR-Alu (SVA) insertion in exon 12 of MSH2 in an individual with early-onset colorectal cancer and a very strong LS family history has been identified (Yang et al. 2021). This insertion was confirmed with RT-PCR analysis, which indicated a larger aberrant MSH2 transcript in one of the family members. The insertion was identified as an SVA element approximately 3 kb in length, belonging to the SVA_F1 family of retrotransposons (Yang et al. 2021).

Inserting a truncated AluSx-like element into MLH1 intron 7 results in aberrant splicing and transcription, causing LS. This study confirms that retrotransposon insertions may be an important mechanism for cancer predisposition (Li et al. 2020).

Using Bionano optical genome mapping, a 39 kb insertion in the MSH2 gene has been detected (Bjørnstad et al. 2024). It was revealed to be a founder event after the precise mapping of the insertion breakpoints and inserted sequence using low-coverage whole-genome sequencing. The same 39 kb insertion was found in two unrelated families with a strong positive family history, suggesting Lynch syndrome. Further and broader analysis found the same insertion in other families from the same region of Norway (Bjørnstad et al. 2024). The roles of TEs in hereditary diseases are detailed in Chap. 9.

10.4 Transposable Elements and Transcriptional Regulation

10.4.1 Transcriptomic Analysis of Cancers and Transposable Elements

The challenge in the TE analysis was analysing the TE expression, which was largely ignored by the conventional workflows. Within the last few years, tools have been developed that allow accurate calling and detection of TEs from the RNA-seq data (Lanciano and Cristofari 2020). The first studies on TE expression in malignant samples have been very informative, showing the specific expression changes in satellite elements, DNA transposons, LINE-1 and, most remarkably, endogenous retroviruses (Lanciano and Cristofari 2020; Jang et al. 2019). The widespread expression of TEs is driving the expression of oncogenes, showing the connection between the insertions of TEs and their functional impact. Indeed, whole transcriptome analysis, in combination with whole-genome sequencing, will undoubtedly improve the diagnostic yield for cancer mutation detection (Reimann et al. 2014; Lin et al. 2019; Choi et al. 2024). This is especially relevant for heterogeneous cancer samples, and we expect to detect multiple pathogenic mutations from the DNA data. Transcriptome analysis technologies are not limited to gene chips or DNA sequencing; more general differential cloning methods are still reliable in studies with non-standard organisms where the reference genomes are unavailable (Koks et al. 2004). Representational difference analysis (RDA) has successfully described the complexity of many cancers and cancer cell lines (Lantto et al. 2016; Hubank and Schatz 1994; Lawson and Berliner 1998; Chen et al. 2005; Zervos et al. 2006).

Whole transcriptome analysis will provide the necessary evidence to determine whether the mutation is expressed and included in the RNA, whether it induces alternative splicing, or whether RNA expression is up- or downregulated (Poudel and Koks 2024; Rothzerg et al. 2020). However, conducting a complete whole transcriptome analysis is not always necessary; targeted quantitative RT-PCR panels are sometimes even more informative, especially if they are designed for specific intracellular pathways (Kingo et al. 2008; Mann et al. 2015). Regarding cancer diagnostics or treatment follow-ups, blood-based transcriptomics is very useful and has proven reliable in different human conditions (Lill et al. 2013; Koks et al. 2015). Blood-based transcriptome analysis can capture repetitive RNA sequences' signatures and provide invaluable clinical care tools (Reggiardo et al. 2023). Whole-genome and exome sequencing will benefit from the community-based standards because different populations have different genetic variant profiles, and using only isolated cases for somatic mutation detection could induce analytical bias (Fang et al. 2021). The ideal reference sample set should have tumour-normal pairs from the same individuals with the blood samples that can be used for the progression monitoring. This combined sample collection would help to verify the pathogenicity of the detected TEs and genomic rearrangements.

10.4.2 Transposable Elements as Activators of Oncogenes

TE-derived enhancers and promoters can significantly contribute to oncogene activation, particularly when epigenetic control is disrupted. Many TEs harbour intrinsic regulatory sequences—including enhancer, promoter and insulator elements—that originally evolved to drive their own expression. These sequences remain inactive when silenced by DNA methylation and repressive histone marks in normal cells. However, in cancer, the loss of TE silencing can expose these latent regulatory elements. Reactivated TE-derived enhancers can upregulate nearby oncogenes by providing aberrant transcription factor binding sites, while TE-derived promoters can initiate chimeric transcripts that include oncogenic sequences. This ectopic or increased expression of oncogenes contributes to tumorigenesis by promoting proliferation, survival and invasion (Jang et al. 2019; Deniz et al. 2019; Babaian and Mager 2016). TE reactivation in cancer not only alters gene expression but also drives widespread chromatin remodelling, profoundly impacting cancer cell plasticity and therapy resistance. Reactivated TEs can introduce new transcription factor binding sites, disrupt insulator elements (such as CTCF boundaries) and reshape topologically associating domains (TADs), leading to large-scale changes in chromatin architecture. This remodelling creates an epigenetic landscape that promotes cellular dedifferentiation, stemness and lineage plasticity—key features of aggressive tumours.

Furthermore, TE-induced chromatin changes can facilitate the activation of alternative signalling pathways, allowing cancer cells to evade targeted therapies. In some contexts, TE-driven innate immune activation through double-stranded RNA-sensing pathways also creates an inflammatory environment that supports immune evasion and resistance. Thus, TE-mediated chromatin remodelling emerges as a central mechanism that fosters both tumour adaptability and therapeutic failure (Brocks et al. 2017; Roulois et al. 2015). Beyond their roles in chromatin remodelling and enhancer reactivation, TEs produce non-coding RNAs (ncRNAs) that influence tumour gene networks. TE-derived ncRNAs, including long non-coding RNAs (lncRNAs) and small RNAs (sRNAs), can modulate gene expression by acting as decoys, scaffolds, or guides for chromatin-modifying complexes. For example, TE-derived lncRNAs can recruit repressive complexes like PRC2 to target loci, altering epigenetic states and silencing tumour suppressor genes (Kapusta et al. 2013; Johnson and Guigo 2014). In other cases, TE-origin RNAs stabilise oncogenic transcriptional programmes by interacting with RNA-binding proteins or miRNAs. A notable example is the lncRNA HERVH, derived from an endogenous retroviral element, which maintains stemness features and has been implicated in promoting therapy resistance and poor prognosis in multiple cancers (Lu et al. 2014; Santoni et al. 2012). Moreover, reactivation of TE-derived double-stranded RNAs can trigger innate immune pathways, influencing tumour microenvironment remodelling and immune escape (Roulois et al. 2015). TE-derived ncRNAs are emerging as key regulators of cancer gene regulatory networks, plasticity and therapeutic outcomes.

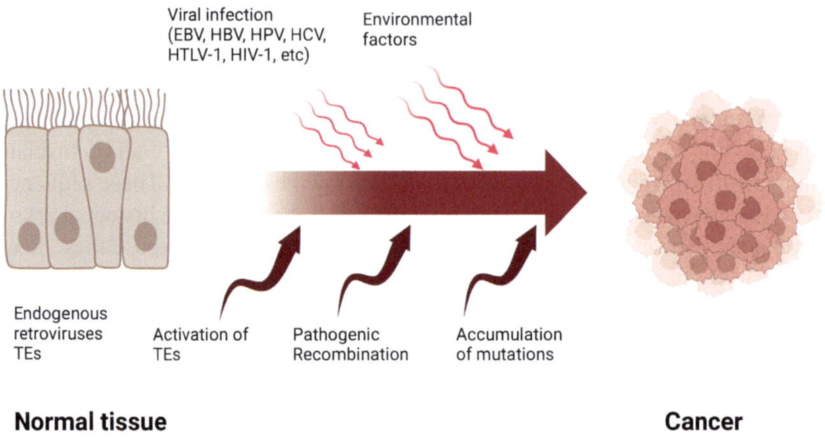

Fig. 10.2 Complexity of carcinogenesis involving the role of transposable elements. Created in BioRender

10.5 Conclusions

Carcinogenesis is a multi-step, complex process that requires many mutations to be introduced into the cell to induce malignant transformation (Fig. 10.2). The standard understanding of this process focuses only on single-nucleotide mutations, either inherited germline or de novo somatic mutations in the target tissues. However, the accumulating evidence indicates that cancers are caused by the TEs and their ability to transform the normally growing cells. While environmental factors, like seasonality and acquired viral infections, impact cancer risks, the complex mutational drive necessary for the malignant transformation is most likely from the endogenous TEs "hidden" in our genomes. The molecular pathology of cancer should focus more on TEs and their pathogenetic role in malignant transformation because the solution to the causative cure for many cancers may lie there.

Acknowledgements The funding and support from the Johanna Sewell Memorial Fund and the Perron Institute are appreciated.

Compliance with Ethical Standards

Conflict of Interest Sulev Kõks is a founder and a director of Genomic Therapeutics Pty Ltd. and Prion OÜ, private companies dedicated to genomic medicine.

Ethical Approval This chapter contains no studies with human/animal participants performed by the author.

References

Ahmed K, Jha S (2023) Oncoviruses: how do they hijack their host and current treatment regimes. Biochim Biophys Acta Rev Cancer 1878(5):188960. https://doi.org/10.1016/j.bbcan.2023.188960

Alkailani MI, Gibbings D (2023) The regulation and immune signature of retrotransposons in cancer. Cancers (Basel) 15(17). https://doi.org/10.3390/cancers15174340

Annapragada AV, Niknafs N, White JR, Bruhm DC, Cherry C, Medina JE, Adleff V, Hruban C, Mathios D, Foda ZH, Phallen J, Scharpf RB, Velculescu VE (2024) Genome-wide repeat landscapes in cancer and cell-free DNA. Sci Transl Med 16(738):eadj9283. https://doi.org/10.1126/scitranslmed.adj9283

Armitage P, Doll R (1954) The age distribution of cancer and a multi-stage theory of carcinogenesis. Br J Cancer 8(1):1–12. https://doi.org/10.1038/bjc.1954.1

Armitage P, Doll R (1957) A two-stage theory of carcinogenesis in relation to the age distribution of human cancer. Br J Cancer 11(2):161–169. https://doi.org/10.1038/bjc.1957.22

Armitage P, Doll R (2004) The age distribution of cancer and a multi-stage theory of carcinogenesis. Int J Epidemiol 33(6):1174–1179. https://doi.org/10.1093/ije/dyh216

Arriagada G, Metzger MJ, Muttray AF, Sherry J, Reinisch C, Street C, Lipkin WI, Goff SP (2014) Activation of transcription and retrotransposition of a novel retroelement, steamer, in neoplastic hemocytes of the mollusk Mya arenaria. Proc Natl Acad Sci USA 111(39):14175–14180. https://doi.org/10.1073/pnas.1409945111

Ashley DJ (1969) The two "hit" and multiple "hit" theories of carcinogenesis. Br J Cancer 23(2):313–328. https://doi.org/10.1038/bjc.1969.41

Babaian A, Mager DL (2016) Endogenous retroviral promoter exaptation in human cancer. Mob DNA 7:24. https://doi.org/10.1186/s13100-016-0080-x

Bartholomew C, Morishita K, Askew D, Buchberg A, Jenkins NA, Copeland NG, Ihle JN (1989) Retroviral insertions in the CB-1/Fim-3 common site of integration activate expression of the Evi-1 gene. Oncogene 4(5):529–534

Belyayev A, Josefiova J, Jandova M, Mahelka V, Krak K, Mandak B (2020) Transposons and satellite DNA: on the origin of the major satellite DNA family in the Chenopodium genome. Mob DNA 11:20. https://doi.org/10.1186/s13100-020-00219-7

Bieuville M, Dujon AM, Raven N, Ujvari B, Pujol P, Eslami SZ, Alix Panabières C, Capp JP, Thomas F (2024) When do Tumours develop? Neoplastic processes across different timescales: age, season and round the circadian clock. Evol Appl 17(10):e70024. https://doi.org/10.1111/eva.70024

Biller LH, Syngal S, Yurgelun MB (2019) Recent advances in Lynch syndrome. Fam Cancer 18(2):211–219. https://doi.org/10.1007/s10689-018-00117-1

Bjørnstad PM, Aaløkken R, Åsheim J, Sundaram AYM, Felde CN, Østby GH, Dalland M, Sjursen W, Carrizosa C, Vigeland MD, Sorte HS, Sheng Y, Ariansen SL, Grindedal EM, Gilfillan GD (2024) A 39 kb structural variant causing Lynch syndrome detected by optical genome mapping and nanopore sequencing. Eur J Hum Genet 32(5):513–520. https://doi.org/10.1038/s41431-023-01494-7

Bouza E, Martin Jimenez M, Alemany L, Arribas J, Banares R, Barragan MB, Eiros Bouza JM, Felip E, Fernandez-Capetillo O, Gracia D, Lopez-Velez R, Bautista Mollar J, Munoz P, Paz-Ares L, Torne A, Tovar J, Valencia E, Palomo E (2021) Overview of virus and cancer relationships. Position paper. Rev Esp Quimioter 34(6):525–555. https://doi.org/10.37201/req/058.2021

Brechot C, Pourcel C, Louise A, Rain B, Tiollais P (1981) Detection of hepatitis B virus DNA sequences in human hepatocellular carcinoma in an integrated form. Prog Med Virol 27:99–102

Brocks D, Schmidt CR, Daskalakis M, Jang HS, Shah NM, Li D, Li J, Zhang B, Hou Y, Laudato S, Lipka DB, Schott J, Bierhoff H, Assenov Y, Helf M, Ressnerova A, Islam MS, Lindroth AM, Haas S, Essers M, Imbusch CD, Brors B, Oehme I, Witt O, Lubbert M, Mallm JP, Rippe K, Will R, Weichenhan D, Stoecklin G, Gerhauser C, Oakes CC, Wang T, Plass C (2017) DNMT

and HDAC inhibitors induce cryptic transcription start sites encoded in long terminal repeats. Nat Genet 49(7):1052–1060. https://doi.org/10.1038/ng.3889
Bruno M, Mahgoub M, Macfarlan TS (2019) The arms race between KRAB-zinc finger proteins and endogenous Retroelements and its impact on mammals. Annu Rev Genet 53:393–416. https://doi.org/10.1146/annurev-genet-112618-043717
Burns KH (2017) Transposable elements in cancer. Nat Rev Cancer 17(7):415–424. https://doi.org/10.1038/nrc.2017.35
Burns MB, Temiz NA, Harris RS (2013) Evidence for APOBEC3B mutagenesis in multiple human cancers. Nat Genet 45(9):977–983. https://doi.org/10.1038/ng.2701
Cairns J (1981) The origin of human cancers. Nature 289(5796):353–357. https://doi.org/10.1038/289353a0
Chen JY, Harrison TJ, Lee CS, Chen DS, Zuckerman AJ (1986) Detection of hepatitis B virus DNA in hepatocellular carcinoma. Br J Exp Pathol 67(2):279–288
Chen H, Wang M, Wang XY, Gao S, Wang J, Guan XM (2005) Identification of differential genes in ovarian cancer using representational difference analysis of cDNA. Chin Med Sci J 20(3):185–189
Chiappinelli KB, Strissel PL, Desrichard A, Li H, Henke C, Akman B, Hein A, Rote NS, Cope LM, Snyder A, Makarov V, Budhu S, Slamon DJ, Wolchok JD, Pardoll DM, Beckmann MW, Zahnow CA, Merghoub T, Chan TA, Baylin SB, Strick R (2015) Inhibiting DNA methylation causes an interferon response in cancer via dsRNA including endogenous retroviruses. Cell 162(5):974–986. https://doi.org/10.1016/j.cell.2015.07.011
Choi Y, Choi SA, Koh EJ, Yun I, Park S, Jeon S, Kim Y, Park S, Woo D, Phi JH, Park SH, Kim DS, Kim SH, Choi JW, Lee JW, Jung TY, Bhak J, Lee S, Kim SK (2024) Comprehensive multiomics analysis reveals distinct differences between pediatric choroid plexus papilloma and carcinoma. Acta Neuropathol Commun 12(1):93. https://doi.org/10.1186/s40478-024-01814-y
Chorazy M (1985) Sequence rearrangements and genome instability. A possible step in carcinogenesis. J Cancer Res Clin Oncol 109(3):159–172. https://doi.org/10.1007/bf00390351
Chuong EB, Elde NC, Feschotte C (2017) Regulatory activities of transposable elements: from conflicts to benefits. Nat Rev Genet 18(2):71–86. https://doi.org/10.1038/nrg.2016.139
Cleveland JL, Weinstein Y, Ihle JN, Askew DS, Rapp UR (1986) Transformation and insertional mutagenesis in vitro of primary hematopoietic stem cell cultures. Curr Top Microbiol Immunol 132:44–54. https://doi.org/10.1007/978-3-642-71562-4_7
de Koning AP, Gu W, Castoe TA, Batzer MA, Pollock DD (2011) Repetitive elements may comprise over two-thirds of the human genome. PLoS Genet 7(12):e1002384. https://doi.org/10.1371/journal.pgen.1002384
Deniz O, Frost JM, Branco MR (2019) Regulation of transposable elements by DNA modifications. Nat Rev Genet 20(7):417–431. https://doi.org/10.1038/s41576-019-0106-6
Deniz Ö, Ahmed M, Todd CD, Rio-Machin A, Dawson MA, Branco MR (2020) Endogenous retroviruses are a source of enhancers with oncogenic potential in acute myeloid leukaemia. Nat Commun 11(1):3506. https://doi.org/10.1038/s41467-020-17206-4
Dopico XC, Evangelou M, Ferreira RC, Guo H, Pekalski ML, Smyth DJ, Cooper N, Burren OS, Fulford AJ, Hennig BJ, Prentice AM, Ziegler AG, Bonifacio E, Wallace C, Todd JA (2015) Widespread seasonal gene expression reveals annual differences in human immunity and physiology. Nat Commun 6:7000. https://doi.org/10.1038/ncomms8000
Fang LT, Zhu B, Zhao Y, Chen W, Yang Z, Kerrigan L, Langenbach K, de Mars M, Lu C, Idler K, Jacob H, Zheng Y, Ren L, Yu Y, Jaeger E, Schroth GP, Abaan OD, Talsania K, Lack J, Shen TW, Chen Z, Stanbouly S, Tran B, Shetty J, Kriga Y, Meerzaman D, Nguyen C, Petitjean V, Sultan M, Cam M, Mehta M, Hung T, Peters E, Kalamegham R, Sahraeian SME, Mohiyuddin M, Guo Y, Yao L, Song L, Lam HYK, Drabek J, Vojta P, Maestro R, Gasparotto D, Kõks S, Reimann E, Scherer A, Nordlund J, Liljedahl U, Jensen RV, Pirooznia M, Li Z, Xiao C, Sherry ST, Kusko R, Moos M, Donaldson E, Tezak Z, Ning B, Tong W, Li J, Duerken-Hughes P, Catalanotti C, Maheshwari S, Shuga J, Liang WS, Keats J, Adkins J, Tassone E, Zismann V, McDaniel T, Trent J, Foox J, Butler D, Mason CE, Hong H, Shi L, Wang C, Xiao W (2021) Establishing

community reference samples, data and call sets for benchmarking cancer mutation detection using whole-genome sequencing. Nat Biotechnol 39(9):1151–1160. https://doi.org/10.1038/s41587-021-00993-6

Fröhlich A, Pfaff AL, Bubb VJ, Koks S, Quinn JP (2022) Characterisation of the function of a SINE-VNTR-Alu retrotransposon to modulate isoform expression at the MAPT locus. Front Mol Neurosci 15:815695. https://doi.org/10.3389/fnmol.2022.815695

Gasparotto E, Burattin FV, Di Gioia V, Panepuccia M, Ranzani V, Marasca F, Bodega B (2023) Transposable elements co-option in genome evolution and gene regulation. Int J Mol Sci 24:3. https://doi.org/10.3390/ijms24032610

Hamilton DG, Jones ME, Cameron EZ, Kerlin DH, McCallum H, Storfer A, Hohenlohe PA, Hamede RK (2020) Infectious disease and sickness behaviour: tumour progression affects interaction patterns and social network structure in wild Tasmanian devils. Proc Biol Sci 287(1940):20202454. https://doi.org/10.1098/rspb.2020.2454

Hanahan D, Weinberg RA (2000) The hallmarks of cancer. Cell 100(1):57–70. https://doi.org/10.1016/s0092-8674(00)81683-9

Hanahan D, Weinberg RA (2011) Hallmarks of cancer: the next generation. Cell 144(5):646–674. https://doi.org/10.1016/j.cell.2011.02.013

Hauptmann S, Schmitt WD (2006) Transposable elements–is there a link between evolution and cancer? Med Hypotheses 66(3):580–591. https://doi.org/10.1016/j.mehy.2005.08.051

Hecht F, Morgan R, Schrier SL, Adams J, Sandberg AA (1985) The Philadelphia (Ph) chromosome in leukemia. I. A new mechanism due to interstitial deletion and insertion in chronic myelocytic leukemia. Cancer Genet Cytogenet 14(1–2):3–10. https://doi.org/10.1016/0165-4608(85)90209-2

Hehir-Kwa JY, Marschall T, Kloosterman WP, Francioli LC, Baaijens JA, Dijkstra LJ, Abdellaoui A, Koval V, Thung DT, Wardenaar R, Renkens I, Coe BP, Deelen P, de Ligt J, Lameijer EW, van Dijk F, Hormozdiari F, Genome of the Netherlands C, Uitterlinden AG, van Duijn CM, Eichler EE, de Bakker PI, Swertz MA, Wijmenga C, van Ommen GB, Slagboom PE, Boomsma DI, Schonhuth A, Ye K, Guryev V (2016) A high-quality human reference panel reveals the complexity and distribution of genomic structural variants. Nat Commun 7:12989. https://doi.org/10.1038/ncomms12989

Hishinuma F, DeBona PJ, Astrin S, Skalka AM (1981) Nucleotide sequence of acceptor site and termini of integrated avian endogenous provirus ev1: integration creates a 6 bp repeat of host DNA. Cell 23(1):155–164. https://doi.org/10.1016/0092-8674(81)90280-4

Ho XD, Nguyen HG, Trinh LH, Reimann E, Prans E, Kõks G, Maasalu K, Le VQ, Nguyen VH, Le NTN, Phung P, Märtson A, Lattekivi F, Kõks S (2017a) Analysis of the expression of repetitive DNA elements in osteosarcoma. Front Genet 8:193. https://doi.org/10.3389/fgene.2017.00193

Ho XD, Phung P, Le VQ, Nguyen VH, Reimann E, Prans E, Kõks G, Maasalu K, Le NT, Trinh LH, Nguyen HG, Märtson A, Kõks S (2017b) Whole transcriptome analysis identifies differentially regulated networks between osteosarcoma and normal bone samples. Exp Biol Med (Maywood) 242(18):1802–1811. https://doi.org/10.1177/1535370217736512

Hsu TC (1983) Genetic instability in the human population: a working hypothesis. Hereditas 98(1):1–9. https://doi.org/10.1111/j.1601-5223.1983.tb00574.x

Hu Z, Guo X, Li Z, Meng Z, Huang S (2024) The neoantigens derived from transposable elements – a hidden treasure for cancer immunotherapy. Biochim Biophys Acta Rev Cancer 1879(5):189126. https://doi.org/10.1016/j.bbcan.2024.189126

Huang X, Park H, Greene J, Pao J, Mulvey E, Zhou SX, Albert CM, Moy F, Sachdev D, Yee D, Rader C, Hamby CV, Loeb DM, Cairo MS, Zhou X (2015) IGF1R- and ROR1-specific CAR T cells as a potential therapy for high risk sarcomas. PLoS One 10(7):e0133152. https://doi.org/10.1371/journal.pone.0133152

Hubank M, Schatz DG (1994) Identifying differences in mRNA expression by representational difference analysis of cDNA. Nucleic Acids Res 22(25):5640–5648. https://doi.org/10.1093/nar/22.25.5640

Huff V (1996) Genotype/phenotype correlations in Wilms' tumor. Med Pediatr Oncol 27(5):408–414. https://doi.org/10.1002/(sici)1096-911x(199611)27:5<408::Aid-mpo4>3.0.Co;2-q

Hysaj O, Karavasiloglou N, Limam M, Wanner M, Korol D, Rohrmann S (2022) Is season of diagnosis a predictor of cancer survival? Results from the Zurich cancer registry. Nutrients 14(20). https://doi.org/10.3390/nu14204291

Ihle JN, Smith-White B, Sisson B, Parker D, Blair DG, Schultz A, Kozak C, Lunsford RD, Askew D, Weinstein Y et al (1989) Activation of the c-H-ras proto-oncogene by retrovirus insertion and chromosomal rearrangement in a Moloney leukemia virus-induced T-cell leukemia. J Virol 63(7):2959–2966. https://doi.org/10.1128/jvi.63.7.2959-2966.1989

Ishizuka JJ, Manguso RT, Cheruiyot CK, Bi K, Panda A, Iracheta-Vellve A, Miller BC, Du PP, Yates KB, Dubrot J, Buchumenski I, Comstock DE, Brown FD, Ayer A, Kohnle IC, Pope HW, Zimmer MD, Sen DR, Lane-Reticker SK, Robitschek EJ, Griffin GK, Collins NB, Long AH, Doench JG, Kozono D, Levanon EY, Haining WN (2019) Loss of ADAR1 in tumours overcomes resistance to immune checkpoint blockade. Nature 565(7737):43–48. https://doi.org/10.1038/s41586-018-0768-9

Jang HS, Shah NM, Du AY, Dailey ZZ, Pehrsson EC, Godoy PM, Zhang D, Li D, Xing X, Kim S, O'Donnell D, Gordon JI, Wang T (2019) Transposable elements drive widespread expression of oncogenes in human cancers. Nat Genet 51(4):611–617. https://doi.org/10.1038/s41588-019-0373-3

Jiang Y, Zhang H, Wang J, Chen J, Guo Z, Liu Y, Hua H (2023) Exploiting RIG-I-like receptor pathway for cancer immunotherapy. J Hematol Oncol 16(1):8. https://doi.org/10.1186/s13045-023-01405-9

Johnson R, Guigo R (2014) The RIDL hypothesis: transposable elements as functional domains of long noncoding RNAs. RNA 20(7):959–976. https://doi.org/10.1261/rna.044560.114

Jones PA, Baylin SB (2002) The fundamental role of epigenetic events in cancer. Nat Rev Genet 3(6):415–428. https://doi.org/10.1038/nrg816

Jones ME, Cockburn A, Hamede R, Hawkins C, Hesterman H, Lachish S, Mann D, McCallum H, Pemberton D (2008) Life-history change in disease-ravaged Tasmanian devil populations. Proc Natl Acad Sci USA 105(29):10023–10027. https://doi.org/10.1073/pnas.0711236105

Ju G, Skalka AM (1980) Nucleotide sequence analysis of the long terminal repeat (LTR) of avian retroviruses: structural similarities with transposable elements. Cell 22(2 Pt 2):379–386. https://doi.org/10.1016/0092-8674(80)90348-7

Kapusta A, Kronenberg Z, Lynch VJ, Zhuo X, Ramsay L, Bourque G, Yandell M, Feschotte C (2013) Transposable elements are major contributors to the origin, diversification, and regulation of vertebrate long noncoding RNAs. PLoS Genet 9(4):e1003470. https://doi.org/10.1371/journal.pgen.1003470

Katzir N, Rechavi G, Cohen JB, Unger T, Simoni F, Segal S, Cohen D, Givol D (1985) "Retroposon" insertion into the cellular oncogene c-myc in canine transmissible venereal tumor. Proc Natl Acad Sci USA 82(4):1054–1058. https://doi.org/10.1073/pnas.82.4.1054

Katzir N, Arman E, Cohen D, Givol D, Rechavi G (1987) Common origin of transmissible venereal tumors (TVT) in dogs. Oncogene 1(4):445–448

Kingo K, Aunin E, Karelson M, Ratsep R, Silm H, Vasar E, Koks S (2008) Expressional changes in the intracellular melanogenesis pathways and their possible role in the pathogenesis of vitiligo. J Dermatol Sci 52(1):39–46. https://doi.org/10.1016/j.jdermsci.2008.03.013

Klein G, Lenoir G (1982) Translocations involving Ig-locus-carrying chromosomes: a model for genetic transposition in carcinogenesis. Adv Cancer Res 37:381–387. https://doi.org/10.1016/s0065-230x(08)60887-8

Knudson AG Jr (1971) Mutation and cancer: statistical study of retinoblastoma. Proc Natl Acad Sci USA 68(4):820–823. https://doi.org/10.1073/pnas.68.4.820

Kobayashi T, Urakami S, Hirayama Y, Yamamoto T, Nishizawa M, Takahara T, Kubo Y, Hino O (1997) Intragenic Tsc2 somatic mutations as Knudson's second hit in spontaneous and chemically induced renal carcinomas in the Eker rat model. Jpn J Cancer Res 88(3):254–261. https://doi.org/10.1111/j.1349-7006.1997.tb00375.x

Kõks S, Kõks G (2018) Chapter 22: The role of human endogenous retroviruses (HERVs) in the pathologies of the nervous system. In: Gerlai RT (ed) Molecular-genetic and statistical techniques for behavioral and neural research. Academic, San Diego, pp 519–533. https://doi.org/10.1016/B978-0-12-804078-2.00022-2

Koks S, Mannisto PT, Bourin M, Shlik J, Vasar V, Vasar E (2000) Cholecystokinin-induced anxiety in rats: relevance of pre-experimental stress and seasonal variations. J Psychiatry Neurosci 25(1):33–42

Koks S, Luuk H, Nelovkov A, Areda T, Vasar E (2004) A screen for genes induced in the amygdaloid area during cat odor exposure. Genes Brain Behav 3(2):80–89. https://doi.org/10.1046/j.1601-183x.2003.00047.x

Koks G, Uudelepp ML, Limbach M, Peterson P, Reimann E, Koks S (2015) Smoking-induced expression of the GPR15 gene indicates its potential role in chronic inflammatory pathologies. Am J Pathol 185(11):2898–2906. https://doi.org/10.1016/j.ajpath.2015.07.006

Kõks S, Singleton LM, Quinn JP, Bubb VJ, Pfaff AL (2022) Analysis of the retrotransposon SINE-VNTR-Alu (SVA) polymorphisms in the genetics and pathophysiology of complex diseases. In: Proukakis C (ed) Genomic structural variants in nervous system disorders. Springer US, New York, pp 63–77. https://doi.org/10.1007/978-1-0716-2357-2_4

Krishnan VS, Kõks S (2023) Transcriptional landscape of repetitive elements in psoriatic skin from large cohort studies: relevance to psoriasis pathophysiology. Int J Mol Sci 24(23). https://doi.org/10.3390/ijms242316725

Kulski JK, Pfaff AL, Marney LD, Fröhlich A, Bubb VJ, Quinn JP, Koks S (2023) Regulation of expression quantitative trait loci by SVA retrotransposons within the major histocompatibility complex. Exp Biol Med (Maywood) 248(23):2304–2318. https://doi.org/10.1177/15353702231209411

Kulski JK, Pfaff AL, Koks S (2024a) SVA regulation of transposable element clustered transcription within the major histocompatibility complex genomic class II region of the Parkinson's progression markers initiative. Genes (Basel) 15(9). https://doi.org/10.3390/genes15091185

Kulski JK, Suzuki S, Shiina T, Pfaff AL, Kõks S (2024b) Regulatory SVA retrotransposons and classical HLA genotyped-transcripts associated with Parkinson's disease. Front Immunol 15:1349030. https://doi.org/10.3389/fimmu.2024.1349030

Kulski JK, Pfaff AL, Koks S (2025) The transcription of transposable elements differentially regulated by SVAs in the major histocompatibility complex class I region of a Parkinson's progression markers initiative cohort. J Mol Pathol 6(1):1

Lambe M, Blomqvist P, Bellocco R (2003) Seasonal variation in the diagnosis of cancer: a study based on national cancer registration in Sweden. Br J Cancer 88(9):1358–1360. https://doi.org/10.1038/sj.bjc.6600901

Lanciano S, Cristofari G (2020) Measuring and interpreting transposable element expression. Nat Rev Genet 21(12):721–736. https://doi.org/10.1038/s41576-020-0251-y

Lander ES, Linton LM, Birren B, Nusbaum C, Zody MC, Baldwin J, Devon K, Dewar K, Doyle M, FitzHugh W, Funke R, Gage D, Harris K, Heaford A, Howland J, Kann L, Lehoczky J, LeVine R, McEwan P, McKernan K, Meldrim J, Mesirov JP, Miranda C, Morris W, Naylor J, Raymond C, Rosetti M, Santos R, Sheridan A, Sougnez C, Stange-Thomann N, Stojanovic N, Subramanian A, Wyman D, Rogers J, Sulston J, Ainscough R, Beck S, Bentley D, Burton J, Clee C, Carter N, Coulson A, Deadman R, Deloukas P, Dunham A, Dunham I, Durbin R, French L, Grafham D, Gregory S, Hubbard T, Humphray S, Hunt A, Jones M, Lloyd C, McMurray A, Matthews L, Mercer S, Milne S, Mullikin JC, Mungall A, Plumb R, Ross M, Shownkeen R, Sims S, Waterston RH, Wilson RK, Hillier LW, McPherson JD, Marra MA, Mardis ER, Fulton LA, Chinwalla AT, Pepin KH, Gish WR, Chissoe SL, Wendl MC, Delehaunty KD, Miner TL, Delehaunty A, Kramer JB, Cook LL, Fulton RS, Johnson DL, Minx PJ, Clifton SW, Hawkins T, Branscomb E, Predki P, Richardson P, Wenning S, Slezak T, Doggett N, Cheng J-F, Olsen A, Lucas S, Elkin C, Uberbacher E, Frazier M, Gibbs RA, Muzny DM, Scherer SE, Bouck JB, Sodergren EJ, Worley KC, Rives CM, Gorrell JH, Metzker ML, Naylor SL, Kucherlapati RS, Nelson DL, Weinstock GM, Sakaki Y, Fujiyama A, Hattori M, Yada T, Toyoda A, Itoh T, Kawagoe C, Watanabe H, Totoki Y, Taylor T, Weissenbach J, Heilig R, Saurin W, Artiguenave F, Brottier P, Bruls T, Pelletier E, Robert C, Wincker P, Rosenthal A, Platzer M, Nyakatura G,

Taudien S, Rump A, Smith DR, Doucette-Stamm L, Rubenfield M, Weinstock K, Lee HM, Dubois J, Yang H, Yu J, Wang J, Huang G, Gu J, Hood L, Rowen L, Madan A, Qin S, Davis RW, Federspiel NA, Abola AP, Proctor MJ, Roe BA, Chen F, Pan H, Ramser J, Lehrach H, Reinhardt R, McCombie WR, de la Bastide M, Dedhia N, Blöcker H, Hornischer K, Nordsiek G, Agarwala R, Aravind L, Bailey JA, Bateman A, Batzoglou S, Birney E, Bork P, Brown DG, Burge CB, Cerutti L, Chen H-C, Church D, Clamp M, Copley RR, Doerks T, Eddy SR, Eichler EE, Furey TS, Galagan J, Gilbert JGR, Harmon C, Hayashizaki Y, Haussler D, Hermjakob H, Hokamp K, Jang W, Johnson LS, Jones TA, Kasif S, Kaspryzk A, Kennedy S, Kent WJ, Kitts P, Koonin EV, Korf I, Kulp D, Lancet D, Lowe TM, McLysaght A, Mikkelsen T, Moran JV, Mulder N, Pollara VJ, Ponting CP, Schuler G, Schultz J, Slater G, Smit AFA, Stupka E, Szustakowki J, Thierry-Mieg D, Thierry-Mieg J, Wagner L, Wallis J, Wheeler R, Williams A, Wolf YI, Wolfe KH, Yang S-P, Yeh R-F, Collins F, Guyer MS, Peterson J, Felsenfeld A, Wetterstrand KA, Myers RM, Schmutz J, Dickson M, Grimwood J, Cox DR, Olson MV, Kaul R, Raymond C, Shimizu N, Kawasaki K, Minoshima S, Evans GA, Athanasiou M, Schultz R, Patrinos A, Morgan MJ, International Human Genome Sequencing C, Whitehead Institute for Biomedical Research CfGR, The Sanger C, Washington University Genome Sequencing C, Institute UDJG, Baylor College of Medicine Human Genome Sequencing C, Center RGS, Genoscope, Cnrs UMR, Department of Genome Analysis IoMB, Center GTCS, Beijing Genomics Institute/Human Genome C, Multimegabase Sequencing Center TIfSB, Stanford Genome Technology C, University of Oklahoma's Advanced Center for Genome T, Max Planck Institute for Molecular G, Cold Spring Harbor Laboratory LAHGC, Biotechnology GBGRCf, *Genome Analysis G, Scientific management: National Human Genome Research Institute USNIoH, Stanford Human Genome C, University of Washington Genome C, Department of Molecular Biology KUSoM, University of Texas Southwestern Medical Center at D, Office of Science USDoE, The Wellcome T (2001) Initial sequencing and analysis of the human genome. Nature 409(6822):860–921. https://doi.org/10.1038/35057062

Lantto TA, Laakso I, Dorman HJ, Mauriala T, Hiltunen R, Kõks S, Raasmaja A (2016) Cellular stress and p53-associated apoptosis by Juniperus communis L. berry extract treatment in the human SH-SY5Y Neuroblastoma cells. Int J Mol Sci 17(7). https://doi.org/10.3390/ijms17071113

Lawson ND, Berliner N (1998) Representational difference analysis of a committed myeloid progenitor cell line reveals evidence for bilineage potential. Proc Natl Acad Sci USA 95(17):10129–10133. https://doi.org/10.1073/pnas.95.17.10129

Leonova KI, Brodsky L, Lipchick B, Pal M, Novototskaya L, Chenchik AA, Sen GC, Komarova EA, Gudkov AV (2013) p53 cooperates with DNA methylation and a suicidal interferon response to maintain epigenetic silencing of repeats and noncoding RNAs. Proc Natl Acad Sci USA 110(1):E89–E98. https://doi.org/10.1073/pnas.1216922110

Li Y, Salo-Mullen E, Varghese A, Trottier M, Stadler ZK, Zhang L (2020) Insertion of an Alu-like element in MLH1 intron 7 as a novel cause of Lynch syndrome. Mol Genet Genomic Med 8(12):e1523. https://doi.org/10.1002/mgg3.1523

Lichtenberg U, Zock C, Doerfler W (1987) Insertion of adenovirus type 12 DNA in the vicinity of an intracisternal a particle genome in Syrian hamster tumor cells. J Virol 61(9):2719–2726. https://doi.org/10.1128/jvi.61.9.2719-2726.1987

Ligtenberg MJ, Kuiper RP, Chan TL, Goossens M, Hebeda KM, Voorendt M, Lee TY, Bodmer D, Hoenselaar E, Hendriks-Cornelissen SJ, Tsui WY, Kong CK, Brunner HG, van Kessel AG, Yuen ST, van Krieken JH, Leung SY, Hoogerbrugge N (2009) Heritable somatic methylation and inactivation of MSH2 in families with Lynch syndrome due to deletion of the 3′ exons of TACSTD1. Nat Genet 41(1):112–117. https://doi.org/10.1038/ng.283

Lill M, Koks S, Soomets U, Schalkwyk LC, Fernandes C, Lutsar I, Taba P (2013) Peripheral blood RNA gene expression profiling in patients with bacterial meningitis. Front Neurosci 7:33. https://doi.org/10.3389/fnins.2013.00033

Lim JK (1979) Site-specific instability in Drosophila melanogaster: the origin of the mutation and cytogenetic evidence for site specificity. Genetics 93(3):681–701. https://doi.org/10.1093/genetics/93.3.681
Lin X, Xia Y, Hu D, Mao Q, Yu Z, Zhang H, Li C, Chen G, Liu F, Zhu W, Shi Y, Zhang H, Zheng J, Sun T, Xu J, Chao HH, Zheng X, Luo X (2019) Transcriptome-wide piRNA profiling in human gastric cancer. Oncol Rep 41(5):3089–3099. https://doi.org/10.3892/or.2019.7073
Livezey KW, Simon D (1997) Accumulation of genetic alterations in a human hepatoma cell line transfected with hepatitis B virus. Mutat Res 377(2):187–198. https://doi.org/10.1016/s0027-5107(97)00068-7
Lu X, Sachs F, Ramsay L, Jacques PE, Goke J, Bourque G, Ng HH (2014) The retrovirus HERVH is a long noncoding RNA required for human embryonic stem cell identity. Nat Struct Mol Biol 21(4):423–425. https://doi.org/10.1038/nsmb.2799
Mann MB, Black MA, Jones DJ, Ward JM, Yew CC, Newberg JY, Dupuy AJ, Rust AG, Bosenberg MW, McMahon M, Print CG, Copeland NG, Jenkins NA (2015) Transposon mutagenesis identifies genetic drivers of Braf(V600E) melanoma. Nat Genet 47(5):486–495. https://doi.org/10.1038/ng.3275
Mardis ER (2019) Neoantigens and genome instability: impact on immunogenomic phenotypes and immunotherapy response. Genome Med 11(1):71. https://doi.org/10.1186/s13073-019-0684-0
Martins F, Rosspopoff O, Carlevaro-Fita J, Forey R, Offner S, Planet E, Pulver C, Pak H, Huber F, Michaux J, Bassani-Sternberg M, Turelli P, Trono D (2024) A cluster of evolutionarily recent KRAB zinc finger proteins protects cancer cells from replicative stress-induced inflammation. Cancer Res 84(6):808–826. https://doi.org/10.1158/0008-5472.Can-23-1237
Märtson A, Kõks S, Reimann E (2013) Transcriptome analysis of osteosarcoma identifies suppression of wnt pathway and up-regulation of adiponectin as potential biomarker. Genomics Discov 1(1):3–9. https://doi.org/10.7243/2052-7993-1-3
Meštrović N, Mravinac B, Pavlek M, Vojvoda-Zeljko T, Šatović E, Plohl M (2015) Structural and functional liaisons between transposable elements and satellite DNAs. Chromosom Res 23(3):583–596. https://doi.org/10.1007/s10577-015-9483-7
Mitreiter K, Schmidt J, Luz A, Atkinson MJ, Höfler H, Erfle V, Strauss PG (1994) Disruption of the murine p53 gene by insertion of an endogenous retrovirus-like element (ETn) in a cell line from radiation-induced osteosarcoma. Virology 200(2):837–841. https://doi.org/10.1006/viro.1994.1253
Moan J, Lagunova Z, Bruland O, Juzeniene A (2010) Seasonal variations of cancer incidence and prognosis. Dermatoendocrinol 2(2):55–57. https://doi.org/10.4161/derm.2.2.12664
Mombach DM, Fontoura Gomes T, Silva MM, Loreto ÉLS (2022) Molecular and biological effects of cisplatin in drosophila. Comp Biochem Physiol C Toxicol Pharmacol 252:109229. https://doi.org/10.1016/j.cbpc.2021.109229
Mombach DM, Mercuri RLV, da Fontoura Gomes TMF, Galante PAF, Loreto ELS (2024) Transposable elements alter gene expression and may impact response to cisplatin therapy in ovarian cancer. Carcinogenesis 45(9):685–695. https://doi.org/10.1093/carcin/bgae029
Morris CM, Heisterkamp N, Kennedy MA, Fitzgerald PH, Groffen J (1990) Ph-negative chronic myeloid leukemia: molecular analysis of ABL insertion into M-BCR on chromosome 22. Blood 76(9):1812–1818
Nelson PN, Carnegie PR, Martin J, Davari Ejtehadi H, Hooley P, Roden D, Rowland-Jones S, Warren P, Astley J, Murray PG (2003) Demystified. Human endogenous retroviruses. Mol Pathol 56(1):11–18. https://doi.org/10.1136/mp.56.1.11
Nelson PN, Hooley P, Roden D, Davari Ejtehadi H, Rylance P, Warren P, Martin J, Murray PG (2004) Human endogenous retroviruses: transposable elements with potential? Clin Exp Immunol 138(1):1–9. https://doi.org/10.1111/j.1365-2249.2004.02592.x
Nguyen LL, Watson ZL, Ortega R, Woodruff ER, Jordan KR, Iwanaga R, Yamamoto TM, Bailey CA, To F, Jeong AD, Guntupalli SR, Behbakht K, Gibaja V, Arnoult N, Cocozaki A, Chuong EB, Bitler BG (2024) Combining EHMT and PARP inhibition: a strategy to diminish therapy-

resistant ovarian cancer tumor growth while stimulating immune activation. Mol Cancer Ther:of1–of16. https://doi.org/10.1158/1535-7163.Mct-23-0613

Nolano A, Medugno A, Trombetti S, Liccardo R, De Rosa M, Izzo P, Duraturo F (2022) Hereditary colorectal cancer: state of the art in Lynch syndrome. Cancers (Basel) 15(1). https://doi.org/10.3390/cancers15010075

Oprandy JJ, Chang PW (1983) 5-bromodeoxyuridine induction of hematopoietic neoplasia and retrovirus activation in the soft-shell clam, Mya arenaria. J Invertebr Pathol 42(2):196–206. https://doi.org/10.1016/0022-2011(83)90062-9

Ostrander EA, Davis BW, Ostrander GK (2016) Transmissible tumors: breaking the cancer paradigm. Trends Genet 32(1):1–15. https://doi.org/10.1016/j.tig.2015.10.001

Panda A, de Cubas AA, Stein M, Riedlinger G, Kra J, Mayer T, Smith CC, Vincent BG, Serody JS, Beckermann KE, Ganesan S, Bhanot G, Rathmell WK (2018) Endogenous retrovirus expression is associated with response to immune checkpoint blockade in clear cell renal cell carcinoma. JCI Insight 3(16). https://doi.org/10.1172/jci.insight.121522

Pasquesi GIM, Allen H, Ivancevic A, Barbachano-Guerrero A, Joyner O, Guo K, Simpson DM, Gapin K, Horton I, Nguyen LL, Yang Q, Warren CJ, Florea LD, Bitler BG, Santiago ML, Sawyer SL, Chuong EB (2024) Regulation of human interferon signaling by transposon exonization. Cell. https://doi.org/10.1016/j.cell.2024.11.016

Peck S, Corkrey R, Hamede R, Jones M, Canfield P (2015) Hematologic and serum biochemical reference intervals for wild Tasmanian devils (Sarcophilus harrisii). Vet Clin Pathol 44(4):519–529. https://doi.org/10.1111/vcp.12304

Pfaff AL, Bubb VJ, Quinn JP, Koks S (2023) A genome-wide screen for the Exonisation of reference SINE-VNTR-Alus and their expression in CNS tissues of individuals with amyotrophic lateral sclerosis. Int J Mol Sci 24(14). https://doi.org/10.3390/ijms241411548

Poudel BH, Koks S (2024) The whole transcriptome analysis using FFPE and fresh tissue samples identifies the molecular fingerprint of osteosarcoma. Exp Biol Med (Maywood) 249:10161. https://doi.org/10.3389/ebm.2024.10161

Pritchard DJ, Finkel MP, Reilly CA Jr (1975) The etiology of osteosarcoma. A review of current considerations. Clin Orthop Relat Res 111:14–22. https://doi.org/10.1097/00003086-197509000-00003

Ramel C (1989) The nature of spontaneous mutations. Mutat Res 212(1):33–42. https://doi.org/10.1016/0027-5107(89)90020-1

Ramel C, Cederberg H, Magnusson J, Vogel E, Natarajan AT, Mullender LH, Nivard JM, Parry JM, Leyson A, Comendador MA, Sierra LM, Ferreiro JA, Consuegra S (1996) Somatic recombination, gene amplification and cancer. Mutat Res 353(1–2):85–107. https://doi.org/10.1016/0027-5107(95)00243-x

Rassool F, Martiat P, Taj A, Klisak I, Goldman J (1990) Interstitial insertion of varying amounts of ABL-containing genetic material into chromosome 22 in Ph-negative CML. Leukemia 4(4):273–277

Raven N, Klaassen M, Madsen T, Thomas F, Hamede RK, Ujvari B (2022) Transmissible cancer influences immune gene expression in an endangered marsupial, the Tasmanian devil (Sarcophilus harrisii). Mol Ecol 31(8):2293–2311. https://doi.org/10.1111/mec.16408

Reddy EP, Smith MJ, Canaani E, Robbins KC, Tronick SR, Zain S, Aaronson SA (1980) Nucleotide sequence analysis of the transforming region and large terminal redundancies of Moloney murine sarcoma virus. Proc Natl Acad Sci USA 77(9):5234–5238. https://doi.org/10.1073/pnas.77.9.5234

Reddy EP, Smith MJ, Aaronson SA (1981) Complete nucleotide sequence and organization of the Moloney murine sarcoma virus genome. Science 214(4519):445–450. https://doi.org/10.1126/science.6170110

Reggiardo RE, Maroli SV, Peddu V, Davidson AE, Hill A, LaMontagne E, Aaraj YA, Jain M, Chan SY, Kim DH (2023) Profiling of repetitive RNA sequences in the blood plasma of patients with cancer. Nat Biomed Eng 7(12):1627–1635. https://doi.org/10.1038/s41551-023-01081-7

Reimann E, Kõks S, Ho XD, Maasalu K, Märtson A (2014) Whole exome sequencing of a single osteosarcoma case–integrative analysis with whole transcriptome RNA-seq data. Hum Genomics 8(1):20. https://doi.org/10.1186/s40246-014-0020-0

Riccardi VM, Sujansky E, Smith AC, Francke U (1978) Chromosomal imbalance in the Aniridia-Wilms' tumor association: 11p interstitial deletion. Pediatrics 61(4):604–610

Ricceri F, Trevisan M, Fiano V, Grasso C, Fasanelli F, Scoccianti C, De Marco L, Tos AG, Vineis P, Sacerdote C (2014) Seasonality modifies methylation profiles in healthy people. PLoS One 9(9):e106846. https://doi.org/10.1371/journal.pone.0106846

Rodriguez-Martin B, Alvarez EG, Baez-Ortega A, Zamora J, Supek F, Demeulemeester J, Santamarina M, Ju YS, Temes J, Garcia-Souto D, Detering H, Li Y, Rodriguez-Castro J, Dueso-Barroso A, Bruzos AL, Dentro SC, Blanco MG, Contino G, Ardeljan D, Tojo M, Roberts ND, Zumalave S, Edwards PA, Weischenfeldt J, Puiggros M, Chong Z, Chen K, Lee EA, Wala JA, Raine KM, Butler A, Waszak SM, FCP N, Schumacher SE, Monlong J, Maura F, Bolli N, Bourque G, Gerstein M, Park PJ, Wedge DC, Beroukhim R, Torrents D, Korbel JO, Martincorena I, Fitzgerald RC, Van Loo P, Kazazian HH, Burns KH, Group PSVW, Campbell PJ, Tubio JMC, Consortium P (2020) Pan-cancer analysis of whole genomes identifies driver rearrangements promoted by LINE-1 retrotransposition. Nat Genet 52(3):306–319. https://doi.org/10.1038/s41588-019-0562-0

Rothzerg E, Ho XD, Xu J, Wood D, Märtson A, Maasalu K, Kõks S (2020) Alternative splicing of leptin receptor overlapping transcript in osteosarcoma. Exp Biol Med (Maywood) 245(16):1437–1443. https://doi.org/10.1177/1535370220949139

Roulois D, Loo Yau H, Singhania R, Wang Y, Danesh A, Shen SY, Han H, Liang G, Jones PA, Pugh TJ, O'Brien C, De Carvalho DD (2015) DNA-Demethylating agents target colorectal cancer cells by inducing viral mimicry by endogenous transcripts. Cell 162(5):961–973. https://doi.org/10.1016/j.cell.2015.07.056

Rowe HM, Trono D (2011) Dynamic control of endogenous retroviruses during development. Virology 411(2):273–287. https://doi.org/10.1016/j.virol.2010.12.007

Rowley JD (1979) Chromosome abnormalities in leukemia. Haematol Blood Transfus 23:43–52. https://doi.org/10.1007/978-3-642-67057-2_5

Sager R (1979) Transposable elements and chromosomal rearrangements in cancer–a possible link. Nature 282(5738):447–449. https://doi.org/10.1038/282447a0

Samia AM, Nenow J (2022) Seasonal variation in the diagnosis of skin cancers from 1983 to 2017 in Greenville, North Carolina. Cureus 14(3):e23254. https://doi.org/10.7759/cureus.23254

Santoni FA, Guerra J, Luban J (2012) HERV-H RNA is abundant in human embryonic stem cells and a precise marker for pluripotency. Retrovirology 9:111. https://doi.org/10.1186/1742-4690-9-111

Savage AL, Bubb VJ, Breen G, Quinn JP (2013) Characterisation of the potential function of SVA retrotransposons to modulate gene expression patterns. BMC Evol Biol 13:101. https://doi.org/10.1186/1471-2148-13-101

Schöbel A, Nguyen-Dinh V, Schumann GG, Herker E (2021) Hepatitis C virus infection restricts human LINE-1 retrotransposition in hepatoma cells. PLoS Pathog 17(4):e1009496. https://doi.org/10.1371/journal.ppat.1009496

Schulz WA, Steinhoff C, Florl AR (2006) Methylation of endogenous human retroelements in health and disease. Curr Top Microbiol Immunol 310:211–250. https://doi.org/10.1007/3-540-31181-5_11

Schumann GG (2007) APOBEC3 proteins: major players in intracellular defence against LINE-1-mediated retrotransposition. Biochem Soc Trans 35(Pt 3):637–642. https://doi.org/10.1042/bst0350637

Shah NM, Jang HJ, Liang Y, Maeng JH, Tzeng SC, Wu A, Basri NL, Qu X, Fan C, Li A, Katz B, Li D, Xing X, Evans BS, Wang T (2023) Pan-cancer analysis identifies tumor-specific antigens derived from transposable elements. Nat Genet 55(4):631–639. https://doi.org/10.1038/s41588-023-01349-3

Sharma A, Wolfgruber TK, Presting GG (2013) Tandem repeats derived from centromeric retrotransposons. BMC Genomics 14:142. https://doi.org/10.1186/1471-2164-14-142
Sherman M, Shafritz DA (1984) Hepatitis B virus and hepatocellular carcinoma: molecular biology and mechanistic considerations. Semin Liver Dis 4(2):98–112. https://doi.org/10.1055/s-2008-1040650
Shinnick TM, Lerner RA, Sutcliffe JG (1981) Nucleotide sequence of Moloney murine leukaemia virus. Nature 293(5833):543–548. https://doi.org/10.1038/293543a0
Silver MJ, Kessler NJ, Hennig BJ, Dominguez-Salas P, Laritsky E, Baker MS, Coarfa C, Hernandez-Vargas H, Castelino JM, Routledge MN, Gong YY, Herceg Z, Lee YS, Lee K, Moore SE, Fulford AJ, Prentice AM, Waterland RA (2015) Independent genomewide screens identify the tumor suppressor VTRNA2-1 as a human epiallele responsive to periconceptional environment. Genome Biol 16(1):118. https://doi.org/10.1186/s13059-015-0660-y
Skalka A, Ju G, Hishinuma F, DeBona PJ, Astrin S (1981) Structural analogies among avian retroviral DNAs and transposable elements. Cold Spring Harb Symp Quant Biol 45(Pt 2):739–746. https://doi.org/10.1101/sqb.1981.045.01.092
Smida J, Xu H, Zhang Y, Baumhoer D, Ribi S, Kovac M, von Luettichau I, Bielack S, O'Leary VB, Leib-Mösch C, Frishman D, Nathrath M (2017) Genome-wide analysis of somatic copy number alterations and chromosomal breakages in osteosarcoma. Int J Cancer 141(4):816–828. https://doi.org/10.1002/ijc.30778
Stammnitz MR, Gori K, Kwon YM, Harry E, Martin FJ, Billis K, Cheng Y, Baez-Ortega A, Chow W, Comte S, Eggertsson H, Fox S, Hamede R, Jones M, Lazenby B, Peck S, Pye R, Quail MA, Swift K, Wang J, Wood J, Howe K, Stratton MR, Ning Z, Murchison EP (2023) The evolution of two transmissible cancers in Tasmanian devils. Science 380(6642):283–293. https://doi.org/10.1126/science.abq6453
Steffen D (1984) Proviruses are adjacent to c-myc in some murine leukemia virus-induced lymphomas. Proc Natl Acad Sci USA 81(7):2097–2101. https://doi.org/10.1073/pnas.81.7.2097
Stopera SA, Davie JR, Ray M (1990) Transposition of the abl proto-oncogene in Philadelphia--negative chronic myeloid leukaemia and acute lymphocytic leukaemia. Cytobios 61(246–247):161–170
Storer J, Hubley R, Rosen J, Wheeler TJ, Smit AF (2021) The Dfam community resource of transposable element families, sequence models, and genome annotations. Mob DNA 12(1):2. https://doi.org/10.1186/s13100-020-00230-y
Strong LC, Riccardi VM, Ferrell RE, Sparkes RS (1981) Familial retinoblastoma and chromosome 13 deletion transmitted via an insertional translocation. Science 213(4515):1501–1503. https://doi.org/10.1126/science.7280668
Su Y, Ghodke PP, Egli M, Li L, Wang Y, Guengerich FP (2019) Human DNA polymerase η has reverse transcriptase activity in cellular environments. J Biol Chem 294(15):6073–6081. https://doi.org/10.1074/jbc.RA119.007925
Sutcliffe JG, Shinnick TM, Verma IM, Lerner RA (1980) Nucleotide sequence of Moloney leukemia virus: 3′ end reveals details of replications, analogy to bacterial transposons, and an unexpected gene. Proc Natl Acad Sci USA 77(6):3302–3306. https://doi.org/10.1073/pnas.77.6.3302
Sutcliffe JG, Shinnick TM, Lerner RA (1981) Moloney murine leukemia virus is a transposon: nucleotide sequence analysis identifies genes and replication details. Cold Spring Harb Symp Quant Biol 45(Pt 2):707–710. https://doi.org/10.1101/sqb.1981.045.01.088
Telesnitsky A, Goff SP (1997) Reverse transcriptase and the generation of retroviral DNA. In: Coffin JM, Hughes SH, Varmus HE (eds) Retroviruses. Cold Spring Harbor Laboratory Press, Cold Spring Harbor
Temin HM (1970) Malignant transformation of cells by viruses. Perspect Biol Med 14(1):11–26. dx.doi.org. https://doi.org/10.1353/pbm.1970.0006
Temin HM (1974) On the origin of RNA tumor viruses. Annu Rev Genet 8:155–177. https://doi.org/10.1146/annurev.ge.08.120174.001103

Temin HM (1980) Origin of retroviruses from cellular moveable genetic elements. Cell 21(3):599–600. https://doi.org/10.1016/0092-8674(80)90420-1
Temin HM (1986) Retroviruses and evolution. Cell Biophys 9(1–2):9–16. https://doi.org/10.1007/bf02797372
Timp W, Feinberg AP (2013) Cancer as a dysregulated epigenome allowing cellular growth advantage at the expense of the host. Nat Rev Cancer 13(7):497–510. https://doi.org/10.1038/nrc3486
Tubio JMC, Li Y, Ju YS, Martincorena I, Cooke SL, Tojo M, Gundem G, Pipinikas CP, Zamora J, Raine K, Menzies A, Roman-Garcia P, Fullam A, Gerstung M, Shlien A, Tarpey PS, Papaemmanuil E, Knappskog S, Van Loo P, Ramakrishna M, Davies HR, Marshall J, Wedge DC, Teague JW, Butler AP, Nik-Zainal S, Alexandrov L, Behjati S, Yates LR, Bolli N, Mudie L, Hardy C, Martin S, McLaren S, O'Meara S, Anderson E, Maddison M, Gamble S, Foster C, Warren AY, Whitaker H, Brewer D, Eeles R, Cooper C, Neal D, Lynch AG, Visakorpi T, Isaacs WB, Veer LV, Caldas C, Desmedt C, Sotiriou C, Aparicio S, Foekens JA, Eyfjörd JE, Lakhani SR, Thomas G, Myklebost O, Span PN, Børresen-Dale AL, Richardson AL, Van de Vijver M, Vincent-Salomon A, Van den Eynden GG, Flanagan AM, Futreal PA, Janes SM, Bova GS, Stratton MR, McDermott U, Campbell PJ (2014) Mobile DNA in cancer. Extensive transduction of nonrepetitive DNA mediated by L1 retrotransposition in cancer genomes. Science 345(6196):1251343. https://doi.org/10.1126/science.1251343
Tulchiner G, Staudacher N, Fritz J, Hackl M, Pichler M, Seles M, Shariat SF, D'Andrea D, Gust K, Albrecht W, Grubmüller K, Madersbacher S, Graf S, Lusuardi L, Augustin H, Berger A, Loidl W, Horninger W, Pichler R (2021) Seasonal variations in the diagnosis of testicular germ cell tumors: a National Cancer Registry Study in Austria. Cancers (Basel) 13(21). https://doi.org/10.3390/cancers13215377
Turelli P, Playfoot C, Grun D, Raclot C, Pontis J, Coudray A, Thorball C, Duc J, Pankevich EV, Deplancke B, Busskamp V, Trono D (2020) Primate-restricted KRAB zinc finger proteins and target retrotransposons control gene expression in human neurons. Sci Adv 6(35):eaba3200. https://doi.org/10.1126/sciadv.aba3200
Twu JS, Schloemer RH (1987) Transcriptional trans-activating function of hepatitis B virus. J Virol 61(11):3448–3453. https://doi.org/10.1128/jvi.61.11.3448-3453.1987
Valdebenito-Maturana B (2024) The spatial and cellular portrait of transposable element expression during gastric cancer. Sci Rep 14(1):22727. https://doi.org/10.1038/s41598-024-73744-7
van Haaften G, Plasterk RH, Tijsterman M (2004) Genomic instability and cancer: scanning the Caenorhabditis elegans genome for tumor suppressors. Oncogene 23(51):8366–8375. https://doi.org/10.1038/sj.onc.1208011
Vijaya S, Steffen DL, Kozak C, Robinson HL (1987) Dsi-1, a region with frequent proviral insertions in Moloney murine leukemia virus-induced rat thymomas. J Virol 61(4):1164–1170. https://doi.org/10.1128/jvi.61.4.1164-1170.1987
Voskarides K (2019) The "cancer-cold" hypothesis and possible extensions for the Nordic populations. Scand J Public Health 47(5):477–481. https://doi.org/10.1177/1403494819831905
Walker C, Bottger SA, Mulkern J, Jerszyk E, Litvaitis M, Lesser M (2009) Mass culture and characterization of tumor cells from a naturally occurring invertebrate cancer model: applications for human and animal disease and environmental health. Biol Bull 216(1):23–39. https://doi.org/10.1086/BBLv216n1p23
Weinstein ME, Grossman A, Perle MA, Wilmot PL, Verma RS, Silver RT, Arlin Z, Allen SL, Amorosi E, Waintraub SE et al (1988) The karyotype of Philadelphia chromosome-negative, bcr rearrangement-positive chronic myeloid leukemia. Cancer Genet Cytogenet 35(2):223–229. https://doi.org/10.1016/0165-4608(88)90244-0
Win AK, Jenkins MA, Dowty JG, Antoniou AC, Lee A, Giles GG, Buchanan DD, Clendenning M, Rosty C, Ahnen DJ, Thibodeau SN, Casey G, Gallinger S, Le Marchand L, Haile RW, Potter JD, Zheng Y, Lindor NM, Newcomb PA, Hopper JL, MacInnis RJ (2017) Prevalence and penetrance of major genes and polygenes for colorectal cancer. Cancer Epidemiol Biomarkers Prev 26(3):404–412. https://doi.org/10.1158/1055-9965.EPI-16-0693

Wolf NK, Largaespada DA, Moriarity BS (2015) Coping with cancer genes altered by copy number. Oncotarget 6(34):35155–35156. https://doi.org/10.18632/oncotarget.6215

Yang L, Emerman M, Malik HS, McLaughlin RNJ (2020) Retrocopying expands the functional repertoire of APOBEC3 antiviral proteins in primates. elife 9. https://doi.org/10.7554/eLife.58436

Yang C, Li Y, Trottier M, Farrell MP, Rai VK, Salo-Mullen EE, Gallagher DJ, Stadler ZK, van der Klift HM, Zhang L (2021) Insertion of an SVA element in MSH2 as a novel cause of Lynch syndrome. Genes Chromosomes Cancer 60(8):571–576. https://doi.org/10.1002/gcc.22950

Zapatka M, Borozan I, Brewer DS, Iskar M, Grundhoff A, Alawi M, Desai N, Sultmann H, Moch H, Pathogens P, Cooper CS, Eils R, Ferretti V, Lichter P, Consortium P (2020) The landscape of viral associations in human cancers. Nat Genet 52(3):320–330. https://doi.org/10.1038/s41588-019-0558-9

Zervos EE, Tanner SM, Osborne DA, Bloomston M, Rosemurgy AS, Ellison EC, Melvin WS, de la Chapelle A (2006) Differential gene expression in patients genetically predisposed to pancreatic cancer. J Surg Res 135(2):317–322. https://doi.org/10.1016/j.jss.2006.03.022

Zhang T, Tan S, Tang N, Li Y, Zhang C, Sun J, Guo Y, Gao H, Cai Y, Sun W, Wang C, Fu L, Ma H, Wu Y, Hu X, Zhang X, Gee P, Yan W, Zhao Y, Chen Q, Guo B, Wang H, Zhang YE (2024) Heterologous survey of 130 DNA transposons in human cells highlights their functional divergence and expands the genome engineering toolbox. Cell 187(14):3741–3760.e3730. https://doi.org/10.1016/j.cell.2024.05.007

Chapter 11
Transposable Elements as Drivers of Age-Related Neurodegenerative Disease

Chiara Scopa, Megan Muench, and Davide Trotti

Abstract Transposable elements (TEs) constitute a large part of the genome and are increasingly recognized for their contribution to age-related neurodegeneration. TEs are typically silenced through epigenetic and RNA-based mechanisms. However, with age, repression mechanisms can falter, leading to aberrant TE activation and potentially contributing to the pathogenesis of neurodegenerative diseases (NDs). There is increasing evidence that aberrant activation of TEs contributes to several prominent features of neurodegenerative diseases (NDs), including genomic instability, mitochondrial dysfunction, inflammation, impaired autophagy, and telomere attrition. TE mobilization and resulting de novo insertions can disrupt neuronal gene regulation globally, adversely affecting mitochondrial function and autophagy pathways. In turn, the genomic instability and mitochondrial damage induced by these disruptions may further exacerbate TE activation, creating a detrimental feedback loop. Also, TE-derived double-stranded RNA, double-stranded DNA, and RNA::DNA hybrids can trigger the innate immune response. Aberrant TE activation has been observed in several NDs, such as Alzheimer's disease (AD), amyotrophic lateral sclerosis/frontotemporal dementia (ALS/FTD), Parkinson's disease (PD), and Huntington's disease (HD). Therefore, targeting TE mobilization pathways could represent a promising therapeutic avenue for mitigating ND progression.

Keywords Transposable elements · Aging · Neurodegenerative diseases · ALS · Mitochondria · Innate immunity · Neuroinflammation

---

C. Scopa · M. Muench · D. Trotti (✉)
Weinberg ALS Center, Vickie and Jack Farber Institute for Neuroscience, Department of Neuroscience, Thomas Jefferson University, Philadelphia, PA, USA
e-mail: Davide.Trotti@jefferson.edu

L. M. Vaschetto (ed.), *Transposable Elements, Transcriptomics, and Diseases*, Epigenetics and Human Health 14, https://doi.org/10.1007/978-3-032-04125-8_11

Abbreviations

AD	Alzheimer's disease
ALS	Amyotrophic lateral sclerosis
ALS/FTD	Amyotrophic lateral sclerosis/frontotemporal dementia
AMD	Age-related macular degeneration
ASO	Antisense oligonucleotide
AT	Ataxia telangiectasia
ATM	Ataxia telangiectasia mutated
CDP	X-linked dystonia-parkinsonism
CJD	Creutzfeldt–Jakob disease
CSF	Cerebrospinal fluid
DNMTs	DNA methyltransferases
dsRNA	Double-stranded RNA
Env	Envelope
ERV	Endogenous retrotransposon
G4S	G-quadruplex structures
GWAS	genome-wide association studies
HGPS	Hutchinson-Gilford progeria syndrome
hMSCs	Human mesenchymal stem cells
HP1	Heterochromatin protein 1
HR	Homologous recombination
IAP	Intracisternal A-particle
INF I	Type I interferon
LINE	Long interspersed nuclear element
lncRNA	L long non-coding RNA
LTR	Long terminal repeat
NDs	Neurodegenerative diseases
NRTIs	Nucleoside reverse transcriptase inhibitors
PD	Parkinson's disease
piRNAs	PIWI-interacting RNA
PPRs	Pattern recognition receptors
PrP^C	Cellular prion protein
PSP	Progressive supranuclear palsy
RIPs	Retrotransposon insertion polymorphisms
RNAi	RNA interference
RTEs	Retrotransposons
SINEs	Small interspersed nuclear elements
siRNAs	Small interfering RNAs
SNP	single-nucleotide polymorphism
SSB	Single-strand breaks
ssDNA	Single-stranded DNA
ssRNA	Single-stranded RNA

SVA	SINE-VNTR-Alu
SVs	Structural variations
TERT	Telomerase reverse transcriptase
TEs	Transposable elements
TET	Ten-eleven translocation
TLR8	Toll-like receptor 8
TREX1	Three-prime repair exonuclease 1
TSE	Transmissible spongiform encephalopathy
UTR	Untranslated region
WRN	Werner syndrome
5hmC	5-hydroxymethylcytosine

11.1 Introduction

11.1.1 The Role of Transposable Elements in the Genome

Making up almost half of the human genome, TEs are mobile DNA sequences that can insert themselves into different locations within the genome (Chalopin et al. 2015). TEs are classified into two main groups based on their mechanism of transposition. Class I elements, or retrotransposons, utilize a "copy-and-paste" mechanism to transpose within the genome. This involves the generation of an RNA intermediate, which is subsequently reverse-transcribed into a DNA copy that integrates into a new genomic location (Bourque et al. 2018; Gebrie 2023). In contrast, Class II elements, or DNA transposons, mobilize their DNA directly, either through a 'cut-and-paste' or a 'peel-and-paste' mechanism (Bourque et al. 2018; Gebrie 2023). Given the potential of TEs to disrupt genome integrity and induce gene dysfunction, mammalian genomes have evolved a suite of defense mechanisms to restrict TE activity (Platt et al. 2018). A key mechanism for controlling TE activity involves their silencing through epigenetic modifications, incorporating them into heterochromatin. In fact, mechanisms of de novo TE insertions are tightly restricted by inhibiting the expression of TE mRNA. However, there is evidence for both evolutionary benefit and cost in TE derepression. According to Friedli and Trono (2015), it is the increased endovirome, or the collection of all TEs within the genome, that gives way to human-specific cognitive function. In the brain, there are significant levels of TE mRNA, and resulting somatic insertions are reported. The retrotransposon class of long interspersed nuclear element 1 or LINE-1 (L1) was found to be highly expressed in brain and immune cells (He et al. 2021) and accounts for 20% of both human and mouse genomes (Toda et al. 2024). The milder regulation of TE insertions in some specific cell type or in a specific time frame might be linked to their capability to modulate the expression of neighboring genes and thus regulate gene expression. For example, TE plays a crucial role in proper neuronal differentiation, and during pregnancy, TEs influence the expression of thousands of genes (Lynch et al. 2015). LINE-1 has been indicated to have a functional, non-random process in somatic cell programming. Neural progenitor cells abundantly express L1 during differentiation (Muotri et al. 2005). Events that are associated with TE derepression, such as chromatin opening and long non-coding RNA (lncRNA) production, occur in the trans-differentiation of fibroblasts into dopaminergic neurons, suggesting that cell fate can be influenced by retrotransposon dynamics (Della Valle et al. 2020). Further, the SOX2 transcription factor can repress L1 expression in neural stem cells (Muotri et al. 2005). As Sox2 decreased in cell fate determination, L1 expression and L1 promoter derepression increased. This correlation suggests that retrotransposons can contribute to neuronal diversity (Erwin et al., 2014). Additionally, new somatic L1 insertions may produce neuronal genomic plasticity due to variations in DNA sequences and to the alteration of the transcriptome. Thus, L1-induced variation may play a role in neuronal plasticity (Singer et al. 2010).

The distribution of TEs within the genome is rarely random. Instead, TEs exhibit distinct preferences for insertion within specific genomic features or compartments shaped by opposing selective pressures, reflecting the delicate balance between promoting TE mobilization and minimizing their detrimental impacts (Bourque et al. 2018). Recent evidence suggests that in the prefrontal cortex, a zinc-finger protein (ZFP189) coordinates complex social behaviors by controlling transposon-mediated immune function (Truby et al. 2024). This association is in line with co-evolutionary pressure for both social behavior and immune defense and suggests a novel pathway for TEs to act as *cis*-regulators in neurons.

11.1.2 Regulation of TE Silencing

To maintain genome integrity and ensure proper development, organisms have evolved a complex defense system to tightly regulate the expression of TEs. This defense system includes epigenetic regulation, RNA interference (RNAi), and other protein-based mechanisms.

11.1.2.1 DNA Methylation

DNA methylation, specifically the conversion of cytosine to 5-methylcytosine (5mC), is crucial for silencing TEs across various eukaryotic organisms. In vertebrates, DNA methylation primarily occurs at CpG dinucleotide sequences, while some plants and other eukaryotes also exhibit frequent methylation at non-CpG sites (Romero et al. 2006). 5mC is established by DNA methyltransferases (DNMTs), which are classified as DNMT1 or DNMT3 in eukaryotes. DNMT1 maintains 5mC across cell division by preferentially methylating hemi-methylated CpG dinucleotides following DNA replication, whereas DNMT3 enzymes catalyze de novo deposition of 5mC (Di Stefano 2022). 5mC can be further modified by ten-eleven translocation (TET) enzymes to generate 5-hydroxymethylcytosine (5hmC) (Di Stefano 2022).

Studies utilizing loss-of-function mutants of DNA methyltransferase (DNMT) genes provide compelling evidence for the importance of DNA methylation in TE control. Disruption of DNMT function in *Arabidopsis thaliana* (plant), *Neurospora crassa* (fungus), and zebrafish leads to the accumulation of TE transcripts and increased transposition activity (Tsukahara et al. 2009; Zhou et al. 2001; Chernyavskaya et al. 2017). Similarly, targeted knockouts of Dnmt1 and Dnmt3C in mice result in significant derepression of intracisternal A-particle (IAP) retrotransposons and evolutionarily young TE families, respectively (Walsh et al. 1998; Barau et al. 2016). Disruption of DNMT3L, a DNMT cofactor in male murine germ cells, leads to reactivation of IAP and LINE1 elements and male sterility (Cuellar et al. 2017). These findings collectively demonstrate that the depletion of 5mC is associated with a marked increase in TE upregulation and genomic instability (Di Stefano 2022).

In somatic cells, DNA methylation patterns at TEs are generally stable, suggesting it is the primary silencing mechanism (Smith and Meissner 2013). This translates to limited TE expression in most somatic tissues. However, exceptions exist, with LINE1 expression and retrotransposition observed in the mouse and human brain (Muotri et al. 2005; Coufal et al. 2011). Interestingly, the brain exhibits similar global 5mC levels compared to other somatic tissues but higher levels of 5hmC, a marker of active demethylation by TET enzymes (Globisch et al. 2010). Furthermore, studies have shown that site-specific methylation of the first seven CpG dinucleotides within the LINE1 promoter effectively suppresses somatic transposition activity (Hata and Sakaki 1997).

In conclusion, DNA methylation, particularly 5mC at specific genomic regions, serves as a crucial line of defense against TE activity across diverse eukaryotes. Disruption of DNA methylation pathways leads to increased TE expression and highlights the importance of this epigenetic mechanism in maintaining genomic integrity (Jansz 2019).

11.1.2.2 Histone Modifications

In addition to DNA methylation, histone modifications play a pivotal role in regulating TEs (Di Stefano 2022). These modifications often cooperate with DNA methylation, reinforcing TE silencing (Li et al. 2021). Histone methylation, primarily occurring on lysine (K) and arginine (R) residues, plays a key role in this process. H3K9 and H3K27 methylation are generally associated with transcriptional repression. H3K9 methylation, particularly H3K9me2/3, is a hallmark of constitutive heterochromatin and is enriched at inactive genes and TEs. It is proposed that H3K9me2/3 recruits heterochromatin protein 1 (HP1), which then multimerizes to induce chromatin condensation and transcriptional repression (Bannister et al. 2001; Nakayama et al. 2001).

Trimethylated histone H4 lysine 20 (H4K20me3) is another important mark associated with heterochromatin and is enriched in genomic regions that harbor LINEs, endogenous retroviruses (ERVs), and satellite DNA in human cells (Ozturk et al. 2021). SUV420H1 and SUV420H2 are the primary enzymes responsible for catalyzing H4K20me3 in mice. SUV420H2 knockout in mouse embryonic stem cells leads to the derepression of repetitive DNA elements (Kurup et al. 2020). H4K20me3 often co-localizes with H3K9 methylation and DNA methylation at repetitive elements (Ren et al. 2021). While H3K9 methylation is generally upstream of H4K20 methylation, H4K20me3 can also occur independently of H3K9me3 in specific contexts (van Kruijsbergen et al. 2017).

SUMOylation pathways also play a critical role in heterochromatin formation and TE silencing from yeast to mammals (Stielow et al. 2008; Shin et al. 2005; Maison et al. 2011).

11.1.2.3 RNA-Based Silencing Mechanisms

Several RNA-based mechanisms also contribute to TE silencing, including small non-coding RNAs and proteins involved in RNA interference (RNAi) pathways. Small RNA-mediated mechanisms are crucial for establishing repressive histone modifications at TE loci in many organisms. Small non-coding RNAs, such as piRNAs, play a pivotal role in TE silencing. In Drosophila, piRNAs associate with the Piwi protein complex, recognizing nascent TE transcripts through RNA-RNA interactions (Sienski et al. 2012, 2015; Donertas et al. 2013; Muerdter et al. 2013). This piRNA-guided complex recruits chromatin-modifying enzymes, including Eggless (H3K9 methyltransferase) and its cofactor Windel, to TE loci, leading to their transcriptional silencing (Osumi et al. 2019). Additionally, the SUMO E3 ligase, Su (var)2–10, contributes to this process by facilitating the recruitment of Eggless and Windel (Andreev et al. 2022).

While Piwi-mediated H3K9 methylation is crucial for TE silencing, additional mechanisms also contribute. Piwi depletion can lead to decreased H3K9 methylation and an increase in the activating H3K4me2 mark at some TE loci, suggesting that Piwi may also inhibit H3K4me2 (Ilyin et al. 2017). This is further supported by the involvement of the histone demethylase, dLsd1, in TE silencing (Di Stefano 2022).

Proteins involved in RNA interference (RNAi) pathways, such as Dicer and APOBEC family proteins, also contribute to TE silencing. Dicer is an enzyme that processes double-stranded RNA into small interfering RNAs (siRNAs). Studies in mouse embryonic stem cells and oocytes have shown that Dicer deficiency leads to increased expression of various TEs, including ERVs, LINEs, and small interspersed nuclear elements (SINEs) (Calabrese et al. 2007; Bodak et al. 2017; Berrens et al. 2017).

Human APOBEC3 (hA3) proteins, known for their role in antiviral defense, have also been implicated in TE restriction. These proteins deaminate viral minus-strand DNA during reverse transcription, leading to hypermutation. Recent findings suggest that hA3G can also restrict the transposition of specific retrotransposons, including MusD and intracisternal A-particle LTR retrotransposons (Esnault et al. 2005). Human APOBEC3 family members, from hA3A to hA3H, differentially reduce the level of L1 retrotransposition. In particular, hA3A, hA3B, hA3F, and hA3G protein levels strongly decrease the transposition of L1 (Kinomoto et al. 2007).

11.1.2.4 Other Regulatory Pathways

Additional pathways contribute to TE regulation (Yushkova and Moskalev 2023), including the lack of a polyadenylation signal and the sequestration of TE products in stress granules (Slotkin and Martienssen 2007; Hwang et al. 2019). SIRT6, a protein involved in lifespan regulation, has also been shown to repress LINE1 transcription in stem cells and brain cells by promoting heterochromatin formation at the 5′ untranslated region (UTR) of the retrotransposon (Van Meter et al. 2016). A recently discovered mechanism involves TREX1 (three-prime repair exonuclease 1) in suppressing retrotransposons. TREX1 eliminates single-stranded cDNA derived from retrotransposons, thereby preventing DNA damage and potentially autoimmune diseases (Stetson et al. 2008; Volkman and Stetson 2014). Studies suggest that TREX1 can suppress LINE1 elements via both exonuclease-dependent and -independent pathways (Li et al. 2017; Thomas et al. 2017).

11.1.3 Aging and Neurodegenerative Diseases: An Overview

Aging is a natural process characterized by a progressive decline in biological functions. This decline is associated with an increased risk of various age-related diseases, including neurodegenerative diseases (NDs). NDs are characterized by progressive dysfunction and loss of neurons, leading to a decline in cognitive and motor function. Age is considered the most significant risk factor for NDs. The brain accumulates damage from various sources, including oxidative stress, inflammation, and protein misfolding during the aging process. Moreover, as we age, cells lose their ability to repair themselves and divide, leading to a decline in cellular function resulting in cellular senescence, which can contribute to the deterioration of brain tissue. For instance, post-mitotic cells, such as neurons, are increasingly vulnerable to the accumulation of age-related damage throughout a lifetime (Hou et al. 2019).

Finally, the resilience of the brain to pathological damage or changes, called brain and cognitive reserve, diminishes with age, making individuals more susceptible to the effects of neurodegeneration (Stern et al. 2019). In fact, brain and cognitive reserve is greater in individuals with a higher premorbid IQ, education, and similar characteristics who also have a reduced risk of developing dementia (Stern et al. 2019) and may have slower rates of age-related cognitive decline (Zahodne et al. 2015). In this context, the age-associated loss in neural stem cell number and/

or activity could participate in the brain reserve and, thus, in the decline in brain function (Culig et al. 2022).

11.1.4 The Link Between TEs, Aging, and Neurodegeneration

The derepression of TEs has been increasingly linked to the development and progression of age-related disorders and NDs. Since the number of individuals living with neurodegenerative diseases (NDs) is expected to double by 2050, researching the etiology of these diseases, such as Alzheimer's, Parkinson's, and amyotrophic lateral sclerosis (ALS), is of great importance (Wyss-Coray 2016). Understanding the intricate relationship between TEs and NDs may lead to the discovery of novel biomarkers and potential therapeutic interventions.

Interestingly, studying social insects that have a vast difference in lifespan within a population, such as the queen bee or king termite, revealed a connection between TEs and longevity (Elsner et al. 2018). The reproducing queen or king can live decades, while the worker population only lives a few weeks, highlighting a physiological difference between the organisms. The majority of older worker insects, but not the reproductive ones, had a downregulation of PIWI-interacting RNA (piRNA), which is known to repress TEs. Increased piRNA defense is typically found in the germline, so it appears that somatic expression coincides with the social evolution of reproductive queens or kings. However, there is no equivalent of somatic upregulation of piRNAs to promote longevity in humans.

Despite years of investigation, the link between aging, age-related disease, and TEs is still not completely understood. Here, we review the mounting evidence for TE contribution in all the major hallmarks of NDs, including genomic instability, mitochondrial dysfunction, inflammation, impaired autophagy, and telomere attrition. In addition, we summarize the evidence that aberrant TE activation is directly linked to patients with AD, ALS/FTD, PD, HD, and other neurodegenerative proteinopathies.

11.2 TEs in Aging

11.2.1 TE Reactivation in Aging

The concept that biological aging is intricately linked to the progressive deterioration of genetic integrity has been extensively discussed in scientific literature (Lopez-Otin et al. 2013; Guo et al. 2022; Tenchov et al. 2024). Aging is associated with extensive remodeling of the epigenome (Gorbunova et al. 2021). While these changes are generally cell-type specific, a consistent trend is the loss of heterochromatin and associated transcriptional derepression, which has been observed across species from yeast to humans (Sen et al. 2016).

Cytosine methylation tends to decrease with aging over large, slowly evolving, gene-poor regions of the genome that are heterochromatic and enriched for repetitive sequences, whereas methylation at gene-associated CpG islands can increase or decrease with pronounced tissue-specific effects (Horvath and Raj 2018).

Dynamic changes in chromatin organization are one of the hallmarks of aging. Chromatin domains characterized by specific histone modifications may be affected by dynamic changes in their spatial organization and interactions within the nucleus. Additionally, alterations in the spatial organization of chromatin domains, such as changes in their associations with the nuclear lamina and disruptions in nuclear architecture, are likely to contribute to the aging process. Given the increasing recognition of the RNA interference (RNAi) machinery's role in establishing and maintaining heterochromatin and silencing, dysregulation of this pathway may also contribute to age-related dysfunction (Wood and Helfand 2013).

Studies in *C. elegans* have shown a causal link between aging and histone alterations. In fact, genetic interventions to directly modulate histone marks can lead to changes in lifespan (Greer et al. 2010; McColl et al. 2008). A study on the aging *Drosophila* genome showed several chromatin changes that occurred with age (Wood et al. 2010). The most notable change with age was a significant decrease in the enrichment of the repressive heterochromatin mark, H3K9me3, as well as HP1 (heterochromatin protein 1, a component of heterochromatin) at pericentric heterochromatin *loci* (Larson et al. 2012). In addition, older flies showed an overall decrease in heterochromatic regions as well as nuclear reorganization of HP1. Age-related chromatin changes have also been observed in mammalian model systems. A reduction in histone acetylation and an increase in histone methylation, specifically acetyl-histone H3 Lys9 and dimethyl-histone H3 Lys9, in aged mice (Watanabe and Bloch 2013) and differences in DNA methylation patterns (Maegawa et al. 2010) have been observed between young and old mice.

Since TEs are concentrated in heterochromatic regions, and changes in the heterochromatic state are linked to aging, researchers have investigated the potential cause-and-effect relationship between these two phenomena. Studies have shown that the mechanisms that normally suppress TE activity weaken with age, leading to increased TE expression in aged organisms (Gorbunova et al. 2021; Jonsson et al. 2020). Furthermore, age-related reductions in DNA methylation and heterochromatin loss contribute to this increase in TE expression and transposition (Mosaddeghi et al. 2023). In yeast, for example, the retrotransposon Ty1 had increased mobility with age, and this was correlated with chromosomal rearrangements (Maxwell et al. 2011). Transposition in mitochondrial DNA also causes senescence in Neurospora (Bertrand et al. 1985). In C. elegans, expression of the endogenous retrotransposon Cer1 (gypsy/Ty3family) increases with age (Dennis et al. 2012), and somatic excision of the DNA transposon Tc1 increases by more than 14-fold during the organism's lifespan (Egilmez and Shmookler Reis 1994). In Drosophila, a significant age-dependent transposition rate of retrotransposon copies was also documented (Filatov et al. 1998). Several TEs were active in the Drosophila brain during normal aging. Using a GFP-engineered gypsy element (the gypsy-TRAP reporter), researchers were able to detect somatic transposition in the brain that increased with age. Finally, disrupting the RNAi protein Ago2 caused an increase in transposition, resulting in age-related memory impairment and an overall shorter lifespan (Li et al. 2013a).

In mammals, the murine intracisternal A-particle (IAP) retrotransposon becomes active upon aging and is associated with progressive demethylation of the IAP promoter during the lifespan (Barbot et al. 2002). Altered DNA methylation and histone modifications at repetitive DNA elements are reported in aging mouse brains (Ryu et al. 2011). Specifically, overall histone acetylation of both H3 and H4 decreased with age at several TEs as assayed by ChIP-qPCR. Additionally, a decrease in histone methylation marks, such as H3K9me3 and H4K20me3, correlates with aging (Ryu et al. 2011).

In humans, a significant decline in average Alu methylation over the lifespan was detected (Jintaridth and Mutirangura 2010), and transcription of the retrotransposon LINE1, detected by L1 ORF1 mRNA levels, increased with age (Roberson et al. 2019). Using multiple RNA-seq datasets generated from human fibroblasts, LaRocca and colleagues demonstrated that transcripts of repetitive elements accumulate with aging. Interestingly, the level of this accumulation was sufficient to predict age accurately, confirming the important role played by repetitive element transcripts in aging (LaRocca et al. 2020).

Cellular senescence, a state of irreversible cell cycle arrest, significantly contributes to the aging process. Therefore, it is not surprising that elevated TE expression has been observed in senescent human cells (Wang et al. 2011). Transcriptome profiling has shown that senescent human stem and progenitor cells (s-HSPCs) robustly express genomic TEs (Colombo et al. 2018). Senescent human fibroblasts exhibit altered chromatin architecture, characterized by a global compaction of euchromatin-rich regions and a concomitant opening of constitutive heterochromatin in gene-poor regions (De Cecco et al. 2013a, b). This chromatin remodeling is associated with increased expression of young Alu, SINE-VNTR-Alu (SVA), and LINE-1 subfamilies, suggesting increased TE activity, including potential LINE-1 retrotransposition. These findings indicate that TEs may play a significant role in driving cellular senescence. Indeed, in human mesenchymal stem cells (hMSCs) derived from patients with Hutchinson-Gilford progeria syndrome (HGPS) or Werner syndrome (WRN), treatment with an L1 antisense oligonucleotide (ASO) rescued the heterochromatic state and mitigated senescent phenotypes (Della Valle et al. 2022).

The reactivation of TEs during aging can have significant consequences on gene expression. TE transposition poses a risk for insertional mutagenesis, leading to the potential disruption of gene function and the production of aberrant proteins (Bravo et al. 2020). Moreover, TEs can influence the expression of neighboring genes by harboring regulatory elements (Elbarbary et al. 2016).

11.2.2 TEs in Age-Related Diseases

The aberrant activation of TEs with age suggests a potential role for TEs in the pathogenesis of age-related diseases. However, their precise roles in disease pathogenesis remain mostly unclear.

Age-related macular degeneration (AMD) is a progressive chronic disease of the central retina leading to blindness (Kaarniranta et al. 2020). A study demonstrated the crucial role of a TE, Alu, in the pathogenesis of this disease (Kaneko et al. 2011). In detail, AMD is associated with decreased levels of DICER1, leading to an accumulation of Alu RNA in retinal pigment epithelial cells. This leads to NLRP3 inflammasome activation, cellular damage, inflammation, and degeneration (Fukuda et al. 2021; Tarallo et al. 2012). The degeneration of retinal pigment epithelium is blocked by antisense oligonucleotides targeting Alu and Alu-like RNAs, demonstrating the pathogenic role of this TE in an AMD context (Kaneko et al. 2011).

TEs have also been implicated in the pathogenesis of premature aging disorders, including Hutchinson-Gilford progeria syndrome (HGPS) and Werner syndrome (WRN). HGPS is caused by mutations in the LMNA gene, leading to the production of a truncated protein called progerin (Eriksson et al. 2003; Noda et al. 2015). Progerin accumulates in the inner nuclear lamina of cells, making the nucleus unstable. This can lead to nuclear blebbing, which is a characteristic of progeria. In contrast, WRN is associated with mutations in the WRN gene, which encodes a RecQ helicase (Yu et al. 1996).

Depletion of lamin A/C in mammalian cells has been shown to induce the upregulation of LINE-1 elements, which is also seen in transcriptomic data obtained from HGPS patient-derived fibroblasts (Vazquez et al. 2019; LaRocca et al. 2020). Furthermore, mesenchymal stem cells derived from individuals with HGPS and WRN showed an increased expression of L1 elements (Le Breton et al. 2024). A recent study also reported an upregulation of the HERV-K retrotransposon family in cellular models of HGPS and WRN (Liu et al. 2023).

The impact of TE activity extends beyond the age-related disorders described here. There is growing evidence linking changes in TE activity to neurodegenerative diseases, which can be considered a major class of age-related conditions.

11.3 TEs in Neurodegenerative Disease

Aberrant activation of TEs has also been observed in neurodegenerative diseases such as Alzheimer's disease (AD) and amyotrophic lateral sclerosis (ALS). Growing evidence demonstrates that TE reactivation causes detrimental effects, including genomic instability, mitochondrial dysfunction, and chronic inflammation—all hallmarks of neurodegenerative disease (Singh et al. 2024). In this section, we will discuss the relationship between TE reactivation and neurodegenerative diseases.

11.3.1 TEs and Neurodegenerative Disease Hallmarks

11.3.1.1 Genomic Instability

DNA damage, which can be induced by many endogenous and exogenous factors that alter the chemical properties of one or both DNA strands, is a continuous challenge for cells and threatens the stability of the genome. Post-mitotic neurons are particularly vulnerable to impaired DNA damage repair (Caldecott et al. 2022). Genomic instability, or increased accumulation of DNA damage, is a shared hallmark of normal aging and neurodegenerative diseases (Fuchs et al. 2022). The DNA damage response (DDR) is highly conserved and involves complex signaling pathways to prevent aberrant activation of genomic instability, epigenetic alterations, telomere attrition, mitochondrial dysfunction, and cellular senescence. DNA repair pathways can inhibit L1 retrotransposition events. For example, BRCA1, a homologous recombination (HR) factor, can directly control L1 ORF2 protein translation (Mita et al. 2020). However, a decline in repair by DDR will lead to DNA damage accumulation, which could be a trigger of neuronal death.

Since TEs can destabilize their surroundings after mobilization and insertion into the genome, they can directly contribute to single-strand breaks (SSB) and, thus, genomic instability and DDR (Balachandran et al. 2022). Neurons have a reduced capacity for SSB repair pathways (Iyama and Wilson 2013). L1 retrotransposons can alter genome structure by inducing DNA rearrangements and ultimately contribute to neurodegenerative disease pathways. A cellular sensor of double-stranded DNA breaks, named the ataxia telangiectasia mutated (*ATM*) gene, controls the damage associated with L1-induced genomic instability. Ataxia telangiectasia (AT) is a genetic disorder that results in inactivated ATM, and patients experience progressive neurodegeneration. In AT patients' post-mortem tissue, L1 DNA is significantly increased. This supports cell and mouse models that show loss of ATM results in longer or more L1 retrotransposition events (Coufal et al. 2011). Therefore, the lack of DNA damage repair is associated with retrotransposon modulation and results in progressive neurodegeneration.

Topoisomerase-induced "programmed" single-strand and double-strand breaks require repair under normal cell conditions. However, in neurons, defective repair is apparent in genetic diseases, such as spinocerebellar ataxia (Zagnoli-Vieira et al. 2018). One proposed mechanism of why neurons are more impacted by topoisomerase-induced DNA breaks compared to other post-mitotic cells is due to the intrinsic high transcriptional demand, leading to the possibility of sequestered topoisomerase (Caldecott et al. 2022). Interestingly, topoisomerases are known to promote heterochromatin organization and, therefore, impact the abundance of transposons (Lee and Wang 2019). Heterochromatin is essential in TE silencing, and heterochromatin alteration has been linked to tau-mediated neurodegeneration (Frost et al. 2014). Therefore, the "trapping" of topoisomerases in neurons could be a possible mechanism in which TEs are aberrantly activated in neurodegenerative disease, although this requires further investigation.

Additionally, causal candidate insertions of *Alu*, named LN15 and 16, are located in *KANSL1*, a trait-associated single-nucleotide polymorphism (SNP) identified by GWAS (genome-wide association studies) as increasing Parkinson's disease risk (Ahn et al. 2023). This gene encodes a protein that modifies chromatin accessibility, thus further highlighting the importance of genomic instability in the risk of neurodegenerative disease. It was also associated with increased expression of *MAPT*, an adjacent gene that encodes for tau, also a known risk factor for PD.

11.3.1.2 Mitochondrial Dysfunction

Highly energy-dependent neurons require functional mitochondria to meet their metabolic demands. Mitochondria have long been implicated as players in neurodegenerative disease (Swerdlow and Khan 2004). The dysfunction of the mitochondrial network and the subsequent oxidative stress are important shared hallmarks of neurodegenerative diseases that often precede neuronal loss. There are no TEs present among the 13 genes that make up the mitochondrial genome. However, mitochondrial function relies on 1500 nuclear-encoded genes, which can be impacted by retrotransposons. De novo insertions or hypomethylation events within or near nuclear mitochondria genes have the potential to contribute to mRNA instability and downstream mitochondrial dysfunction. The translocase of outer mitochondrial membrane 40 (TOMM40) gene is an essential part of the TOM complex, which is responsible for transporting pre-proteins into the mitochondria (Shiota et al. 2015). Disrupting the TOM complex leads to an induced stress response and mitophagy. A potential source of TOM40 dysfunction is the 16 Alu insertions in *TOMM40* introns 6 and 9 (Payton et al. 2016). Such Alu insertions create a region of genome instability, which can lead to increased mRNA degradation (An et al. 2004) or the formation of alternate TOM40 isoforms (Larsen et al. 2017). Hypothetically, a defective TOM40 (TOM40') production would destabilize the TOM complex and restrict the transport of pre-proteins, which could have cell-type-specific effects on the metabolic homeostasis of neurons. However, this mechanism has not been completely solidified and requires further research.

Alternatively, mitochondrial dysfunction has been shown to be the trigger of aberrant TE activation, instead of the other way around (Baeken et al. 2020). By using pharmacological agents to inhibit the mitochondrial respiratory chain complex I and induce ROS in vitro, Baeken et al. found an increased expression of ORF1 accompanied by a loss of DNA methylation. They showed that this finding of increased L1 can be replicated in both human dopaminergic LUHMES cells and in mouse models. Importantly, they were able to reverse this effect by treating it with a mitochondrial antioxidant, including the decrease of global methylation. When mitochondria require an emergency response, they utilize the epigenetic mechanism of demethylation to control their numerous nuclear-encoded genes (Mayorga et al. 2019). Therefore, mitochondrial distress might be the upstream trigger of demethylation, and TE mobilization might be a consequence.

In support of mitochondrial stress as a trigger, ionizing radiation and pharmacologically induced oxidative stress cause a combined total of 18% of TEs in Drosophila to be activated (de Oliveira et al. 2021). However, the same study found that some TEs are repressed under stress conditions, and the activation was correlated with the severity of the stressor. Additionally, some TEs that were activated only had partial sequences and, therefore, did not have the ability to mobilize, which further highlights the variability between TE families in activation and likely downstream effects.

11.3.1.3 Inflammation

Mounting evidence suggests a critical role for retrotransposons (RTEs) in inflammation. Endogenous retrotransposon (ERV) transcription can directly upregulate the expression of interferon-stimulated genes because some long terminal repeat (LTR) sequences within ERVs function as promoters or enhancers for these genes (Burns 2020).

Furthermore, during their transposition, cytosolic TE-derived single-stranded RNA (ssRNA) can serve as a template for cDNA synthesis, resulting in the formation of ssDNA, dsRNA, and RNA:DNA hybrids. The presence of these nucleic acids in the cytosol is detected by pattern recognition receptors (PRRs), triggering an interferon response (Jonsson et al. 2020). This is a key mechanism of ERV-mediated inflammation, where cytoplasmic dsRNA can be recognized by RNA sensors such as RIG-I, MDA5, and endosomal TLRs, while cytosolic DNA is sensed through the cGAS/STING pathway, leading to IFN-γ production (Li et al. 2013b; Stetson and Medzhitov 2006; Lim et al. 2015; Thomas et al. 2017; Chalmers and Wu 2020). For instance, in SIRT6-deficient mice, a derepression of long interspersed nuclear element 1 (LINE1) retrotransposons was observed in various tissues, including the brain. SIRT6 deficiency resulted in the accumulation of cytoplasmic LINE-1 cDNA, which triggered an inflammatory response through the cGAS-STING pathway (Simon et al. 2019).

RTE nucleic acids can also trigger innate immunity through recognition by both transmembrane and cytosolic PRRs (Grandi and Tramontano 2018; Chesnokova et al. 2022). For instance, HERV ssRNA can be sensed by TLR7 and TLR8 (Heil et al. 2004), while ERV RNA:DNA hybrids are recognized by TLR9 (Rigby et al. 2014). Activation of transmembrane PRRs initiates an intracellular signaling cascade leading to the expression of pro-inflammatory cytokines, chemokines, and type I IFN (IFN-I) (Grandi and Tramontano 2018; Jonsson et al. 2020). Additionally, cytosolic PRR activation leads to IRF3 activation and translocation to the nucleus, resulting in IFN-I production. MDA5 has been shown to sense HERV and LINE-1 dsRNA (Chiappinelli et al. 2015; Tunbak et al. 2020), and ERV cDNA can be sensed by cGAS.

11.3.1.4 Dysregulated Autophagy

The collective decline of protein clearance pathways, such as autophagy, occurs as a common denominator of neurodegenerative disease and aging. The loss of proteostasis is highly associated with cellular senescence and closely related to other neurodegenerative disease hallmarks, such as genomic instability and neuroinflammation (Meller and Shalgi 2021).

Selective autophagy eliminates stress granules through SQSMT1, which is implicated in ALS/FTD (Yang et al. 2023; Deng et al. 2020). These findings corroborate previous research on Ty1, the most abundant retrotransposon in yeast, which forms virus-like particles that are targeted for degradation by selective autophagy, leading to a decrease in Ty1 transposition (Suzuki et al. 2011). Further, NDP52 and p62-dependent autophagy is the mechanism by which retrotransposable element RNA is degraded in the cytoplasm, thus preventing further insertion into the genome (Guo et al. 2014). Therefore, the loss of autophagy during aging could contribute to TE activation.

Autophagy is also becoming increasingly appreciated for its role in genomic stability. By influencing both the modulation of DNA repair and responding to DNA damage, the autophagic pathway controls genomic stability directly and indirectly (Hewitt and Korolchuk 2017). Therefore, insults to the autophagic pathway could lead to both the derepression of TEs and a reduced degradation of their mRNA.

11.3.1.5 Telomere Attrition

Shorter telomeres are a marker of aging, and telomere length is crucial for genomic stability. Telomeres suppress chromatin accessibility, so when telomeres are shortened, TEs can induce more retrotransposition events. In mouse embryonic stem cells, shortened telomeres were found to de-repress TEs, such as LINE-1s and HERVs (Zhao et al. 2023). Telomere attrition also plays a role in neurodegenerative diseases, as evident by a mouse model of short telomeres that shows neurodegeneration (Whittemore et al. 2019).

Furthermore, telomerase reverse transcriptase (TERT) has been shown to improve and reduce oxidative stress, another hallmark of neurodegenerative disease, making telomere maintenance a potential therapeutic target (Yu et al. 2023). In an iPSC model of telomere shortening, differentiated neurons, and astrocytes displayed increased DNA damage and inflammation (Harley et al. 2024). Additional studies in the context of neurodegeneration are required to show a direct link between telomere attrition and TE reactivation in neurodegenerative disease.

Having explored the various mechanisms by which TEs contribute to the neurodegeneration phenotype (Fig. 11.1), in the following sections, we will delve deeper into their contributions in the context of specific neurodegenerative diseases.

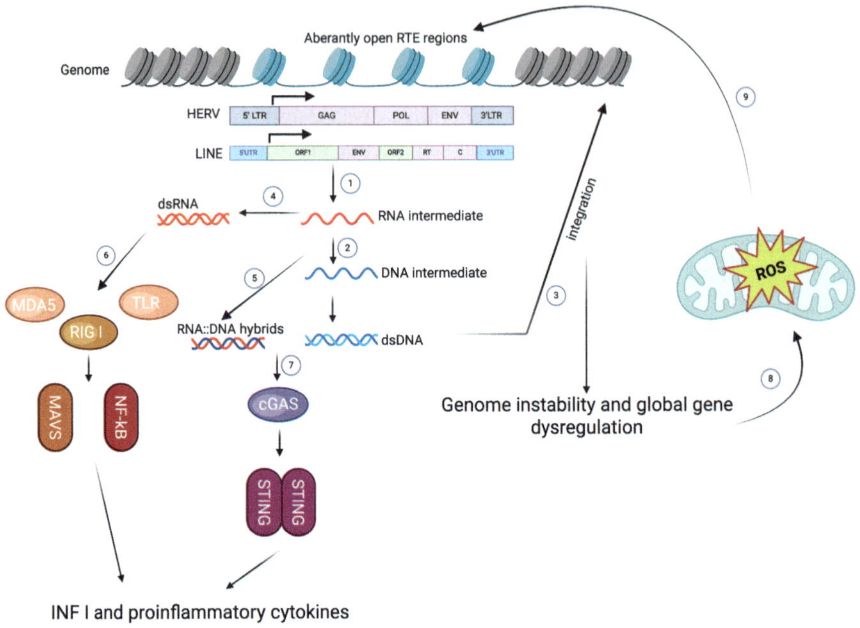

Fig. 11.1 TE mechanism in neurodegeneration. TE mobilization involves transcription into intermediate RNA (1) followed by reverse transcription and integration of the resulting cDNA into the genome (2). TE insertion can destabilize the surrounding genomic region (3), contributing to single-strand breaks and, thus, genomic instability and DDR. During mobilization, cytosolic TE-derived ssRNA can be reverse transcribed, leading to the formation of ssDNA, dsRNA (4), and RNA::DNA hybrids (5). These non-canonical nucleic acids are recognized by PRRs, triggering an interferon response. Specifically, dsRNA is sensed by RNA receptors such as RIG-I and MDA5, as well as endosomal TLRs (6), while cytosolic DNA and RNA::DNA hybrids activate the cGAS/STING pathway (7). Furthermore, TE-derived genomic instability can impair mitochondrial function, resulting in oxidative stress (8). Increased ROS levels can, in turn, trigger aberrant TE activation (9). (Figure made with BioRender.com)

11.3.2 Alzheimer's Disease

Alzheimer's disease (AD) is the most common cause of dementia associated with memory loss and progressive cognitive decline. Post-mortem AD brains showed increased chromatin opening (Frost et al. 2014) and increased expression of retrotransposons, specifically HERV-L, HERV-K, and HERV-H families, (Guo et al. 2018; Sun et al. 2018). Indeed, during the clinical phenoconversion to late-onset AD, increased activity of retrotransposons in the brain, named "retrotransposon storm," has been reported (Macciardi et al. 2022).

Increasing evidence supports the crucial role of Tau in the TE reactivation in AD. Expression of a disease-associated human tau protein in Drosophila neurons induced chromatin relaxation, retrotransposon activation, and neuronal cell death (Sun et al. 2018). Furthermore, it has been demonstrated that the overexpression of

Tau can impair the post-transcriptional suppression of TEs in fly brains by reducing the levels of PIWI proteins (Sun et al. 2018). Analysis of the human cortex revealed a positive association between retrotransposon expression and the burden of neurofibrillary tangles. Moreover, an enrichment of the H3K9ac histone modification was observed at some HERV *loci* (Guo et al. 2018). More recently, similar observations were obtained in murine models of tauopathy. TE (predominantly ERV) transcription levels were increased in the CNS of tauopathy mice compared to control animals. In detail, both LINE1 and ERV copy numbers were increased in the cortex of tauopathy (Ramirez et al. 2022). No significant differences in LINE1 copy number were found between the brains and blood samples of AD patients and age-matched controls, suggesting that aging and the neurodegenerative processes associated with AD may not be the primary factors driving the retrotransposition activity of LINE1 elements (Chesnokova et al. 2022). Several studies have demonstrated a causal link between TE reactivation and neuroinflammation in AD. In a mouse model, disruption of Trim28, a protein that regulates the nuclear accumulation and toxicity of tau (Rousseaux et al. 2016), led to the activation of ERVs in neural progenitor cells. Furthermore, ERV activation in neurons was associated with the formation of ERV protein aggregates and microglial activation (Jonsson et al. 2021). In another mouse model, single-stranded RNA derived from HERV-K was shown to induce neuronal injury through Toll-like receptor 8 (TLR8) signaling. Additionally, elevated expression of both HERV-K and TLR8 was observed in cortical samples from AD patients (Evering et al. 2023). RTE-derived dsRNA and dsRNA PPRs are significantly elevated in human and mouse models (Saleh et al. 2019). As described in the paragraph above, TE expression can activate the innate immune system through the cGAS-STING pathway (Gorbunova et al. 2021). In this context, our recent paper demonstrated that upregulation of the AP-1 subunit, c-JUN, triggers the decondensation of genomic regions containing HERVs in in vitro AD iPSCs-derived models. This leads to the cytoplasmic accumulation of HERV-K-derived RNA::DNA hybrids, activation of the cGAS-STING cascade, and increased cleaved caspase-3, suggesting the initiation of programmed cell death (Scopa et al. 2023). Notably, transcripts of HERV-K were also detected in the cerebrospinal fluid (CSF) of a significant percentage of AD patients (56%) compared to controls (8%), further supporting a link between HERV-K activation and AD (Dembny et al. 2020). Interestingly, RIG-I was found to be significantly upregulated in the temporal cortex and plasma of AD patients with mild cognitive impairment (de Rivero Vaccari et al. 2014).

11.3.3 Amyotrophic Lateral Sclerosis/Frontotemoral Dementia (ALS/FTD)

The initial link between TEs and ALS/FTD arose from observations of elevated reverse transcriptase activity in the blood of patients and their relatives (Steele et al. 2005; McCormick et al. 2008). This was followed by the discovery of upregulated

HERV-K expression in post-mortem brain tissue from ALS patients compared to controls (Douville et al. 2011). Notably, pol transcripts (indicative of active transcription) were most abundant in the motor cortex, and reverse transcriptase was localized to motor neurons in these patients (Douville et al. 2011). The presence of HERV-K in cortical and spinal neurons of ALS patients was further confirmed (Li et al. 2015). In vitro and mouse studies demonstrated the specific neurotoxic potential of the HERV-K envelope (env) protein, suggesting an important role of HERV-K in neurodegeneration and disease pathogenesis. In detail, in human neurons, env protein expression caused neurite retraction, while in transgenic animals, it led to the development of motor dysfunction (Li et al. 2015). Importantly, Li et al. revealed that the RNA-binding protein TDP-43 can regulate HERV-K expression. Supporting the role of TDP-43 and TE interplay in ALS/FTD pathology, loss of nuclear TDP-43 was shown to directly influence genomic instability and subsequent LINE-1 element expression (Liu et al. 2019). The mislocalization of TDP-43 triggered nuclear transcriptional changes in genes associated with DNA damage and repair, leading to decreased genome integrity and potentially allowing for the derepression of LINE-1 elements (Liu et al. 2019). However, another study suggests that TDP-43 pathology burden alone may not be sufficient to explain TE changes observed in ALS/FTD post-mortem tissue (Prudencio et al. 2017). This last study confirmed TE upregulation in C9orf72 mutation carriers, the most common genetic form of ALS/FTD, but attributed this overexpression to increased RNA polymerase II-mediated transcription in post-mortem frontal cortex tissues (Prudencio et al. 2017).

11.3.4 Parkinson's Disease

Parkinson's disease (PD) is the second most common neurodegenerative disorder, characterized by the gradual loss of dopaminergic neurons in the substantia nigra (Copley and Shorter 2023; Chesnokova et al. 2022). A longitudinal study of 423 PD patients found that non-reference TEs significantly correlated with the progression of PD and clinically important traits (Koks et al. 2022). Additionally, several SVA Retrotransposon Insertion Polymorphisms (RIPs) are associated with PD progression. In particular, seven SVAs were linked to significant differences in disease severity scores and changes in dopaminergic degeneration (Pfaff et al. 2021). Furthermore, retroelement recombination events are implicated in early-onset Parkinson's disease. Whole-genome sequencing analysis of three families with early-onset PD revealed the presence of five different PRKN structural variations (SVs) and the presence of retrotransposon sequences within two kilobases of the deletion breakpoints, suggesting a crucial role of retrotransposition events in originating these deletions (Bravo et al. 2018). Finally, multiple PD models exhibit increased L1 expression observed in both human (Baeken et al. 2020) and mouse PD models (Blaudin de The et al. 2018). In detail, increased L1 expression was observed in SNpc dopaminergic neurons of 6-week-old PD mice compared to WT, and this overexpression triggered oxidative stress-induced DNA damage (Blaudin de The et al. 2018). Experiments with cell cultures showed that respiratory chain complex I inhibitors cause a decrease in overall DNA methylation level, suggesting a potential mechanism of TE mobilization (Baeken et al. 2020).

11.3.5 X-Linked Dystonia-Parkinsonism

X-linked dystonia-parkinsonism (XDP) is a genetic disorder caused by a rare polymorphic SVA-family retrotransposon antisense insertion within intron 32 of the *TAF1* gene, which plays a crucial role in transcriptional control (Singh et al. 2024). XDP is a neurodegenerative movement disorder that is characterized by atrophy in the striatum of the basal ganglia, similar to Parkinson's disease (Arasaratnam et al. 2021). Interestingly, the length of the variable number tandem repeat within the SVA element appears to correlate with disease severity, with longer repeats associated with an earlier age of onset (Bragg et al. 2017). Removal of the SVA insertion from the intron restores *TAF1* expression levels to their typical range (Makino et al. 2007; Aneichyk et al. 2018). Patient-derived XDP iPSCs showed phenotypic rescue after excision of the SVA (Rakovic et al. 2018). However, although iPSC-derived neurons also showed a deficit in TAF1 expression, removal of the retrotransposon insertion did not restore TAF1 expression levels. The inserted repeat is guanine-rich and has the ability to form G-quadruplex structures (G4S). A recent study found that, in patient-derived fibroblasts and neural progenitor cells, G4S in the SVA region causes aberrant TAF1 expression by inhibiting its binding partner RNA polymerase II (Nicoletto et al. 2024).

11.3.6 Huntington's Disease

Huntington disease (HD) is caused by a CAG repeat expansion in the HTT gene, encoding huntingtin (Bates et al. 2015). In a Drosophila model of HD, an increase in TE expression and mobilization was observed. Notably, inhibiting TE activity rescued eye neurodegeneration and genomic instability in larval brains (Casale et al. 2022). Further studies in a murine model of HD suggest a potential role for L1 mobilization in the pathogenesis of the disease. An increase in L1 DNA copy number was observed in the caudate region of HD mouse brains compared to wild-type controls, and this elevated L1 transcription was associated with decreased methylation within L1 regions (Tan et al. 2018). Conversely, RNA-seq data from two HD mouse models revealed decreased expression of certain L1 RNAs in the striatum of transgenic mice expressing mutant huntingtin protein compared to those expressing the wild-type protein. Consistent with these findings, human HD post-mortem brain tissue showed a trend toward downregulation of L1 expression compared to normal controls (Floreani et al. 2021).

11.3.7 Creutzfeldt–Jakob Disease

Creutzfeldt–Jakob disease (CJD) is a transmissible spongiform encephalopathy (TSE) found in humans. CJD is a fatal neurodegenerative disorder that is characterized by vacuolization of the brain and aberrant protein aggregation. The key molecular event in the pathogenesis of CJD is the accumulation of the misfolded isoform of the cellular prion protein (PrP^{C}), named PrP^{Sc}, which leads to progressive neurodegeneration and astrogliosis (Sitammagari and Masood 2025; Chesnokova et al. 2022). PrP^{Sc} can bind to PrP^{C} and promote misfolding, leading to the amplification of toxic protein aggregation and disease progression (Lathe and Darlix 2020). The "protein-only" hypothesis suggests that PrPC refolds into PrPSc, which then acts as a template to convert more PrPC into PrPSc, leading to disease spread without the presence of nucleic acids, such as a viral genome (Lathe and Darlix 2020). However, as Lathe and Darlix (2020) describe, the protein-only hypothesis does not explain the extensive variation of TSE strains and evident strain competition. Additionally, there is evidence that just the misfolded protein itself may not be enough to propagate disease, indicating an endogenous nucleic acid intermediate. In fact, in cells that have low propensity for TSE propagation, retrovirus infection increased levels of PrP^{Sc} and infectivity (Leblanc et al. 2006; Kim et al. 2016). In scrapie-infected hamsters, TEs such as LINE-1 were upregulated, suggesting an interaction with host RTEs as the potential nucleic acid involved in prion disease (Lathe and Harris 2009). Further, elevated levels of HERV families were observed in the CSF of individuals with sporadic CJD compared to controls. Concentrations of HERV-W and HERV-L were higher in CJD samples than in the control group (Jeong et al. 2010). Therefore, while the pathogenesis of CJD and other TSEs was originally thought to be controlled only by the propagation of misfolded prions, it is possible that TEs also play a role.

11.3.8 TE Targeting in Neurodegenerative Diseases

As highlighted in this chapter, aging weakens silencing mechanisms of TEs, leading to their derepression and detrimental effects in the pathogenesis of age-related neurodegenerative diseases. Therefore, TEs represent a potential therapeutic target.

Several strategies have been explored to target the effects of aberrant reactivated TE. One approach consists of targeting TE regulators, such as transcription factors or RNA-binding proteins, to restore proper TE expression. Engrailed, a homeoprotein, exemplifies this approach. It represses LINE-1 elements by directly binding to their promoter, protecting dopaminergic neurons from DNA damage *caused by* oxidative stress and, thereby, from degeneration (Blaudin de The et al. 2018). Due to its ability to be internalized by cells, Engrailed holds promise as a therapeutic protein capable of modulating LINE-1 activity and conferring protection against oxidative stress (Blaudin de The et al. 2018).

Another approach is to develop therapies that specifically target and inhibit the activity of TEs. In this context, nucleoside reverse transcriptase inhibitors (NRTIs) have demonstrated consistent efficacy across various models (Gorbunova et al. 2021). Notably, many NRTIs are already FDA-approved for HIV treatment and exhibit good tolerability profiles. Building on this existing knowledge, clinical trials are currently underway to evaluate the therapeutic potential of antiretroviral therapies for AD (Sullivan et al. 2024) and ALS (Gold et al. 2019). Among the FDA-approved NRTIs, stavudine reduces DNA damage and mitigates neurodegeneration in dopaminergic neurons under acute oxidative stress (Blaudin de The et al. 2018), and lamivudine inhibits the LINE-1 cDNA synthesis, resulting in decreased expression of genes associated with the IFN-1 response and the neuroinflammation linked to LINE-1 elements (De Cecco et al. 2013a). Clinical trials assessing the safety and tolerability of a combination of NRTIs, namely Triumeq (abacavir, lamivudine, and dolutegravir), are being conducted in the context of ALS (Gold et al. 2019), paving the way for a phase 3 clinical trial. Additionally, Transposon Therapeutics developed a drug, TPN-101, that inhibits LINE-1 reverse transcriptase. Phase-2 clinical trials showed that both ALS patients with *C9orf72* mutations and progressive supranuclear palsy (PSP) patients receiving TPN-101 showed a reduction of biomarkers such as neurofilament light and osteopontin (National Library of Medicine [NLM], NCT04993755). Thus, targeting retrotransposons could be impactful across diverse neurodegenerative diseases.

The most direct approach to inhibit retrotransposons would be to strengthen their silencing epigenetic mechanisms, such as through SIRT6. Male mice overexpressing Sirt6 display an extended lifespan (Kanfi et al. 2012), potentially due to enhanced L1 element silencing and reduced inflammation. The development of small-molecule SIRT6 activators holds promise for a range of health benefits, including improved retrotransposon control.

Finally, caloric restriction, by reducing the metabolic rate, leads to a decrease in reactive oxidative species production and inflammation, and an increase in DNA repair, all of which are relevant for retrotransposon activation (Heydari et al. 2007; Lim et al. 2024). In mice, caloric restriction significantly reduces TE expression (De Cecco et al. 2013b).

11.4 Current Challenges and Open Questions

Despite significant advances in understanding the role of TEs in the pathogenesis of neurodegenerative disease, several conceptual and technical challenges remain unresolved. Addressing these barriers is critical to moving the field forward and translating the basic insights into therapeutic breakthroughs.

One major hurdle is the accurate and reliable detection of TE mobilization and transcriptional activation in aged tissues, especially post-mortem brain samples. Due to the repetitive nature of TEs and the abundance of inactive, truncated copies, standard sequencing and bioinformatic methods often fall short in detecting TEs mobilization events. Additionally, detecting somatic TE insertions, particularly those that occur at low frequency within neuronal populations, poses significant methodological challenges. Thus, the development of more sensitive techniques, including improved long-read sequencing approaches, single-cell technologies, and more robust bioinformatic pipelines, is essential.

Current model systems, including cell lines and animal models, have limitations in fully capturing the complexity of human aging and neurodegeneration. Murine models and simpler organisms like *Drosophila* provide valuable insights, but species-specific differences in TE regulation complicate translational interpretation. The application of human iPSC-derived neurons, brain organoids, and multi-cellular co-culture systems could bridge these gaps, offering a more faithful representation of age-related TE dynamics and disease mechanisms. Yet, even these human-derived models currently lack the aging context, making methods to artificially age these systems or induce aging-associated TE activation critical research priorities.

Moreover, TE activation in neurodegeneration might arise secondarily to other upstream pathological events, such as mitochondrial dysfunction or protein aggregation. Differentiating whether TE reactivation is a primary pathogenic event or a secondary epiphenomenon triggered by other cellular stresses remains challenging. Delineating these hierarchies is crucial, as it significantly affects therapeutic strategies aimed at targeting TE activity or its upstream triggers.

11.4.1 Open Questions

Several critical open questions remain, providing valuable directions for future research on the role of TEs in neurodegenerative diseases. First, longitudinal studies are needed to clarify how TE expression and mobilization evolve throughout life in individuals predisposed to neurodegeneration, and whether specific patterns of TE activation could serve as biomarkers for disease onset or progression. Second, the interplay between genetic variations and environmental factors such as diet, toxins, infections, and lifestyle in modulating TE activity warrants further investigation, particularly to identify why some individuals are more susceptible to TE-induced neurotoxicity. Additionally, understanding why specific neuronal populations are uniquely vulnerable to TE-driven genomic instability and identifying the cellular and molecular factors underpinning neuronal resilience or susceptibility remain important goals. Finally, therapeutic approaches targeting TE activation pathways must be explored carefully to ensure they specifically inhibit harmful TE activity without compromising essential physiological roles. Addressing these challenges will require collaborative efforts, combining advances from genetics, molecular biology, bioinformatics, and clinical neuroscience. Resolving these issues is crucial to harnessing our knowledge of TEs in aging and neurodegeneration, paving the way toward effective diagnostic and therapeutic solutions for age-related neurodegenerative diseases.

11.5 Conclusions

TEs play a diverse role in neurodegenerative diseases, and further research is necessary to fully uncover the intricate pathways that they are involved with. They have evolved in our genomes to play a pivotal role in neurodevelopment. However, aberrant TE derepression is associated with aging, which is a major risk factor for neurodegeneration. This is likely due to a collapse of the regulatory mechanisms that keep TEs in place, specifically in highly vulnerable neurons. In this chapter, we discussed the growing evidence for specific TE contribution in all the major hallmarks of neurodegenerative diseases: genomic instability, mitochondrial dysfunction, neuroinflammation, impaired autophagy, and telomere attrition. In addition, we summarized the evidence that aberrant TE activation is directly linked to patients with AD, ALS/FTD, PD, HD, and other neurodegenerative proteinopathies.

Acknowledgments We thank the members of the Weinberg ALS Center for providing critical feedback on the manuscript. The following sources supported this work: National Institute of Health grants RO1-NS136624 (D.T.), RO3-AG088868 (C.S.); Family Strong for ALS Foundation, Farber Family Foundation, and the Aldrich Foundation (Weinberg ALS Center).

Refined with the assistance of artificial intelligence (AI) tools, Gemini, 2.5 flash. Every section was carefully reviewed, revised, and approved by the authors to ensure that the text is appropriate concerning content and scientific correctness and maintains the integrity of the original content.

Compliance with Ethical Standards

Conflict of Interest The authors declare no conflict of interest.

Ethical Approval This chapter contains no studies with human/animal participants performed by the author.

References

Ahn HW, Worman ZF, Lechsinska A et al (2023) Retrotransposon insertions associated with risk of neurologic and psychiatric diseases. EMBO Rep 24(1):e55197. https://doi.org/10.15252/embr.202255197

An HJ, Lee D, Lee KH et al (2004) The association of Alu repeats with the generation of potential AU-rich elements (ARE) at 3′ untranslated regions. BMC Genomics 5(1):97. https://doi.org/10.1186/1471-2164-5-97

Andreev VI, Yu C, Wang J et al (2022) Panoramix SUMOylation on chromatin connects the piRNA pathway to the cellular heterochromatin machinery. Nat Struct Mol Biol 29(2):130–142. https://doi.org/10.1038/s41594-022-00721-x

Aneichyk T, Hendriks WT, Yadav R et al (2018) Dissecting the causal mechanism of X-linked dystonia-parkinsonism by integrating genome and transcriptome assembly. Cell 172(5):897–909 e821. https://doi.org/10.1016/j.cell.2018.02.011

Arasaratnam CJ, Singh-Bains MK, Waldvogel HJ et al (2021) Neuroimaging and neuropathology studies of X-linked dystonia parkinsonism. Neurobiol Dis 148:105186. https://doi.org/10.1016/j.nbd.2020.105186

Baeken MW, Moosmann B, Hajieva P (2020) Retrotransposon activation by distressed mitochondria in neurons. Biochem Biophys Res Commun 525(3):570–575. https://doi.org/10.1016/j.bbrc.2020.02.106
Balachandran P, Walawalkar IA, Flores JI et al (2022) Transposable element-mediated rearrangements are prevalent in human genomes. Nat Commun 13(1):7115. https://doi.org/10.1038/s41467-022-34810-8
Bannister AJ, Zegerman P, Partridge JF et al (2001) Selective recognition of methylated lysine 9 on histone H3 by the HP1 chromo domain. Nature 410(6824):120–124. https://doi.org/10.1038/35065138
Barau J, Teissandier A, Zamudio N et al (2016) The DNA methyltransferase DNMT3C protects male germ cells from transposon activity. Science 354(6314):909–912. https://doi.org/10.1126/science.aah5143
Barbot W, Dupressoir A, Lazar V et al (2002) Epigenetic regulation of an IAP retrotransposon in the aging mouse: progressive demethylation and de-silencing of the element by its repetitive induction. Nucleic Acids Res 30(11):2365–2373. https://doi.org/10.1093/nar/30.11.2365
Bates GP, Dorsey R, Gusella JF et al (2015) Huntington disease. Nat Rev Dis Primers 1:15005. https://doi.org/10.1038/nrdp.2015.5
Berrens RV, Andrews S, Spensberger D et al (2017) An endosiRNA-based repression mechanism counteracts transposon activation during global DNA demethylation in embryonic stem cells. Cell Stem Cell 21(5):694–703 e697. https://doi.org/10.1016/j.stem.2017.10.004
Bertrand H, Chan BS, Griffiths AJ (1985) Insertion of a foreign nucleotide sequence into mitochondrial DNA causes senescence in Neurospora intermedia. Cell 41(3):877–884. https://doi.org/10.1016/s0092-8674(85)80068-4
Blaudin de The FX, Rekaik H, Peze-Heidsieck E et al (2018) Engrailed homeoprotein blocks degeneration in adult dopaminergic neurons through LINE-1 repression. EMBO J 37(15). https://doi.org/10.15252/embj.201797374
Bodak M, Cirera-Salinas D, Yu J et al (2017) Dicer, a new regulator of pluripotency exit and LINE-1 elements in mouse embryonic stem cells. FEBS Open Bio 7(2):204–220. https://doi.org/10.1002/2211-5463.12174
Bourque G, Burns KH, Gehring M et al (2018) Ten things you should know about transposable elements. Genome Biol 19(1):199. https://doi.org/10.1186/s13059-018-1577-z
Bragg DC, Mangkalaphiban K, Vaine CA et al (2017) Disease onset in X-linked dystonia-parkinsonism correlates with expansion of a hexameric repeat within an SVA retrotransposon in TAF1. Proc Natl Acad Sci USA 114(51):E11020–E11028. https://doi.org/10.1073/pnas.1712526114
Bravo P, Darvish H, Tafakhori A et al (2018) Molecular characterization of PRKN structural variations identified through whole-genome sequencing. Mol Genet Genomic Med 6(6):1243–1248. https://doi.org/10.1002/mgg3.482
Bravo JI, Nozownik S, Danthi PS et al (2020) Transposable elements, circular RNAs and mitochondrial transcription in age-related genomic regulation. Development 147(11). https://doi.org/10.1242/dev.175786
Burns KH (2020) Our conflict with transposable elements and its implications for human disease. Annu Rev Pathol 15:51–70. https://doi.org/10.1146/annurev-pathmechdis-012419-032633
Calabrese JM, Seila AC, Yeo GW et al (2007) RNA sequence analysis defines dicer's role in mouse embryonic stem cells. Proc Natl Acad Sci USA 104(46):18097–18102. https://doi.org/10.1073/pnas.0709193104
Caldecott KW, Ward ME, Nussenzweig A (2022) The threat of programmed DNA damage to neuronal genome integrity and plasticity. Nat Genet 54(2):115–120. https://doi.org/10.1038/s41588-021-01001-y
Casale AM, Liguori F, Ansaloni F et al (2022) Transposable element activation promotes neurodegeneration in a drosophila model of Huntington's disease. iScience 25(1):103702. https://doi.org/10.1016/j.isci.2021.103702

Chalmers TJ, Wu LE (2020) Transposable elements cross kingdom boundaries and contribute to inflammation and ageing: somatic Acquisition of Foreign Transposable Elements as a catalyst of genome instability, epigenetic dysregulation, inflammation, senescence, and ageing. BioEssays 42(3):e1900197. https://doi.org/10.1002/bies.201900197

Chalopin D, Naville M, Plard F et al (2015) Comparative analysis of transposable elements highlights mobilome diversity and evolution in vertebrates. Genome Biol Evol 7(2):567–580. https://doi.org/10.1093/gbe/evv005

Chernyavskaya Y, Mudbhary R, Zhang C et al (2017) Loss of DNA methylation in zebrafish embryos activates retrotransposons to trigger antiviral signaling. Development 144(16):2925–2939. https://doi.org/10.1242/dev.147629

Chesnokova E, Beletskiy A, Kolosov P (2022) The role of transposable elements of the human genome in neuronal function and pathology. Int J Mol Sci 23(10). https://doi.org/10.3390/ijms23105847

Chiappinelli KB, Strissel PL, Desrichard A et al (2015) Inhibiting DNA methylation causes an interferon response in cancer via dsRNA including endogenous retroviruses. Cell 162(5):974–986. https://doi.org/10.1016/j.cell.2015.07.011

Colombo AR, Elias HK, Ramsingh G (2018) Senescence induction universally activates transposable element expression. Cell Cycle 17(14):1846–1857. https://doi.org/10.1080/15384101.2018.1502576

Copley KE, Shorter J (2023) Repetitive elements in aging and neurodegeneration. Trends Genet 39(5):381–400. https://doi.org/10.1016/j.tig.2023.02.008

Coufal NG, Garcia-Perez JL, Peng GE et al (2011) Ataxia telangiectasia mutated (ATM) modulates long interspersed element-1 (L1) retrotransposition in human neural stem cells. Proc Natl Acad Sci USA 108(51):20382–20387. https://doi.org/10.1073/pnas.1100273108

Cuellar TL, Herzner AM, Zhang X et al (2017) Silencing of retrotransposons by SETDB1 inhibits the interferon response in acute myeloid leukemia. J Cell Biol 216(11):3535–3549. https://doi.org/10.1083/jcb.201612160

Culig L, Sahbaz BD, Bohr VA (2022) Effects of lifespan-extending interventions on cognitive healthspan. Expert Rev Mol Med 25:e2. https://doi.org/10.1017/erm.2022.36

De Cecco M, Criscione SW, Peckham EJ et al (2013a) Genomes of replicatively senescent cells undergo global epigenetic changes leading to gene silencing and activation of transposable elements. Aging Cell 12(2):247–256. https://doi.org/10.1111/acel.12047

De Cecco M, Criscione SW, Peterson AL et al (2013b) Transposable elements become active and mobile in the genomes of aging mammalian somatic tissues. Aging (Albany NY) 5(12):867–883. https://doi.org/10.18632/aging.100621

de Oliveira DS, Rosa MT, Vieira C et al (2021) Oxidative and radiation stress induces transposable element transcription in Drosophila melanogaster. J Evol Biol 34(4):628–638. https://doi.org/10.1111/jeb.13762

de Rivero Vaccari JP, Brand FJ 3rd, Sedaghat C et al (2014) RIG-1 receptor expression in the pathology of Alzheimer's disease. J Neuroinflammation 11:67. https://doi.org/10.1186/1742-2094-11-67

Della Valle F, Thimma MP, Caiazzo M et al (2020) Transdifferentiation of mouse embryonic fibroblasts into dopaminergic neurons reactivates LINE-1 repetitive elements. Stem Cell Rep 14(1):60–74. https://doi.org/10.1016/j.stemcr.2019.12.002

Della Valle F, Reddy P, Yamamoto M et al (2022) LINE-1 RNA causes heterochromatin erosion and is a target for amelioration of senescent phenotypes in progeroid syndromes. Sci Transl Med 14(657):eabl6057. https://doi.org/10.1126/scitranslmed.abl6057

Dembny P, Newman AG, Singh M, H et al. (2020) Human endogenous retrovirus HERV-K(HML-2) RNA causes neurodegeneration through toll-like receptors. JCI Insight 5 (7). https://doi.org/10.1172/jci.insight.131093

Deng Z, Lim J, Wang Q et al (2020) ALS-FTLD-linked mutations of SQSTM1/p62 disrupt selective autophagy and NFE2L2/NRF2 anti-oxidative stress pathway. Autophagy 16(5):917–931. https://doi.org/10.1080/15548627.2019.1644076

Dennis S, Sheth U, Feldman JL et al (2012) C. elegans germ cells show temperature and age-dependent expression of Cer1, a Gypsy/Ty3-related retrotransposon. PLoS Pathog 8(3):e1002591. https://doi.org/10.1371/journal.ppat.1002591
Di Stefano L (2022) All quiet on the TE front? The role of chromatin in transposable element silencing. Cells 11(16). https://doi.org/10.3390/cells11162501
Donertas D, Sienski G, Brennecke J (2013) Drosophila Gtsf1 is an essential component of the Piwi-mediated transcriptional silencing complex. Genes Dev 27(15):1693–1705. https://doi.org/10.1101/gad.221150.113
Douville R, Liu J, Rothstein J et al (2011) Identification of active loci of a human endogenous retrovirus in neurons of patients with amyotrophic lateral sclerosis. Ann Neurol 69(1):141–151. https://doi.org/10.1002/ana.22149
Egilmez NK, Shmookler Reis RJ (1994) Age-dependent somatic excision of transposable element Tc1 in Caenorhabditis elegans. Mutat Res 316(1):17–24. https://doi.org/10.1016/0921-8734(94)90004-3
Elbarbary RA, Lucas BA, Maquat LE (2016) Retrotransposons as regulators of gene expression. Science 351(6274):aac7247. https://doi.org/10.1126/science.aac7247
Elsner D, Meusemann K, Korb J (2018) Longevity and transposon defense, the case of termite reproductives. Proc Natl Acad Sci USA 115(21):5504–5509. https://doi.org/10.1073/pnas.1804046115
Eriksson M, Brown WT, Gordon LB et al (2003) Recurrent de novo point mutations in Lamin a cause Hutchinson-Gilford progeria syndrome. Nature 423(6937):293–298. https://doi.org/10.1038/nature01629
Esnault C, Heidmann O, Delebecque F et al (2005) APOBEC3G cytidine deaminase inhibits retrotransposition of endogenous retroviruses. Nature 433(7024):430–433. https://doi.org/10.1038/nature03238
Evering TH, Marston JL, Gan L et al (2023) Transposable elements and Alzheimer's disease pathogenesis. Trends Neurosci 46(3):170–172. https://doi.org/10.1016/j.tins.2022.12.003
Filatov DA, Morozova TV, Pasyukova EG (1998) Age dependence of the copia transposition rate is positively associated with copia transcript abundance in a Drosophila melanogaster isogenic line. Mol Gen Genet 258(6):646–654. https://doi.org/10.1007/s004380050778
Floreani L, Ansaloni F, Mangoni D et al (2021) Analysis of LINE1 retrotransposons in Huntington's disease. Front Cell Neurosci 15:743797. https://doi.org/10.3389/fncel.2021.743797
Friedli M, Trono D (2015) The developmental control of transposable elements and the evolution of higher species. Annu Rev Cell Dev Biol 31:429–451. https://doi.org/10.1146/annurev-cellbio-100814-125514
Frost B, Hemberg M, Lewis J et al (2014) Tau promotes neurodegeneration through global chromatin relaxation. Nat Neurosci 17(3):357–366. https://doi.org/10.1038/nn.3639
Fuchs J, Schwer B, El-Khamisy SF (2022) Editorial: genomic instability and neurodegeneration. Front Aging Neurosci 14:940459. https://doi.org/10.3389/fnagi.2022.940459
Fukuda S, Varshney A, Fowler BJ et al (2021) Cytoplasmic synthesis of endogenous Alu complementary DNA via reverse transcription and implications in age-related macular degeneration. Proc Natl Acad Sci USA 118(6). https://doi.org/10.1073/pnas.2022751118
Gebrie A (2023) Transposable elements as essential elements in the control of gene expression. Mob DNA 14(1):9. https://doi.org/10.1186/s13100-023-00297-3
Globisch D, Munzel M, Muller M et al (2010) Tissue distribution of 5-hydroxymethylcytosine and search for active demethylation intermediates. PLoS One 5(12):e15367. https://doi.org/10.1371/journal.pone.0015367
Gold J, Rowe DB, Kiernan MC et al (2019) Safety and tolerability of Triumeq in amyotrophic lateral sclerosis: the lighthouse trial. Amyotroph Lateral Scler Frontotemporal Degener 20(7–8):595–604. https://doi.org/10.1080/21678421.2019.1632899
Gorbunova V, Seluanov A, Mita P et al (2021) The role of retrotransposable elements in ageing and age-associated diseases. Nature 596(7870):43–53. https://doi.org/10.1038/s41586-021-03542-y

Grandi N, Tramontano E (2018) Human endogenous retroviruses are ancient acquired elements still shaping innate immune responses. Front Immunol 9:2039. https://doi.org/10.3389/fimmu.2018.02039

Greer EL, Maures TJ, Hauswirth AG et al (2010) Members of the H3K4 trimethylation complex regulate lifespan in a germline-dependent manner in C. elegans. Nature 466(7304):383–387. https://doi.org/10.1038/nature09195

Guo H, Chitiprolu M, Gagnon D et al (2014) Autophagy supports genomic stability by degrading retrotransposon RNA. Nat Commun 5:5276. https://doi.org/10.1038/ncomms6276

Guo C, Jeong HH, Hsieh YC et al (2018) Tau activates transposable elements in Alzheimer's disease. Cell Rep 23(10):2874–2880. https://doi.org/10.1016/j.celrep.2018.05.004

Guo J, Huang X, Dou L et al (2022) Aging and aging-related diseases: from molecular mechanisms to interventions and treatments. Signal Transduct Target Ther 7(1):391. https://doi.org/10.1038/s41392-022-01251-0

Harley J, Santosa MM, Ng CY et al (2024) Telomere shortening induces aging-associated phenotypes in hiPSC-derived neurons and astrocytes. Biogerontology 25(2):341–360. https://doi.org/10.1007/s10522-023-10076-5

Hata K, Sakaki Y (1997) Identification of critical CpG sites for repression of L1 transcription by DNA methylation. Gene 189(2):227–234. https://doi.org/10.1016/s0378-1119(96)00856-6

He J, Babarinde IA, Sun L et al (2021) Identifying transposable element expression dynamics and heterogeneity during development at the single-cell level with a processing pipeline scTE. Nat Commun 12(1):1456. https://doi.org/10.1038/s41467-021-21808-x

Heil F, Hemmi H, Hochrein H et al (2004) Species-specific recognition of single-stranded RNA via toll-like receptor 7 and 8. Science 303(5663):1526–1529. https://doi.org/10.1126/science.1093620

Hewitt G, Korolchuk VI (2017) Repair, reuse, recycle: The expanding role of autophagy in genome maintenance. Trends Cell Biol 27(5):340–351. https://doi.org/10.1016/j.tcb.2016.11.011

Heydari AR, Unnikrishnan A, Lucente LV et al (2007) Caloric restriction and genomic stability. Nucleic Acids Res 35(22):7485–7496. https://doi.org/10.1093/nar/gkm860

Horvath S, Raj K (2018) DNA methylation-based biomarkers and the epigenetic clock theory of ageing. Nat Rev Genet 19(6):371–384. https://doi.org/10.1038/s41576-018-0004-3

Hou Y, Dan X, Babbar M et al (2019) Ageing as a risk factor for neurodegenerative disease. Nat Rev Neurol 15(10):565–581. https://doi.org/10.1038/s41582-019-0244-7

Hwang YE, Baek YM, Baek A et al (2019) Oxidative stress causes Alu RNA accumulation via PIWIL4 sequestration into stress granules. BMB Rep 52(3):196–201. https://doi.org/10.5483/BMBRep.2019.52.3.146

Ilyin AA, Ryazansky SS, Doronin SA et al (2017) Piwi interacts with chromatin at nuclear pores and promiscuously binds nuclear transcripts in drosophila ovarian somatic cells. Nucleic Acids Res 45(13):7666–7680. https://doi.org/10.1093/nar/gkx355

Iyama T, Wilson DM 3rd (2013) DNA repair mechanisms in dividing and non-dividing cells. DNA Repair (Amst) 12(8):620–636. https://doi.org/10.1016/j.dnarep.2013.04.015

Jansz N (2019) DNA methylation dynamics at transposable elements in mammals. Essays Biochem 63(6):677–689. https://doi.org/10.1042/EBC20190039

Jeong BH, Lee YJ, Carp RI et al (2010) The prevalence of human endogenous retroviruses in cerebrospinal fluids from patients with sporadic Creutzfeldt-Jakob disease. J Clin Virol 47(2):136–142. https://doi.org/10.1016/j.jcv.2009.11.016

Jintaridth P, Mutirangura A (2010) Distinctive patterns of age-dependent hypomethylation in interspersed repetitive sequences. Physiol Genomics 41(2):194–200. https://doi.org/10.1152/physiolgenomics.00146.2009

Jonsson ME, Garza R, Johansson PA et al (2020) Transposable elements: a common feature of neurodevelopmental and neurodegenerative disorders. Trends Genet 36(8):610–623. https://doi.org/10.1016/j.tig.2020.05.004

Jonsson ME, Garza R, Sharma Y et al (2021) Activation of endogenous retroviruses during brain development causes an inflammatory response. EMBO J 40(9):e106423. https://doi.org/10.15252/embj.2020106423

Kaarniranta K, Pawlowska E, Szczepanska J et al (2020) DICER1 in the pathogenesis of age-related macular degeneration (AMD) – Alu RNA accumulation versus miRNA dysregulation. Aging Dis 11(4):851–862. https://doi.org/10.14336/AD.2019.0809

Kaneko H, Dridi S, Tarallo V et al (2011) DICER1 deficit induces Alu RNA toxicity in age-related macular degeneration. Nature 471(7338):325–330. https://doi.org/10.1038/nature09830

Kanfi Y, Naiman S, Amir GP et al (2012) The sirtuin SIRT6 regulates lifespan in male mice. Nature 483(7388):218–221. https://doi.org/10.1038/nature10815

Kim BH, Shin HY, Goto JJ, C et al. (2016) Cellular prion protein combined with Galectin-3 and -6 affects the infectivity titer of an endogenous retrovirus assayed in hippocampal neuronal cells. PLoS One 11 (12):e0167293. https://doi.org/10.1371/journal.pone.0167293

Kinomoto M, Kanno T, Shimura M et al (2007) All APOBEC3 family proteins differentially inhibit LINE-1 retrotransposition. Nucleic Acids Res 35(9):2955–2964. https://doi.org/10.1093/nar/gkm181

Koks S, Pfaff AL, Singleton LM et al (2022) Non-reference genome transposable elements (TEs) have a significant impact on the progression of the Parkinson's disease. Exp Biol Med (Maywood) 247(18):1680–1690. https://doi.org/10.1177/15353702221117147

Kurup JT, Han Z, Jin W et al (2020) H4K20me3 methyltransferase SUV420H2 shapes the chromatin landscape of pluripotent embryonic stem cells. Development 147(23). https://doi.org/10.1242/dev.188516

LaRocca TJ, Cavalier AN, Wahl D (2020) Repetitive elements as a transcriptomic marker of aging: evidence in multiple datasets and models. Aging Cell 19(7):e13167. https://doi.org/10.1111/acel.13167

Larsen PA, Lutz MW, Hunnicutt KE et al (2017) The Alu neurodegeneration hypothesis: a primate-specific mechanism for neuronal transcription noise, mitochondrial dysfunction, and manifestation of neurodegenerative disease. Alzheimers Dement 13(7):828–838. https://doi.org/10.1016/j.jalz.2017.01.017

Larson K, Yan SJ, Tsurumi A et al (2012) Heterochromatin formation promotes longevity and represses ribosomal RNA synthesis. PLoS Genet 8(1):e1002473. https://doi.org/10.1371/journal.pgen.1002473

Lathe R, Darlix JL (2020) Prion protein PrP nucleic acid binding and mobilization implicates retroelements as the replicative component of transmissible spongiform encephalopathy. Arch Virol 165(3):535–556. https://doi.org/10.1007/s00705-020-04529-2

Lathe R, Harris A (2009) Differential display detects host nucleic acid motifs altered in scrapie-infected brain. J Mol Biol 392(3):813–822. https://doi.org/10.1016/j.jmb.2009.07.045

Le Breton A, Bettencourt MP, Gendrel AV (2024) Navigating the brain and aging: exploring the impact of transposable elements from health to disease. Front Cell Dev Biol 12:1357576. https://doi.org/10.3389/fcell.2024.1357576

Leblanc P, Alais S, Porto-Carreiro I et al (2006) Retrovirus infection strongly enhances scrapie infectivity release in cell culture. EMBO J 25(12):2674–2685. https://doi.org/10.1038/sj.emboj.7601162

Lee SK, Wang W (2019) Roles of topoisomerases in heterochromatin, aging, and diseases. Genes (Basel) 10(11). https://doi.org/10.3390/genes10110884

Li W, Prazak L, Chatterjee N, Gruninger S et al (2013a) Activation of transposable elements during aging and neuronal decline in Drosophila. Nat Neurosci 16(5):529–531. https://doi.org/10.1038/nn.3368

Li XD, Wu J, Gao D et al (2013b) Pivotal roles of cGAS-cGAMP signaling in antiviral defense and immune adjuvant effects. Science 341(6152):1390–1394. https://doi.org/10.1126/science.1244040

Li W, Lee MH, Henderson L et al (2015) Human endogenous retrovirus-K contributes to motor neuron disease. Sci Transl Med 7(307):307ra153. https://doi.org/10.1126/scitranslmed.aac8201

Li P, Du J, Goodier JL et al (2017) Aicardi-Goutieres syndrome protein TREX1 suppresses L1 and maintains genome integrity through exonuclease-independent ORF1p depletion. Nucleic Acids Res 45(8):4619–4631. https://doi.org/10.1093/nar/gkx178

Li Y, Chen X, Lu C (2021) The interplay between DNA and histone methylation: molecular mechanisms and disease implications. EMBO Rep 22(5):e51803. https://doi.org/10.15252/embr.202051803

Lim YW, Sanz LA, Xu X et al (2015) Genome-wide DNA hypomethylation and RNA:DNA hybrid accumulation in Aicardi-Goutieres syndrome. elife 4. https://doi.org/10.7554/eLife.08007

Lim GM, Maharajan N, Cho GW (2024) How calorie restriction slows aging: an epigenetic perspective. J Mol Med (Berl) 102(5):629–640. https://doi.org/10.1007/s00109-024-02430-y

Liu EY, Russ J, Cali CP et al (2019) Loss of nuclear TDP-43 is associated with Decondensation of LINE retrotransposons. Cell Rep 27(5):1409–1421. e1406. https://doi.org/10.1016/j.celrep.2019.04.003

Liu X, Liu Z, Wu Z et al (2023) Resurrection of endogenous retroviruses during aging reinforces senescence. Cell 186(2):287–304 e226. https://doi.org/10.1016/j.cell.2022.12.017

Lopez-Otin C, Blasco MA, Partridge L et al (2013) The hallmarks of aging. Cell 153(6):1194–1217. https://doi.org/10.1016/j.cell.2013.05.039

Lynch VJ, Nnamani MC, Kapusta A et al (2015) Ancient transposable elements transformed the uterine regulatory landscape and transcriptome during the evolution of mammalian pregnancy. Cell Rep 10(4):551–561. https://doi.org/10.1016/j.celrep.2014.12.052

Macciardi F, Giulia Bacalini M, Miramontes R et al (2022) A retrotransposon storm marks clinical phenoconversion to late-onset Alzheimer's disease. Geroscience 44(3):1525–1550. https://doi.org/10.1007/s11357-022-00580-w

Maegawa S, Hinkal G, Kim HS et al (2010) Widespread and tissue specific age-related DNA methylation changes in mice. Genome Res 20(3):332–340. https://doi.org/10.1101/gr.096826.109

Maison C, Bailly D, Roche D et al (2011) SUMOylation promotes de novo targeting of HP1alpha to pericentric heterochromatin. Nat Genet 43(3):220–227. https://doi.org/10.1038/ng.765

Makino S, Kaji R, Ando S et al (2007) Reduced neuron-specific expression of the TAF1 gene is associated with X-linked dystonia-parkinsonism. Am J Hum Genet 80(3):393–406. https://doi.org/10.1086/512129

Maxwell PH, Burhans WC, Curcio MJ (2011) Retrotransposition is associated with genome instability during chronological aging. Proc Natl Acad Sci USA 108(51):20376–20381. https://doi.org/10.1073/pnas.1100271108

Mayorga L, Salassa BN, Marzese DM, Loos MA, Eiroa HD, Lubieniecki F, Garcia Samartino C, Romano PS, Roque M (2019) Mitochondrial stress triggers a pro-survival response through epigenetic modifications of nuclear DNA. Cell Mol Life Sci 76(7):1397–1417. https://doi.org/10.1007/s00018-019-03008-5

McColl G, Killilea DW, Hubbard AE et al (2008) Pharmacogenetic analysis of lithium-induced delayed aging in Caenorhabditis elegans. J Biol Chem 283(1):350–357. https://doi.org/10.1074/jbc.M705028200

McCormick AL, Brown RH Jr, Cudkowicz ME et al (2008) Quantification of reverse transcriptase in ALS and elimination of a novel retroviral candidate. Neurology 70(4):278–283. https://doi.org/10.1212/01.wnl.0000297552.13219.b4

Meller A, Shalgi R (2021) The aging proteostasis decline: from nematode to human. Exp Cell Res 399(2):112474. https://doi.org/10.1016/j.yexcr.2021.112474

Mita T, Harabuchi Y, Maeda S (2020) Discovery of a synthesis method for a difluoroglycine derivative based on a path generated by quantum chemical calculations. Chem Sci 11(29):7569–7577. https://doi.org/10.1039/d0sc02089c

Mosaddeghi P, Farahmandnejad M, Zarshenas MM (2023) The role of transposable elements in aging and cancer. Biogerontology 24(4):479–491. https://doi.org/10.1007/s10522-023-10028-z

Muerdter F, Guzzardo PM, Gillis J et al (2013) A genome-wide RNAi screen draws a genetic framework for transposon control and primary piRNA biogenesis in drosophila. Mol Cell 50(5):736–748. https://doi.org/10.1016/j.molcel.2013.04.006

Muotri AR, Chu VT, Marchetto MC et al (2005) Somatic mosaicism in neuronal precursor cells mediated by L1 retrotransposition. Nature 435(7044):903–910. https://doi.org/10.1038/nature03663

Nakayama J, Rice JC, Strahl BD et al (2001) Role of histone H3 lysine 9 methylation in epigenetic control of heterochromatin assembly. Science 292(5514):110–113. https://doi.org/10.1126/science.1060118

Nicoletto G, Terreri M, Maurizio I et al (2024) G-quadruplexes in an SVA retrotransposon cause aberrant TAF1 gene expression in X-linked dystonia parkinsonism. Nucleic Acids Res 52(19):11571–11586. https://doi.org/10.1093/nar/gkae797

Noda A, Mishima S, Hirai Y et al (2015) Progerin, the protein responsible for the Hutchinson-Gilford progeria syndrome, increases the unrepaired DNA damages following exposure to ionizing radiation. Genes Environ 37:13. https://doi.org/10.1186/s41021-015-0018-4

Osumi K, Sato K, Murano K et al (2019) Essential roles of Windei and nuclear monoubiquitination of Eggless/SETDB1 in transposon silencing. EMBO Rep 20(12):e48296. https://doi.org/10.15252/embr.201948296

Ozturk N, Dansranjavin T, Gies S et al (2021) H4K20me3 marks distal intergenic and repetitive regions in human mature spermatozoa. Development 148(15). https://doi.org/10.1242/dev.196477

Payton A, Sindrewicz P, Pessoa V et al (2016) A TOMM40 poly-T variant modulates gene expression and is associated with vocabulary ability and decline in nonpathologic aging. Neurobiol Aging 39:217 e211–e217. https://doi.org/10.1016/j.neurobiolaging.2015.11.017

Pfaff AL, Bubb VJ, Quinn JP et al (2021) Reference SVA insertion polymorphisms are associated with Parkinson's disease progression and differential gene expression. NPJ Parkinsons Dis 7(1):44. https://doi.org/10.1038/s41531-021-00189-4

Platt RN 2nd, Vandewege MW, Ray DA (2018) Mammalian transposable elements and their impacts on genome evolution. Chromosom Res 26(1–2):25–43. https://doi.org/10.1007/s10577-017-9570-z

Prudencio M, Gonzales PK, Cook CN et al (2017) Repetitive element transcripts are elevated in the brain of C9orf72 ALS/FTLD patients. Hum Mol Genet 26(17):3421–3431. https://doi.org/10.1093/hmg/ddx233

Rakovic A, Domingo A, Grutz K et al (2018) Genome editing in induced pluripotent stem cells rescues TAF1 levels in X-linked dystonia-parkinsonism. Mov Disord 33(7):1108–1118. https://doi.org/10.1002/mds.27441

Ramirez P, Zuniga G, Sun W et al (2022) Pathogenic tau accelerates aging-associated activation of transposable elements in the mouse central nervous system. Prog Neurobiol 208:102181. https://doi.org/10.1016/j.pneurobio.2021.102181

Ren W, Fan H, Grimm SA et al (2021) DNMT1 reads heterochromatic H4K20me3 to reinforce LINE-1 DNA methylation. Nat Commun 12(1):2490. https://doi.org/10.1038/s41467-021-22665-4

Rigby RE, Webb LM, Mackenzie KJ et al (2014) RNA:DNA hybrids are a novel molecular pattern sensed by TLR9. EMBO J 33(6):542–558. https://doi.org/10.1002/embj.201386117

Roberson PA, Romero MA, Osburn SC et al (2019) Skeletal muscle LINE-1 ORF1 mRNA is higher in older humans but decreases with endurance exercise and is negatively associated with higher physical activity. J Appl Physiol (1985) 127(4):895–904. https://doi.org/10.1152/japplphysiol.00352.2019

Romero PR, Zaidi S, Fang YY et al (2006) Alternative splicing in concert with protein intrinsic disorder enables increased functional diversity in multicellular organisms. Proc Natl Acad Sci USA 103(22):8390–8395. https://doi.org/10.1073/pnas.0507916103

Rousseaux MW, de Haro M, Lasagna-Reeves CA et al (2016) TRIM28 regulates the nuclear accumulation and toxicity of both alpha-synuclein and tau. elife 5. https://doi.org/10.7554/eLife.19809

Ryu SH, Kang K, Yoo T et al (2011) Transcriptional repression of repeat-derived transcripts correlates with histone hypoacetylation at repetitive DNA elements in aged mice brain. Exp Gerontol 46(10):811–818. https://doi.org/10.1016/j.exger.2011.07.001

Saleh A, Macia A, Muotri AR (2019) Transposable elements, inflammation, and neurological disease. Front Neurol 10:894. https://doi.org/10.3389/fneur.2019.00894

Scopa C, Barnada SM, Cicardi ME et al (2023) JUN upregulation drives aberrant transposable element mobilization, associated innate immune response, and impaired neurogenesis in Alzheimer's disease. Nat Commun 14(1):8021. https://doi.org/10.1038/s41467-023-43728-8

Sen P, Shah PP, Nativio R et al (2016) Epigenetic mechanisms of longevity and aging. Cell 166(4):822–839. https://doi.org/10.1016/j.cell.2016.07.050

Shin JA, Choi ES, Kim HS et al (2005) SUMO modification is involved in the maintenance of heterochromatin stability in fission yeast. Mol Cell 19(6):817–828. https://doi.org/10.1016/j.molcel.2005.08.021

Shiota T, Imai K, Qiu J et al (2015) Molecular architecture of the active mitochondrial protein gate. Science 349(6255):1544–1548. https://doi.org/10.1126/science.aac6428

Sienski G, Donertas D, Brennecke J (2012) Transcriptional silencing of transposons by Piwi and maelstrom and its impact on chromatin state and gene expression. Cell 151(5):964–980. https://doi.org/10.1016/j.cell.2012.10.040

Sienski G, Batki J, Senti KA et al (2015) Silencio/CG9754 connects the Piwi-piRNA complex to the cellular heterochromatin machinery. Genes Dev 29(21):2258–2271. https://doi.org/10.1101/gad.271908.115

Simon M, Van Meter M, Ablaeva J, et al. (2019) LINE1 Derepression in aged wild-type and SIRT6-deficient mice drives inflammation. Cell Metab 29 (4):871-885 e875. https://doi.org/10.1016/j.cmet.2019.02.014

Singer T, McConnell MJ, Marchetto MC et al (2010) LINE-1 retrotransposons: mediators of somatic variation in neuronal genomes? Trends Neurosci 33(8):345–354. https://doi.org/10.1016/j.tins.2010.04.001

Singh S, Borkar MR, Bhatt LK (2024) Transposable elements: emerging therapeutic targets in neurodegenerative diseases. Neurotox Res 42(1):9. https://doi.org/10.1007/s12640-024-00688-1

Sitammagari KK, Masood W (2025) Creutzfeldt Jakob Disease. In: StatPearls. Treasure Island (FL)

Slotkin RK, Martienssen R (2007) Transposable elements and the epigenetic regulation of the genome. Nat Rev Genet 8(4):272–285. https://doi.org/10.1038/nrg2072

Smith ZD, Meissner A (2013) DNA methylation: roles in mammalian development. Nat Rev Genet 14(3):204–220. https://doi.org/10.1038/nrg3354

Steele AJ, Al-Chalabi A, Ferrante K et al (2005) Detection of serum reverse transcriptase activity in patients with ALS and unaffected blood relatives. Neurology 64(3):454–458. https://doi.org/10.1212/01.WNL.0000150899.76130.71

Stern Y, Barnes CA, Grady C et al (2019) Brain reserve, cognitive reserve, compensation, and maintenance: operationalization, validity, and mechanisms of cognitive resilience. Neurobiol Aging 83:124–129. https://doi.org/10.1016/j.neurobiolaging.2019.03.022

Stetson DB, Medzhitov R (2006) Type I interferons in host defense. Immunity 25(3):373–381. https://doi.org/10.1016/j.immuni.2006.08.007

Stetson DB, Ko JS, Heidmann T et al (2008) Trex1 prevents cell-intrinsic initiation of autoimmunity. Cell 134(4):587–598. https://doi.org/10.1016/j.cell.2008.06.032

Stielow B, Sapetschnig A, Wink C et al (2008) SUMO-modified Sp3 represses transcription by provoking local heterochromatic gene silencing. EMBO Rep 9(9):899–906. https://doi.org/10.1038/embor.2008.127

Sullivan AC, Zuniga G, Ramirez P et al (2024) A pilot study to investigate the safety and feasibility of antiretroviral therapy for Alzheimer's disease (ART-AD). medRxiv. https://doi.org/10.1101/2024.02.26.24303316

Sun W, Samimi H, Gamez M et al (2018) Pathogenic tau-induced piRNA depletion promotes neuronal death through transposable element dysregulation in neurodegenerative tauopathies. Nat Neurosci 21(8):1038–1048. https://doi.org/10.1038/s41593-018-0194-1

Suzuki K, Morimoto M, Kondo C et al (2011) Selective autophagy regulates insertional mutagenesis by the Ty1 retrotransposon in Saccharomyces cerevisiae. Dev Cell 21(2):358–365. https://doi.org/10.1016/j.devcel.2011.06.023

Swerdlow RH, Khan SM (2004) A "mitochondrial cascade hypothesis" for sporadic Alzheimer's disease. Med Hypotheses 63(1):8–20. https://doi.org/10.1016/j.mehy.2003.12.045

Tan H, Wu C, Jin L (2018) A possible role for long interspersed nuclear Elements-1 (LINE-1) in Huntington's disease progression. Med Sci Monit 24:3644–3652

Tarallo V, Hirano Y, Gelfand BD et al (2012) DICER1 loss and Alu RNA induce age-related macular degeneration via the NLRP3 inflammasome and MyD88. Cell 149(4):847–859. https://doi.org/10.1016/j.cell.2012.03.036

Tenchov R, Sasso JM, Wang X et al (2024) Aging hallmarks and progression and age-related diseases: a landscape view of research advancement. ACS Chem Neurosci 15(1):1–30. https://doi.org/10.1021/acschemneuro.3c00531

Thomas CA, Tejwani L, Trujillo CA et al (2017) Modeling of TREX1-dependent autoimmune disease using human stem cells highlights L1 accumulation as a source of Neuroinflammation. Cell Stem Cell 21(3):319–331 e318. https://doi.org/10.1016/j.stem.2017.07.009

Toda T, Bedrosian TA, Schafer ST et al (2024) Long interspersed nuclear elements safeguard neural progenitors from precocious differentiation. Cell Rep 43(2):113774. https://doi.org/10.1016/j.celrep.2024.113774

Truby NL, Kim RK, Silva GM et al (2024) A zinc finger transcription factor enables social behaviors while controlling transposable elements and immune response in prefrontal cortex. Transl Psychiatry 14(1):59. https://doi.org/10.1038/s41398-024-02775-5

Tsukahara S, Kobayashi A, Kawabe A et al (2009) Bursts of retrotransposition reproduced in Arabidopsis. Nature 461(7262):423–426. https://doi.org/10.1038/nature08351

Tunbak H, Enriquez-Gasca R, Tie CHC et al (2020) The HUSH complex is a gatekeeper of type I interferon through epigenetic regulation of LINE-1s. Nat Commun 11(1):5387. https://doi.org/10.1038/s41467-020-19170-5

van Kruijsbergen I, Hontelez S, Elurbe DM, v et al. (2017) Heterochromatic histone modifications at transposons in Xenopus tropicalis embryos. Dev Biol 426 (2):460–471. https://doi.org/10.1016/j.ydbio.2016.08.031

Van Meter M, Simon M, Tombline G et al (2016) JNK phosphorylates SIRT6 to stimulate DNA double-Strand break repair in response to oxidative stress by recruiting PARP1 to DNA breaks. Cell Rep 16(10):2641–2650. https://doi.org/10.1016/j.celrep.2016.08.006

Vazquez BN, Thackray JK, Simonet NG et al (2019) SIRT7 mediates L1 elements transcriptional repression and their association with the nuclear lamina. Nucleic Acids Res 47(15):7870–7885. https://doi.org/10.1093/nar/gkz519

Volkman HE, Stetson DB (2014) The enemy within: endogenous retroelements and autoimmune disease. Nat Immunol 15(5):415–422. https://doi.org/10.1038/ni.2872

Walsh CP, Chaillet JR, Bestor TH (1998) Transcription of IAP endogenous retroviruses is constrained by cytosine methylation. Nat Genet 20(2):116–117. https://doi.org/10.1038/2413

Wang J, Geesman GJ, Hostikka SL et al (2011) Inhibition of activated pericentromeric SINE/Alu repeat transcription in senescent human adult stem cells reinstates self-renewal. Cell Cycle 10(17):3016–3030. https://doi.org/10.4161/cc.10.17.17543

Watanabe K, Bloch W (2013) Histone methylation and acetylation indicates epigenetic change in the aged cochlea of mice. Eur Arch Otorrinolaringol 270(6):1823–1830. https://doi.org/10.1007/s00405-012-2222-1

Whittemore K, Derevyanko A, Martinez P et al (2019) Telomerase gene therapy ameliorates the effects of neurodegeneration associated to short telomeres in mice. Aging (Albany NY) 11(10):2916–2948. https://doi.org/10.18632/aging.101982

Wood JG, Helfand SL (2013) Chromatin structure and transposable elements in organismal aging. Front Genet 4:274. https://doi.org/10.3389/fgene.2013.00274

Wood JG, Hillenmeyer S, Lawrence C et al (2010) Chromatin remodeling in the aging genome of drosophila. Aging Cell 9(6):971–978. https://doi.org/10.1111/j.1474-9726.2010.00624.x

Wyss-Coray T (2016) Ageing, neurodegeneration and brain rejuvenation. Nature 539(7628):180–186. https://doi.org/10.1038/nature20411

Yang C, Wang Z, Kang Y et al (2023) Stress granule homeostasis is modulated by TRIM21-mediated ubiquitination of G3BP1 and autophagy-dependent elimination of stress granules. Autophagy 19(7):1934–1951. https://doi.org/10.1080/15548627.2022.2164427

Yu CE, Oshima J, Fu YH et al (1996) Positional cloning of the Werner's syndrome gene. Science 272(5259):258–262. https://doi.org/10.1126/science.272.5259.258

Yu X, Liu MM, Zheng CY et al (2023) Telomerase reverse transcriptase and neurodegenerative diseases. Front Immunol 14:1165632. https://doi.org/10.3389/fimmu.2023.1165632

Yushkova E, Moskalev A (2023) Transposable elements and their role in aging. Ageing Res Rev 86:101881. https://doi.org/10.1016/j.arr.2023.101881

Zagnoli-Vieira G, Bruni F, Thompson K et al (2018) Confirming TDP2 mutation in spinocerebellar ataxia autosomal recessive 23 (SCAR23). Neurol Genet 4(4):e262. https://doi.org/10.1212/NXG.0000000000000262

Zahodne LB, Stern Y, Manly JJ (2015) Differing effects of education on cognitive decline in diverse elders with low versus high educational attainment. Neuropsychology 29(4):649–657. https://doi.org/10.1037/neu0000141

Zhao N, Yin G, Liu C et al (2023) Critically short telomeres derepress retrotransposons to promote genome instability in embryonic stem cells. Cell Discov 9(1):45. https://doi.org/10.1038/s41421-023-00538-y

Zhou Y, Cambareri EB, Kinsey JA (2001) DNA methylation inhibits expression and transposition of the Neurospora tad retrotransposon. Mol Gen Genomics 265(4):748–754. https://doi.org/10.1007/s004380100472

Chapter 12
Retrotransposons: Silent Architects of Vascular Aging and Disease

Sudip Kumar Paul and Naoya Takayama

Abstract Retrotransposons (RTE) are emerging as critical players in vascular aging and disease. These genetic elements, capable of self-replication through RNA intermediates, remain largely silent under normal conditions but can be activated in response to various stressors, including aging. The double-stranded RNA (dsRNA) derived from reactivated RTEs triggers innate immune responses, contributing to the pathophysiology of vascular diseases such as atherosclerosis. Our recent study has demonstrated that RTE-derived dsRNAs can exacerbate vascular dysfunction by activating inflammatory pathways, disrupting endothelial cell function, promoting vascular smooth muscle cell proliferation, and enhancing vascular smooth muscle cell phenotype switching. These modulations occur within the vascular cells under the influence of other immune cells, like macrophages. In this chapter, we explore the emerging role of RTEs in vascular aging and disease, highlighting their potential as biomarkers for early detection and their therapeutic implications. We discuss the mechanistic insights into RTE activation, their contribution to vascular inflammation, and potential strategies to modulate RTE activity. Targeting these silent genomic elements may lead to novel therapeutic approaches that mitigate vascular aging and improve outcomes in patients with vascular diseases.

Keywords Retrotransposons · Atherosclerosis · Transposable elements · Vascular diseases · LINEs · SINEs · ERVs

Abbreviations

ATF3	Activating transcription factor 3
dsRNA	Double-stranded RNA
ERV	Endogenous retroviruses

S. K. Paul · N. Takayama (✉)
Department of Regenerative Medicine, Graduate School of Medicine, Chiba University, Chiba, Japan

L. M. Vaschetto (ed.), *Transposable Elements, Transcriptomics, and Diseases*, Epigenetics and Human Health 14, https://doi.org/10.1007/978-3-032-04125-8_12

HERV-K	Human endogenous retrovirus K
ICAM-1	Intercellular adhesion molecule 1
IFN	Interferon
IL-6	Interleukin-6
iPSC	Induced pluripotent stem cell
LINE	Long interspersed nuclear elements
LTR	Long terminal repeats
MAVS	Mitochondrial antiviral signaling protein
Mφ	Macrophages
NE	Neutrophil elastase
PAEC	Pulmonary arterial endothelial cell
PAH	Pulmonary arterial hypertension
RTE	Retrotransposons
SA-ERV	Senescence-associated endogenous retrovirus
SINE	Short interspersed nuclear elements
TE	Transposable elements
TNF	Tumor necrosis factor
VEC	Vascular endothelial cell
VSMC	Vascular smooth muscle cell
WS	Werner syndrome

12.1 Introduction: Retrotransposons and Their Hidden Influence

Transposable elements (TEs), often referred to as "jumping genes," are DNA sequences capable of moving from one genomic location to another. This mobility can sometimes lead to mutations, reverse genetic changes, or alterations to a cell's genetic identity and even genome size. These elements were first identified over six decades ago by geneticist Barbara McClintock, who observed their role in hereditary traits, such as the diverse coloration patterns of corn kernels. In subsequent years, it became evident that TEs are not only capable of "jumping," but they are widespread across nearly all organisms, both prokaryotic and eukaryotic, and are present in substantial numbers. For instance, TEs comprise roughly 50% of the human genome and as much as 90% of the maize genome (SanMiguel et al. 1996).

TEs are classified into two main groups: DNA transposons, which utilize a transposase enzyme for insertion and excision, and retrotransposons (RTEs), which replicate via reverse transcription. RTEs are ancient genetic elements propagating by reverse-transcribing RNA into DNA and integrating it into new sites within the genome. They can be divided into two main classes: long terminal repeat (LTR) retrotransposons and non-long terminal repeat (non-LTR) retrotransposons. LTR retrotransposons include endogenous retroviruses (ERVs), whereas non-LTR retrotransposons include long interspersed nuclear elements (LINEs) and short interspersed nuclear elements (SINEs). Typically, RTEs are kept inactive through heterochromatin-mediated repression (de Koning et al. 2011; Goodier 2016; Bulut-Karslioglu et al. 2014). However, under conditions such as aging, epigenetic dysregulation, or disease, RTEs can become reactivated, leading to outcomes such as aberrant transcription, inflammation, alternative splicing, insertional mutagenesis, DNA damage, and genomic instability (Huang et al. 2012; Liu et al. 2023).

Mechanistically, reactivated RTEs can influence pathology through several pathways. First, RTE insertions can induce mutations, which disrupt coding or regulatory regions, potentially activating oncogenes or silencing tumor suppressors, thereby contributing to cancer progression (Iskow et al. 2010; Shukla et al. 2013; Tufarelli et al. 2013). Second, RTE activity can mimic viral infection via "viral mimicry," whereby RTE-derived RNA or DNA is recognized by innate immune sensors, such as RIG-I-like receptors or cyclic GMP-AMP synthase (cGAS), leading to type I interferon (IFN) signaling and chronic inflammation (Roulois et al. 2015; Chiappinelli et al. 2015). This inflammatory response is increasingly implicated in the development of autoimmune diseases and age-related conditions (De Cecco et al. 2019; Liu et al. 2023; Paul et al. 2024; Mao et al. 2024; Grandi and Tramontano 2018; Tokuyama et al. 2021; Freimanis et al. 2010). Third, LTR retrotransposons can act as promoters or enhancers, driving the expression of nearby genes (Hu et al. 2017; Nishihara 2020; Medstrand et al. 2001). For example, the LTR-driven overexpression of proto-oncogenes or immune modulatory genes can disrupt normal cellular functions and exacerbate inflammatory or proliferative disorders (Lamprecht et al. 2010; Mommert et al. 2018). Additionally, persistent RTE activity contributes to genomic instability, which manifests as chromosomal rearrangements, deletions, and copy number variations, further amplifying pathological processes.

Despite being largely silent in modern genomes, the evolutionary significance of RTEs and their impact on genome architecture are profound. Emerging evidence has linked RTEs to a range of pathological conditions, including cardiovascular diseases like atherosclerosis, cancers, aging, and cellular homeostasis (Wang et al. 2011; Chiappinelli et al. 2015; Simon et al. 2019; De Cecco et al. 2019; Saito et al. 2017; Taylor et al. 2022; Liu et al. 2023; Paul et al. 2024; Mao et al. 2024). Elevated ERVs, for example, are known to trigger innate immune responses and induce vascular changes associated with pulmonary arterial hypertension (Taylor et al. 2022; Saito et al. 2017). The involvement of ERVs in immune regulation is further demonstrated by the increased expression of ERVB7-1.LTR-MM, a type of RTE that has been shown to modulate immune responses in cardiomyopathy (Chen et al. 2024). LINEs and SINEs have also been implicated in age-related diseases, including atherosclerosis (Woo et al. 2012; Roy et al. 2024; Sharysh et al. 2020; Wei et al. 2014). Polymorphisms within the sequences of Alu, a type of SINE, such as the ACE −/− genotype, are linked to reduced lifespan, atherosclerosis progression, and gender-specific cardiovascular risks in the elderly Chinese population (Woo et al. 2012). More recent evidence links the dysfunction of vascular homeostasis to RTE reactivation in macrophages (Mφ), leading to the induction of type I interferon (IFN) signaling in age-related cardiovascular conditions such as atherosclerosis (Paul et al. 2024).

This chapter delves into the emerging roles of RTEs as "silent architects" of cardiovascular aging and disease, focusing on their contributions to immune dysregulation and genomic instability. By highlighting the mechanistic insights and pathological consequences of RTE activity, we aim to shed light on their potential as biomarkers and therapeutic targets in cardiovascular health.

12.2 Retrotransposons and Innate Immunity During Aging

Aging is associated with a gradual decline of physiological integrity, which is one of the key contributors to human pathologies, including atherosclerosis. Key mechanisms contributing to aging include telomere shortening, epigenetic dysregulation, genomic instability, disruptions in protein homeostasis, and alterations in signaling pathways. While protein-coding genes have been extensively studied to understand these processes, growing attention has recently turned to the role of RTEs in the human genome, highlighting their significant involvement in the aging process (De Cecco et al. 2019; Simon et al. 2019; Liu et al. 2023; Mao et al. 2024). RTEs usually remain silent by maintaining repressive heterochromatin formation across the genome (Goodier 2016; Bulut-Karslioglu et al. 2014). As we age, the epigenome undergoes extensive reorganization, including alterations in DNA methylation patterns, histone modifications, and chromatin remodeling (López-Otín et al. 2013). With aging, the maintenance of repressive heterochromatin weakens, leading to reduced genomic surveillance and an increased possibility of the derepression of normally silenced RTEs and other protein-coding genes. Derepressed RTEs can produce complementary DNA (cDNA) or RNA transcripts, which later contribute to double-stranded RNA (dsRNA) formation. These cDNAs and dsRNAs are subsequently recognized by DNA sensors (e.g., cGAS, AIM2) and RNA sensors (e.g., RIG-I, MDA5), respectively, resulting in the activation of innate immune responses, primarily through IFN signaling pathways (Fig. 12.1).

Indeed, studies have shown that RTEs, particularly long interspersed nuclear elements-1 (LINE-1), a type of LINE element, become derepressed during cellular senescence, a hallmark of aging, triggering a type I IFN response via the cGAS-STING-dependent cytosolic DNA sensing pathway. This type I IFN response, which is linked to late senescence, sustains the senescence-associated secretory phenotype (SASP) and is driven by cytoplasmic LINE-1 cDNA. Inhibiting LINE-1 reverse transcriptase reduces type I IFN activation and age-related inflammation (inflammaging). Treatment with reverse transcriptase inhibitors, such as lamivudine, has shown promise in mitigating inflammaging and sterile inflammation in aged mice (De Cecco et al. 2019). Additionally, ERVs, particularly human ERV-K (HERV-K), are emerging as important inducers of cGAS-STING-driven type I IFN signaling and have been reported to become reactivated in senescent cells, producing retrovirus-like particles that transmit senescence signals to young cells (Liu

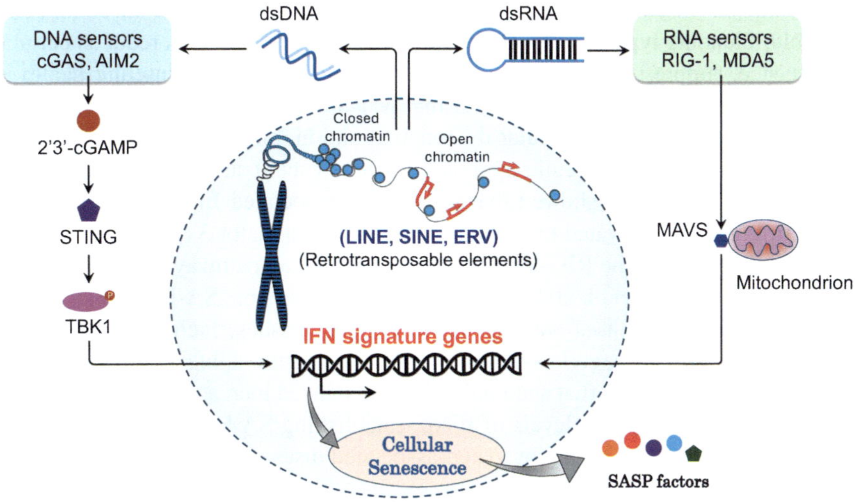

Fig. 12.1 The fate of retrotransposable element resurrection. A schematic representation of the fate of retrotransposable element (RTE) reactivation. Aging-associated epigenetic changes result in an open chromatin state, derepressing RTEs that are typically silenced in closed chromatin. Reactivated RTEs generate double-stranded DNA (dsDNA) or double-stranded RNA (dsRNA) that are sensed by DNA or RNA sensors. This sensing triggers respective pathways that ultimately activate the transcription of interferon (IFN) signature genes, leading to senescence and inflammation. SASP, senescence-associated secretory phenotype

et al. 2023). Therefore, RTEs impact the cells in which they become reactivated and extend their influence to neighboring young cells by transmitting senescence-inducing signals. This ERV resurrection, which is also observed in aged human tissues, primates, and mice, accelerates cellular senescence and tissue aging. Interestingly, blocking ERV activity mitigates senescence, tissue degeneration, and aging, establishing ERV reactivation as both a hallmark and a driver of aging (Liu et al. 2023).

In addition to the cGAS-STING-dependent cytosolic DNA sensing pathway, RTE-derived dsRNAs were reported to trigger a type I IFN response via the RIG-I/MDA5–MAVS signaling pathway during aging, leading to cellular senescence and inflammation (Paul et al. 2024; Mao et al. 2024). Werner syndrome (WS), a premature aging disorder, is caused by mutations in the *WRN* gene, which encodes the Werner protein—a member of the human helicase family essential for maintaining genomic stability, DNA replication, DNA repair, and telomere maintenance. Paul and colleagues showed that WS is associated with the aging-related derepression of RTEs. WS patient-specific induced pluripotent stem cell (iPSC)-derived Mφ (iMφs) exhibited an inverse correlation between H3K9me3 levels and RTE activity, indicative of epigenetic dysregulation. This dysregulation led to the derepression of RTEs followed by the activation of type I IFN signaling via dsRNA-mediated activation of the *DHX58*–MAVS pathway. In iMφs, these derepressed RTEs drove cellular senescence and inflammation and hampered cellular homeostasis. They also compromised vascular health when iMφs were co-cultured with vascular cells. Notably, silencing type I IFN signaling restored iMφ proliferation, reduced cellular senescence, suppressed inflammation, and ultimately improved vascular health in the co-culture. These findings suggest that Mφ-specific RTE reactivation contributes to type I IFN-driven vascular dysfunction and highlight IFN signaling inhibition as a promising therapeutic approach for atherosclerosis in WS patients (Paul et al. 2024). Mao et al. showed that senescence-associated ERVs (SA-ERVs) are transcriptionally reactivated by activating transcription factor 3 (ATF3), generating dsRNAs that activate the RIG-I/MDA5–MAVS signaling pathway, triggering a type I IFN response (Mao et al. 2024). Elevated ATF3 expression, SA-ERV reactivation, and type I IFN responses were observed in aged tissues, including those from Hutchinson–Gilford progeria syndrome, also a premature aging disorder, patients. Additionally, they found that aged individuals displayed increased SA-ERV-derived dsRNAs and higher blood levels of IFNβ, establishing SA-ERVs as a hallmark of aging and a potential contributor to age-associated diseases.

12.3 Impact of Retrotransposons on Vascular Aging

A well-established risk factor for cardiovascular diseases is aging, which is closely linked to progressive structural and functional alterations in the vasculature. These changes include circulation anomalies driven by elevated oxidative stress, premature cellular senescence, and disruptions in the synthesis and secretion of endothelium-derived vasoactive molecules. Such impairments collectively compromise vascular homeostasis, contributing to the pathogenesis of age-related vascular disorders such as atherosclerosis. Besides dietary and lifestyle changes, aging-related biological processes, such as vascular aging, also contribute to the increased risk of developing atherosclerosis. Traditionally, studies on cardiovascular diseases have focused predominantly on determinants derived from protein-coding genes. However, increasing attention is being directed toward non-coding regions of the genome, such as RTEs, which hold critical insights into the molecular mechanisms driving vascular aging and its associated pathologies (Lu et al. 2000; Lu and Ramos 2003; Banaz-Yaşar et al. 2010; Lacey et al. 2015; Saito et al. 2017; Hueso et al. 2018; Taylor et al. 2022; Paul et al. 2024).

Reactivated RTEs have been shown to influence key cell types involved in atherosclerosis. Myeloid cells, such as Mφ, and vascular cells, including vascular endothelial cells (VECs) and vascular smooth muscle cells (VSMCs), are critical for maintaining vascular homeostasis. Dysfunction in any of these cell types can compromise vascular integrity and health (Paul et al. 2024).

In this chapter, we describe the association between RTE and specific diseases.

12.3.1 Atherosclerosis

RTEs are emerging as significant contributors to atherosclerosis. Studies indicate that RTE reactivation in Mφ can exacerbate inflammation by generating dsRNA, which triggers type I IFN responses through RNA-sensing pathways such as *DHX58* (Paul et al. 2024). Recently, we established an in vitro co-culture system for iMφs, iPSC-derived VECs (iVECs), and iPSC-derived VSMCs (iVSMCs) to uncover the mechanisms underlying atherosclerosis in WS patients, which was previously unresolved due to the failure of WS mouse models to exhibit the atherosclerotic phenotype observed in human patients, limiting their utility in studying disease progression (Chang et al. 2004). In addition, a recent study indicates that *WRN*-deficient embryonic stem cells differentiated into endothelial cells do not display signs of premature aging (Wu et al. 2018). Consistently, our WS-iVECs and WS-iVSMCs did not exhibit abnormalities in monoculture conditions. However, our co-culture system revealed that WS-iMφs induce endothelial dysfunction in WS-iVECs and a synthetic phenotype in WS-iVSMCs, mimicking the early stages of atherosclerotic changes and highlighting Mφ as central players in WS-associated vascular pathology (Paul et al. 2024).

One of the key features of early atherosclerosis is the increased adhesion of Mφ to VECs, along with inflammation and breakdown of the extracellular matrix (Lee et al. 2019). In our study, WS-iMφs adhered more strongly to WS-iVECs compared to the interaction between healthy iMφs and healthy iVECs. This enhanced adhesion was associated with higher expression of the adhesion molecule ICAM1 and increased secretion of inflammatory cytokines such as IL-6 and TNF-α in WS-iVECs after the co-culture. Additionally, WS-iVECs showed elevated activity of matrix metalloproteinase-1 (MMP-1), indicating that inflammation in these cells may drive degradation of the extracellular matrix (Paul et al. 2024). In addition, using a porcine aortic endothelial cell model, de novo retrotransposition events by LINE-1 were found to reduce endothelial cell proliferation and migration, causing G0/G1 cell cycle arrest. Further analysis revealed that LINE-1 can selectively suppress key angiogenic factors, including vascular endothelial growth factor and angiopoietin receptor Tie-2, suggesting that LINE-1 activity inhibits angiogenesis and tumor vascularization by impairing endothelial cell function (Banaz-Yaşar et al. 2010).

Traditionally, VSMCs have been viewed as major contributors to atherosclerosis due to their aberrant proliferation, which promotes plaque formation. A hallmark of atherosclerosis is the phenotypic switching of VSMCs from a contractile to an adverse synthetic phenotype, characterized by increased proliferation and extracellular matrix production (Rensen et al. 2007). Benzo(a)pyrene and its oxidative metabolites activate the gene L1Md, which codes for a type of LINE element, in a concentration-dependent manner in murine VSMCs, leading to switching from a quiescent to proliferative phenotype both in vivo and in vitro (Lu et al. 2000). In addition, we showed WS-iMφs promoted a switch from the contractile to the synthetic phenotype in WS-iVSMCs, which was evident by the increased cell proliferation and decreased gene expression of contractile markers (e.g., *CNN1*, *ACTA2*).

The phenotype switch was further validated by reduced calponin-1 protein levels in the co-cultured WS-iVSMCs (Paul et al. 2024). Synthetic VSMCs contribute to vascular calcification by differentiating into osteoblast-like cells, depositing calcium in vessel walls, accelerating plaque instability, and increasing the risk of thrombosis and plaque rupture (Durham et al. 2018). Also, they are prone to apoptosis, necrosis, and senescence, all of which contribute to inflammation and plaque vulnerability (Bennett et al. 2016). These findings highlight the potential role of RTEs in regulating vascular dysfunction by reducing the angiogenic capacity of VECs and phenotype switching of VSMCs.

Additional cross-co-culture experiments revealed that WS-iMφs are the primary drivers of vascular inflammation, independent of the vascular cell type (iVECs and iVSMCs). Direct Mφ-vascular cell interactions, rather than secreted factors alone, were necessary to induce inflammatory and phenotypic changes, underscoring the orchestrating role of WS-iMφs in vascular dysfunction (Paul et al. 2024). This effect was attributed to heightened type I IFN signaling and RTE-derived dsRNA activating inflammatory pathways. However, these findings are yet to be validated in mouse models; such epigenetic dysregulation is expected to amplify inflammatory signaling and contribute to endothelial injury and plaque formation, hallmark events in atherosclerosis progression.

12.3.2 Pulmonary Arterial Hypertension-Induced Vascular Remodeling

Pulmonary arterial hypertension (PAH)-induced vascular remodeling involves structural changes in the arterial wall, such as increased vascular stiffness and reduced compliance. Emerging evidence links RTE activation to these pathological processes. A recent study identified HERV-K reactivation as a significant contributor to immune dysregulation and vascular remodeling in PAH. Elevated HERV-K expression was observed in lung immune complexes of PAH patients, particularly in endothelial cells, dendritic cells, and Mφs. SAM domain and HD domain-containing protein 1 (SAMHD1) functions as an innate immune regulator and is known for its role in restricting HIV replication. Saito and colleagues reported the presence of SAMHD1 immune complexes in the lungs of patients diagnosed with PAH (Saito et al. 2017). This unexpected finding highlighted elevated levels of HERV-K gene products, specifically HERV-K envelope proteins and deoxyuridine triphosphate nucleotide hydrolase (dUTPase), in PAH-affected lungs compared to control lungs. A further analysis demonstrated an increased expression of HERV-K dUTPase in circulating monocytes derived from PAH patients and in iPSCs induced from the fibroblasts of PAH patients. Notably, treating monocytes and pulmonary arterial endothelial cells (PAECs) with HERV-K dUTPase triggered SAMHD1 upregulation and elevated cytokine production, including IL-6. This cascade not only exacerbated inflammation but also promoted apoptosis in PAECs, contributing to vascular remodeling—a hallmark of PAH. In vivo studies reinforced these findings, as rats exposed to HERV-K dUTPase developed pulmonary hypertension, mirroring the pathological features observed in PAH patients (Saito et al. 2017). These results underscore the critical role of HERV-K elements in vascular dysfunction and suggest that targeting HERV-K dUTPase may represent a novel therapeutic avenue for managing PAH.

Taylor et al. observed that ERVs can become transcriptionally active in neutrophils, triggering innate immune responses through type I IFN signaling in PAH (Taylor et al. 2022). Neutrophils from PAH patients show an increased production and release of neutrophil elastase (NE), thereby promoting the enhanced formation of neutrophil extracellular traps. These neutrophils exhibit heightened adhesion and impaired migration, which are correlated with elevated β1-integrin and vinculin levels. Proteomic and transcriptomic analyses identified an IFN signature in PAH neutrophils that was driven by the upregulated expression of HERV-K envelope proteins. The overexpression of the HERV-K envelope in neutrophil-like cells amplified NE and IFN gene expression, while HERV-K dUTPase enhanced vinculin levels. CD66b-specific extracellular vesicles (EVs) from PAH-patient neutrophils were enriched in NE and HERV-K envelope proteins. When administered to mice, these EVs induced PAH, an effect that was reduced by elafin, an NE inhibitor. Overall, Taylor et al. suggest that elevated levels of human ERVs and elastase are associated with a neutrophil-mediated innate immune response in PAH.

12.3.3 Hypertension

Blood pressure, a well-established intermediary biomarker, is recognized as a predictor of cardiovascular events (Fuchs and Whelton 2020). Chronic arsenic exposure has been linked to LINE-1 hypomethylation and increased blood pressure, suggesting a connection between environmental toxins, epigenetic dysregulation, and vascular health. A cross-sectional study conducted in Bangladesh investigated the relationship between arsenic exposure, LINE-1 methylation levels, and blood pressure in a population with varying degrees of exposure to arsenic-contaminated groundwater (Hossain et al. 2017). The study revealed that higher arsenic exposure was associated with reduced LINE-1 DNA methylation at CpG sites, which correlated with elevated blood pressure. Another study examining DNA methylation patterns of retrotransposons and pro-inflammatory genes in an elderly cohort, drawn from the Normative Aging Study, found a positive correlation between DNA methylation at Alu sites and high blood pressure (Alexeeff et al. 2013). This association suggests that alterations in gene expression may contribute to the mechanistic pathways affecting cardiovascular health during aging.

12.3.4 Aortic Aneurysm and Heart Valve Disease

An aortic aneurysm is a dilation or bulge in the aorta, the primary artery responsible for transporting oxygen-rich blood from the heart. Aneurysms can form in either the chest (thoracic aortic aneurysm) or the abdomen (abdominal aortic aneurysm). One study confirmed a significant association between SNPs rs10757278 and rs1333049 on chromosome 9p21.3 with abdominal aortic aneurysm (AAA) (Biros et al. 2010). Additionally, it reported the upregulation of LINE-1 RTE at the site of AAA, suggesting the potential involvement of LINE-1 RTE in the disease process, warranting further investigation to elucidate its role in the pathophysiology of AAA. Lastly, TNF plays a key role in various inflammatory diseases, including arthritis and psoriasis. A spontaneous retrotransposon insertion in the 3′UTR of Tnf has been shown to disrupt its normal regulation, leading to polyarthritis and heart valve disease (Lacey et al. 2015). This study identified CCCH-containing zinc-finger proteins as new regulators of Tnf expression and revealed a novel regulatory element within the TNF 3′UTR. The study also showed the mutations caused severe joint and heart complications in BPSM1 mice, a model for rheumatoid arthritis and ankylosing spondylitis.

12.4 Biomarkers, Therapeutic Potential, and Future

The contributions of RTEs to vascular diseases have shed light on their potential to function as biomarkers for disease progression and as therapeutic targets.

12.4.1 RTE-Derived dsRNAs as Biomarkers for Vascular Disease

The correlation between heightened dsRNA production and activation of innate immunity via the type I IFN signaling pathway underscores the central role of RTEs in the pathophysiology of vascular diseases (Paul et al. 2024). Furthermore, several studies suggest dsRNA as a potential biomarker of aging (Liu et al. 2023; Mao et al. 2024). These observations highlight RTE-derived dsRNAs as potential biomarkers for diagnosing vascular disorders. Specifically, the presence of RTE-derived dsRNAs could serve as a molecular signature of vascular inflammation and injury, allowing for early detection and monitoring of disease progression. In fact, circulating levels of these dsRNAs may reflect the severity of vascular damage, providing a non-invasive means of assessing disease status. Moreover, the detection of specific RTE transcripts could enable the identification of individuals at high risk for developing vascular diseases, including those with underlying genetic predispositions to such conditions.

12.4.2 *Therapeutic Implications of Targeting Retrotransposon Activity*

Given the emerging evidence linking RTE activation to vascular disease, targeting RTE activity holds great promise as a therapeutic strategy. Several studies have demonstrated that modulating RTE activity can reduce inflammation (Mao et al. 2024), mitigate vascular remodeling, and even prevent thrombosis (Taylor et al. 2022). For example, inhibitors of reverse transcriptase, the enzyme responsible for converting retrotransposon RNA into DNA, have shown potential in reducing the effects of RTE-induced inflammation (De Cecco et al. 2019; Taylor et al. 2022). Reverse transcriptase inhibitors, which are already used in the treatment of HIV, could be repurposed to target RTEs and their inflammatory consequences in vascular diseases. Such interventions could potentially halt or reverse the progression of vascular damage, particularly in diseases driven by chronic inflammation such as atherosclerosis.

The therapeutic potential of targeting RTEs extends beyond reverse transcriptase inhibitors. Other molecular strategies, such as the use of small interfering RNAs (siRNAs) or CRISPR-based technologies, could be employed to specifically silence RTE activity (Mao et al. 2024). These strategies could be tailored to reduce the expression of particular RTEs or their RNA products, thereby preventing the initiation of inflammatory cascades that lead to vascular dysfunction.

Understanding the molecular mechanisms that govern RTE activation is critical for the development of more targeted therapeutic interventions. For instance, modulating chromatin accessibility or regulating the host cellular machinery that triggers RTE activation could offer new avenues for treatment. By selectively blocking the pathways that lead to RTE expression, it may be possible to prevent the harmful effects of RTE-derived dsRNAs on vascular health without disrupting normal cellular processes.

As our understanding of RTEs in vascular diseases deepens, several emerging interventions hold promise for clinical application. The use of reverse transcriptase inhibitors, as mentioned earlier, represents one potential strategy. Other approaches, such as the inhibition of specific inflammatory pathways activated by RTEs, could also prove effective in reducing vascular damage. Additionally, targeting the cellular machinery that controls RTE silencing, such as the use of epigenetic modulators, may offer a more precise and durable solution for managing vascular diseases associated with RTE activity.

12.5 Conclusion: Silent Architects Shaping Vascular Health

RTE-derived dsRNAs offer significant potential as biomarkers for vascular diseases and as therapeutic targets. Their ability to exacerbate inflammation and vascular dysfunction positions them as crucial players in disease progression. The therapeutic implications of targeting RTE activity, particularly through the use of reverse transcriptase inhibitors and other molecular interventions, provide exciting new avenues for the treatment of vascular diseases. As research continues, future studies should focus on refining these therapeutic strategies and exploring their efficacy in clinical settings, ultimately providing new tools to combat vascular diseases and improve patient outcomes.

Compliance with Ethical Standards

Conflict of Interest The authors declare that they have no conflict of interest.

Ethical Approval This chapter contains no studies with human/animal participants performed by the authors.

References

Alexeeff SE, Baccarelli AA, Halonen J et al (2013) Association between blood pressure and DNA methylation of retrotransposons and pro-inflammatory genes. Int J Epidemiol 42:270–280

Banaz-Yaşar F, Steffen G, Hauschild J et al (2010) LINE-1 retrotransposition events affect endothelial proliferation and migration. Histochem Cell Biol 134:581–589

Bennett MR, Sinha S, Owens GK (2016) Vascular smooth muscle cells in atherosclerosis. Circ Res 118:692–702

Biros E, Cooper M, Palmer LJ et al (2010) Association of an allele on chromosome 9 and abdominal aortic aneurysm. Atherosclerosis 212:539–542

Bulut-Karslioglu A, De La Rosa-Velázquez IA, Ramirez F et al (2014) Suv39h-dependent H3K9me3 marks intact retrotransposons and silences LINE elements in mouse embryonic stem cells. Mol Cell 55:277–290

Chang S, Multani AS, Cabrera NG et al (2004) Essential role of limiting telomeres in the pathogenesis of Werner syndrome. Nat Genet 36:877–882

Chen S, Jiang F, Wu J et al (2024) Identification of key TE associated with myocarditis based on RNA and single-cell sequencing data mining. Reprod Breed 4:102–109

Chiappinelli KB, Strissel PL, Desrichard A et al (2015) Inhibiting DNA methylation causes an interferon response in cancer via dsRNA including endogenous retroviruses. Cell 162:974–986

De Cecco M, Ito T, Petrashen AP et al (2019) L1 drives IFN in senescent cells and promotes age-associated inflammation. Nature 566:73–78

de Koning APJ, Gu W, Castoe TA et al (2011) Repetitive elements may comprise over two-thirds of the human genome. PLoS Genet 7:e1002384

Durham AL, Speer MY, Scatena M et al (2018) Role of smooth muscle cells in vascular calcification: implications in atherosclerosis and arterial stiffness. Cardiovasc Res 114:590–600

Freimanis G, Hooley P, Ejtehadi HD et al (2010) A role for human endogenous retrovirus-K (HML-2) in rheumatoid arthritis: investigating mechanisms of pathogenesis. Clin Exp Immunol 160:340–347

Fuchs FD, Whelton PK (2020) High blood pressure and cardiovascular disease. Hypertension 75:285–292

Goodier JL (2016) Restricting retrotransposons: a review. Mob DNA 7:16

Grandi N, Tramontano E (2018) HERV envelope proteins: physiological role and pathogenic potential in cancer and autoimmunity. Front Microbiol 9:462

Hossain K, Suzuki T, Hasibuzzaman MM et al (2017) Chronic exposure to arsenic, LINE-1 hypomethylation, and blood pressure: a cross-sectional study in Bangladesh. Environ Health 16:20. https://doi.org/10.1186/s12940-017-0231-7

Hu T, Zhu X, Pi W et al (2017) Hypermethylated LTR retrotransposon exhibits enhancer activity. Epigenetics 12:226–237

Huang CRL, Burns KH, Boeke JD (2012) Active transposition in genomes. Annu Rev Genet 46:651–675

Hueso M, Cruzado JM, Torras J, Navarro E (2018) ALU minating the path of atherosclerosis progression: chaos theory suggests a role for Alu repeats in the development of atherosclerotic vascular disease. Int J Mol Sci 19:1734. https://doi.org/10.3390/ijms19061734

Iskow RC, McCabe MT, Mills RE et al (2010) Natural mutagenesis of human genomes by endogenous retrotransposons. Cell 141:1253–1261

Lacey D, Hickey P, Arhatari BD et al (2015) Spontaneous retrotransposon insertion into TNF 3’UTR causes heart valve disease and chronic polyarthritis. Proc Natl Acad Sci USA 112:9698–9703

Lamprecht B, Bonifer C, Mathas S (2010) Repeat-element driven activation of proto-oncogenes in human malignancies. Cell Cycle 9:4276–4281

Lee SJ, Baek SE, Jang MA, Kim CD (2019) SIRT1 inhibits monocyte adhesion to the vascular endothelium by suppressing Mac-1 expression on monocytes. Exp Mol Med 51:1–12

Liu X, Liu Z, Wu Z et al (2023) Resurrection of endogenous retroviruses during aging reinforces senescence. Cell 186:287–304.e26

López-Otín C, Blasco MA, Partridge L et al (2013) The hallmarks of aging. Cell 153:1194–1217

Lu KP, Ramos KS (2003) Redox regulation of a novel L1Md-A2 retrotransposon in vascular smooth muscle cells. J Biol Chem 278:28201–28209

Lu KP, Hallberg LM, Tomlinson J, Ramos KS (2000) Benzo(a)pyrene activates L1Md retrotransposon and inhibits DNA repair in vascular smooth muscle cells. Mutat Res 454:35–44

Mao J, Zhang Q, Zhuang Y et al (2024) Reactivation of senescence-associated endogenous retroviruses by ATF3 drives interferon signaling in aging. Nat Aging 4:1794–1812

Medstrand P, Landry JR, Mager DL (2001) Long terminal repeats are used as alternative promoters for the endothelin B receptor and apolipoprotein C-I genes in humans. J Biol Chem 276:1896–1903

Mommert M, Tabone O, Oriol G et al (2018) LTR-retrotransposon transcriptome modulation in response to endotoxin-induced stress in PBMCs. BMC Genomics 19:522

Nishihara H (2020) Transposable elements as genetic accelerators of evolution: contribution to genome size, gene regulatory network rewiring and morphological innovation. Genes Genet Syst 94:269–281

Paul SK, Oshima M, Patil A et al (2024) Retrotransposons in Werner syndrome-derived macrophages trigger type I interferon-dependent inflammation in an atherosclerosis model. Nat Commun 15:4772

Rensen SSM, Doevendans PAFM, van Eys GJJM (2007) Regulation and characteristics of vascular smooth muscle cell phenotypic diversity. Neth Heart J 15:100–108

Roulois D, Loo Yau H, Singhania R et al (2015) DNA-demethylating agents target colorectal cancer cells by inducing viral mimicry by endogenous transcripts. Cell 162:961–973

Roy N, Haq I, Ngo JC et al (2024) Elevated expression of the retrotransposon LINE-1 drives Alzheimer’s disease-associated microglial dysfunction. Acta Neuropathol 148:75

Saito T, Miyagawa K, Chen S-Y et al (2017) Upregulation of human endogenous retrovirus-K is linked to immunity and inflammation in pulmonary arterial hypertension. Circulation 136:1920–1935

SanMiguel P, Tikhonov A, Jin YK et al (1996) Nested retrotransposons in the intergenic regions of the maize genome. Science 274:765–768

Sharysh DV, Markov AV, Sleptcov AA et al (2020) LINE-1 retrotransposon demethylation is associated with atherosclerotic plaque destabilization. Med Genet 19(5):50–51

Shukla R, Upton KR, Muñoz-Lopez M et al (2013) Endogenous retrotransposition activates oncogenic pathways in hepatocellular carcinoma. Cell 153:101–111

Simon M, Van Meter M, Ablaeva J et al (2019) LINE1 derepression in aged wild-type and SIRT6-deficient mice drives inflammation. Cell Metab 29:871–885.e5

Taylor S, Isobe S, Cao A et al (2022) Endogenous retroviral elements generate pathologic neutrophils in pulmonary arterial hypertension. Am J Respir Crit Care Med 206(8):1019–1034

Tokuyama M, Gunn BM, Venkataraman A et al (2021) Antibodies against human endogenous retrovirus K102 envelope activate neutrophils in systemic lupus erythematosus. J Exp Med 218:e20191766. https://doi.org/10.1084/jem.20191766

Tufarelli C, Cruickshanks HA, Meehan RR (2013) LINE-1 activation and epigenetic silencing of suppressor genes in cancer: causally related events? Mob Genet Elements 3:e26832

Wang J, Geesman GJ, Hostikka SL et al (2011) Inhibition of activated pericentromeric SINE/Alu repeat transcription in senescent human adult stem cells reinstates self-renewal. Cell Cycle 10:3016–3030

Wei L, Liu S, Su Z et al (2014) LINE-1 hypomethylation is associated with the risk of coronary heart disease in the Chinese population. Arq Bras Cardiol 102:481–488

Woo J, Tang NLS, Leung J, Kwok T (2012) The Alu polymorphism of angiotensin I converting enzyme (ACE) and atherosclerosis, incident chronic diseases and mortality in an elderly Chinese population. J Nutr Health Aging 16:262–268

Wu Z, Zhang W, Song M et al (2018) Differential stem cell aging kinetics in Hutchinson-Gilford progeria syndrome and Werner syndrome. Protein Cell 9:333–350

Chapter 13
The Role of Transposable Elements in Therapeutics

Kathryn E. Howe and Tara T. Doucet-O'Hare

Abstract Transposable elements (TEs) are pervasive genomic sequences that have significantly shaped genome evolution. Initially discovered by Barbara McClintock in maize, TEs are now understood to influence both coding and noncoding regions of the genome, contributing to physiological processes and, when dysregulated, to disease. In humans, TEs such as LINE-1 (L1), Alu elements, and human endogenous retroviruses (HERVs) are differentially regulated throughout development and cellular differentiation. While TE expression is tightly controlled under normal conditions, aberrant reactivation has been associated with a range of pathologies, including neurodegenerative disorders, autoimmune diseases, and malignancies. Recent advances have highlighted the potential of targeting TE activity for therapeutic purposes. Strategies under investigation include the use of small molecules, gene therapies to modulate TE expression, and immunotherapeutic approaches to eliminate cells expressing aberrant TE products. Antiretroviral drugs originally designed to inhibit HIV replication are also being repurposed to target HERV-mediated diseases. In contrast, some studies are attempting to induce expression of TEs to promote an immune response through viral mimicry and target TE-expressing cells for destruction in cancer. This chapter reviews the biological roles of TEs in disease, evaluates current and emerging therapeutic strategies aimed at modulating TE activity, and explores the potential of leveraging TE biology for the development of targeted and personalized therapies.

Keywords Transposable elements · Disease · Cancer · Viral mimicry · Gene therapy · Epigenetics · Anti-cancer drugs · Therapeutics

K. E. Howe · T. T. Doucet-O'Hare (✉)
Department of Genetics and Biochemistry and Center for Human Genetics, Clemson University, Greenwood, SC, USA
e-mail: tdoucet@clemson.edu

L. M. Vaschetto (ed.), *Transposable Elements, Transcriptomics, and Diseases*, Epigenetics and Human Health 14, https://doi.org/10.1007/978-3-032-04125-8_13

Abbreviations

3TC	An HIV drug cocktail containing zidovudine dolutegravir, and abacavir
AD	Alzheimer's disease
AIDS	Acquired immunodeficiency syndrome
AIF	Anoxia inducible factor
Alu	TE is named for its original characterization in *Arthrobacter luteus (Alu)*
AMP	Adenosine 3′,5′-monophosphate
ALS	Amyotrophic lateral sclerosis
APC	Adenomatous polyposis coli
AP-1	Activator protein 1
ART	Antiretroviral therapy
AT/RT	Atypical teratoid rhabdoid tumor
ASRGL1	Asparaginase and isoaspartyl peptidase 1
BBB	Blood brain barrier
B cells	B lymphocyte
Cas9	CRISPR-associated protein 9
ccRCC	Clear cell renal cell carcinoma
CD3	Cluster of differentiation 3
CD4	Cluster of differentiation 4
CD19	Cluster of differentiation 19
cDNA	Complementary deoxyribonucleic acid
c-myc	Cellular myelocytomatosis
cGAMP	Cyclic GMP-AMP
cGAS	Cyclic GMP-AMP synthase
cGAS-STING	Cyclic GMP-AMP synthase-stimulator of interferon genes
c-Jun	c-Jun N-terminal kinase (JNK)
CRISPR	Clustered regularly interspaced short palindromic repeats
DAMPs	Damage-associated molecular patterns
DHX58	DExH-box helicase 58
DJ-1	Parkinsonism-associated deglycase gene, which encodes PARK-7 protein
DNA	Deoxyribonucleic acid
DNMT	DNA methyltransferase
DNMTi	DNA methyltransferase inhibitor
DRD2	Dopamine receptor D2
Ds	Dissociation element
dsRNA	Double-stranded RNA
Dux	Family of transcription factors encoded by double homeobox
EBV	Epstein Barr virus
Env	Envelope
EN	Endonuclease

ERV	Endogenous retrovirus
ESC	Embryonic stem cell
FI	Fusion inhibitor
G4	G (guanine) quadruplex
Gag	Group-specific antigen
GMP	Guanosine 3′,5′-cyclic monophosphate
HDAC	Histone deacetylase
HDACi	Histone deacetylase inhibitor
HERV	Human endogenous retrovirus
HERV-E	Human endogenous retrovirus E
HERV-H	Human endogenous retrovirus H
HERV-K	Human endogenous retrovirus K
HERV-W	Human endogenous retrovirus W
HIF	Hypoxia-induced transcription factors
HIV	Human immunodeficiency virus
HIV-1	Human immunodeficiency virus type 1
HLA	Human leukocyte antigen
HML-2	Human endogenous MMTV-like
HNPCC	Hereditary non-polyposis colorectal cancer
ICW	In-cell western
INF-α	Interferon-α
INF-I	Interferon I
iPSC	Induced pluripotent stem cells
KO	Knock out
Kap1	Krüppel-associated box (KRAB)-associated protein 1
KS	Kaposi sarcoma
L1	LINE1
LINE	Long interspersed element
lncRNA	Long noncoding RNA
mAbs	Monoclonal antibodies
MAPT	Microtubule-associated protein tau
miRNA	Micro RNA
MMTV	Mouse mammary tumor virus
mRNA	Messenger RNA
MRI	Magnetic resonance imaging
MS	Multiple sclerosis
MSRV	Multiple sclerosis associated retrovirus
NEAT	Nuclear enriched abundant transcript 1
NK	Natural killer cells
Notch	Neurogenic locus notch homolog protein
Np-9	Nuclear protein of 9 KDa
NRTI	Nucleoside reverse transcriptase inhibitor
NNRTI	Non-nucleoside reverse transcriptase inhibitor
ORF	Open reading frame
ORF0	LINE-1 open reading frame protein 0

ORF1p	LINE-1 open reading frame protein 1
ORF2p	LINE-1 open reading frame protein 2
PAMPs	Pathogen-associated molecular patterns
PD	Parkinson's disease
PB	PiggyBac transposable element
PBMC	Peripheral blood mononuclear cell
PGBD5	PiggyBac transposable element derived 5
PI	Protease inhibitor
PIWI	P-element induced wimpy testis
piRNA	PIWI-interacting RNA
Pol	Polymerase
PLZF	Promyelocytic leukemia zinc finger protein
Ras	Rat sarcoma
RCC	Renal cell carcinoma
Rec	HERV functional homolog of HIV-1 Rev. and HTLV Rev.
RNA	Ribonucleic acid
RT	Reverse transcriptase
SARE	Secondary anoxia response element
SB	Sleeping beauty transposon
SET	Su(var)3–9, Enhancer-of-zeste and trithorax (name derived from *Drosophila*)
SETDB1	SET domain bifurcated histone lysine methyltransferase 1
SINE	Short interspersed nuclear element
SIRT6	Sirtuin 6
SLE	Systemic lupus erythematosus
SNCA	α-synuclein
STING	STimulator of INterferon Genes
STK19	Serine/threonine kinase 19
SVA	SINE-VNTR-Alu
Tat	Transactivator of transcription (from HIV)
T cell	Lymphocyte produced by the thymus
TLR-4	Toll-like receptor 4
TLR-8	Toll-like receptor 8
TAF1	Transcription initiation factor TFIID subunit 1, TBP-associated factor 250 kDa
TDP-43	Transactive response DNA binding protein 43 kDa
TE	Transposable element
Trim28	Tripartite motif-containing protein 28
VHL	von Hippel-Lindau tumor suppressor
VNTR	Variable number tandem repeat
XDP	X-linked dystonia parkinsonism

13.1 Introduction

Transposable elements (TEs) were first discovered by Barbara McClintock in 1944, as she worked in maize and observed that breakage occurred at specific sites on maize chromosomes (McClintock 1950). Through a series of genetic crosses, McClintock revealed chromosomal breakage at a locus she named *Dissociation (Ds)*. She observed that the breakage occurred repeatedly on chromosome 9 and led to variable pigmentation of the maize kernels (McClintock 1950). These findings were the first indication that the mobilization of these genetic elements could lead to disparate physiological outcomes and challenged the contemporaneous concept of a gene (Crow and Dove 2000). We now know that the process of transposition occurs in both noncoding and coding sections of the genome, including areas of the genome important for genetic regulation (Kazazian Jr. et al. 1988; Azlan et al. 2016; Yuan et al. 2010). In addition, TEs do not necessarily need to insert into a new region of the genome and disrupt a coding region, enhancer, or splice junction to cause a phenotype or pathology. For example, the expression of human endogenous retrovirus K (HERV-K) envelope (env) in amyotrophic lateral sclerosis (ALS) directly contributes to neurodegeneration (Li et al. 2015). Furthermore, the presence, prevalence, and activity of transposons can vary within a population, affect gene expression, and sometimes lead to disease (Bourgeois and Boissinot 2019; Marchi et al. 2014). Regardless of whether TE expression directly causes diseases and acts as the primary insult, or if TE expression is a secondary effect following other changes like chromatin remodeling and subsequent abnormal transcription, targeting cells that overexpress certain pathogenic TE transcripts or proteins is still a valuable therapeutic avenue. In this chapter, we will explore how therapeutic approaches targeting TE expression can be beneficial.

TEs have been incorporated into the genomes of many species throughout evolution and those that contain open reading frames and resultant proteins might be co-opted for other host functions. Although some TEs are specific to a single species (Wells and Feschotte 2020), there are also many examples of TEs that have been conserved in unrelated species, like vertebrates, for millions of years (Bejerano et al. 2006). There are two classes of TEs, DNA transposons and retrotransposons. Within the retrotransposon class there are several types of TEs, long interspersed elements (LINEs), short interspersed nuclear elements (SINEs), Alu elements, named after the restriction enzyme AluI, which cuts at specific DNA sequences and was originally characterized in *Arthrobacter luteus (Alu)*, and endogenous retroviruses (ERVs) (Schmid and Deininger 1975; Jay and Wu 1976). Two of the major distinctions between TE types are (1) the ability to self-mobilize (e.g., autonomous or non-autonomous elements) and (2) the presence of open reading frames that can lead to protein expression. LINE 1 (L1) elements are the only known potentially active and autonomous retrotransposons in the human genome, and even these elements are primarily inactive in humans due to 5′ truncations, inversions, or point mutations that have been acquired over evolution. Retrotransposons use a copy-and-paste method to move around the genome and go through an RNA intermediate using reverse transcriptase. However, several full-length active L1 elements, i.e., capable of copying and pasting themselves into new locations in the human genome, have been identified (Brouha et al. 2003). Non-autonomous elements, like Alu and SINE-VNTR (variable number tandem repeat) -Alu (SVA) elements, rely on the proteins encoded by L1 elements to move (Dewannieux et al. 2003; Ostertag et al. 2003). *Alu* elements were originally named in the mid-1970s; however, SVA elements were discovered in the 1990s and specifically identified in the *STK19* gene, where Shen et al. (1994) reported SINE-R and VNTR sequences were also associated with *Alu*-like sequences, and the entire structure was flanked by target site duplications (TSDs) (Shen et al. 1994). It was later posited that the SVA originated from an alternatively spliced transcript that included all of the elements now contained within its sequence; however, the possibility that the incorporation of all of its parts (the CCCTCT hexamer, first *Alu*, SVA-U, second *Alu*, VNTR, and the SINE-R sequence) could all have occurred independently cannot be completely ruled out (Hancks and Kazazian Jr. 2010). Both *Alu* elements and SVA elements are restricted to the primate lineage, with SVA only present in *hominids*, and are absent from old-world monkeys (Chuong et al. 2017).

LINE elements were technically first identified in 1980 by J. Adams; however, these elements were not named "LINEs" until the following decades, as more sequencing revealed the presence of both long and short interspersed elements, and a need arose for a more concise term such as "LINE" or "L1" (Adams et al. 1980). L1 elements have three open reading frames, ORF0p, ORF1p, and ORF2p, some of which have been detected in embryonic stem cells (ESC) and have been shown to play essential roles in ESC cell maintenance and/or transition to differentiated states. (Percharde et al. 2018; Garcia-Perez et al. 2007). For example, Percharde et al. showed that ORF1p acts as a nuclear scaffold that recruits nucleolin and Kap1/Trim28 to repress Dux, a master activator for transcriptional programming in the 2-cell embryo (Percharde et al. 2018). Some SINE elements, on the other hand, are expressed in certain stress conditions but otherwise are epigenetically silenced (Dunker et al. 2017; Cheng et al. 2020). ERV sequences are found across many different but related species, like old-world monkeys and humans, suggesting that a common ancestor was infected and that the virus became fixed in the genome before the species diverged (Pontis et al. 2022). In human endogenous retroviruses (HERVs), most loci are not full-length proviruses; however, there are still many that are transcribed and several that can produce proteins (Durnaoglu et al. 2021; Doucet-O'Hare et al. 2021; Garcia-Montojo et al. 2018). Some HERV elements have been shown to play an integral role in pre-implantation embryos and in embryonic development in general (Grow et al. 2015; Macfarlan et al. 2012). Furthermore, different HERV families are expressed in a specific temporal manner; for example, HERV-W is expressed in peripheral blood mononuclear cells (PBMCs) in children under the age of one and in adults over the age of 60, while HERV-H elements are typically expressed in children under the age of four, and HERV-K element expression peaks in children under the age of one (Balestrieri et al. 2015). Although many TEs have important roles in developmental processes, their continued expression after certain developmental timepoints and/or their reactivation in mature differentiated cell types has been suggested to contribute to many different human pathologies (Li et al. 2015; Doucet-O'Hare et al. 2021; Shah et al. 2022; Shah et al. 2023). It is now clear that proper regulation of TE expression is paramount to both human development and human disease.

Due to the aberrant TE promoter-initiated expression in certain pathological conditions and the absence of TE expression in many differentiated normal tissues, TE expression provides a potential opportunity for therapeutic targeting. TE expression has been observed in many diseases that do not have targeted or effective treatments, such as ALS, multiple sclerosis (MS), atypical teratoid rhabdoid tumors (AT/RT), and glioblastoma, which increases the appeal of developing TE-targeted treatments (Perron et al. 1989; Douville et al. 2011; Li et al. 2015; Doucet-O'Hare et al. 2021; Shah et al. 2022, 2023; Serra et al. 2001). Several types of strategies have been suggested to target TE expression in disease, including small molecules to inhibit their protein function, targeting the expression of genes that modulate TE expression, and enhancing the ability of the immune system to target cells expressing aberrant levels of TEs (Singh et al. 2024). In cancers that express HERVs or L1s, drug therapies could deactivate retroelement transcription, leading to a decrease in HERV/L1 protein expression and impairing the ability of tumor cells to retain their pluripotent state and continually divide (Doucet-O'Hare et al. 2021; Shah et al. 2023; Zehrbach et al. 2024). Antiretroviral therapies, which were originally designed to inhibit the proteins of Human Immunodeficiency Virus (HIV) (Table 13.1), are being repurposed in current in vitro and in vivo models and there are also clinical trials to explore their therapeutic potential in HERV-mediated disease (Table 13.2) (Shah et al. 2022, 2023; Gold et al. 2019; Garcia-Montojo et al. 2021).

Furthermore, some TEs can be used as a delivery system for gene therapies. Gene therapy is the manipulation of gene expression to target a specific cellular response; for example, it can be used to restore function of a gene with a deletion in a monogenic disorder, confer new abilities to immune cells to target cancer cells with specific expression hallmarks, or induce expression of suicide genes in a targeted cell population (Morgan et al. 2021). Suicide gene therapy is a therapeutic option for some cancers, including colon, lung, breast, gliomas and others, that utilize a non-toxic prodrug that is converted into a cytotoxic drug by a therapeutic transgene when the prodrug enters the tumor cells (Zarogoulidis et al. 2013). Some TEs can stably integrate into the genome and can be synthesized using specific sequences with the machinery needed to transcribe a newly integrated open reading frame for a protein. The two most common approaches for this TE-based gene therapy are the Sleeping Beauty (SB) and piggyBac (PB) transposons (Tipanee et al. 2017). Both SB and PB vectors are used in preclinical and clinical trials to try and establish potential therapeutics in patients with cancer as a form of CAR-T therapy (Sandoval-Villegas et al. 2021). SB has also been used in an Alzheimer's disease (AD) clinical trial where the authors tried to increase the amount of nerve growth factor in a patient's brain (Eyjolfsdottir et al. 2016). In contrast, a strategy has been proposed to induce 'viral mimicry' by increasing the expression of some HERVs to induce an immune response as a therapeutic strategy (Seetharam et al. 2025). The interplay between TEs and cancer is discussed in Chap. 10. In this chapter, we will

Table 13.1 ART drugs approved for HIV treatment

Drug Class	Generic Name	Brand Name	FDA Approval Date	Clinical Trial PubMed ID	ClinicalTrials. gov ID	Abbreviation
NRTIs	Zidovudine	Retrovir	March 19, 1987	2869302	NA	AZT/ZDV
	Lamivudine	Epivir	November 17, 1995	7477217	NA	3TC
	Abacavir	Ziagen	December 17, 1998	10103339	NA	ABC
	Tenofovir disoproxil fumarate	Vired	October 26, 2001	11557462	NA	TDF
	Emtricitabine	Emtriva	July 2, 2003	15249567	NA	FTC
NNRTIs	Nevirapine	Viramune	June 21, 1996	8452345	NA	EFV
	Efavirenz	Sustiva	September 17, 1998	10601505	NA	ETR
	Etravirine	Intelence	January 18, 2008	20051805	NA	NVP
	Rilpivirine	Edurant	May 20, 2011	21763936	NCT00540449	RPV
FI	Enfuvirtide	Fuzeon	March 13, 2003	12167274	NA	T-20
PI	Ritonavir	Norvir	March 1, 1996	9492772	NA	ATV
	Atazanavir	Reyataz	June 20, 2003	14685054	NA	DRV
	Tipranavir	Aptivus	June 22, 2005	15097154	NA	RTV
	Darunavir	Prezista	June 23, 2006	19195454	NA	TPV
INSTI	Raltegravir	Isentress	October 12, 2007	17133211	NA	RAL
	Dolutegravir	Tiviccay	August 12, 2013	23306000	NCT01227824	DTG

List of HIV ART drugs separated by class, with generic name, brand, and FDA approval date. The PubMed ID trial number is linked to the first clinical trial in which the drug was tested. ClinicalTrials.gov was established in 2000, but it was not a requirement to register a clinical trial until 2007. Any clinicaltrial.gov references that are available have been provided

Table 13.2 ART clinical trials in TE-associated diseases

Disease	Drug(s)	Drug Class	Trial Acronym	Phase	Status	Objectives	Results	ClinicalTrial. gov ID
Kaposi's sarcoma	ART	All	NA	2	Unknown, begain 1998	Evaluate the response of patients with slowly progressing Kaposi's sarcoma to antiretroviral therapy (ART)	NA	NCT00003419
Lupus	Abacavir/ lamivudine	NRTI	PENCIL	2	Not yet recruting	Effect of NRTIs on IFN signature in low- activity systemic lupus patients	NA	NCT06356740
Alzheimer's disease	Lamivudine	NRTI	ART-AD	1/2	Completed November 2023	CNS penetration, neuroprotection, and safety of 3TC	Well-tolerated drug with beneficial biomarker effects; larger studies with controls needed to assess effectiveness	NCT04552795
Alzheimer's disease	Emtricitabine	NRTI	LINE-AD	1	Recruiting	Reduction of inflammatory disease state	NA	NCT04500847
Parkinson's disease	TDF	NRTI	NA	1	Recruiting	Assess the safety and efficacy	NA	NCT06356662
Multiple sclerosis	Raltegravir	INSTI	INSPIRE	2	Completed September 2014	Efficacy in prevention of progression of relapsing and remitting MS	No impact on MS disease activity	NCT01767701
ALS	Triumeq	NRTI and INSTI	LIGHTHOUSE	2	Completed December 2018	Assess the safety and efficacy	Drug was well-tolerated with indications of a biological response; larger trial with controls needed to assess effectiveness	NCT02868580

(continued)

Table 13.2 (continued)

Disease	Drug(s)	Drug Class	Trial Acronym	Phase	Status	Objectives	Results	ClinicalTrial.gov ID
Brain cancer	Indinavir and ritonavir	PI	NA	2	Unknown, began in 2007	Efficacy of radiation therapy with or without ART	NA	NCT00637637
Glioma	Nelfinavir	PI	NA	1	Terminated (Insuffient accrual)	Dose determination of nelfinavir with radiations and temozolimide	NA	NCT00915694
Glioma	Abacavir and lamivudine, ritonavir	NRTI, NRTI and PI	STARLITE	1	Recruiting	Dose determination, safety, and tolerability of drug combination for non-surgical tumors	NA	NCT06428045
Breast cancer	Ritonavir	PI	NA	1	Completed July 2014	Determine optimal dose of ritonavir and its impact on biomarkers after surgical resection	No results have been published yet	NCT01009437
Prostate cancer	Nelfinavir	PI	COAST	1/2	Recruiting	Safety and efficacy of hydroxychloroquine, metformin, sirolimus, dasatinib, and nelfinavir in prostate cancer	NA	NCT05036226

Clinical trials utilizing ART therapy in other diseases. Examples of clinical trials that are trying to repurpose ART therapy for other diseases along with their status. Any publication that accompanies the clinical trial results have been included

look at different diseases and how TEs are related to their development and treatment, highlighting a few examples of TE-targeted therapeutic applications. We will also explore the pathways and mechanisms by which targeting TE products can be used to achieve a therapeutic effect.

13.2 AIDS/Autoimmune Disorders

13.2.1 HIV-1 and Antiretroviral Therapy

First identified in 1981, human immunodeficiency virus type 1 (HIV-1) is a retrovirus capable of integrating into the host genome and transmitting between individuals through blood contact, sexual activity, and from mother to child (Landers et al. 2021; Gartner et al. 2020; Gottlieb et al. 1981). In the early stages of infection, the virus causes a sore throat, muscle aches, fever, chills, and fatigue, but if it progresses without treatment, it will turn into acquired immunodeficiency syndrome (AIDS) (Schacker et al. 1996). HIV-1 also increases the risk of developing certain cancers like Kaposi sarcoma (KS) and non-Hodgkins's lymphoma (Grulich et al. 2007). HIV destroys CD4+ cells (helper T cells, lymphocyte cells produced by the thymus), which are a critical part of the adaptive immune response in protecting the body from infection (Workowski et al. 2021). Currently, HIV-1 infections are treated with one of many drug cocktails known as antiretroviral therapy (ART) drugs that include nucleoside reverse transcriptase inhibitor (NRTIs), non-nucleoside reverse transcriptase inhibitors (NNRTIs), protease inhibitors (PIs), integrase stand transfer inhibitors (INSTIs), and fusion inhibitors (FI), all of which work in different ways to limit the ability of the retrovirus to reverse-transcribe its RNA, cleave the polyprotein into functional protein segments, integrate into the genome, or fuse with cellular receptors, respectively (Fischl et al. 1987; Tseng et al. 2015). Many ARTs, like tenofovir, alafenamide, and lamivudine, are NRTIs that either directly inhibit reverse transcription or provide inhibitory nucleosides that prevent subsequent protein expression (Gandhi et al. 2023). Capsid inhibitors, like lenacapavir, are used to prevent the capsid core from opening, preventing the release of the HIV enzymes and RNA, which could infect new cells (Link et al. 2020). Integrase inhibitors prevent the HIV DNA from being inserted into the host DNA (Metifiot et al. 2013). PIs, like Atazanavir and Ritonavir, prevent the cleavage of the viral polyprotein into mature proteins, thus inhibiting viral replication for which these proteins are necessary (Clinic 2023). There are several additional inhibitors that are used in antiretroviral cocktails that work together to limit the activity and function of HIV (Table 13.1). HIV and TE rely on similar proteins like reverse transcriptase and integrase to incorporate into the genome; the drugs developed to fight HIV are promising therapeutic agents for other disease types that have a TE signature. In this chapter, we highlight the use of these antiviral therapeutics as experimental treatments for ERV expression.

HIV infections have also been associated with an increase in HERV-K expression both in an antibody response and in a T-cell response when comparing HIV-positive patients to uninfected donors (Lower et al. 1996; Garrison et al. 2007; SenGupta et al. 2011; Tandon et al. 2011; Vogetseder et al. 1993). Several reasons for the increase in HERV-K expression in individuals with HIV have been proposed, including (1) a decrease in global methylation regulation, (2) activation of HERV-K long terminal repeat (LTR), the HERV-K promoter, by the HIV transactivator of transcription (tat) protein, (3), or loss of immune regulation (Contreras-Galindo et al. 2007). Beyond HERVs, L1 retrotransposition has also been found to be increased in HIV-infected Jurkat cells, and another study uncovered an accumulation of extrachromosomal L1 DNA in HIV-infected CD4+ T cells (Jones et al. 2013). Furthermore, RNA levels of HERV-K are found to be increased in plasma samples and T cells from HIV-infected patients (Contreras-Galindo et al. 2006; Garrison et al. 2007). In one study, the activation of certain HERV-K MMTV-like loci (HML-2) in HIV-infected cells was found to be cell-type specific (Vincendeau et al. 2015). Interestingly, several studies have found intracellular HML-2 transcripts were elevated in PBMCs from individuals with HIV as compared with controls, even when HIV transcripts were undetectable in plasma (Vincendeau et al. 2015; Sugimoto et al. 2001; Brinzevich et al. 2014). Together, these findings suggest that HIV infection induces broad and multifaceted dysregulation of retroelement activity, with potential implications for both viral pathogenesis and host immune responses.

There have been several studies that have looked at how an immune response to HML-2 influences HIV pathogenesis. In patients with HIV, high levels of HML-2 pol RNA were detected in almost all patients, while it was nearly undetectable in most control samples, suggesting that HIV induces HML-2 expression (Contreras-Galindo et al. 2006). HML-2-specific antibodies have been detected in the blood of HIV-infected individuals at higher titers than in uninfected controls (Michaud et al. 2014). There are HIV-infected individuals classified as "elite controllers" who can control the transcription and reintegration of HIV in the absence of ART treatment, and these individuals had even higher titers of the HML-2 antibodies compared to those treated with ART (Michaud et al. 2014). This finding allows for the possibility that an immune response against HML-2 may correlate with HIV suppression. In addition, some groups have reported a T-cell response to HML-2 in HIV-infected patients that was correlated with better control of viral titers; further, higher T-cell counts suggest that HML-2 transcripts induced by HIV infection could produce viral proteins able to generate a humoral and/or cytotoxic T-cell response (SenGupta et al. 2011). In summary, the evidence to date demonstrates a significant interplay between HIV infection and endogenous retroelement activation, particularly involving HERV-K elements. However, the precise mechanisms and consequences of this interaction remain incompletely understood. Future research aimed at delineating these pathways may not only deepen our understanding of HIV pathogenesis but also uncover novel immunological targets and therapeutic interventions.

13.2.2 Lupus

Systemic lupus erythematosus (SLE) is a chronic autoimmune disease without a definitive cure. In autoimmune diseases, the body's immune system mistakenly recognizes and attacks its own proteins and tissues. In SLE patients, there is an increase in apoptosis and a decreased ability to effectively clear apoptotic cells (Mistry and Kaplan 2017). This triggers an interferon-alpha (IFN-α) response, activating dendritic cells and natural killer (NK) cells, which subsequently induce apoptosis (Ronnblom and Pascual 2008). This process establishes a feed-forward loop that perpetually activates the innate immune response (Pan et al. 2020). TE expression is linked to IFN-α levels in individuals with SLE. LTR retrotransposons, DNA transposons, LINEs, and SINEs were all found to be upregulated in SLE patients, and higher TE expression correlated with a higher innate immune response (Wang et al. 2024). Different TE types are expressed in a cell-specific manner across different SLE phenotypes (Cutts et al. 2024). In the neutrophils, there is an increase in L1 retrotransposons that is caused by a reduction in epigenetic silencing in SLE patients (Ukadike et al. 2023). The INF response can be inhibited in cells when treated with a reverse transcriptase inhibitor (Ukadike et al. 2023). HERV-K antibodies were also detected in plasma samples of SLE patients, which form immune complexes that activate neutrophils (Table 13.2) (Tokuyama et al. 2021).

In terms of therapeutics, most SLE treatments are designed for symptom management and do not target the cause of the disease; as such, there have not been many therapeutic approaches toward TE expression in this disease. However, there have been studies to try to establish why patients with SLE have TE expression. One theory concerns the regulation of heat shock protein levels. Heat shock protein 90 (Hsp-90) is a chaperone protein that can silence the post-transcriptional expression of TEs using the P-element-induced Wimpy testis (PIWI)-interacting RNAs (piRNAs) mechanism of epigenetic silencing (Hummel et al. 2017). Heat shock proteins were downregulated in patient-derived CD3 T cells, CD19 B cells, and monocytes; this is a possible mechanism that could explain TEs overexpression (Zhang et al. 2018). However, it is also worth noting that Hsp-90 protein is overexpressed in most patients with SLE (Shukla and Pitha 2012). Further studies are needed to link a subset of SLE patients with decreased Hsp-90 protein levels and increased levels of TE expression; for that subset of patients, an Hsp-90 activator could serve as a potential therapeutic.

TE activation has emerged as a contributing factor to the persistent innate immune activation characteristic of SLE. Elevated expression of various TEs, including LTR retrotransposons and LINEs, has been correlated with increased interferon responses and disease severity. Despite current therapies focusing primarily on symptom management, modulation of TE expression offers a potential avenue for more targeted treatment strategies. Dysregulation of Hsp-90 and its role in post-transcriptional TE silencing highlights a novel therapeutic target, although further studies are needed to define patient subsets that would most benefit from such an approach.

13.3 Neurological Disorders

Neurological disorders can be categorized based on when they appear during a person's life. For instance, some are neurodevelopmental and emerge in the beginning of life, others are neurodegenerative and emerge later in life, and some arise unexpectedly, such as cognitive impairments following a viral infection (Collaborators 2024). Neurodevelopmental disorders generally manifest due to disruption of brain growth, damage to the brain, spinal cord, or peripheral nerves, or an impairment in cognitive, sensory, socioemotional, or motor function (Collaborators 2024). Some neurological disorders can be managed with lifestyle changes, while others require constant and comprehensive care throughout life; however, most of these challenging illnesses do not have a cure. A recent study estimated that nearly a third of the world's population had some kind of condition affecting nervous system function (Collaborators 2024). Targeting treatments for neurological disorders where TE elements are initiating or perpetuating the disease could be beneficial to many people (Collaborators 2024)..

Many common neurological disorders manifest relatively early in human development and in childhood-like attention deficit hyperactivity disorder (ADHD) and autism spectrum disorder (ASD). ASD development generally begins early during childhood, with diagnoses appearing even at 6 months of age (Association 2013). Individuals in families where one or more members are diagnosed with ASD are also at a higher risk of having ADHD, a common neurological disorder in children and adults (Septier et al. 2019). There is evidence that genes involved with postsynaptic density, neural projections, and synaptic transmission are implicated in both disorders (Lotan et al. 2014; Liu et al. 2017). In *Drosophila melanogaster*, it has been demonstrated that TEs have a significant role in synaptic plasticity, cognition, tissue development, and morphogenesis (Nikolaienko et al. 2018; Miao et al. 2020), suggesting that TEs could be involved in ADHD and ASD development. TE involvement in neurodifferentiation has also been explored in vitro in human cells, where it was found that regulation of HML-2, a human-specific HERV-K element, is essential for neurodevelopment (Wang et al. 2020). Furthermore, studies on both ADHD and ASD patients revealed dysregulated transcription of the *env* gene from several HERVs in the PBMCs of patients compared to age and sex matched unaffected controls (Balestrieri et al. 2012, 2014, 2016). Genetic variants that influence or are associated with TR expression can also contribute to the development of disorders such as ADHD (Luo et al. 2020). TE-targeted treatments could be beneficial for individuals with ADHD or ASD, but this requires additional research. Advancing the development of TE-targeted therapies for these disorders will require a deeper understanding of the specific roles of TEs and the mechanisms through which their transcription or protein expression may contribute to ADHD and ASD.

As the brain ages, neurodegeneration becomes a concern. Aging can be accompanied by activation of TEs like LINEs and ERVs that can contribute to age-related disease by increasing stress, inflammation, and genomic and cellular damage (Wahl et al. 2021; Li et al. 2013a). One study looked at the derepression of LINE-1 elements during cellular senescence and demonstrated that their increased transcription activates a type-I interferon (IFN-I) response (De Cecco et al. 2019). The authors further showed that treatment of aged mice with NRTIs, which inhibit LINE-1 reverse transcriptase, downregulated IFN-I activation and age-associated inflammation in multiple tissues (De Cecco et al. 2019). In one study of rapid aging and its putative effects on TE expression, the authors created a sirtuin 6 (SIRT6) knockout (KO) mouse model that exhibited a shortened lifespan, growth retardation, and highly elevated L1 activity (Simon et al. 2019). The SIRT6 knockout mice accumulated large amounts of L1 complementary deoxyribonucleic acid (cDNA) in their cytoplasm, triggering an IFN-I response via activation of cyclic GMP-AMP Synthase (cGAS) (Simon et al. 2019). In humans, cGAS recognizes double-stranded DNA, and upon sufficient binding, forms dimers arranged in a ladder-like structure along the DNA (Li et al. 2013b; Andreeva et al. 2017). The ladder structure signals for the production of cyclic guanosine 3′,5′-monophosphate (GMP)-cyclic adenosine 3′,5′-monophosphate (AMP) or (cGAMP), a secondary messaging molecule that will then be detected by the stimulator of interferon genes (STING) (Ishikawa et al. 2009; Ergun et al. 2019). The STING complex then undergoes a series of conformational shifts that trigger an immune response (Ergun et al. 2019). The cGAS-STING pathway helps to protect the body against viral infection and has been a therapeutic target for inflammatory diseases. In the same study, which used the SIRT6 KO mice, advanced-age wild-type mice also showed accumulation of L1 cDNA and IFN-I responses, demonstrating that this trend is consistent with normal aging as well (Simon et al. 2019). Reverse transcriptase inhibitors rescued the IFN-I response and prevented DNA damage while extending the lifespan of SIRT6 KO mice (Simon et al. 2019).

Studies in mutated TE-overexpressing mice have shown that treatment with NRTIs could extend their life span when paired with an antiaging agent (Gorbunova et al. 2021). In aging-related disease, chronic inflammation and senescence pathways can activate HERVs and elevated levels of HERVs lead to continued inflammation and rapid senescence, when cells are alive but no longer dividing (Liu et al. 2023b). A recent study observed that human mesenchymal progenitor cells in a senescent state released elevated levels of HERV-K env protein into the culture supernatant in vitro (Liu et al. 2023b). When this conditioned media was applied to young, non-senescent cells, it triggered rapid induction of senescence (Liu et al. 2023b). The introduction of an anti-env antibody blocked this effect, and the cells continued to divide (Liu et al. 2023b). Recently, in vitro co-culture systems with macrophages, vascular endothelial cells, and vascular smooth muscle cells derived from patients with Werner syndrome, a rare genetic disease that causes rapid premature aging, showed an activation of type-I interferon signaling and changes in chromatin accessibility that affected the cellular homeostasis (Paul et al. 2024). This study further established that TE-derived double-stranded RNA (dsRNA) activated the DExH-box helicase 58 (DHX58)-dependent cytoplasmic RNA sensing pathway in Werner syndrome patient-derived macrophages; however, silencing type-I interferon signaling rescued cell proliferation and decreased senescence and inflammation markers (Paul et al. 2024). In an exciting advance, some researchers are now developing a small molecule approach to inhibit LINE-1 endonuclease and have shown that treatment with these inhibitors decreases L1-induced damage, inflammation, and senescence in cell culture (D'Ordine et al. 2024). Developing ways to maintain TE suppression during aging could extend both longevity and the quality of life. The roles of TEs in aging pathways are discussed in detail in Chap. 11.

Neurological disorders can manifest throughout the human lifespan and are influenced by both genetic and environmental factors. Emerging evidence indicates that TEs, including LINEs and ERVs, play critical roles in both neurodevelopmental and neurodegenerative processes. TEs have been implicated in early-onset conditions such as ASD and ADHD, with studies revealing dysregulation of HERV transcripts and TE-linked genes involved in synaptic function. In aging and neurodegeneration, TE reactivation contributes to chronic inflammation, genomic instability, and cellular senescence, in part through the induction of interferon responses. Clinical trials targeting TE activity, including RT inhibitors and anti-env antibodies, have shown promise in mitigating these pathological effects and restoring cellular homeostasis (Tables 13.2 and 13.3) (Sullivan et al. 2024; Gold et al. 2019; Diebold and Derfuss 2019). These findings underscore the potential for TE-targeted therapies to benefit a broad spectrum of neurological disorders across the lifespan.

13.3.1 Alzheimer's Disease (AD)

AD is one of the most common neurodegenerative disorders characterized by progressive memory loss and cognitive dysfunction. AD is associated with plaques that are caused by an aggregation of amyloid beta (Aβ) proteins. One study revealed that numerous members of the HERV-K superfamily exhibited expression correlating with the expression of genes commonly dysregulated in AD. Furthermore, their elements were located in close physical proximity to genes that drive these pathways (Dawson et al. 2023). The mechanisms that underlie TE dysregulation in AD are still under active investigation. One study combined functional genomics with the use of both familial and sporadic AD patient-derived induced pluripotent stem cells (iPSCs) to evaluate how changes in chromatin leading to TE expression may occur (Scopa et al. 2023). This study demonstrated that upregulation of activator protein 1 (AP-1) subunit and c-Jun N-terminal kinase (JNK) (c-Jun) initiates the decondensation of TE regions, leading to cytoplasmic accumulation of HERV-K-derived RNA-DNA hybrids and the activation of the cyclic GMP-AMP Synthase-STimulator of INterferon Genes (cGAS-STING) cascade (Scopa et al. 2023). Inhibition of c-Jun blocked all downstream molecular consequences, reducing neuronal death and the impairment of neurogenesis (Scopa et al. 2023). A targeted treatment that could inhibit c-Jun in the correct cellular population could be a good therapeutic target for future development.

In one recent study, widespread dysregulation of TE expression was observed in the brains of individuals with human herpes virus infection (Feng et al. 2025). The

Table 13.3 HDACi, DNMTi, and Monoclonal antibodies used for treatment of TE-associated disease

HDAC Inhibitors												
				Current Clinical Trials					Status of Clinical Trials			
Inhibitor name	Brand name	Target	Status	Phase 1	Phase 2	Phase 3	Phase 4	Other	Active/ recruiting	Complete	Terminated	Current disease groups
Belinostat	Beleodaq	Pan	Approved in 2014	35	38	2	0	0	11	31	8	Prtipheral T-cell lymphoma, malignant thyomoma
Givinostat	Duvyzat	Pan	Approved in 2024	10	16	4	0	1	6	12	6	Duchenne muscular dystrophy, juvenile idiopathic arthritis, and polycythaemia vera
Entinostat	N/A	Class 1	Not approved	43	37	2	0	2	10	40	16	Breast cancer, non-small cell lung cancer
SAHA	Vorinostat	Class 1, 2, and 4	Approved in 2006	168	151	9	0	9	26	162	60	Cutansous T-cell lymphoma, gliomas, ovarian, Alzheimer disease
Panobinostat	Farydak	Non-selective	Approved in 2015	90	80	3	1	2	5	85	43	Multiple myeloma, prostate cancer, breast cancer, recurrent glioblastoma
Romidepsin	Istodax	Class 2	Approved in 2009	57	59	5	0	3	14	61	17	Cutaneous T-cell lymphoma
Rocilinostat	N/A	HDAC6	Not approved	12	4	0	0	0	1	8	3	Multiple myeloma, breast cancer

(continued)

Table 13.3 (continued)

DNMT inhibitors

				Current clinical trials					Status of clinical trials			
Inhibitor name	Brand name	Target	Status	Phase 1	Phase 2	Phase 3	Phase 4	Other	Active/ recruiting	Complete	Terminated	Current disease groups
Resveratrol	N/A	DNMT1 and DNMT3a/b	Natural product	32	37	17	7	122	24	142	12	Colerectal cancer, Alzheimer disease
Fludarabine	Fludara	DNMT1 and DNMT3a/b	Approved in 1991	1074	1209	171	22	220	845	772	317	B-cell chronic lymphocytic leukemia, Multpile sclerosis, acute myeloid leukemia
Genistein	N/A	DNMT1	Natural product	18	38	8	5	23	11	51	5	Colorectal cancer, breast cancer, cardiomyopathy, bladder cancer
Curcumin	N/A	DNMT1 and DNMT3a/b	Natural product	57	96	37	15	167	54	188	17	Breast and cervical cancer, depression
Disulfiram	Anrabuse	DNMT1	Approved 1951	20	28	2	5	15	8	35	10	Addiction, Alcholism, breast cancer,

Antibody treatments

				Current clinical trials					Status of clinical trials			
Antibody	Other names	Target	Status	Phase 1	Phase 2	Phase 3	Phase 4	Other	Active/ recruiting	Complete	Terminated	
GNbAC1	Temelimab	HERV-W env	Unknown	3	7	0	0	0	0	10	2	Multiple sclerosis
KO1	N/A	HERV-K env	N/A	N/A	N/A	N/A	N/A	N/A	N/A	N/A	N/A	Amyotrophic lateral sclerosis (ALS)

Select examples of HDACi and DNMTi drugs and a summary of the clinical trials by numbers that are recorded on clinicaltrials.gov. The inhibitor name, the brand name, and the target of the drug are included along with approval status by the FDA. Further details include the number of different trial phases performed for each drug, the status of the clinical trials, and examples of current diseases being treated with them

study further demonstrated an astrocyte-specific upregulation of LINE1 which they suggest targets Nuclear Enriched Abundant Transcript 1 (*NEAT1*) expression through long-range enhancer-promoter chromatin interactions (Feng et al. 2025). These results could indicate that treatments that target chromatin remodeling may be effective in AD patients with TE dysregulation. LINE-1 upregulation has also been observed in late-onset AD and may be connected to microglial dysregulation and contribute to disease pathogenesis (Roy et al. 2024). Small molecules that inhibit LINE-1 protein function or therapies targeted at downregulation of LINE-1 transcription and protein expression in specific cell types could be useful for individuals with TE-mediated AD.

Tau aggregates that cause the development of neurofibrillary tangles are also characteristic of AD. Studies have shown that tau is associated with retrotransposons and linked to TE activation (Guo et al. 2018). In addition, Tau propagation is increased by the overexpression of Gag/Pol and Env from HERV-K, R, H, and W (Liu et al. 2023a). High levels of tau prevent the suppression of TE post-transcriptionally, which creates a positive feedback loop, where tau activates TEs and TEs cause tau to spread (Guo et al. 2018). In murine studies, Tau also caused an increase in LINE-1 elements (Ramirez et al. 2022). Using an NRTI is a potential therapeutic approach to limit the ability of L1 DNA to insert back into the genome. There are current clinical trials investigating the safety of the NRTI, lamivudine, also known as 3TC, which has shown early promise in Drosophila, mouse, and cell models of tauopathy (Table 13.2) (Sullivan et al. 2024).

A correlation has been observed between the expression levels of HERVs and the activation of Toll-like receptor 8 (TLR8) (Dembny et al. 2020). TLR8 is an endosomal pattern recognition receptor that plays a crucial role in the innate immune system by detecting pathogen-associated molecular patterns (PAMPs) and damage-associated molecular patterns (DAMPs) (Ma et al. 2006; Jefferies et al. 2003). Upon activation, TLR8 initiates downstream signaling cascades that lead to the production of proinflammatory cytokines and the activation of innate immune defenses (Akira and Takeda 2004). The interaction between elevated HERV expression and TLR8 signaling suggests that HERV-derived nucleic acids or proteins may act as endogenous ligands, contributing to chronic inflammation and immune dysregulation in disease contexts. TLR signaling is known to help reduce amyloid plaques but also produce inflammatory cytokines in AD patients. HERV-K env has been shown to bind with TLR8 and cause neurodegeneration in mouse models (Dembny et al. 2020). In the same study, HERV-K transcripts were more commonly found in human CSF samples from patients with AD than in the controls (Dembny et al. 2020). Taken together, these findings indicate that the targeting of HERVs in AD may be a potentially fruitful endeavor to develop new AD therapies.

13.3.2 Parkinson's Disease (PD)

Parkinson's disease (PD) is the second most common neurodegenerative disorder in individuals over the age of 65 (Poewe et al. 2017). In PD, the neuropathological hallmarks of the disease include neuronal loss in the substantia nigra, leading to dopamine deficiency and intracellular aggregation of α-synuclein (Poewe et al. 2017). Although this disease can be characterized by the depletion of dopaminergic neurons and resulting decline in mobility, there are many non-motor symptoms that can add to patient disability, such as mitochondrial dysfunction, oxidative stress, and neuroinflammation (Poewe et al. 2017). One of the earliest links between TE activity and PD was reported in a 2003 study, which found that the promoter region of the Parkinsonism-associated deglycase (*DJ-1* gene, which encodes PARK7)—deleted in patients with early-onset PD—was flanked by Alu elements on both sides of the deletion (Bonifati et al. 2003). This pattern suggested that the deletion may have resulted from an unequal crossing-over event mediated by the Alu elements (Bonifati et al. 2003).

In another example of TE involvement in a variant of PD, X-linked dystonia parkinsonism (XDP), an insertion of an SVA into an intron of the TATA-Box Binding Protein Associated Factor 1 (TAF1) gene decreased the expression of the dopamine receptor D2 (DRD2) gene and the TAF1 gene (Makino et al. 2007). This study also identified an abnormal pattern of DNA methylation in the SVA within the 34th *TAF1* intron that could account for the decrease in its expression (Makino et al. 2007). The D2 receptor gene in the caudate nucleus is important for motor control and function, and in PD patients, dopamine creation and uptake are severely decreased (Ramesh and Arachchige 2023). In another study of XDP, the authors identified an SVA in intron 32 of *TAF1*, performed excision of the SVA element with CRISPR/Cas 9 in cells, and rescued the expression of TAF1 (Aneichyk et al. 2018). CRISPR/Cas 9 could be used in a therapeutic context. G-quadruplexes were recently investigated as a potential source of *TAF1* silencing in individuals with XDP who have SVA insertions in their introns (Nicoletto et al. 2024). The data indicated that G-quadruplex formation occurring at the site of the SVA contributes to aberrant *TAF1* expression, providing a mechanistic explanation for this interaction between a TE and a disease-associated gene (Nicoletto et al. 2024).

Currently, many treatments for PD focus on the restoration of striatal dopamine, deep-brain stimulation, and the mitigation of the aggregation of α-synuclein. The overexpression of α-synuclein has been attributed to both missense substitutions and to multiplication of the *SNCA* locus, which encodes α-synuclein, in families with autosomal dominant Parkinsonism (Farrer 2006; Ross et al. 2008). In a study by Ross and colleagues, the authors found that there were many Alu subtypes at both the 5′ and the 3′ ends of the *SNCA* multiplication regions, which were sequenced in multiple families (Ross et al. 2008). The authors suggest these Alu sequences could have led to genomic instability, unequal recombination, and copy number variations of this genomic region (Ross et al. 2008). Additionally, somatic retrotransposition may influence the abundance of different brain metabolites, including neurotransmitters like those affected in PD (Abrusan 2012). A metabolic flux analysis suggested that somatic transposition alters the biosynthesis of metabolites, including dopamine, serotonin, and glutamate, and thus could contribute to the development of PD (Abrusan 2012).

Some TE loci are present in only a subset of the population, and it is important to consider this in ongoing research regarding TEs and disease. Non-reference TEs are elements that are absent from the reference genome but present in the human population and sometimes restricted to a single genetic ancestry. One group found that the presence of non-reference Alu, LINE1, and SVA loci correlates with disease progression of PD, although the presence or expression of these elements can contribute to disease in a myriad of ways (Koks et al. 2022). This study systematically evaluated the presence of non-reference TEs in individuals with PD as compared to unaffected individuals and showed that the TEs are expressed in and contribute to the regionally specific dopaminergic neurodegeneration of the putamen (Koks et al. 2022). Reference SVA insertion polymorphisms, which are found in the human reference genome, have also been associated with PD disease progression. One study suggests they should be considered as regulatory domains that affect processes that occur in PD progression (Pfaff et al. 2021). In a follow-up study, the same group suggested that the presence of SVA_67 in the intron of the microtubule-associated protein tau (*MAPT*) is polymorphic for its variable number tandem repeat domain and that this variation correlated with PD progression (Frohlich et al. 2024). With all the available evidence presented herein, TE-targeted therapeutics may be a fertile area of research for improved PD patient outcomes. There are current active studies funded by The Michael J. Fox Foundation focused on the connection between TEs, inflammation, and PD (Table 13.2).

13.3.3 Multiple Sclerosis (MS)

MS is a neuroinflammatory disorder characterized by the accumulation of plaques in the central nervous system, composed of inflammatory cells, demyelinated axons, and astrogliosis (Ghasemi et al. 2017). The plaque aggregation can cause neuronal dysfunction, ataxia, and cognitive impairments. MS has been associated with multiple types of viruses, including Epstein-Barr virus (EBV), in fact, a recent article used longitudinal data to reveal that individuals who are diagnosed with MS have a very high prevalence of EBV (Bjornevik et al. 2022). Approximately 95% of individuals have been infected with EBV during their lifetime, which made it difficult to connect infection with MS diagnosis without a well-documented longitudinal cohort; however, Bjornevik et al. used data from the United States military to assess individuals who had EBV and then received an MS diagnosis later in life (Bjornevik et al. 2022). They found the risk of MS increased 32-fold after infection with EBV but was not increased after infection with other viruses (Bjornevik et al. 2022). The authors note that the increase in the neurofilament light chain, used as an indicator of neuronal degradation, increased in their cohort only after EBV seroconversion

(Bjornevik et al. 2022). They conclude that their work suggests EBV is the leading cause of MS (Bjornevik et al. 2022).

Interestingly, it has been suggested that EBV may also stimulate the expression of endogenous retroviruses (Wieland et al. 2022). Further supporting this possible connection, Perron et al. found that individuals with MS also have higher levels of HERV expression when compared with healthy controls (Perron et al. 1997). HERV-W env proteins can be found in the active white matter of the brain, where plaque lesions cause a breakdown of the blood–brain barrier (BBB) (Frischer et al. 2009). Among the upregulated HERVs in MS patients, the most commonly found group is HERV-W. MS-associated retrovirus (MSRV) is a member of HERV-W and was previously known as LM7 before being found in several sample types from MS patients (Perron et al. 1997). Furthermore, MSRV expression has been detected in cerebrospinal fluid of MS patients, and its presence appears to correlate with both their clinical progression and prognosis (Sotgiu et al. 2010).

One strategy for targeting the effects of TEs, especially ERVs, in patients is through blocking the effects of the env protein. The MSRV env, like many other HERVs, is made up of three domains, including the signal peptide (SP), the surface domain (SU), and a transmembrane domain (TM). One group has demonstrated that the MSRV env SU is an agonist of the Toll-like receptor 4 (TLR4) and thus induces release of proinflammatory cytokines from PBMCs and microglial cells from the CNS (Rolland et al. 2005; Rolland et al. 2006). Monoclonal antibodies (mAbs) have been a therapeutic strategy to block the interaction between ERV env proteins and cell receptors. The humanized IgG4 mAb GNbAC1 was tested to prevent damage incurred due to MSRV env expression (Curtin et al. 2015). Once GNbAC1, also known as temelimab, is bound to the MSRV env, the protein is then unable to initiate the proinflammatory cascade (Curtin et al. 2015). Preventing the cascade protects the myelin repair process and arrests the neuronal degradation that occurs in MS patients (Curtin et al. 2015). Early clinical trial results have been promising and showed a decrease in expression of MSRV, accompanied by neuroprotective and regenerative effects in patients with MS, with no impairment of the immune system (Table 13.3) (Curtin et al. 2015).

In the Phase 2B clinical trial using temelimab, the study design included a double-blind 48-week trial in MS patients with relapsing and remitting disease with the addition of another 48-week extension phase to test the efficacy and safety of the drug (Hartung et al. 2022). The study had a primary endpoint to reduce the cumulative gadolinium-enhancing T1-lesions in the brain magnetic resonance imaging (MRI) scans at week 24 of the trial. Although the primary endpoint of the trial was not met, at week 48, participants treated with temelimab had fewer new T1-hypointense lesions; however, there was no statistically significant difference in reductions in overall atrophy in the brain compared with the placebo group (Hartung et al. 2022). The authors concluded that temelimab did not show an effect on acute inflammation in the patients but did demonstrate enough efficacy that this treatment, or others like it, should be further investigated for progressive MS (Hartung et al. 2022). In a reanalysis of many trials conducted with temelimab, pooled results for MS treatment show statistically significant differences in outcomes for those in the placebo versus the treatment arms of these studies (Irfan et al. 2022). These authors conclude that temelimab is safe and a good treatment option for patients with MS (Irfan et al. 2022).

In summary, the involvement of HERV-W expression—particularly MSRV env—in the pathogenesis of multiple sclerosis underscores the potential of targeting endogenous retroviral proteins as a therapeutic strategy. Temelimab, a monoclonal antibody directed against the MSRV env protein, has shown encouraging neuroprotective effects in early trials despite not meeting all primary endpoints. These findings support continued exploration of TE-targeted therapies as a promising approach in the management of progressive MS and highlight the importance of refining treatment strategies to address chronic neuroinflammation linked to endogenous retroviral activity.

13.3.4 Amyotrophic Lateral Sclerosis (ALS)

ALS is a rare degenerative motor neuron disease in which a patient slowly loses the ability to move, eat, and talk. HERV-K proteins are more highly transcribed in individuals with ALS in their motor cortex, and this expression has been targeted in clinical trials with NRTIs and PIs (Douville et al. 2011; Garcia-Montojo et al. 2021). The first connection between ALS and reverse transcriptase occurred in a work published in the 1970s, where nervous system tissue from two patients with Guamanian ALS showed 'RNase-sensitive DNA polymerase activity' which was absent from the brains of two unaffected individuals from the United States (Viola et al. 1975). This concept bore out in studies performed in the early 2000s, where one group found that both individuals with ALS and their blood relatives had significantly higher levels of reverse transcriptase in serum compared to non-blood-related and spousal controls (Steele et al. 2005). In 2011, active loci of ERVs were identified in the neurons of individuals with ALS (Douville et al. 2011). This study also showed that the HML-2 expression correlated strongly with Transactive response DNA binding protein 43 kDa (TDP-43) levels, which were already known to be dysregulated in ALS (Douville et al. 2011). A critical step in solidifying the role of HERVs in ALS disease pathogenesis came with the creation of a transgenic mouse model that overexpressed the HML-2 env protein in neurons (Li et al. 2015). These animals displayed a progressive motor dysfunction, loss of volume in the motor cortex, and many other pathological characteristics regularly observed in the brains of deceased individuals with ALS (Li et al. 2015). This study further revealed that the expression of HML-2 was regulated by TDP-43, which binds to the HML-2 promoter DNA and increases its transcription (Li et al. 2015). TDP-43 is known to act both through binding of DNA promoters and through direct binding of RNA molecules and thus may have disparate effects on expression of TEs depending on cell type specificity and epigenetic context within a cell.

The studies that followed sought to determine how pervasive HERV-K involvement was in individuals with ALS and whether these individuals' diseases could be better stratified to identify the most effective treatments. In 2019, it was demonstrated in an analysis of 148 postmortem individuals with ALS that three major expression profiles, including ALS-TE, ALS-Ox (oxidative stress), and ALS-Glia, could be identified, and aberrant pathways detected in this analysis pointed to stress, inflammation, and TDP-43-associated pathology (Tam et al. 2019). In this study, the authors demonstrated that TDP-43 directly binds to retrotransposon transcripts with a UGUGU repeat motif, such as LINE-1 s, SINEs, and some LTRs from ERVs, and contributes to their silencing in an RNA-mediated manner (Tam et al. 2019). Thus, it seems that the role of TDP-43 in TE-mediated ALS pathology is complex and may affect TE expression and its potential contribution to disease in a myriad of ways. These results emphasize the need to identify patients with a TE-linked pathology as those who would benefit from TE-targeted treatments in the future. A 2023 study of 208 individuals with ALS also stratified the patients into three groups, including two of which are the same (ALS-Glia and ALS-Ox) and one that is different (ALS-TD) (Eshima et al. 2023). ALS-TD is characterized by transcriptional dysregulation. ALS-Ox and ALS-TD both have heightened levels of TE expression compared to unaffected individuals and ALS-Glia (Eshima et al. 2023). By looking at TE locus-specific expression in subtypes of ALS, it appears there may be a subset of individuals with ALS who would benefit from a TE-targeted treatment and should be recruited for clinical trials.

Normally, TDP-43 is in the nuclei of glia and neurons, whereas in ALS patients, the protein aggregates in the cytoplasm and is a hallmark of the disease (Simula et al. 2021). Mislocalization and aggregation of TDP-43 impair its ability to bind RNA and suppress HERV-K expression effectively. Although the connection between HERV-K and TDP-43 in ALS-related neurodegeneration is still being clarified, multiple studies—including a recent one revealing a novel pathogenic role for TDP-43—have reported elevated levels of env, pol, and TDP-43 proteins in the brain tissue of ALS patients (Garcia-Montojo et al. 2024). In this study, TDP-43 and asparaginase and isoaspartyl peptidase 1 (ASRGL1) expression co-localized in unaffected individuals; however, in individuals with ALS who had lower expression of ASRGL1, TDP-43 aggregated in the cytoplasm of neurons (Garcia-Montojo et al. 2024). The *ASGRL1* gene contains a HERV-K (subtype HML-2) locus in the opposite orientation of the gene (Garcia-Montojo et al. 2024). When the HERV-K locus was overexpressed, a significant decrease in ASRGL1 expression occurred and, in a mouse model, led to an accumulation of misfolded, fragmented, phosphorylated, and mislocalized TDP-43 in the murine motor cortex (Garcia-Montojo et al. 2024). Targeted downregulation of HERV-K transcription restored normal expression of ASRGL1 and could be a potential avenue for treatment in patients with this ALS etiology. In addition, HERV-K env has been found to be neurotoxic as it is released extracellularly (Steiner et al. 2022).

The connection between TEs and TDP-43 extends beyond findings in human and in mouse models and has been demonstrated in *Drosophila melanogaster* models as well (Chang and Dubnau 2019). In 2019, Chang and Dubnau showed in a *Drosophila* model that TDP-43 overexpression in glial cells leads to the death of adjacent neurons (Chang and Dubnau 2019). The authors uncovered that the spreading of cell death is caused by an ERV, gypsy, expressed in the glial cells, which leads to DNA damage and neuronal cell death (Chang and Dubnau 2019). In later work, the same authors uncovered that gypsy ERV expression and TDP-43 proteinopathy are "mutually reinforcing" (Chang and Dubnau 2023). The authors demonstrated that either expression of the *Drosophila* mdg4-ERV (gypsy) or the human HERV-K subtype HML-2 is sufficient to stimulate aggregation of human TDP-43 in the cytoplasm of glial cells in vivo (Chang and Dubnau 2023).

Early cell culture models show that a HERV-K env antibody, GN-KO1, can reduce the neuronal death caused by CSF with high env levels in cultured human neurons (Table 13.3) (Li et al. 2022). Due to the accumulating evidence indicating that TE expression may contribute to ALS pathology, clinical trials using drugs and treatments that target these elements should be prioritized in the future, especially due to the lack of available treatments for individuals with ALS. There have been some clinical trials that assessed the effectiveness of NRTI and PRs in ALS patients, but these did not produce positive results despite in vitro experiments that showed early promise (Westarp et al. 1993). However, in a Phase 2A clinical trial with Triumeq, a combined mixture of dolutegravir, abacavir, and lamivudine that is widely successful in HIV, the authors showed a decrease in HERV-K expression and in the rate of disease progression (Table 13.2) (Gold et al. 2019). These results were confirmed by another study in 2022 that looked at HERV-K levels during and after the end of treatment, which also found an improved clinical response (Garcia-Montojo et al. 2021). The key to bringing TE-targeted treatments to individuals with ALS will be clearly identifying which patients have disease mediated through TE mechanisms and finding the best way to evaluate efficacy for these treatments. For some patients who have a more advanced disease before treatment, perhaps simply slowing the progression of symptoms would be a worthwhile reason to offer treatment. For individuals who are recently diagnosed, research is needed to understand whether inhibiting TE can prevent the progression of the disease. In future studies, it would be useful to see if there are clinical characteristics that correlate with individuals that have ALS who have TE involvement, as this could increase the likelihood these patients would be identified quickly and have immediate access to treatment.

13.4 Cancer

Reverse transcriptase activity in cancer cells was first described in the 1970s (Sarngadharan et al. 1972). In 1986, Ono and his team completed endogenous retroviral sequencing and identified the human Mouse Mammary Tumor Virus (MMTV)-like virus (including subtypes HML-1-10) as part of the HERV-K superfamily (Ono et al. 1986). Subsequently, HERV-K expression was found in many types of cancer, including melanoma, leukemia, prostate, colorectal, and brain cancer (Buscher et al. 2005; Brodsky et al. 1993; Manca et al. 2022). Besides HERVs, other TEs are upregulated in cancers, including LINEs in epithelial, ovarian, and colon cancers (Solyom et al. 2012; Perkio et al. 2024; McKerrow et al. 2023). Another study observed that a higher rate of retrotransposition in colorectal tumors correlated with a poorer outcome for patients (Cajuso et al. 2019). Alu elements are found to be upregulated in some patients with breast, lung, pancreatic, and thyroid cancer (Shaban et al. 2022; Stricker et al. 2024; Bouras et al. 2021; Sun et al. 2024; Daskalos et al. 2009). While this is not an exhaustive list of all the cancers that have upregulated TE expression, it does showcase the broad nature of TE expression in tumors and their potential role in cancer development. Studies that work to identify the presence of TE activity or expression in cancer have the potential to generate new insights for therapeutic interventions. In this section, a few examples of current therapeutic approaches will be highlighted.

HERVs are associated with cancer through two mechanisms: expression of oncogenic HERV proteins and indirect transcriptional regulation of oncogenes (Shah et al. 2021). For example, nuclear protein of 9 KDa (np9), a splice variant from the env gene of HERV-K101 and a homolog of HIV-1 Rev. (Rec) from HERV-K type 2 can bind with the zinc finger domain of the tumor suppressor Promyelocytic leukemia zinc finger protein (PLZF) (Denne et al. 2007). This interaction removes transcriptional repression of the cellular myelocytomatosis (c-myc) proto-oncogene promoter and induces cell growth and proliferation. This process has been associated with germline and leukemia cancers. Np9 can also bind with a ligand of the Numb protein involved in the Numb/Notch pathway, which has been linked with breast cancers, leukemias, and germline cancers (Denne et al. 2007). The HERV-K env (subtype HML-2) has also been associated with cell stemness and cellular proliferation in pediatric atypical teratoid rhabdoid tumors (AT/RT) (Doucet-O'Hare et al. 2021). Targeting the expression of env in AT/RT resulted in tumor cell death and a decrease in rat sarcoma (*Ras)* signaling (Doucet-O'Hare et al. 2021). These studies and many others suggest targeting specific HERV proteins may be a treatment option in tumors where the current standard of care is ineffective (Dang et al. 2023).

Current therapeutic approaches include the regulation of expression through transcriptase inhibitors and epigenetic modifications. Viral mimicry is a therapeutic strategy that leverages TE expression to stimulate the immune system in the fight against cancer. By inducing the expression of viral-like proteins within tumor cells, this approach aims to trigger an immune response capable of recognizing and targeting the tumor. The strategy holds particular promise for immunologically 'cold' tumors, where enhancing immune infiltration could significantly improve therapeutic outcomes. The induction of viral mimicry activates expression of HERVs by disarming the normal silencing mechanisms, and the subsequent increase in dsRNA elicits an innate and adaptive immune response (Petrizzo et al. 2021). The viral mimicry response is triggered by epigenetic modifications. DNA methyltransferases (DNMTs) methylate cytosines at unmethylated CpG sites, preventing transcription factors from binding to LTRs and thereby silencing TE expression. Inhibition of DNMT activity with a DNMT inhibitor (DNMTi) can reverse this effect and induce TE expression, which can then trigger the viral mimic response (Jin and Robertson 2013). Other epigenetic regulation includes histone methylation or acetylation with H3K9me2, H3K9me3, and H3K27ac, which are key regulators for many types of HERVs (Liu et al. 2018). Collectively, these findings emphasize the critical need for continued research into the mechanisms underlying TE activation in cancer, as well as the development of targeted therapies that exploit TE expression to improve patient outcomes. Future research should focus on identifying the specific contexts in which modulating TE activity can be most effectively leveraged for therapeutic benefit.

13.4.1 Brain Cancer (Gliomas)—and Viral Mimicry

Gliomas are heterogeneous tumors and often healthy brain tissue is contained within the tumor microenvironment (Neftel et al. 2019). The cellular heterogeneity within the tumor types allows for a broad range of mutations and different epigenetic modifications from tumor to tumor. In gliomas, HERVs contribute to tumor development and pathogenesis through their involvement in regulating oncogenes and tumor suppressors, as well as through the expression of oncogenic HERV proteins (Shah et al. 2022; Shah et al. 2023). The HERV-K envelope protein contains alternatively spliced transcripts, and the HERV-K proteins Np9 and Rec can activate the c-MYC proto-oncogene, which is associated with tumorigenesis in a variety of cancers (Kessler et al. 2014). Notably, c-MYC expression is also regulated by the transcriptional repressor PLZF, suggesting a point of convergence in HERV-mediated oncogenic pathways (McConnell et al. 2003). In gliomas, PLZF concentrations were three times higher than in normal controls (Choi et al. 2014). Under normal conditions, PLZF regulates transcription by repressing proliferative genes, leading to cell cycle arrest at the G1 to S phase transition (Shaknovich et al. 1998). However, paradoxically, in cell culture experiments, PLZF has been shown to promote proliferation in NIH3T3 cells and stimulate cancer cell growth (Choi et al. 2014). This dysregulation may be explained by the ability of HERV proteins to bind PLZF, thereby interfering with its normal repression of c-MYC expression.

One emerging clinical trial strategy involves the use of DNMT inhibitors in glioma treatment. Gliomas are characterized by aberrant DNA methylation patterns, including hypermethylation of CpG islands near tumor suppressor and apoptotic genes, and concurrent hypomethylation of regions normally silenced, such as TEs (Choudhury et al. 2015). This dysregulation contributes to unchecked tumor growth and evasion of apoptosis. DNMT inhibitors aim to reverse these effects by demethylating and reactivating silenced tumor suppressor genes, thereby restoring growth regulation and apoptotic pathways (Dammann et al. 2017). In addition to gene reactivation, DNMT inhibitors promote the expression of TEs, including HERVs, which can generate viral-like transcripts that trigger an innate immune response. Given that gliomas are typically immunologically "cold" tumors, the re-expression of HERVs may elicit a viral mimicry response, activating $CD8^+$ T-cell—mediated cytotoxicity against tumor cells (Chiappinelli et al. 2015). While this therapeutic approach has shown promise in acute myeloid leukemia and myelodysplastic syndromes, results in solid tumors such as lung, ovarian, and gliomas have been less robust, underscoring the need for further refinement and combination strategies (Kagan et al. 2023).

Histone deacetylase inhibitors (HDACi) are another drug therapeutic option currently being explored in glioma treatment. HDAC inhibitors work to decrease the amount of histone acetylation, create a more open chromatin structure, and increase the expression of genes typically silenced by the cancer (Li and Seto 2016). There are several different types of histone deacetylases (HDACs) and identifying which ones are overexpressed can help guide the inhibitor choice. The inhibitors are small enough to cross the BBB, and several have been tested in glioma clinical trials. One of the drugs in Phase 1 and Phase 2 clinical trials is vorinostat. In one of its Phase 2 trials, the efficacy of vorinostat in patients that did or did not undergo surgery was studied. Among patients who received oral vorinostat and surgical resection, 17% were free of progression in the first 6 months and at a median of 11.2 months showed stable disease. Unfortunately, it can be difficult to determine the proper dose of vorinostat, like many other therapeutics, at which the treatment has a high efficacy and minimal toxic side effects. There are ongoing trials to find optimal drug combinations with vorinostat (Table 13.3) (Minami et al. 2014).

Gliomas are complex and heterogeneous tumors that present significant treatment challenges due to their variable genetic and epigenetic landscapes. HERVs, particularly HERV-K, contribute to glioma pathogenesis by interacting with regulatory proteins like PLZF and promoting oncogene activation, such as c-MYC. Therapeutically, epigenetic drugs like DNMT inhibitors and HDAC inhibitors are being explored to reverse aberrant gene silencing and induce immune-mediated tumor responses. While DNMT inhibitors can re-express silenced tumor suppressors and trigger viral mimicry pathways, HDAC inhibitors such as vorinostat aim to remodel chromatin structure and restore gene expression. Though promising, further studies are needed to optimize dosing and combination strategies for maximum therapeutic benefit.

13.4.2 Colorectal Cancer

HERV-H is one of the largest HERV families. HERV-H elements rarely exist as full-length proviruses and instead manifest as solo LTRs or partial proviral sequences lacking *pol* and *env* genes (Hirose et al. 1993). HERV-H transcripts are found in several tumor cell lines, including lung cancers, bladder carcinomas, and testicular tumors (Hirose et al. 1993; Park et al. 2023; Gimenez et al. 2010). HERV-H is also upregulated in colorectal cancer tissue and cell lines and can be used as a biomarker for colorectal cancer (Wentzensen et al. 2007). Colorectal cancer patients that have high HERV-H expression are often poor responders to chemotherapy, and the preferred treatment in these individuals is an immune checkpoint inhibitor (Golkaram et al. 2021). Tumors with HERV-H expression are associated with immunoevasion along with an increase in CD19 production and elimination of either is followed by tumor regression (Kudo-Saito et al. 2014). The role of HERV-H in colorectal cancers must be further elucidated before effective treatments can be developed.

In addition to HERV expression in colorectal cancer, other TE activity has been demonstrated as well (Solyom et al. 2012). An L1 insertion into the *APC* gene, responsible for familial adenomatous polyposis, was detected in 1992 (Miki et al. 1992). The L1 inserted into the last exon of the adenomatous polyposis coli (APC) gene and resulted in complete disruption of *APC* expression; this was the first report of a tumor suppressor gene being disrupted by a somatic insertion of a mobile element (Miki et al. 1992). This phenomenon is not unique to L1 elements, as somatic insertions of Alu elements and SVA elements also occur and have led to hereditary non-polyposis colorectal cancer (HNPCC) and Lynch syndrome, respectively (Kloor et al. 2004). Using an L1-targeted sequencing approach, one group found 69 instances of somatic L1 retrotransposition in 16 colorectal tumors and showed that numerous genes with a putative role in tumorigenesis had insertions (Solyom et al. 2012). In a study focused on assessing the pattern of somatic retrotransposition of L1 and its typical activity in normal colorectal epithelium, the authors found 1708 somatic L1 retrotransposition events, and the frequency of these events showed a positive correlation with increased age (Nam et al. 2023). This study further analyzed the rate of retrotransposition in matched colorectal cancers and found that the L1 retrotransposition rate is substantially increased as colorectal tumors form (Nam et al. 2023). A similar trend has been seen in esophageal cancers and the precursor cells, which may develop into tumors like Barrett's esophagus (Doucet-O'Hare et al. 2015).

As groups explore potential therapeutics for colorectal tumors that express TE transcripts and proteins, it will be crucial to identify patients who will benefit from these types of treatments quickly. A recent study showcased a method for detecting ORF1p, a L1 protein, in plasma from patients with many types of cancer, which provides an excellent opportunity for clinicians to identify the patients who may benefit from TE-targeted therapies (Taylor et al. 2023). In a different study, an 'In-Cell Western' (ICW) assay was developed in which compounds could be quickly screened for their impact on the expression of the L1 ORF1p (Kou et al. 2022). The authors found three compounds (hydroxyprogesterone, 2,2′:5′,2″-terthiophene, and ethynyl estradiol) that all decreased ORF1p expression via affecting L1 RNA transcription (Kou et al. 2022).

These findings highlight the significant role that TEs, particularly HERVs and L1 elements, play in the development and progression of colorectal cancer. HERV-H expression has been associated with immune evasion and resistance to chemotherapy, suggesting its potential as both a biomarker and a therapeutic target. Similarly, the disruption of tumor suppressor genes by somatic insertions of L1, Alu, and SVA elements demonstrates the broader genomic instability driven by TE activity. Recent advances in diagnostic techniques, such as the detection of L1 ORF1p protein in patient plasma, offer promising avenues for the early identification of patients who may benefit from TE-targeted therapies. Moreover, high-throughput screening methods, including the ICW assay, have enabled the rapid identification of small molecules capable of suppressing TE expression. These developments underscore the importance of continued investigation into TE biology in colorectal cancer, with the goal of improving diagnostic precision and expanding treatment options through personalized therapeutic strategies.

13.4.3 Hypoxia Activation of TEs

Hypoxia-inducible transcription factors (HIFs) are proteins that regulate the level of oxygen under oxidative and glycolytic conditions. Both anoxia and hypoxia occur in tumors, and understanding how oxygen level affects TE expression in malignancies is important for developing targeted therapeutics (Vaupel et al. 1989). In cancer, many tumors have regions of hypoxia, which activate HIF proteins and are associated with higher mortality (Fig. 13.1) (Wicks and Semenza 2022). In 1989, Anderson, Stoler, and Scarcello were the first to show that the absence of oxygen (anoxia) activated a retrotransposon-like VL30 element in rat fibroblasts (Anderson et al. 1989). The rat VL30 elements are endogenous retroviral-like sequences that have long LTR elements very similar to retroviral *gag* and *pol* genes and specific tRNA primer binding sites that are used by retroviral RTs (Manly et al. 1988; Keshet et al. 1980). Growing the rat fibroblasts in the absence of oxygen led to the increase of VL30 element transcription and the induction and secretion of several cellular

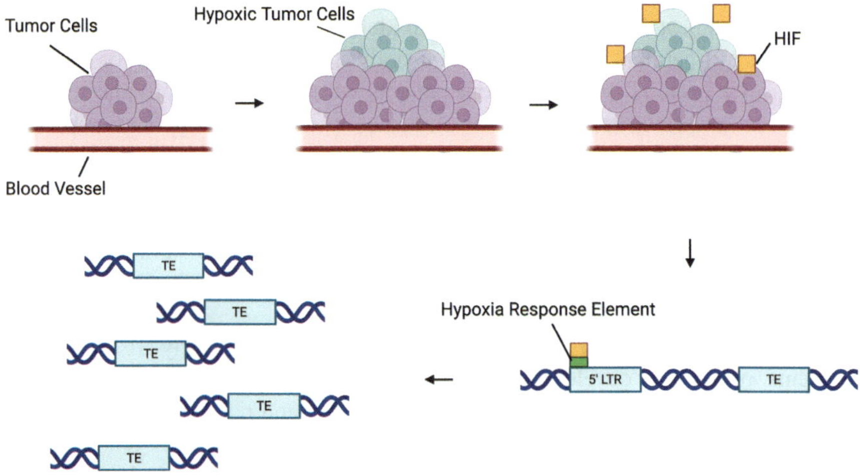

Fig. 13.1 Overview of hypoxic activation of TEs. As tumors grow (tumor cells in purple and green), more oxygen is necessary for the larger cell mass; however, eventually demand outpaces supply from existing blood vessels (bolded red lines). This leads to hypoxic regions of the tumor where cells release signaling molecules such as hypoxia inducible factor 1 (HIF-1), shown as a yellow square, which binds to hypoxia response elements, shown as green rectangles, within the 5′ end of the LTRs, triggering TE transcription. This results in elevated TE expression in the tumor microenvironment. Created in BioRender

proteins, including a protease cathepsin L (Anderson et al. 1989). Anderson et al. (1989) concluded that the fibroblast response to anoxia did not constitute a survival mechanism, but rather resembled cellular behaviors characteristic of the early phases of wound healing and processes observed in malignant cancer cells (Anderson et al. 1989). Estes et al. later identified a 14-base pair sequence in the VL30 element, named secondary anoxia response element (SARE), that was similar to the binding domain in HIF-1. The SARE element is able to bind to HIF-1 in both hypoxic and anoxic conditions. They were able to identify a secondary binding location, which was more predominant in anoxic fibroblasts, called anoxia inducible factor (AIF)(Estes et al. 1995).

In 2010, aberrant methylation and subsequent mobilization of Alu elements were found to contribute to the genomic instability present in hypoxic conditions in U87MG glioblastoma cells (Pal et al. 2010). The authors showed a significant increase in SINE and LINE transcripts under hypoxic conditions and used bisulfite sequencing to demonstrate the change of methylation at SINE loci (Pal et al. 2010). The authors suggested that hypoxic conditions may contribute to TE-derived genomic instability in other tumors as well (Pal et al. 2010). Renal cell carcinoma (RCC) is the most common and lethal kidney cancer with constitutive stabilization of HIF factors 1α and 2α (Siebenthall et al. 2019). The von Hippel-Lindau (VHL) tumor suppressor gene is inactive in most clear cell RCC (ccRCC), which allows for stabilization of both HIF-1α and HIF2α to stabilize and bind their targets (Cherkasova et al. 2011). These authors further showed HERV-E expression in most ccRCC tumors, but not in non-clear cell kidney tumors and that loss of VHL leads to HIF stability that then binds the hypoxia response element in the 5′ LTR and activates HERV-E transcription. In a recent study on ccRCC, Jiang et al. showed that HIFs commonly induce ERV expression in kidney cancers (Jiang et al. 2025). These ERVs have proteins that present like human leukocyte antigen (HLA) bound peptides on the surface of macrophages, which induce a CD8+ T-cell response (Jiang et al. 2025). The researchers further demonstrated that the expression of ERVs can be modulated—either upregulated or downregulated—by HIF stabilizers or repressors, suggesting the potential to induce this response in other models. They substantiated this finding by successfully inducing an ERV response in a non-ccRCC tumor through the administration of an HIF-stabilizing agent (Jiang et al. 2025).

In summary, HIFs are central regulators of oxygen-sensitive gene expression, and their activation under hypoxic and anoxic conditions has profound implications for tumor biology. Early studies established the responsiveness of retrotransposon-like elements to oxygen deprivation, while subsequent research has shown that hypoxia can destabilize genomic integrity via TE activation. In renal cancers, HIF stabilization due to VHL loss promotes ERV expression, which then may be leveraged to stimulate tumor-specific immune responses. These findings underscore the potential of targeting hypoxia-responsive elements for therapeutic benefit in cancer.

13.5 Conclusion

TEs, once considered "junk DNA," are now recognized as key regulators in human health and disease. Across diverse pathological landscapes—including viral infections, autoimmune disorders, neurodegenerative diseases, and cancers—TEs, particularly HERVs and LINE-1 elements, emerge as critical modulators of gene expression, immune activation, and genomic stability. Their aberrant activation contributes to chronic inflammation, proteinopathies, and tumorigenesis, revealing shared molecular mechanisms underlying seemingly disparate conditions.

In neurodegenerative diseases such as Alzheimer's disease, Parkinson's disease, ALS, and MS, dysregulated TE activity not only exacerbates neuroinflammation and neuronal death but also interacts with hallmark pathological proteins like TDP-43, tau, and α-synuclein. Likewise, in autoimmune diseases like SLE and infections such as HIV, TE reactivation amplifies immune responses, often leading to tissue damage and persistent disease activity. In cancer, TEs can drive oncogenesis through insertional mutagenesis, epigenetic deregulation, and activation of immune evasion pathways, while also offering unique vulnerabilities for therapeutic targeting.

Therapeutically, targeting TEs has transitioned from concept to clinical exploration. Strategies including RT inhibitors, monoclonal antibodies against TE-derived proteins, and epigenetic modulators have shown preclinical and early clinical promise (Fig. 13.2). Stratification of patients based on TE expression profiles—such as in

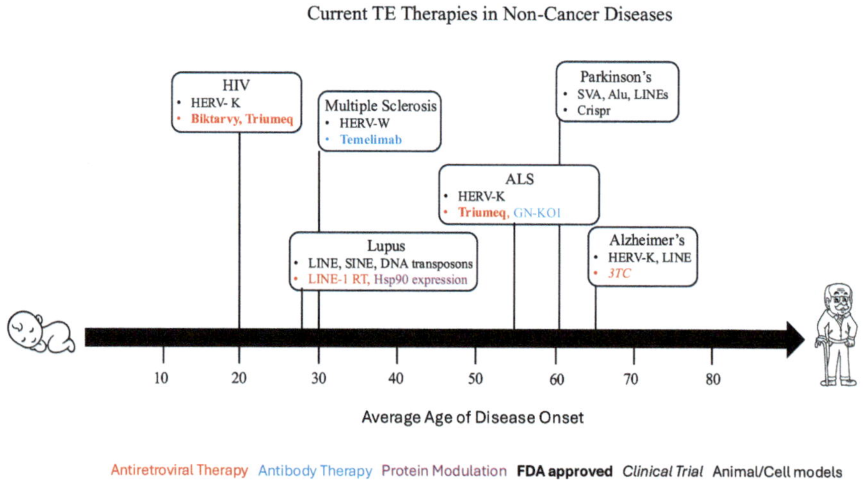

Fig. 13.2 Current TE Therapies in Non-Cancer Diseases. A timeline of ages at which diseases typically manifest, along with information about which TEs are expressed in those conditions and treatments that are currently being used to target them. Antiretroviral therapies are highlighted in red, antibody therapies in blue, and protein modulation therapies in purple. Treatments that have been FDA-approved are bolded, treatments currently under a clinical trial are in italics, and treatments only tested in animal or cell models are in plain text

ALS and glioma—highlights the potential of precision medicine approaches to optimize outcomes. Furthermore, novel diagnostics capable of detecting TE-related biomarkers in accessible biofluids are enhancing early disease identification and monitoring treatment response.

As our understanding of TE biology continues to deepen, the integration of TE-targeted therapies into standard clinical practice holds substantial promise. Future research must focus on identifying patient subsets most likely to benefit from these treatments, refining delivery methods, and minimizing off-target effects. By embracing TEs as both pathogenic drivers and therapeutic targets, we open new frontiers in the treatment of complex, currently intractable diseases.

Acknowledgments Kathryn E. Howe and Dr. Tara T. Doucet-O'Hare wrote and edited this chapter together. They would like to thank Dr. Robert Anholt and Dr. Lela Lackey for their helpful review and comments, which greatly improved the chapter.

The text of this chapter was refined with the assistance of artificial intelligence (AI) tools-OpenAI's ChatGPT, GPT-4-turbo, used in June of 2025. Every section was carefully reviewed, revised, and approved by the author/s to ensure that the text is appropriate in terms of content and scientific correctness and maintains the integrity of the original content.

Compliance with Ethical Standards

Funding This study was funded by start-up funds provided to Dr. Tara T. Doucet-O'Hare by Clemson University.

Conflict of Interest Dr. Tara T. Doucet-O'Hare declares that she has no conflict of interest. Kathryn E. Howe declares that he has no conflict of interest.

Ethical Approval This chapter is a review of previously published literature; as such, no animal or human studies were performed.

References

Abrusan G (2012) Somatic transposition in the brain has the potential to influence the biosynthesis of metabolites involved in Parkinson's disease and schizophrenia. Biol Direct 7:41. https://doi.org/10.1186/1745-6150-7-41

Adams JW, Kaufman RE, Kretschmer PJ, Harrison M, Nienhuis AW (1980) A family of long reiterated DNA sequences, one copy of which is next to the human beta globin gene. Nucleic Acids Res 8(24):6113–6128. https://doi.org/10.1093/nar/8.24.6113

Akira S, Takeda K (2004) Toll-like receptor signalling. Nat Rev Immunol 4(7):499–511. https://doi.org/10.1038/nri1391

Anderson GR, Stoler DL, Scarcello LA (1989) Normal fibroblasts responding to anoxia exhibit features of the malignant phenotype. J Biol Chem 264(25):14885–14892

Andreeva L, Hiller B, Kostrewa D, Lassig C, de Oliveira Mann CC, Jan Drexler D, Maiser A, Gaidt M, Leonhardt H, Hornung V, Hopfner KP (2017) cGAS senses long and HMGB/TFAM-bound U-turn DNA by forming protein-DNA ladders. Nature 549(7672):394–398. https://doi.org/10.1038/nature23890

Aneichyk T, Hendriks WT, Yadav R, Shin D, Gao D, Vaine CA, Collins RL, Domingo A, Currall B, Stortchevoi A, Multhaupt-Buell T, Penney EB, Cruz L, Dhakal J, Brand H, Hanscom C, Antolik

C, Dy M, Ragavendran A, Underwood J, Cantsilieris S, Munson KM, Eichler EE, Acuna P, Go C, Jamora RDG, Rosales RL, Church DM, Williams SR, Garcia S, Klein C, Muller U, Wilhelmsen KC, Timmers HTM, Sapir Y, Wainger BJ, Henderson D, Ito N, Weisenfeld N, Jaffe D, Sharma N, Breakefield XO, Ozelius LJ, Bragg DC, Talkowski ME (2018) Dissecting the causal mechanism of X-linked dystonia-parkinsonism by integrating genome and transcriptome assembly. Cell 172(5):897–909 e821. https://doi.org/10.1016/j.cell.2018.02.011

Association AP (2013) Diagnostic and statistical manual of mental disorders: DSM-5. American psychiatric association

Azlan A, Dzaki N, Azzam G (2016) Argonaute: the executor of small RNA function. J Genet Genomics 43(8):481–494. https://doi.org/10.1016/j.jgg.2016.06.002

Balestrieri E, Arpino C, Matteucci C, Sorrentino R, Pica F, Alessandrelli R, Coniglio A, Curatolo P, Rezza G, Macciardi F, Garaci E, Gaudi S, Sinibaldi-Vallebona P (2012) HERVs expression in autism Spectrum disorders. PLoS One 7(11):e48831. https://doi.org/10.1371/journal.pone.0048831

Balestrieri E, Pitzianti M, Matteucci C, D'Agati E, Sorrentino R, Baratta A, Caterina R, Zenobi R, Curatolo P, Garaci E, Sinibaldi-Vallebona P, Pasini A (2014) Human endogenous retroviruses and ADHD. World J Biol Psychiatry 15(6):499–504. https://doi.org/10.3109/15622975.2013.862345

Balestrieri E, Pica F, Matteucci C, Zenobi R, Sorrentino R, Argaw-Denboba A, Cipriani C, Bucci I, Sinibaldi-Vallebona P (2015) Transcriptional activity of human endogenous retroviruses in human peripheral blood mononuclear cells. Biomed Res Int 2015:164529. https://doi.org/10.1155/2015/164529

Balestrieri E, Cipriani C, Matteucci C, Capodicasa N, Pilika A, Korca I, Sorrentino R, Argaw-Denboba A, Bucci I, Miele MT, Coniglio A, Alessandrelli R, Vallebona PS (2016) Transcriptional activity of human endogenous retrovirus in Albanian children with autism spectrum disorders. New Microbiol 39(3):228–231

Bejerano G, Lowe CB, Ahituv N, King B, Siepel A, Salama SR, Rubin EM, Kent WJ, Haussler D (2006) A distal enhancer and an ultraconserved exon are derived from a novel retroposon. Nature 441(7089):87–90. https://doi.org/10.1038/nature04696

Bjornevik K, Cortese M, Healy BC, Kuhle J, Mina MJ, Leng Y, Elledge SJ, Niebuhr DW, Scher AI, Munger KL, Ascherio A (2022) Longitudinal analysis reveals high prevalence of Epstein-Barr virus associated with multiple sclerosis. Science 375(6578):296–301. https://doi.org/10.1126/science.abj8222

Bonifati V, Rizzu P, van Baren MJ, Schaap O, Breedveld GJ, Krieger E, Dekker MC, Squitieri F, Ibanez P, Joosse M, van Dongen JW, Vanacore N, van Swieten JC, Brice A, Meco G, van Duijn CM, Oostra BA, Heutink P (2003) Mutations in the DJ-1 gene associated with autosomal recessive early-onset parkinsonism. Science 299(5604):256–259. https://doi.org/10.1126/science.1077209

Bouras A, Leone M, Bonadona V, Lebrun M, Calender A, Boutry-Kryza N (2021) Identification and characterization of new Alu element insertion in the BRCA1 exon 14 associated with hereditary breast and ovarian cancer. Genes (Basel) 12(11):1736. https://doi.org/10.3390/genes12111736

Bourgeois Y, Boissinot S (2019) On the population dynamics of junk: a review on the population genomics of transposable elements. Genes (Basel) 10(6):419. https://doi.org/10.3390/genes10060419

Brinzevich D, Young GR, Sebra R, Ayllon J, Maio SM, Deikus G, Chen BK, Fernandez-Sesma A, Simon V, Mulder LC (2014) HIV-1 interacts with human endogenous retrovirus K (HML-2) envelopes derived from human primary lymphocytes. J Virol 88(11):6213–6223. https://doi.org/10.1128/JVI.00669-14

Brodsky I, Foley B, Haines D, Johnston J, Cuddy K, Gillespie D (1993) Expression of HERV-K proviruses in human leukocytes. Blood 81(9):2369–2374

Brouha B, Schustak J, Badge RM, Lutz-Prigge S, Farley AH, Moran JV, Kazazian HH Jr (2003) Hot L1s account for the bulk of retrotransposition in the human population. Proc Natl Acad Sci USA 100(9):5280–5285. https://doi.org/10.1073/pnas.0831042100

Buscher K, Trefzer U, Hofmann M, Sterry W, Kurth R, Denner J (2005) Expression of human endogenous retrovirus K in melanomas and melanoma cell lines. Cancer Res 65(10):4172–4180. https://doi.org/10.1158/0008-5472.CAN-04-2983

Cajuso T, Sulo P, Tanskanen T, Katainen R, Taira A, Hanninen UA, Kondelin J, Forsstrom L, Valimaki N, Aavikko M, Kaasinen E, Ristimaki A, Koskensalo S, Lepisto A, Renkonen-Sinisalo L, Seppala T, Kuopio T, Bohm J, Mecklin JP, Kilpivaara O, Pitkanen E, Palin K, Aaltonen LA (2019) Retrotransposon insertions can initiate colorectal cancer and are associated with poor survival. Nat Commun 10(1):4022. https://doi.org/10.1038/s41467-019-11770-0

Chang YH, Dubnau J (2019) The gypsy endogenous retrovirus drives non-cell-autonomous propagation in a drosophila TDP-43 model of neurodegeneration. Curr Biol 29(19):3135–3152 e3134. https://doi.org/10.1016/j.cub.2019.07.071

Chang YH, Dubnau J (2023) Endogenous retroviruses and TDP-43 proteinopathy form a sustaining feedback driving intercellular spread of drosophila neurodegeneration. Nat Commun 14(1):966. https://doi.org/10.1038/s41467-023-36649-z

Cheng Y, Saville L, Gollen B, Isaac C, Belay A, Mehla J, Patel K, Thakor N, Mohajerani MH, Zovoilis A (2020) Increased processing of SINE B2 ncRNAs unveils a novel type of transcriptome deregulation in amyloid beta neuropathology. elife 9:e61265. https://doi.org/10.7554/eLife.61265

Cherkasova E, Malinzak E, Rao S, Takahashi Y, Senchenko VN, Kudryavtseva AV, Nickerson ML, Merino M, Hong JA, Schrump DS, Srinivasan R, Linehan WM, Tian X, Lerman MI, Childs RW (2011) Inactivation of the von Hippel-Lindau tumor suppressor leads to selective expression of a human endogenous retrovirus in kidney cancer. Oncogene 30(47):4697–4706. https://doi.org/10.1038/onc.2011.179

Chiappinelli KB, Strissel PL, Desrichard A, Li H, Henke C, Akman B, Hein A, Rote NS, Cope LM, Snyder A, Makarov V, Budhu S, Slamon DJ, Wolchok JD, Pardoll DM, Beckmann MW, Zahnow CA, Merghoub T, Chan TA, Baylin SB, Strick R (2015) Inhibiting DNA methylation causes an interferon response in cancer via dsRNA including endogenous retroviruses. Cell 162(5):974–986. https://doi.org/10.1016/j.cell.2015.07.011

Choi WI, Kim MY, Jeon BN, Koh DI, Yun CO, Li Y, Lee CE, Oh J, Kim K, Hur MW (2014) Role of promyelocytic leukemia zinc finger (PLZF) in cell proliferation and cyclin-dependent kinase inhibitor 1A (p21WAF/CDKN1A) gene repression. J Biol Chem 289(27):18625–18640. https://doi.org/10.1074/jbc.M113.538751

Choudhury SR, Cui Y, Milton JR, Li J, Irudayaraj J (2015) Selective increase in subtelomeric DNA methylation: an epigenetic biomarker for malignant glioma. Clin Epigenetics 7:107. https://doi.org/10.1186/s13148-015-0140-y

Chuong EB, Elde NC, Feschotte C (2017) Regulatory activities of transposable elements: from conflicts to benefits. Nat Rev Genet 18(2):71–86. https://doi.org/10.1038/nrg.2016.139

Clinic C (2023) Antiretroviral Therapy.. https://my.clevelandclinic.org/health/treatments/antiretroviral-therapy

Collaborators GBDNSD (2024) Global, regional, and national burden of disorders affecting the nervous system, 1990-2021: a systematic analysis for the global burden of disease study 2021. Lancet Neurol 23(4):344–381. https://doi.org/10.1016/S1474-4422(24)00038-3

Contreras-Galindo R, Kaplan MH, Markovitz DM, Lorenzo E, Yamamura Y (2006) Detection of HERV-K(HML-2) viral RNA in plasma of HIV type 1-infected individuals. AIDS Res Hum Retrovir 22(10):979–984. https://doi.org/10.1089/aid.2006.22.979

Contreras-Galindo R, Almodovar-Camacho S, Gonzalez-Ramirez S, Lorenzo E, Yamamura Y (2007) Comparative longitudinal studies of HERV-K and HIV-1 RNA titers in HIV-1-infected patients receiving successful versus unsuccessful highly active antiretroviral therapy. AIDS Res Hum Retrovir 23(9):1083–1086. https://doi.org/10.1089/aid.2007.0054

Crow JF, Dove WF (2000) Perspectives on genetics: anecdotal, historical, and critical commentaries, 1987–1998. University of Wisconsin Press, Madison

Curtin F, Perron H, Kromminga A, Porchet H, Lang AB (2015) Preclinical and early clinical development of GNbAC1, a humanized IgG4 monoclonal antibody targeting endogenous retroviral MSRV-Env protein. MAbs 7(1):265–275. https://doi.org/10.4161/19420862.2014.985021

Cutts Z, Patterson S, Maliskova L, Taylor KE, Ye CJ, Dall'Era M, Yazdany J, Criswell LA, Fragiadakis GK, Langelier C, Capra JA, Sirota M, Lanata CM (2024) Cell-specific transposable element and gene expression analysis across systemic lupus erythematosus phenotypes. ACR Open Rheumatol 6(11):769–779. https://doi.org/10.1002/acr2.11713

Dammann RH, Richter AM, Jimenez AP, Woods M, Kuster M, Witharana C (2017) Impact of natural compounds on DNA methylation levels of the tumor suppressor gene RASSF1A in cancer. Int J Mol Sci 18(10). https://doi.org/10.3390/ijms18102160

Dang DD, Rosenblum JS, Shah AH, Zhuang Z, Doucet-O'Hare TT (2023) Epigenetic regulation in primary CNS tumors: an opportunity to bridge old and new WHO classifications. Cancers (Basel) 15(9):2511. https://doi.org/10.3390/cancers15092511

Daskalos A, Nikolaidis G, Xinarianos G, Savvari P, Cassidy A, Zakopoulou R, Kotsinas A, Gorgoulis V, Field JK, Liloglou T (2009) Hypomethylation of retrotransposable elements correlates with genomic instability in non-small cell lung cancer. Int J Cancer 124(1):81–87. https://doi.org/10.1002/ijc.23849

Dawson T, Rentia U, Sanford J, Cruchaga C, Kauwe JSK, Crandall KA (2023) Locus specific endogenous retroviral expression associated with Alzheimer's disease. Front Aging Neurosci 15:1186470. https://doi.org/10.3389/fnagi.2023.1186470

De Cecco M, Ito T, Petrashen AP, Elias AE, Skvir NJ, Criscione SW, Caligiana A, Brocculi G, Adney EM, Boeke JD, Le O, Beausejour C, Ambati J, Ambati K, Simon M, Seluanov A, Gorbunova V, Slagboom PE, Helfand SL, Neretti N, Sedivy JM (2019) L1 drives IFN in senescent cells and promotes age-associated inflammation. Nature 566(7742):73–78. https://doi.org/10.1038/s41586-018-0784-9

Dembny P, Newman AG, Singh M, Hinz M, Szczepek M, Kruger C, Adalbert R, Dzaye O, Trimbuch T, Wallach T, Kleinau G, Derkow K, Richard BC, Schipke C, Scheidereit C, Stachelscheid H, Golenbock D, Peters O, Coleman M, Heppner FL, Scheerer P, Tarabykin V, Ruprecht K, Izsvak Z, Mayer J, Lehnardt S (2020) Human endogenous retrovirus HERV-K(HML-2) RNA causes neurodegeneration through toll-like receptors. JCI. Insight 5(7):e131093. https://doi.org/10.1172/jci.insight.131093

Denne M, Sauter M, Armbruester V, Licht JD, Roemer K, Mueller-Lantzsch N (2007) Physical and functional interactions of human endogenous retrovirus proteins Np9 and rec with the promyelocytic leukemia zinc finger protein. J Virol 81(11):5607–5616. https://doi.org/10.1128/JVI.02771-06

Dewannieux M, Esnault C, Heidmann T (2003) LINE-mediated retrotransposition of marked Alu sequences. Nat Genet 35(1):41–48. https://doi.org/10.1038/ng1223

Diebold M, Derfuss T (2019) The monoclonal antibody GNbAC1: targeting human endogenous retroviruses in multiple sclerosis. Ther Adv Neurol Disord 12:1756286419833574. https://doi.org/10.1177/1756286419833574

D'Ordine AM, Jogl G, Sedivy JM (2024) Identification and characterization of small molecule inhibitors of the LINE-1 retrotransposon endonuclease. Nat Commun 15(1):3883. https://doi.org/10.1038/s41467-024-48066-x

Doucet-O'Hare TT, Rodic N, Sharma R, Darbari I, Abril G, Choi JA, Young Ahn J, Cheng Y, Anders RA, Burns KH, Meltzer SJ, Kazazian HH Jr (2015) LINE-1 expression and retrotransposition in Barrett's esophagus and esophageal carcinoma. Proc Natl Acad Sci USA 112(35):E4894–E4900. https://doi.org/10.1073/pnas.1502474112

Doucet-O'Hare TT, DiSanza BL, DeMarino C, Atkinson AL, Rosenblum JS, Henderson LJ, Johnson KR, Kowalak J, Garcia-Montojo M, Allen SJ, Orr BA, Santi M, Wang T, Fathi S, Lee MH, Sampson K, Li W, Zhuang Z, Nath A (2021) SMARCB1 deletion in atypical teratoid

rhabdoid tumors results in human endogenous retrovirus K (HML-2) expression. Sci Rep 11(1):12893. https://doi.org/10.1038/s41598-021-92223-x
Douville R, Liu J, Rothstein J, Nath A (2011) Identification of active loci of a human endogenous retrovirus in neurons of patients with amyotrophic lateral sclerosis. Ann Neurol 69(1):141–151. https://doi.org/10.1002/ana.22149
Dunker W, Zhao Y, Song Y, Karijolich J (2017) Recognizing the SINEs of infection: regulation of retrotransposon expression and modulation of host cell processes. Viruses 9(12). https://doi.org/10.3390/v9120386
Durnaoglu S, Lee SK, Ahnn J (2021) Human endogenous retroviruses as gene expression regulators: insights from animal models into human diseases. Mol Cells 44(12):861–878. https://doi.org/10.14348/molcells.2021.5016
Ergun SL, Fernandez D, Weiss TM, Li L (2019) STING polymer structure reveals mechanisms for activation, Hyperactivation, and inhibition. Cell 178(2):290–301 e210. https://doi.org/10.1016/j.cell.2019.05.036
Eshima J, O'Connor SA, Marschall E, Consortium NA, Bowser R, Plaisier CL, Smith BS (2023) Molecular subtypes of ALS are associated with differences in patient prognosis. Nat Commun 14(1):95. https://doi.org/10.1038/s41467-022-35494-w
Estes SD, Stoler DL, Anderson GR (1995) Anoxic induction of a sarcoma virus-related VL30 retrotransposon is mediated by a cis-acting element which binds hypoxia-inducible factor 1 and an anoxia-inducible factor. J Virol 69(10):6335–6341. https://doi.org/10.1128/JVI.69.10.6335-6341.1995
Eyjolfsdottir H, Eriksdotter M, Linderoth B, Lind G, Juliusson B, Kusk P, Almkvist O, Andreasen N, Blennow K, Ferreira D, Westman E, Nennesmo I, Karami A, Darreh-Shori T, Kadir A, Nordberg A, Sundstrom E, Wahlund LO, Wall A, Wiberg M, Winblad B, Seiger A, Wahlberg L, Almqvist P (2016) Targeted delivery of nerve growth factor to the cholinergic basal forebrain of Alzheimer's disease patients: application of a second-generation encapsulated cell biodelivery device. Alzheimer's Res Ther 8(1):30. https://doi.org/10.1186/s13195-016-0195-9
Farrer MJ (2006) Genetics of Parkinson disease: paradigm shifts and future prospects. Nat Rev Genet 7(4):306–318. https://doi.org/10.1038/nrg1831
Feng Y, Cao SQ, Shi Y, Sun A, Flanagan ME, Leverenz JB, Pieper AA, Jung JU, Cummings J, Fang EF, Zhang P, Cheng F (2025) Human herpesvirus-associated transposable element activation in human aging brains with Alzheimer's disease. Alzheimers Dement 21(2):e14595. https://doi.org/10.1002/alz.14595
Fischl MA, Richman DD, Grieco MH, Gottlieb MS, Volberding PA, Laskin OL, Leedom JM, Groopman JE, Mildvan D, Schooley RT et al (1987) The efficacy of azidothymidine (AZT) in the treatment of patients with AIDS and AIDS-related complex. A double-blind, placebo-controlled trial. N Engl J Med 317(4):185–191. https://doi.org/10.1056/NEJM198707233170401
Frischer JM, Bramow S, Dal-Bianco A, Lucchinetti CF, Rauschka H, Schmidbauer M, Laursen H, Sorensen PS, Lassmann H (2009) The relation between inflammation and neurodegeneration in multiple sclerosis brains. Brain 132(Pt 5):1175–1189. https://doi.org/10.1093/brain/awp070
Frohlich A, Pfaff AL, Middlehurst B, Hughes LS, Bubb VJ, Quinn JP, Koks S (2024) Deciphering the role of a SINE-VNTR-Alu retrotransposon polymorphism as a biomarker of Parkinson's disease progression. Sci Rep 14(1):10932. https://doi.org/10.1038/s41598-024-61753-5
Gandhi RT, Bedimo R, Hoy JF, Landovitz RJ, Smith DM, Eaton EF, Lehmann C, Springer SA, Sax PE, Thompson MA, Benson CA, Buchbinder SP, Del Rio C, Eron JJ Jr, Gunthard HF, Molina JM, Jacobsen DM, Saag MS (2023) Antiretroviral drugs for treatment and prevention of HIV infection in adults: 2022 recommendations of the international antiviral society-USA panel. JAMA 329(1):63–84. https://doi.org/10.1001/jama.2022.22246
Garcia-Montojo M, Doucet-O'Hare T, Henderson L, Nath A (2018) Human endogenous retrovirus-K (HML-2): a comprehensive review. Crit Rev Microbiol 44(6):715–738. https://doi.org/10.1080/1040841X.2018.1501345
Garcia-Montojo M, Fathi S, Norato G, Smith BR, Rowe DB, Kiernan MC, Vucic S, Mathers S, van Eijk RPA, Santamaria U, Rogers ML, Malaspina A, Lombardi V, Mehta PR, Westeneng

HJ, van den Berg LH, Al-Chalabi A, Gold J, Nath A (2021) Inhibition of HERV-K (HML-2) in amyotrophic lateral sclerosis patients on antiretroviral therapy. J Neurol Sci 423:117358. https://doi.org/10.1016/j.jns.2021.117358

Garcia-Montojo M, Fathi S, Rastegar C, Simula ER, Doucet-O'Hare T, Cheng YHH, Abrams RPM, Pasternack N, Malik N, Bachani M, Disanza B, Maric D, Lee MH, Wang H, Santamaria U, Li W, Sampson K, Lorenzo JR, Sanchez IE, Mezghrani A, Li Y, Sechi LA, Pineda S, Heiman M, Kellis M, Steiner J, Nath A (2024) TDP-43 proteinopathy in ALS is triggered by loss of ASRGL1 and associated with HML-2 expression. Nat Commun 15(1):4163. https://doi.org/10.1038/s41467-024-48488-7

Garcia-Perez JL, Marchetto MC, Muotri AR, Coufal NG, Gage FH, O'Shea KS, Moran JV (2007) LINE-1 retrotransposition in human embryonic stem cells. Hum Mol Genet 16(13):1569–1577. https://doi.org/10.1093/hmg/ddm105

Garrison KE, Jones RB, Meiklejohn DA, Anwar N, Ndhlovu LC, Chapman JM, Erickson AL, Agrawal A, Spotts G, Hecht FM, Rakoff-Nahoum S, Lenz J, Ostrowski MA, Nixon DF (2007) T cell responses to human endogenous retroviruses in HIV-1 infection. PLoS Pathog 3(11):e165. https://doi.org/10.1371/journal.ppat.0030165

Gartner MJ, Roche M, Churchill MJ, Gorry PR, Flynn JK (2020) Understanding the mechanisms driving the spread of subtype C HIV-1. EBioMedicine 53:102682. https://doi.org/10.1016/j.ebiom.2020.102682

Ghasemi N, Razavi S, Nikzad E (2017) Multiple sclerosis: pathogenesis, symptoms, Diagnoses and Cell-Based Therapy. Cell J 19(1):1–10. https://doi.org/10.22074/cellj.2016.4867

Gimenez J, Montgiraud C, Pichon JP, Bonnaud B, Arsac M, Ruel K, Bouton O, Mallet F (2010) Custom human endogenous retroviruses dedicated microarray identifies self-induced HERV-W family elements reactivated in testicular cancer upon methylation control. Nucleic Acids Res 38(7):2229–2246. https://doi.org/10.1093/nar/gkp1214

Gold J, Rowe DB, Kiernan MC, Vucic S, Mathers S, van Eijk RPA, Nath A, Garcia Montojo M, Norato G, Santamaria UA, Rogers ML, Malaspina A, Lombardi V, Mehta PR, Westeneng HJ, van den Berg LH, Al-Chalabi A (2019) Safety and tolerability of Triumeq in amyotrophic lateral sclerosis: the lighthouse trial. Amyotroph Lateral Scler Frontotemporal Degener 20(7–8):595–604. https://doi.org/10.1080/21678421.2019.1632899

Golkaram M, Salmans ML, Kaplan S, Vijayaraghavan R, Martins M, Khan N, Garbutt C, Wise A, Yao J, Casimiro S, Abreu C, Macedo D, Costa AL, Alvim C, Mansinho A, Filipe P, Marques da Costa P, Fernandes A, Borralho P, Ferreira C, Aldeia F, Malaquias J, Godsey J, So A, Pawlowski T, Costa L, Zhang S, Liu L (2021) HERVs establish a distinct molecular subtype in stage II/III colorectal cancer with poor outcome. NPJ Genom Med 6(1):13. https://doi.org/10.1038/s41525-021-00177-w

Gorbunova V, Seluanov A, Mita P, McKerrow W, Fenyo D, Boeke JD, Linker SB, Gage FH, Kreiling JA, Petrashen AP, Woodham TA, Taylor JR, Helfand SL, Sedivy JM (2021) The role of retrotransposable elements in ageing and age-associated diseases. Nature 596(7870):43–53. https://doi.org/10.1038/s41586-021-03542-y

Gottlieb MS, Schroff R, Schanker HM, Weisman JD, Fan PT, Wolf RA, Saxon A (1981) Pneumocystis-Carinii pneumonia and mucosal candidiasis in previously healthy homosexual men – evidence of a new acquired cellular immunodeficiency. New Engl J Med 305(24):1425–1431. https://doi.org/10.1056/Nejm198112103052401

Grow EJ, Flynn RA, Chavez SL, Bayless NL, Wossidlo M, Wesche DJ, Martin L, Ware CB, Blish CA, Chang HY, Pera RA, Wysocka J (2015) Intrinsic retroviral reactivation in human preimplantation embryos and pluripotent cells. Nature 522(7555):221–225. https://doi.org/10.1038/nature14308

Grulich AE, van Leeuwen MT, Falster MO, Vajdic CM (2007) Incidence of cancers in people with HIV/AIDS compared with immunosuppressed transplant recipients: a meta-analysis. Lancet 370(9581):59–67. https://doi.org/10.1016/S0140-6736(07)61050-2

Guo C, Jeong HH, Hsieh YC, Klein HU, Bennett DA, De Jager PL, Liu Z, Shulman JM (2018) Tau activates transposable elements in Alzheimer's disease. Cell Rep 23(10):2874–2880. https://doi.org/10.1016/j.celrep.2018.05.004

Hancks DC, Kazazian HH Jr (2010) SVA retrotransposons: evolution and genetic instability. Semin Cancer Biol 20(4):234–245. https://doi.org/10.1016/j.semcancer.2010.04.001

Hartung HP, Derfuss T, Cree BA, Sormani MP, Selmaj K, Stutters J, Prados F, MacManus D, Schneble HM, Lambert E, Porchet H, Glanzman R, Warne D, Curtin F, Kornmann G, Buffet B, Kremer D, Kury P, Leppert D, Ruckle T, Barkhof F (2022) Efficacy and safety of temelimab in multiple sclerosis: results of a randomized phase 2b and extension study. Mult Scler 28(3):429–440. https://doi.org/10.1177/13524585211024997

Hirose Y, Takamatsu M, Harada F (1993) Presence of env genes in members of the RTVL-H family of human endogenous retrovirus-like elements. Virology 192(1):52–61. https://doi.org/10.1006/viro.1993.1007

Hummel B, Hansen EC, Yoveva A, Aprile-Garcia F, Hussong R, Sawarkar R (2017) The evolutionary capacitor HSP90 buffers the regulatory effects of mammalian endogenous retroviruses. Nat Struct Mol Biol 24(3):234–242. https://doi.org/10.1038/nsmb.3368

Irfan SA, Murtaza M, Ahmed A, Altaf H, Ali AA, Shabbir N, Baig MMA (2022) Promising role of temelimab in multiple sclerosis treatment. Mult Scler Relat Disord 61:103743. https://doi.org/10.1016/j.msard.2022.103743

Ishikawa H, Ma Z, Barber GN (2009) STING regulates intracellular DNA-mediated, type I interferon-dependent innate immunity. Nature 461(7265):788–792. https://doi.org/10.1038/nature08476

Jay E, Wu R (1976) Arthrobacter luteus restriction endonuclease recognition sequence and its cleavage map of SV40 DNA. Biochemistry 15(16):3612–3620. https://doi.org/10.1021/bi00661a032

Jefferies CA, Doyle S, Brunner C, Dunne A, Brint E, Wietek C, Walch E, Wirth T, O'Neill LA (2003) Bruton's tyrosine kinase is a toll/interleukin-1 receptor domain-binding protein that participates in nuclear factor kappaB activation by toll-like receptor 4. J Biol Chem 278(28):26258–26264. https://doi.org/10.1074/jbc.M301484200

Jiang Q, Braun DA, Clauser KR, Ramesh V, Shirole NH, Duke-Cohan JE, Nabilsi N, Kramer NJ, Forman C, Lippincott IE, Klaeger S, Phulphagar KM, Chea V, Kim N, Vanasse AP, Saad E, Parsons T, Carr-Reynolds M, Carulli I, Pinjusic K, Jiang Y, Li R, Syamala S, Rachimi S, Verzani EK, Stevens JD, Lane WJ, Camp SY, Meli K, Pappalardi MB, Herbert ZT, Qiu X, Cejas P, Long HW, Shukla SA, Van Allen EM, Choueiri TK, Churchman LS, Abelin JG, Gurer C, MacBeath G, Childs RW, Carr SA, Keskin DB, Wu CJ, Kaelin WG Jr (2025) HIF regulates multiple translated endogenous retroviruses: implications for cancer immunotherapy. Cell 188(7):1807–1827. e1834. https://doi.org/10.1016/j.cell.2025.01.046

Jin B, Robertson KD (2013) DNA methyltransferases, DNA damage repair, and cancer. Adv Exp Med Biol 754:3–29. https://doi.org/10.1007/978-1-4419-9967-2_1

Jones RB, Song H, Xu Y, Garrison KE, Buzdin AA, Anwar N, Hunter DV, Mujib S, Mihajlovic V, Martin E, Lee E, Kuciak M, Raposo RA, Bozorgzad A, Meiklejohn DA, Ndhlovu LC, Nixon DF, Ostrowski MA (2013) LINE-1 retrotransposable element DNA accumulates in HIV-1-infected cells. J Virol 87(24):13307–13320. https://doi.org/10.1128/JVI.02257-13

Kagan AB, Garrison DA, Anders NM, Webster JA, Baker SD, Yegnasubramanian S, Rudek MA (2023) DNA methyltransferase inhibitor exposure-response: challenges and opportunities. Clin Transl Sci 16(8):1309–1322. https://doi.org/10.1111/cts.13548

Kazazian HH Jr, Wong C, Youssoufian H, Scott AF, Phillips DG, Antonarakis SE (1988) Haemophilia a resulting from de novo insertion of L1 sequences represents a novel mechanism for mutation in man. Nature 332(6160):164–166. https://doi.org/10.1038/332164a0

Keshet E, Shaul Y, Kaminchik J, Aviv H (1980) Heterogeneity of "virus-like" genes encoding retrovirus-associated 30S RNA and their organization within the mouse genome. Cell 20(2):431–439. https://doi.org/10.1016/0092-8674(80)90629-7

Kessler AF, Wiesner M, Denner J, Kammerer U, Vince GH, Linsenmann T, Lohr M, Ernestus RI, Hagemann C (2014) Expression-analysis of the human endogenous retrovirus HERV-K in human astrocytic tumors. BMC Res Notes 7:159. https://doi.org/10.1186/1756-0500-7-159

Kloor M, Sutter C, Wentzensen N, Cremer FW, Buckowitz A, Keller M, von Knebel DM, Gebert J (2004) A large MSH2 Alu insertion mutation causes HNPCC in a German kindred. Hum Genet 115(5):432–438. https://doi.org/10.1007/s00439-004-1176-9

Koks S, Pfaff AL, Singleton LM, Bubb VJ, Quinn JP (2022) Non-reference genome transposable elements (TEs) have a significant impact on the progression of the Parkinson's disease. Exp Biol Med (Maywood) 247(18):1680–1690. https://doi.org/10.1177/15353702221117147

Kou Y, Wang S, Ma Y, Zhang N, Zhang Z, Liu Q, Mao Y, Zhou R, Yi D, Ma L, Zhang Y, Li Q, Wang J, Wang J, Zhou X, He C, Ding J, Cen S, Li X (2022) A high throughput cell-based screen assay for LINE-1 ORF1p expression inhibitors using the in-cell Western technique. Front Pharmacol 13:881938. https://doi.org/10.3389/fphar.2022.881938

Kudo-Saito C, Yura M, Yamamoto R, Kawakami Y (2014) Induction of immunoregulatory CD271+ cells by metastatic tumor cells that express human endogenous retrovirus H. Cancer Res 74(5):1361–1370. https://doi.org/10.1158/0008-5472.CAN-13-1349

Landers S, Kapadia F, Bowleg L (2021) 1981-2021: HIV and our world. Am J Public Health 111(7):1180–1182. https://doi.org/10.2105/AJPH.2021.306360

Li Y, Seto E (2016) HDACs and HDAC inhibitors in cancer development and therapy. Cold Spring Harb Perspect Med 6(10):a026831. https://doi.org/10.1101/cshperspect.a026831

Li W, Prazak L, Chatterjee N, Gruninger S, Krug L, Theodorou D, Dubnau J (2013a) Activation of transposable elements during aging and neuronal decline in drosophila. Nat Neurosci 16(5):529–531. https://doi.org/10.1038/nn.3368

Li X, Shu C, Yi G, Chaton CT, Shelton CL, Diao J, Zuo X, Kao CC, Herr AB, Li P (2013b) Cyclic GMP-AMP synthase is activated by double-stranded DNA-induced oligomerization. Immunity 39(6):1019–1031. https://doi.org/10.1016/j.immuni.2013.10.019

Li W, Lee MH, Henderson L, Tyagi R, Bachani M, Steiner J, Campanac E, Hoffman DA, von Geldern G, Johnson K, Maric D, Morris HD, Lentz M, Pak K, Mammen A, Ostrow L, Rothstein J, Nath A (2015) Human endogenous retrovirus-K contributes to motor neuron disease. Sci Transl Med 7(307):307ra153. https://doi.org/10.1126/scitranslmed.aac8201

Li W, Pandya D, Pasternack N, Garcia-Montojo M, Henderson L, Kozak CA, Nath A (2022) Retroviral elements in pathophysiology and as therapeutic targets for amyotrophic lateral sclerosis. Neurotherapeutics 19(4):1085–1101. https://doi.org/10.1007/s13311-022-01233-8

Link JO, Rhee MS, Tse WC, Zheng J, Somoza JR, Rowe W, Begley R, Chiu A, Mulato A, Hansen D, Singer E, Tsai LK, Bam RA, Chou CH, Canales E, Brizgys G, Zhang JR, Li J, Graupe M, Morganelli P, Liu Q, Wu Q, Halcomb RL, Saito RD, Schroeder SD, Lazerwith SE, Bondy S, Jin D, Hung M, Novikov N, Liu X, Villasenor AG, Cannizzaro CE, Hu EY, Anderson RL, Appleby TC, Lu B, Mwangi J, Liclican A, Niedziela-Majka A, Papalia GA, Wong MH, Leavitt SA, Xu Y, Koditek D, Stepan GJ, Yu H, Pagratis N, Clancy S, Ahmadyar S, Cai TZ, Sellers S, Wolckenhauer SA, Ling J, Callebaut C, Margot N, Ram RR, Liu YP, Hyland R, Sinclair GI, Ruane PJ, Crofoot GE, McDonald CK, Brainard DM, Lad L, Swaminathan S, Sundquist WI, Sakowicz R, Chester AE, Lee WE, Daar ES, Yant SR, Cihlar T (2020) Clinical targeting of HIV capsid protein with a long-acting small molecule. Nature 584(7822):614–618. https://doi.org/10.1038/s41586-020-2443-1

Liu X, Li Z, Fan C, Zhang D, Chen J (2017) Genetics implicate common mechanisms in autism and schizophrenia: synaptic activity and immunity. J Med Genet 54(8):511–520. https://doi.org/10.1136/jmedgenet-2016-104487

Liu M, Thomas SL, DeWitt AK, Zhou W, Madaj ZB, Ohtani H, Baylin SB, Liang G, Jones PA (2018) Dual inhibition of DNA and histone Methyltransferases increases viral mimicry in ovarian cancer cells. Cancer Res 78(20):5754–5766. https://doi.org/10.1158/0008-5472.CAN-17-3953

Liu S, Heumuller SE, Hossinger A, Muller SA, Buravlova O, Lichtenthaler SF, Denner P, Vorberg IM (2023a) Reactivated endogenous retroviruses promote protein aggregate spreading. Nat Commun 14(1):5034. https://doi.org/10.1038/s41467-023-40632-z

Liu X, Liu Z, Wu Z, Ren J, Fan Y, Sun L, Cao G, Niu Y, Zhang B, Ji Q, Jiang X, Wang C, Wang Q, Ji Z, Li L, Esteban CR, Yan K, Li W, Cai Y, Wang S, Zheng A, Zhang YE, Tan S, Cai Y, Song M, Lu F, Tang F, Ji W, Zhou Q, Belmonte JCI, Zhang W, Qu J, Liu GH (2023b) Resurrection of endogenous retroviruses during aging reinforces senescence. Cell 186(2):287–304 e226. https://doi.org/10.1016/j.cell.2022.12.017

Lotan A, Fenckova M, Bralten J, Alttoa A, Dixson L, Williams RW, van der Voet M (2014) Neuroinformatic analyses of common and distinct genetic components associated with major neuropsychiatric disorders. Front Neurosci 8:331. https://doi.org/10.3389/fnins.2014.00331

Lower R, Lower J, Kurth R (1996) The viruses in all of us: characteristics and biological significance of human endogenous retrovirus sequences. Proc Natl Acad Sci USA 93(11):5177–5184. https://doi.org/10.1073/pnas.93.11.5177

Luo X, Guo X, Tan Y, Zhang Y, Garcia-Milian R, Wang Z, Shi J, Yu T, Ji J, Wang X, Xu J, Zhang H, Zuo L, Lu L, Wang K, Li CR (2020) KTN1 variants and risk for attention deficit hyperactivity disorder. Am J Med Genet B Neuropsychiatr Genet 183(4):234–244. https://doi.org/10.1002/ajmg.b.32782

Ma Y, Li J, Chiu I, Wang Y, Sloane JA, Lu J, Kosaras B, Sidman RL, Volpe JJ, Vartanian T (2006) Toll-like receptor 8 functions as a negative regulator of neurite outgrowth and inducer of neuronal apoptosis. J Cell Biol 175(2):209–215. https://doi.org/10.1083/jcb.200606016

Macfarlan TS, Gifford WD, Driscoll S, Lettieri K, Rowe HM, Bonanomi D, Firth A, Singer O, Trono D, Pfaff SL (2012) Embryonic stem cell potency fluctuates with endogenous retrovirus activity. Nature 487(7405):57–63. https://doi.org/10.1038/nature11244

Makino S, Kaji R, Ando S, Tomizawa M, Yasuno K, Goto S, Matsumoto S, Tabuena MD, Maranon E, Dantes M, Lee LV, Ogasawara K, Tooyama I, Akatsu H, Nishimura M, Tamiya G (2007) Reduced neuron-specific expression of the TAF1 gene is associated with X-linked dystonia-parkinsonism. Am J Hum Genet 80(3):393–406. https://doi.org/10.1086/512129

Manca MA, Solinas T, Simula ER, Noli M, Ruberto S, Madonia M, Sechi LA (2022) HERV-K and HERV-H Env proteins induce a humoral response in prostate cancer patients. Pathogens 11(1):95. https://doi.org/10.3390/pathogens11010095

Manly KF, Anderson GR, Stoler DL (1988) Harvey sarcoma virus genome contains no extensive sequences unrelated to those of other retroviruses except ras. J Virol 62(9):3540–3543. https://doi.org/10.1128/JVI.62.9.3540-3543.1988

Marchi E, Kanapin A, Magiorkinis G, Belshaw R (2014) Unfixed endogenous retroviral insertions in the human population. J Virol 88(17):9529–9537. https://doi.org/10.1128/JVI.00919-14

McClintock B (1950) The origin and behavior of mutable loci in maize. Proc Natl Acad Sci USA 36(6):344–355. https://doi.org/10.1073/pnas.36.6.344

McConnell MJ, Chevallier N, Berkofsky-Fessler W, Giltnane JM, Malani RB, Staudt LM, Licht JD (2003) Growth suppression by acute promyelocytic leukemia-associated protein PLZF is mediated by repression of c-myc expression. Mol Cell Biol 23(24):9375–9388. https://doi.org/10.1128/MCB.23.24.9375-9388.2003

McKerrow W, Kagermazova L, Doudican N, Frazzette N, Kaparos EI, Evans SA, Rocha A, Sedivy JM, Neretti N, Carucci J, Boeke JD, Fenyo D (2023) LINE-1 retrotransposon expression in cancerous, epithelial and neuronal cells revealed by 5′ single-cell RNA-Seq. Nucleic Acids Res 51(5):2033–2045. https://doi.org/10.1093/nar/gkad049

Metifiot M, Marchand C, Pommier Y (2013) HIV integrase inhibitors: 20-year landmark and challenges. Adv Pharmacol 67:75–105. https://doi.org/10.1016/B978-0-12-405880-4.00003-2

Miao B, Fu S, Lyu C, Gontarz P, Wang T, Zhang B (2020) Tissue-specific usage of transposable element-derived promoters in mouse development. Genome Biol 21(1):255. https://doi.org/10.1186/s13059-020-02164-3

Michaud HA, de Mulder M, SenGupta D, Deeks SG, Martin JN, Pilcher CD, Hecht FM, Sacha JB, Nixon DF (2014) Trans-activation, post-transcriptional maturation, and induction of antibodies to HERV-K (HML-2) envelope transmembrane protein in HIV-1 infection. Retrovirology 11:10. https://doi.org/10.1186/1742-4690-11-10

Miki Y, Nishisho I, Horii A, Miyoshi Y, Utsunomiya J, Kinzler KW, Vogelstein B, Nakamura Y (1992) Disruption of the APC gene by a retrotransposal insertion of L1 sequence in a colon cancer. Cancer Res 52(3):643–645

Minami J, Suzuki R, Mazitschek R, Gorgun G, Ghosh B, Cirstea D, Hu Y, Mimura N, Ohguchi H, Cottini F, Jakubikova J, Munshi NC, Haggarty SJ, Richardson PG, Hideshima T, Anderson KC (2014) Histone deacetylase 3 as a novel therapeutic target in multiple myeloma. Leukemia 28(3):680–689. https://doi.org/10.1038/leu.2013.231

Mistry P, Kaplan MJ (2017) Cell death in the pathogenesis of systemic lupus erythematosus and lupus nephritis. Clin Immunol 185:59–73. https://doi.org/10.1016/j.clim.2016.08.010

Morgan MA, Galla M, Grez M, Fehse B, Schambach A (2021) Retroviral gene therapy in Germany with a view on previous experience and future perspectives. Gene Ther 28(9):494–512. https://doi.org/10.1038/s41434-021-00237-x

Nam CH, Youk J, Kim JY, Lim J, Park JW, Oh SA, Lee HJ, Park JW, Won H, Lee Y, Jeong SY, Lee DS, Oh JW, Han J, Lee J, Kwon HW, Kim MJ, Ju YS (2023) Widespread somatic L1 retrotransposition in normal colorectal epithelium. Nature 617(7961):540–547. https://doi.org/10.1038/s41586-023-06046-z

Neftel C, Laffy J, Filbin MG, Hara T, Shore ME, Rahme GJ, Richman AR, Silverbush D, Shaw ML, Hebert CM, Dewitt J, Gritsch S, Perez EM, Gonzalez Castro LN, Lan X, Druck N, Rodman C, Dionne D, Kaplan A, Bertalan MS, Small J, Pelton K, Becker S, Bonal D, Nguyen QD, Servis RL, Fung JM, Mylvaganam R, Mayr L, Gojo J, Haberler C, Geyeregger R, Czech T, Slavc I, Nahed BV, Curry WT, Carter BS, Wakimoto H, Brastianos PK, Batchelor TT, Stemmer-Rachamimov A, Martinez-Lage M, Frosch MP, Stamenkovic I, Riggi N, Rheinbay E, Monje M, Rozenblatt-Rosen O, Cahill DP, Patel AP, Hunter T, Verma IM, Ligon KL, Louis DN, Regev A, Bernstein BE, Tirosh I, Suva ML (2019) An integrative model of cellular states, plasticity, and genetics for glioblastoma. Cell 178(4):835–849. e821. https://doi.org/10.1016/j.cell.2019.06.024

Nicoletto G, Terreri M, Maurizio I, Ruggiero E, Cernilogar FM, Vaine CA, Cottini MV, Shcherbakova I, Penney EB, Gallina I, Monchaud D, Bragg DC, Schotta G, Richter SN (2024) G-quadruplexes in an SVA retrotransposon cause aberrant TAF1 gene expression in X-linked dystonia parkinsonism. Nucleic Acids Res 52(19):11571–11586. https://doi.org/10.1093/nar/gkae797

Nikolaienko O, Patil S, Eriksen MS, Bramham CR (2018) Arc protein: a flexible hub for synaptic plasticity and cognition. Semin Cell Dev Biol 77:33–42. https://doi.org/10.1016/j.semcdb.2017.09.006

Ono M, Yasunaga T, Miyata T, Ushikubo H (1986) Nucleotide sequence of human endogenous retrovirus genome related to the mouse mammary tumor virus genome. J Virol 60(2):589–598. https://doi.org/10.1128/JVI.60.2.589-598.1986

Ostertag EM, Goodier JL, Zhang Y, Kazazian HH Jr (2003) SVA elements are nonautonomous retrotransposons that cause disease in humans. Am J Hum Genet 73(6):1444–1451. https://doi.org/10.1086/380207

Pal A, Srivastava T, Sharma MK, Mehndiratta M, Das P, Sinha S, Chattopadhyay P (2010) Aberrant methylation and associated transcriptional mobilization of Alu elements contributes to genomic instability in hypoxia. J Cell Mol Med 14(11):2646–2654. https://doi.org/10.1111/j.1582-4934.2009.00792.x

Pan L, Lu MP, Wang JH, Xu M, Yang SR (2020) Immunological pathogenesis and treatment of systemic lupus erythematosus. World J Pediatr 16(1):19–30. https://doi.org/10.1007/s12519-019-00229-3

Park EG, Lee DH, Kim WR, Lee YJ, Bae WH, Kim JM, Shin HJ, Ha H, Yi JM, Cho SG, Choi YH, Leem SH, Cha HJ, Kim SW, Kim HS (2023) Human endogenous retrovirus-H-derived miR-4454 inhibits the expression of DNAJB4 and SASH1 in non-muscle-invasive bladder cancer. Genes (Basel) 14(7). https://doi.org/10.3390/genes14071410

Paul SK, Oshima M, Patil A, Sone M, Kato H, Maezawa Y, Kaneko H, Fukuyo M, Rahmutulla B, Ouchi Y, Tsujimura K, Nakanishi M, Kaneda A, Iwama A, Yokote K, Eto K, Takayama N (2024) Retrotransposons in Werner syndrome-derived macrophages trigger type I interferon-dependent inflammation in an atherosclerosis model. Nat Commun 15(1):4772. https://doi.org/10.1038/s41467-024-48663-w

Percharde M, Lin CJ, Yin Y, Guan J, Peixoto GA, Bulut-Karslioglu A, Biechele S, Huang B, Shen X, Ramalho-Santos M (2018) A LINE1-Nucleolin partnership regulates early development and ESC identity. Cell 174(2):391–405 e319. https://doi.org/10.1016/j.cell.2018.05.043

Perkio A, Pradhan B, Genc F, Pirttikoski A, Pikkusaari S, Erkan EP, Falco MM, Huhtinen K, Narva S, Hynninen J, Kauppi L, Vaharautio A (2024) Locus-specific LINE-1 expression in clinical ovarian cancer specimens at the single-cell level. Sci Rep 14(1):4322. https://doi.org/10.1038/s41598-024-54113-w

Perron H, Geny C, Laurent A, Mouriquand C, Pellat J, Perret J, Seigneurin JM (1989) Leptomeningeal cell line from multiple sclerosis with reverse transcriptase activity and viral particles. Res Virol 140(6):551–561. https://doi.org/10.1016/s0923-2516(89)80141-4

Perron H, Garson JA, Bedin F, Beseme F, Paranhos-Baccala G, Komurian-Pradel F, Mallet F, Tuke PW, Voisset C, Blond JL, Lalande B, Seigneurin JM, Mandrand B (1997) Molecular identification of a novel retrovirus repeatedly isolated from patients with multiple sclerosis. The collaborative research group on multiple sclerosis. Proc Natl Acad Sci USA 94(14):7583–7588. https://doi.org/10.1073/pnas.94.14.7583

Petrizzo A, Ragone C, Cavalluzzo B, Mauriello A, Manolio C, Tagliamonte M, Buonaguro L (2021) Human endogenous retrovirus reactivation: implications for cancer immunotherapy. Cancers (Basel) 13(9):1999. https://doi.org/10.3390/cancers13091999

Pfaff AL, Bubb VJ, Quinn JP, Koks S (2021) Reference SVA insertion polymorphisms are associated with Parkinson's disease progression and differential gene expression. NPJ Parkinsons Dis 7(1):44. https://doi.org/10.1038/s41531-021-00189-4

Poewe W, Seppi K, Tanner CM, Halliday GM, Brundin P, Volkmann J, Schrag AE, Lang AE (2017) Parkinson disease. Nat Rev Dis Primers 3:17013. https://doi.org/10.1038/nrdp.2017.13

Pontis J, Pulver C, Playfoot CJ, Planet E, Grun D, Offner S, Duc J, Manfrin A, Lutolf MP, Trono D (2022) Primate-specific transposable elements shape transcriptional networks during human development. Nat Commun 13(1):7178. https://doi.org/10.1038/s41467-022-34800-w

Ramesh S, Arachchige A (2023) Depletion of dopamine in Parkinson's disease and relevant therapeutic options: a review of the literature. AIMS Neurosci 10(3):200–231. https://doi.org/10.3934/Neuroscience.2023017

Ramirez P, Zuniga G, Sun W, Beckmann A, Ochoa E, DeVos SL, Hyman B, Chiu G, Roy ER, Cao W, Orr M, Buggia-Prevot V, Ray WJ, Frost B (2022) Pathogenic tau accelerates aging-associated activation of transposable elements in the mouse central nervous system. Prog Neurobiol 208:102181. https://doi.org/10.1016/j.pneurobio.2021.102181

Rolland A, Jouvin-Marche E, Saresella M, Ferrante P, Cavaretta R, Creange A, Marche P, Perron H (2005) Correlation between disease severity and in vitro cytokine production mediated by MSRV (multiple sclerosis associated retroviral element) envelope protein in patients with multiple sclerosis. J Neuroimmunol 160(1–2):195–203. https://doi.org/10.1016/j.jneuroim.2004.10.019

Rolland A, Jouvin-Marche E, Viret C, Faure M, Perron H, Marche PN (2006) The envelope protein of a human endogenous retrovirus-W family activates innate immunity through CD14/TLR4 and promotes Th1-like responses. J Immunol 176(12):7636–7644. https://doi.org/10.4049/jimmunol.176.12.7636

Ronnblom L, Pascual V (2008) The innate immune system in SLE: type I interferons and dendritic cells. Lupus 17(5):394–399. https://doi.org/10.1177/0961203308090020

Ross OA, Braithwaite AT, Skipper LM, Kachergus J, Hulihan MM, Middleton FA, Nishioka K, Fuchs J, Gasser T, Maraganore DM, Adler CH, Larvor L, Chartier-Harlin MC, Nilsson C, Langston JW, Gwinn K, Hattori N, Farrer MJ (2008) Genomic investigation of alpha-synuclein multiplication and parkinsonism. Ann Neurol 63(6):743–750. https://doi.org/10.1002/ana.21380

Roy N, Haq I, Ngo JC, Bennett DA, Teich AF, De Jager PL, Olah M, Sher F (2024) Elevated expression of the retrotransposon LINE-1 drives Alzheimer's disease-associated microglial dysfunction. Acta Neuropathol 148(1):75. https://doi.org/10.1007/s00401-024-02835-6

Sandoval-Villegas N, Nurieva W, Amberger M, Ivics Z (2021) Contemporary transposon tools: a review and guide through mechanisms and applications of sleeping beauty, piggyBac and Tol2 for genome engineering. Int J Mol Sci 22(10):50584. https://doi.org/10.3390/ijms22105084

Sarngadharan MG, Sarin PS, Reitz MS, Gallo RC (1972) Reverse transcriptase activity of human acute leukaemic cells: purification of the enzyme, response to AMV 70S RNA, and characterization of the DNA product. Nat New Biol 240(98):67–72. https://doi.org/10.1038/newbio240067a0

Schacker T, Collier AC, Hughes J, Shea T, Corey L (1996) Clinical and epidemiologic features of primary HIV infection. Ann Intern Med 125(4):257–264. https://doi.org/10.7326/0003-4819-125-4-199608150-00001

Schmid CW, Deininger PL (1975) Sequence organization of the human genome. Cell 6(3):345–358. https://doi.org/10.1016/0092-8674(75)90184-1

Scopa C, Barnada SM, Cicardi ME, Singer M, Trotti D, Trizzino M (2023) JUN upregulation drives aberrant transposable element mobilization, associated innate immune response, and impaired neurogenesis in Alzheimer's disease. Nat Commun 14(1):8021. https://doi.org/10.1038/s41467-023-43728-8

Seetharam D, Chandar J, Ramsoomair CK, Desgraves JF, Alvarado Medina A, Hudson AJ, Amidei A, Castro JR, Govindarajan V, Wang S, Zhang Y, Sonabend AM, Mendez Valdez MJ, Maric D, Govindarajan V, Rivas SR, Lu VM, Tiwari R, Sharifi N, Thomas E, Alexander M, DeMarino C, Johnson K, De La Fuente MI, Alshiekh Nasany R, Noviello TMR, Ivan ME, Komotar RJ, Iavarone A, Nath A, Heiss J, Ceccarelli M, Chiappinelli KB, Figueroa ME, Bayik D, Shah AH (2025) Activating antiviral immune responses potentiates immune checkpoint inhibition in glioblastoma models. J Clin Invest 135(6). https://doi.org/10.1172/JCI183745

SenGupta D, Tandon R, Vieira RG, Ndhlovu LC, Lown-Hecht R, Ormsby CE, Loh L, Jones RB, Garrison KE, Martin JN, York VA, Spotts G, Reyes-Teran G, Ostrowski MA, Hecht FM, Deeks SG, Nixon DF (2011) Strong human endogenous retrovirus-specific T cell responses are associated with control of HIV-1 in chronic infection. J Virol 85(14):6977–6985. https://doi.org/10.1128/JVI.00179-11

Septier M, Peyre H, Amsellem F, Beggiato A, Maruani A, Poumeyreau M, Amestoy A, Scheid I, Gaman A, Bolognani F, Honey G, Bouquet C, Ly-Le Moal M, Bouvard M, Leboyer M, Bourgeron T, Delorme R (2019) Increased risk of ADHD in families with ASD. Eur Child Adolesc Psychiatry 28(2):281–288. https://doi.org/10.1007/s00787-018-1206-0

Serra C, Sotgiu S, Mameli G, Pugliatti M, Rosati G, Dolei A (2001) Multiple sclerosis and multiple sclerosis-associated retrovirus in Sardinia. Neurol Sci 22(2):171–173. https://doi.org/10.1007/s100720170019

Shaban SA, Al-Rahim AM, Suleiman AA (2022) ALU repeat as potential molecular marker in the detection and prognosis of different cancer types: a systematic review. Mol Clin Oncol 16(4):86. https://doi.org/10.3892/mco.2022.2519

Shah AH, Gilbert M, Ivan ME, Komotar RJ, Heiss J, Nath A (2021) The role of human endogenous retroviruses in gliomas: from etiological perspectives and therapeutic implications. Neuro-Oncology 23(10):1647–1655. https://doi.org/10.1093/neuonc/noab142

Shah AH, Govindarajan V, Doucet-O'Hare TT, Rivas S, Ampie L, DeMarino C, Banasavadi-Siddegowda YK, Zhang Y, Johnson KR, Almsned F, Gilbert MR, Heiss JD, Nath A (2022) Differential expression of an endogenous retroviral element [HERV-K(HML-6)] is associated with reduced survival in glioblastoma patients. Sci Rep 12(1):6902. https://doi.org/10.1038/s41598-022-10914-5

Shah AH, Rivas SR, Doucet-O'Hare TT, Govindarajan V, DeMarino C, Wang T, Ampie L, Zhang Y, Banasavadi-Siddegowda YK, Walbridge S, Maric D, Garcia-Montojo M, Suter RK, Lee MH, Zaghloul KA, Steiner J, Elkahloun AG, Chandar J, Seetharam D, Desgraves J, Li W, Johnson K, Ivan ME, Komotar RJ, Gilbert MR, Heiss JD, Nath A (2023) Human endogenous retrovirus K contributes to a stem cell niche in glioblastoma. J Clin Invest 133(13). https://doi.org/10.1172/JCI167929

Shaknovich R, Yeyati PL, Ivins S, Melnick A, Lempert C, Waxman S, Zelent A, Licht JD (1998) The promyelocytic leukemia zinc finger protein affects myeloid cell growth, differentiation, and apoptosis. Mol Cell Biol 18(9):5533–5545. https://doi.org/10.1128/MCB.18.9.5533

Shen L, Wu LC, Sanlioglu S, Chen R, Mendoza AR, Dangel AW, Carroll MC, Zipf WB, Yu CY (1994) Structure and genetics of the partially duplicated gene RP located immediately upstream of the complement C4A and the C4B genes in the HLA class III region. Molecular cloning, exon-intron structure, composite retroposon, and breakpoint of gene duplication. J Biol Chem 269(11):8466–8476

Shukla HD, Pitha PM (2012) Role of hsp90 in systemic lupus erythematosus and its clinical relevance. Autoimmune Dis 2012:728605. https://doi.org/10.1155/2012/728605

Siebenthall KT, Miller CP, Vierstra JD, Mathieu J, Tretiakova M, Reynolds A, Sandstrom R, Rynes E, Haugen E, Johnson A, Nelson J, Bates D, Diegel M, Dunn D, Frerker M, Buckley M, Kaul R, Zheng Y, Himmelfarb J, Ruohola-Baker H, Akilesh S (2019) Integrated epigenomic profiling reveals endogenous retrovirus reactivation in renal cell carcinoma. EBioMedicine 41:427–442. https://doi.org/10.1016/j.ebiom.2019.01.063

Simon M, Van Meter M, Ablaeva J, Ke Z, Gonzalez RS, Taguchi T, De Cecco M, Leonova KI, Kogan V, Helfand SL, Neretti N, Roichman A, Cohen HY, Meer MV, Gladyshev VN, Antoch MP, Gudkov AV, Sedivy JM, Seluanov A, Gorbunova V (2019) LINE1 Derepression in aged wild-type and SIRT6-deficient mice drives inflammation. Cell Metab 29(4):871–885 e875. https://doi.org/10.1016/j.cmet.2019.02.014

Simula ER, Arru G, Zarbo IR, Solla P, Sechi LA (2021) TDP-43 and HERV-K envelope-specific immunogenic epitopes are recognized in ALS patients. Viruses 13(11):2301. https://doi.org/10.3390/v13112301

Singh S, Borkar MR, Bhatt LK (2024) Transposable elements: emerging therapeutic targets in neurodegenerative diseases. Neurotox Res 42(1):9. https://doi.org/10.1007/s12640-024-00688-1

Solyom S, Ewing AD, Rahrmann EP, Doucet T, Nelson HH, Burns MB, Harris RS, Sigmon DF, Casella A, Erlanger B, Wheelan S, Upton KR, Shukla R, Faulkner GJ, Largaespada DA, Kazazian HH Jr (2012) Extensive somatic L1 retrotransposition in colorectal tumors. Genome Res 22(12):2328–2338. https://doi.org/10.1101/gr.145235.112

Sotgiu S, Mameli G, Serra C, Zarbo IR, Arru G, Dolei A (2010) Multiple sclerosis-associated retrovirus and progressive disability of multiple sclerosis. Mult Scler 16(10):1248–1251. https://doi.org/10.1177/1352458510376956

Steele AJ, Al-Chalabi A, Ferrante K, Cudkowicz ME, Brown RH Jr, Garson JA (2005) Detection of serum reverse transcriptase activity in patients with ALS and unaffected blood relatives. Neurology 64(3):454–458. https://doi.org/10.1212/01.WNL.0000150899.76130.71

Steiner JP, Bachani M, Malik N, DeMarino C, Li W, Sampson K, Lee MH, Kowalak J, Bhaskar M, Doucet-O'Hare T, Garcia-Montojo M, Cowen M, Smith B, Reoma LB, Medina J, Brunel J, Pierquin J, Charvet B, Perron H, Nath A (2022) Human endogenous retrovirus K envelope in spinal fluid of amyotrophic lateral sclerosis is toxic. Ann Neurol 92(4):545–561. https://doi.org/10.1002/ana.26452

Stricker E, Peckham-Gregory EC, Lai SY, Sandulache VC, Scheurer ME (2024) Targeted variant assessments of human endogenous retroviral regions in whole genome sequencing data reveal retroviral variants associated with papillary thyroid cancer. Microorganisms 12(12):2435. https://doi.org/10.3390/microorganisms12122435

Sugimoto J, Matsuura N, Kinjo Y, Takasu N, Oda T, Jinno Y (2001) Transcriptionally active HERV-K genes: identification, isolation, and chromosomal mapping. Genomics 72(2):137–144. https://doi.org/10.1006/geno.2001.6473

Sullivan AC, Zuniga G, Ramirez P, Fernandez R, Wang CP, Li J, Davila L, Pelton K, Gomez S, Sohn C, Gonzalez E, Lopez-Cruzan M, Gonzalez DA, Parker A, Zilli E, de Erausquin GA, Seshadri S, Espinoza S, Musi N, Frost B (2024) A pilot study to investigate the safety and feasibility of antiretroviral therapy for Alzheimer's disease (ART-AD). medRxiv. https://doi.org/10.1101/2024.02.26.24303316

Sun S, You E, Hong J, Hoyos D, Del Priore I, Tsanov KM, Mattagajasingh O, Di Gioacchino A, Marhon SA, Chacon-Barahona J, Li H, Jiang H, Hozeifi S, Rosas-Bringas O, Xu KH, Song Y, Lang ER, Rojas AS, Nieman LT, Patel BK, Murali R, Chanda P, Karacay A, Vabret N, De Carvalho DD, Zenklusen D, LaCava J, Lowe SW, Ting DT, Iacobuzio-Donahue CA, Solovyov A, Greenbaum BD (2024) Cancer cells restrict immunogenicity of retrotransposon expression via distinct mechanisms. Immunity 57(12):2879–2894 e2811. https://doi.org/10.1016/j.immuni.2024.10.015

Tam OH, Rozhkov NV, Shaw R, Kim D, Hubbard I, Fennessey S, Propp N, Consortium NA, Fagegaltier D, Harris BT, Ostrow LW, Phatnani H, Ravits J, Dubnau J, Gale Hammell M (2019) Postmortem cortex samples identify distinct molecular subtypes of ALS: retrotransposon activation, oxidative stress, and activated glia. Cell Rep 29(5):1164–1177 e1165. https://doi.org/10.1016/j.celrep.2019.09.066

Tandon R, SenGupta D, Ndhlovu LC, Vieira RG, Jones RB, York VA, Vieira VA, Sharp ER, Wiznia AA, Ostrowski MA, Rosenberg MG, Nixon DF (2011) Identification of human endogenous retrovirus-specific T cell responses in vertically HIV-1-infected subjects. J Virol 85(21):11526–11531. https://doi.org/10.1128/JVI.05418-11

Taylor MS, Wu C, Fridy PC, Zhang SJ, Senussi Y, Wolters JC, Cajuso T, Cheng WC, Heaps JD, Miller BD, Mori K, Cohen L, Jiang H, Molloy KR, Chait BT, Goggins MG, Bhan I, Franses JW, Yang X, Taplin ME, Wang X, Christiani DC, Johnson BE, Meyerson M, Uppaluri R, Egloff AM, Denault EN, Spring LM, Wang TL, Shih IM, Fairman JE, Jung E, Arora KS, Yilmaz OH, Cohen S, Sharova T, Chi G, Norden BL, Song Y, Nieman LT, Pappas L, Parikh AR, Strickland MR, Corcoran RB, Mustelin T, Eng G, Yilmaz OH, Matulonis UA, Chan AT, Skates SJ, Rueda BR, Drapkin R, Klempner SJ, Deshpande V, Ting DT, Rout MP, LaCava J, Walt DR, Burns KH (2023) Ultrasensitive detection of circulating LINE-1 ORF1p as a specific multicancer biomarker. Cancer Discov 13(12):2532–2547. https://doi.org/10.1158/2159-8290.CD-23-0313

Tipanee J, Chai YC, VandenDriessche T, Chuah MK (2017) Preclinical and clinical advances in transposon-based gene therapy. Biosci Rep 37(6). https://doi.org/10.1042/BSR20160614

Tokuyama M, Gunn BM, Venkataraman A, Kong Y, Kang I, Rakib T, Townsend MJ, Costenbader KH, Alter G, Iwasaki A (2021) Antibodies against human endogenous retrovirus K102 envelope activate neutrophils in systemic lupus erythematosus. J Exp Med 218(7). https://doi.org/10.1084/jem.20191766

Tseng A, Seet J, Phillips EJ (2015) The evolution of three decades of antiretroviral therapy: challenges, triumphs and the promise of the future. Br J Clin Pharmacol 79(2):182–194. https://doi.org/10.1111/bcp.12403

Ukadike KC, Najjar R, Ni K, Laine A, Wang X, Bays A, Taylor MS, LaCava J, Mustelin T (2023) Expression of L1 retrotransposons in granulocytes from patients with active systemic lupus erythematosus. Mob DNA 14(1):5. https://doi.org/10.1186/s13100-023-00293-7

Vaupel P, Kallinowski F, Okunieff P (1989) Blood flow, oxygen and nutrient supply, and metabolic microenvironment of human tumors: a review. Cancer Res 49(23):6449–6465

Vincendeau M, Gottesdorfer I, Schreml JM, Wetie AG, Mayer J, Greenwood AD, Helfer M, Kramer S, Seifarth W, Hadian K, Brack-Werner R, Leib-Mosch C (2015) Modulation of human endogenous retrovirus (HERV) transcription during persistent and de novo HIV-1 infection. Retrovirology 12:27. https://doi.org/10.1186/s12977-015-0156-6

Viola MV, Frazier M, White L, Brody J, Spiegelman S (1975) RNA-instructed DNA polymerase activity in a cytoplasmic particulate fraction in brains from Guamanian patients. J Exp Med 142(2):483–494. https://doi.org/10.1084/jem.142.2.483

Vogetseder W, Dumfahrt A, Mayersbach P, Schonitzer D, Dierich MP (1993) Antibodies in human sera recognizing a recombinant outer membrane protein encoded by the envelope gene of the human endogenous retrovirus K. AIDS Res Hum Retrovir 9(7):687–694. https://doi.org/10.1089/aid.1993.9.687

Wahl D, Cavalier AN, Smith M, Seals DR, LaRocca TJ (2021) Healthy aging interventions reduce repetitive element transcripts. J Gerontol A Biol Sci Med Sci 76(5):805–810. https://doi.org/10.1093/gerona/glaa302

Wang T, Medynets M, Johnson KR, Doucet-O'Hare TT, DiSanza B, Li W, Xu Y, Bagnell A, Tyagi R, Sampson K, Malik N, Steiner J, Hadegan A, Kowalak J, O'Malley J, Maric D, Nath A (2020) Regulation of stem cell function and neuronal differentiation by HERV-K via mTOR pathway. Proc Natl Acad Sci USA 117(30):17842–17853. https://doi.org/10.1073/pnas.2002427117

Wang FQ, Dang X, Su H, Lei Y, She CH, Zhang C, Chen X, Yang X, Yang J, Feng H, Yang W (2024) Association of hyperactivated transposon expression with exacerbated immune activation in systemic lupus erythematosus. Mob DNA 15(1):23. https://doi.org/10.1186/s13100-024-00335-8

Wells JN, Feschotte C (2020) A Field guide to eukaryotic transposable elements. Annu Rev Genet 54:539–561. https://doi.org/10.1146/annurev-genet-040620-022145

Wentzensen N, Coy JF, Knaebel HP, Linnebacher M, Wilz B, Gebert J, von Knebel DM (2007) Expression of an endogenous retroviral sequence from the HERV-H group in gastrointestinal cancers. Int J Cancer 121(7):1417–1423. https://doi.org/10.1002/ijc.22826

Westarp ME, Bartmann P, Rossler J, Geiger E, Westphal KP, Schreiber H, Fuchs D, Westarp MP, Kornhuber HH (1993) Antiretroviral therapy in sporadic adult amyotrophic lateral sclerosis. Neuroreport 4(6):819–822. https://doi.org/10.1097/00001756-199306000-00056

Wicks EE, Semenza GL (2022) Hypoxia-inducible factors: cancer progression and clinical translation. J Clin Invest 132(11). https://doi.org/10.1172/JCI159839

Wieland L, Schwarz T, Engel K, Volkmer I, Kruger A, Tarabuko A, Junghans J, Kornhuber ME, Hoffmann F, Staege MS, Emmer A (2022) Epstein-Barr virus-induced genes and endogenous retroviruses in immortalized B cells from patients with multiple sclerosis. Cells 11(22):3619. https://doi.org/10.3390/cells11223619

Workowski KA, Bachmann LH, Chan PA, Johnston CM, Muzny CA, Park I, Reno H, Zenilman JM, Bolan GA (2021) Sexually transmitted infections treatment guidelines, 2021. MMWR Recomm Rep 70(4):1–187. https://doi.org/10.15585/mmwr.rr7004a1

Yuan Z, Sun X, Jiang D, Ding Y, Lu Z, Gong L, Liu H, Xie J (2010) Origin and evolution of a placental-specific microRNA family in the human genome. BMC Evol Biol 10:346. https://doi.org/10.1186/1471-2148-10-346

Zarogoulidis P, Darwiche K, Sakkas A, Yarmus L, Huang H, Li Q, Freitag L, Zarogoulidis K, Malecki M (2013) Suicide gene therapy for cancer – current strategies. J Genet Syndr Gene Ther 4. https://doi.org/10.4172/2157-7412.1000139

Zehrbach NM, Oh N, Ishak CA (2024) Insights into LINE-1 reverse transcription guide therapy development. Trends Cancer 10(4):286–288. https://doi.org/10.1016/j.trecan.2024.02.010

Zhang Z, Shi L, Song L, Maurer K, Petri MA, Sullivan KE (2018) Overall downregulation of mRNAs and enrichment of H3K4me3 change near genome-wide association study signals in systemic lupus erythematosus: cell-specific effects. Front Immunol 9:497. https://doi.org/10.3389/fimmu.2018.00497

GPSR Compliance

The European Union's (EU) General Product Safety Regulation (GPSR) is a set of rules that requires consumer products to be safe and our obligations to ensure this.

If you have any concerns about our products, you can contact us on ProductSafety@springernature.com

In case Publisher is established outside the EU, the EU authorized representative is:

Springer Nature Customer Service Center GmbH
Europaplatz 3
69115 Heidelberg, Germany

Batch number: 10755214

Printed by Printforce, the Netherlands